International Conference on Startup Ventures: Technology Developments and Future Strategies (SV-TDFS-2018)

AA006085

IOP Conference Series: Materials Science and Engineering
Volume 594

Jaipur, India
8 - 9 October 2018

Editors:

Kulwant Singh
Himanshu Chaudhary
Ashish Shrivastava
S.N. Sharan
N.N. Sharma
Jamil Akhtar

ISBN: 978-1-5108-9815-8
ISSN: 1757-8981

Printed from e-media with permission by:

Curran Associates, Inc.
57 Morehouse Lane
Red Hook, NY 12571

Some format issues inherent in the e-media version may also appear in this print version.

Additional copies of this publication are available from:

Curran Associates, Inc.
57 Morehouse Lane
Red Hook, NY 12571 USA
Phone: 845-758-0400
Fax: 845-758-2633
Email: curran@proceedings.com
Web: www.proceedings.com

Table of contents

Volume 594

International Conference on Startup Ventures: Technology Developments and Future Strategies

8–9 October 2018, Manipal University Jaipur, Rajasthan, India

Accepted papers received: 22 July 2019
Published online: 16 September 2019

Preface

Papers

IOP Conf. Series: Materials Science and Engineering **594** (2019) 011001 doi:10.1088/1757-899X/594/1/011001

Preface

An International conference on Startup ventures: Technology Developments and Future Strategies (SV-TDFS) was held at Manipal University Jaipur, Rajasthan, India during 8th-9th October, 2018 with the collaboration of Institute for Engineering Research and Publication (IFERP), Technocrat Research and Development professional association.

The aim of SV-TDFS-2018 is to provide an international open forum for the researchers, academicians, practitioners, startup ventures and technocrats in academia as well as in industries from different parts of the world to interact, exchange concepts, prototypes, innovative research ideas and share the outcomes of their research work which could especially contribute to the academic arena and further benefit the business and industrial community, and as a whole helping the society. SV-TDFS-2018 is the first unique conference of its own kind which is trying to bring academicians, scientists and technocrats from various domains ranging from agriculture to industry, to provide the solution of the common man's basic problem at one platform and also discuss the challenges in their respective domains.

This proceeding contains papers presented at the conference were intended to provide answer of future challenges in the field of material science and engineering for food, green energy, agriculture, water and health applications. In this two-days conference, 5 keynote talks and seven technical sessions were organized in which total number of 76 research papers came from India and abroad. Moreover, 250 delegates have participated from India and various other countries of the world.

Each contributed paper was refereed before being accepted for publication in these proceeding. The papers were accepted for publication based on their interest, relevance, innovation and application to the material science and engineering field.

Sincere thanks to the following individuals: Prof. Bhim Singh (Dean academics, IITD), Prof. Ajay Agrawal (Senior Principal Scientist, CSIR-CEERI Pilani), Dr. Vesna Lavtizar (Kobe University, Japan), Dr. Sugunalakshmi, (Senior Principal Scientist, CSIR-CLRI, Chennai), Prof. G. K. Prabhu (President, MUJ), Prof. Jagannath Korody, (Dean, Faculty of Engineering, MUJ), Prof. H. R. Kamath (Dean, Research & Innovation, MUJ), Prof. Subrata Bandhu Ghosh, (Director, Research, MUJ) for their contribution and support. Special thanks go to our conference sponsors for their continued support: Dr. G. Aren, CEO, Macwin India Pvt. Ltd., New Delhi.

Editors

IOP Conf. Series: Materials Science and Engineering **594** (2019) 011001 doi:10.1088/1757-899X/594/1/011001

International Conference on Start-Up Ventures: Technological Development & Future Strategies (SV-TDFS) during 8th- 9th October 2018 at Manipal University Jaipur

The inauguration of the "International Conference on Start-Up Ventures: Technological Development & Future Strategies (SV-TDFS 2018)" was marked by the lamp-lighting in the presence of academic stalwarts like **Prof. Bhim Singh** (Dean Academics - IIT Delhi), **Prof. Jamil Akhtar** (Chief Scientist, CSIR-CEERI Pilani), **Prof. Ajay Agarwal** (Senior Principal Scientist, CSIR-CEERI Pilani), **Prof. G. K. Prabhu** (President- Manipal University Jaipur), **Prof. Jagannath K.** (Dean, Faculty of Engineering, Manipal University Jaipur), **Prof. H. R. Kamath** (Dean- Research & Innovation, Manipal University Jaipur), **Prof. S. N. Sharan** (Director- SEEC-Manipal University Jaipur) and other dignitaries.

IOP Conf. Series: Materials Science and Engineering **594** (2019) 011001 doi:10.1088/1757-899X/594/1/011001

Prof. Jamil Akhtar (Conference Chair) emphasized the importance of societal benefit of marketability of the Research & Development being carried out in laboratories of scientific and academic institutions.

Prof. Jagannath K. (Dean Engineering, Manipal University Jaipur) pointed out the opportunity for young engineers to utilize the forum for bolstering their confidence as an entrepreneur by discussing the networking capability, technical feasibility, and economic viability of their business models with the experts at this conference.

Prof. G. K. Prabhu (President Manipal University Jaipur) deliberated the importance of Imagination, Ideas, Inspiration, Ideation, Innovation, and Incubation. He urged the students to focus on the cost-effectiveness of their innovation by collaborating with the experts (of various fields like Finance, HR, Legal Aid, Technology, etc.) at Atal Incubation Centre at Manipal University Jaipur – the only one of its kind in Rajasthan.

Prof. Bhim Singh (Chief Guest, SV-TDFS 2018), admitted the reasonable gap between engineering academics and final technological product, gave many inspiring real-time examples of how young entrepreneurs are bridging this gap by their sheer dedication to creating problem-solving products and platforms for the people of their motherland. He cited that innovat ion comes very easy when we focus on what is needed by the countrymen, for them the marketability of the product is quickly taken care of, without any extraneous endeavour. He said that if the young entrepreneurs can correlate with the need and problem solving of their targeted it customers, would herald the growth of very sustainable and scalable economy.

IOP Conf. Series: Materials Science and Engineering **594** (2019) 011001 doi:10.1088/1757-899X/594/1/011001

The inaugural function was concluded with vote of thanks by **Prof. S. N. Sharan** (Conference

Chair). He shared the conception of this unique initiative and urged everyone to utilize this excellent opportunity wholeheartedly.

Everyone unanimously congratulated conference conveners **Dr. Ashish Shrivastava**, **Dr. Himanshu Chaudhary**, **Dr. Kulwant Singh** and their entire team for their pioneering effort of organizing this conference on an innovative and radical theme of start-up and technological entrepreneurship.

In this two-days conference, seven technical sessions were organized and these sessions 76 faculties and scholars presented their research papers. In this conference, 250 delegates were participated from India and various other countries of the world.

IOP Conf. Series: Materials Science and Engineering **594** (2019) 011002 doi:10.1088/1757-899X/594/1/011002

Peer review statement

All papers published in this volume of *IOP Conference Series: Materials Science and Engineering* have been peer reviewed through processes administered by the proceedings Editors. Reviews were conducted by expert referees to the professional and scientific standards expected of a proceedings journal published by IOP Publishing.

IOP Conf. Series: Materials Science and Engineering **594** (2019) 012001 doi:10.1088/1757-899X/594/1/012001

Structural and Optical Properties of Si Nanostructures

Ashish Kumar[1*], Ashok Sharma[1], and Ajay Agarwal[1]

[1]CSIR-Central Electronics Engineering Research Institute, Pilani, 333031, Rajasthan India

*Email: ashishvlsi10@gmail.com

Abstract: Silicon nanostructures were realized using metal assisted chemical etching of silicon wafer (100) for two different etching times: 5 minute and 10 minute. Structural and optical properties of silicon nanostructures were investigated using field emission scanning electron microscopy (FESEM), X-ray diffraction (XRD), and photoluminescence (PL). XRD analysis probes the changes in crystallinity during etching and reveals reflection plane (220) of obtained silicon nanostructures. UV and visible spectra of photoluminescence from Si nanostructures suggest the optically active nature of Si at nanoscale and confinement of carrier. These results show that the structural and optical properties of Si nanostructures strongly depend on etching time of silicon wafer.

Keywords: MACE; Silicon nanostructures; and Quantum confinement

1. Introduction: The famous lecture "There's Plenty of Room at the Bottom: An Invitation to Enter a New Field of Physics" of Richard Feynman in 1959 promoted research community to harvest the unique and structural properties of materials at nanoscale which are absent in bulk structures [1]. However, the visions of Richard Feynman came into the reality with the advancement in technology in the early decade of 21 century. Since then, research community have demonstrated numerous nanostructures of various materials such as Si, ZnO, SnO_2 and etc. to study unique properties of material at nanoscale [2]. Among all these material, silicon is more attractive for nanoelectronics application as well documented properties and easy integration with exciting technology are addition advantage of Si nanostrucutres (NSs) over other materials [3-6]. Since the first report of Si NSs in 1950 by Uhlir [3], Si NSs with control morphology demonstrated potential application in the fields ranging from electronic, sensor, energy storage, flexible electronics photovoltaics, and energy storage electrodes [4-5]. However, with an indirect band gap material Si NSs can hardly be explored for optoelectronic application till the discovery of luminescent in Si NSs due to carrier confinement. Optically active nature of Si NSs or porous Si has attracted significant attention to demonstrate efficient and cost effective light emitting diode and laser.

Despite the various application of Si NSs, current fabrication methods reactive ion etching, other dry etching which are commonly used for micro nanofabrication are not suitable for mass production of Si NSs at large scale with high aspect ratios [4]. With the advancement in catalyst chemistry, an alternative approach metal assist chemical etching (MACE) has attracted a great interest to develop Si NSs because of its simplicity; low cost, and versatility [6-7]. Various types of Si NSs i. e. Si nanowires, porous Si from many type of Si wafer (100) and Si(111) with different doping levels p-type Si or n-type Si wafers, have been demonstrated using this method. The MACE process begins with partial coating of metal nanoparticles over Si surface followed by etching of Si in acidic oxidizing solution of hydrogen peroxide (H_2O_2) and hydrofluoric acid (HF). Here metal nanoparticles works catalyst and enhance the etch rate of Si beneath of NPs. Although realization of Si NSs using MACE is very easy and significant effort have been made to explain etching mechanism of Si in the presence of metal NPs, structural and luminescent properties of Si NSs are yet to address. ,

In this work, we reported the development of vertically aligned Si NSs through etching of lightly doped Si wafer utilizing gold (Au) NPs in MACE process. XRD analysis of silicon NSs showed the variations in the crystal lattice than its bulk wafer which might correspond to oxidation rate of crystal plane. Photoluminescence of Si NSs shows intense blue and red emission at room temperature attributed to quantum confinement carrier at nanoscale.

2. Experimental Procedures:

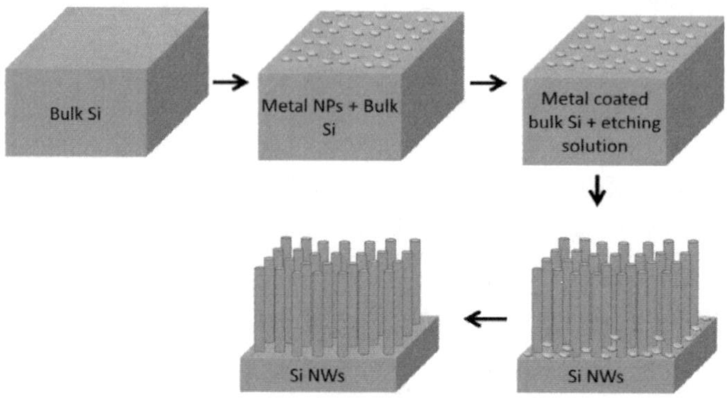

Fig. 1 Processes flow MACE for formation of Si Nanostructures

For experiment, single side polished p-type Si (100) wafers with resistivity (1-10) Ω-cm were considered. Si wafer were cleaned using standard degreasing and RCA cleaning process of semiconductor industry. Cleaned Si wafer were decorated with 10 nm thickness of gold (Au) metal nanoparticles (NPs) by e-beam evaporation system. Further, Au NPs coated wafers were immersed in etchant solution of H_2O_2 and HF (2:1) for 5 and 10 minutes. After developing process, substrates were dipped in etchant solution of Au to remove metal NPs from the Si NSs and followed by bath of NSs in HF solution to remove oxide layer which was produced in Au etchant solution. The process flow for the development of Si NSs figure 1. For further simplicity in discussion, we named the sample S1 and S2 of Si NSs which were obtained by different etching time duration 5 and 10 minutes respectively. Crystal properties of etched wafers were characterized by Powder XRD (Rigaku Smart Lab system® with Cu-Kα radiation operating at 40 kV and 40 mA, with wavelength (λ) = 1.5418 Å. FESEM reveals surface morphology of textured surface of samples. Photoluminescence spectra of samples was recorded with high power 325 nm He-Cd laser excitation power source using Dongwoo optron 80K PL system at room temperature.

3. Results and Discussion:

Textured surface of Si wafer was revealed in FESEM images and shown in the figure 2 (a-b) for sample S1 and S2. Though, surface profile of etched wafer is quite smiler. Etch depth of silicon was not same for different etching time as revealed in cross section images 2(c-d). Etching depth of sample S1 was recorded 22 μm while 33 μm for sample S2. Cross section images exhibited uniform distribution of dense and isolated Si NSs on entire sample. Long NSs as shown in the figures confirmed that MACE process is capable to produce high aspect ratio NSs in less time than dry reactive ion etching process. Vertical alignment of Si NSs shows that electrochemical reaction of HF and H_2O_2 took place at the interface of Si and Au NPs. Au NPs works as catalyst and enhance the anisotropic etching of Si. The electrochemical reaction is similar to two electrode configuration under galvanostatic control. Here Si substrate works as the anode electrode and metal NPs works cathode electrode. During the reaction Si atoms are oxidized in the presence of HF and removed from the surface as intermediate products SiF_4 or SiF_6^{2-} are soluble in water. During the reaction Si atom released electron and oxidize after adsorption of hole. The released electrons are capture by Au NPs. The various chemical reactions have been proposed to describe the MACE analogous anodic etching of Si in HF [5-6].

It is well established that H_2O_2 is reduced at the metal (cathode) and generate hole

$$H_2O_2 + 2H^+ \rightarrow 2H_2O + 2h^+ \quad (1)$$

Generated hole is adsorbed by Si surface and leads oxidation of Si and followed by dissolution of oxide

$$Si + 4h^+ + 4HF \rightarrow SiF_4 + H^+$$

$$SiF^4 + 2HF \rightarrow H_2SiF_6$$

$$Si + 2H_2O \rightarrow SiO_2 + 4H^+ + 4e^-$$

$$SiO_2 + 6HF \rightarrow H_2SiF_6 + 2H_2O \quad (2)$$

Since the reaction take place at the surface of Si, surface states of Si also play an important role to control the morphology of etched Si. Up to now, the role of surface states in the metal assisted chemical etching is not explored. Moreover, morphology of Si NSs i. e. the length, diameter, and interface are strongly functions of catalyst shape, size, concentration of HF and H_2O_2 and etching time.

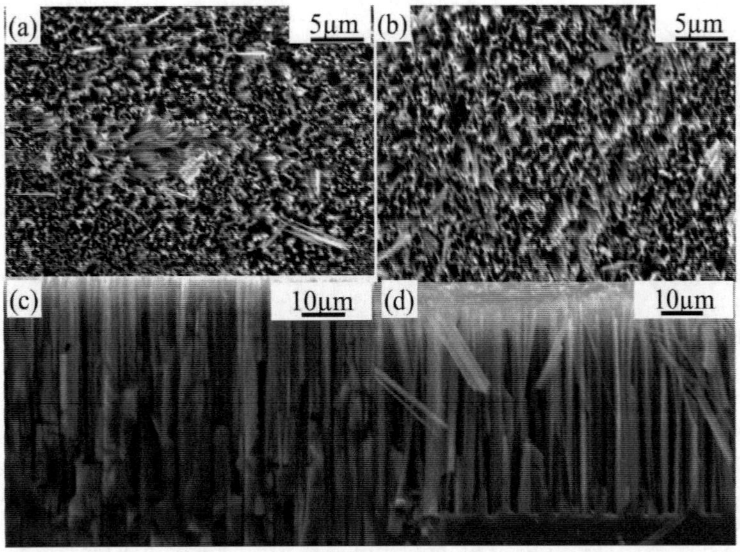

Fig. 2 FESEM images of silicon nanostructures obtain after etching of Si wafer for (a) 5 minutes and (b) 10 minutes

In order to understand the crystal structures of etched Si NSs, XRD plots of the samples (S1 and S2) are presented in figure 3 (a-b) respectively. The common diffraction peak in both figures was observed at diffraction angle of 2θ (44.5°) and can be assigned to diffraction plane of silicon (220) [9]. A distinguish peak for the sample S1 at the value of 2θ (28.53°) is well assigned to Si (111) plane [9]. The presence of plane (111) indicates the Si atom is not etched in (111) direction [10]. The anisotropic etching of Si was described using the back bond breaking theory. According to this theory, bond of Si surface atom connect to underneath atom and break during oxidation and dissolution process of MACE. The stronger the back bond strength the more difficult to remove the atom. Bond strength of surface Si atom is decided by crystal orientation of substrate. Si atom on the surface of a (100) substrate has two back bonds while atom on (111) substrate has three back bond. Therefore, atom on (100) surface plane is easily etched than the atom present on plane (111). The similar result was observed for sample S1. As etching increases, only crystal orientation (220) of Si nanostructures was observed. Here it is worth to mention that etching of Si in MACE process is very

IOP Conf. Series: Materials Science and Engineering **594** (2019) 012001 doi:10.1088/1757-899X/594/1/012001

complicated. The detail etching mechanism of Si with respect to metal NPS is yet to explain. Other than diffraction peaks, a small hump was also observed at lower diffraction angle in both figures and indicated the oxidation of developed Si NSs in open environment. Full width half maximum (FWHM) also gives important information for the crystalline size: Large value of FWHM indicates smaller the crystal size. The values of FWHM corresponding to diffraction plane (220) was calculated via origin peak analyser and found 0.35° and 0.42°for the sample S1 and S2 respectively. Larger etching time produces small crystalline size and responsible for the large value of FWHM.

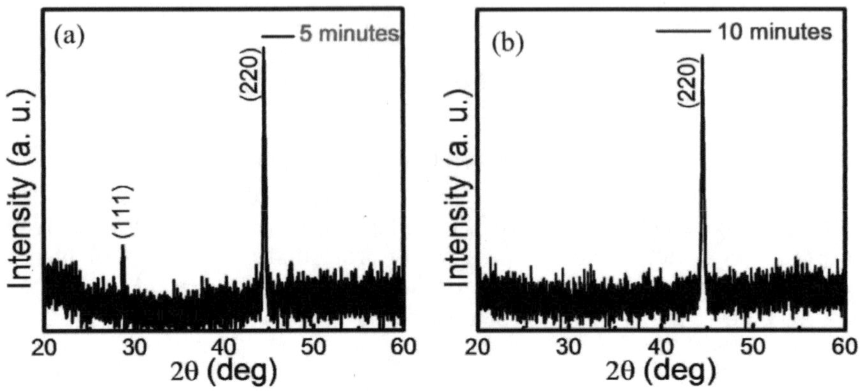

Fig. 3. XRD analysis of etched Si nanostructures obtain after etching of bulk Si wafer

Finally we investigated the luminescent properties of developed Si NSs. Strong luminescent spectra for both samples were recorded from the samples and shown in figures 4 (a-b). Si NSs exhibits visible emission at room temperature without any treatment of surface. However, emission spectrum at the wavelengths 620-650 for sample S1 in figure 4(a) was slightly different from emission spectra of sample S2 where only blue emission centred between 350-450 nm was observed in figure 4 (b). The image of room temperature emission of the samples were also captured and shown in the inset of graph corresponding to each sample. Visible emission of Si NSs is strongly depends on structural properties of Si NSs and therefore PL can provide the information about size or morphology present in samples. Multicolour emission covering entire visible spectra from individual Si nanowire is reported by K. Seo and his research group [11]. Therefore it is concluded that a range of diameters of Si NSs were present in the sample S1 as also confirmed in XRD analysis by the presence of (111) crystal plane. However, as the etching time increased, plane (111) dissolved and only single plane of NSs could be responsible for single peak of luminescent for sample S2. Additionally small particles sizes enhance carrier confinement and might be responsible for blue shift in emission. Here, small particles size could be achieved by long etching time i.e. 10 minutes for sample S_2. These results are similar to reported photoluminescence spectra of Si NWs produced in longer etching time corresponding to carrier confinement at Bohr radius (≈ 5nm) of Si NSs. However diameters range (≈ 50nm to 100nm) of Si NSs in this experiment was measured using SEM within its resolution limit. Therefore, surface morphology of Si NSs could be responsible lower wavelength emission instead of quantum confinement of carrier. Origin of red emission is expected from the carrier confinement corresponding to bandgap of silicon. However the blue, orange-red emission is greater than the band gap of bulk Si can be assign to combine effect of carrier confinement and surface states. The present blue emission in the fabricated Si NSs could be attributed to low doping level of Si wafer and presence of dangling bond at the surface. A reasonable mechanism for explanation of luminescent in Si NSs is subject of debate.

IOP Conf. Series: Materials Science and Engineering **594** (2019) 012001 doi:10.1088/1757-899X/594/1/012001

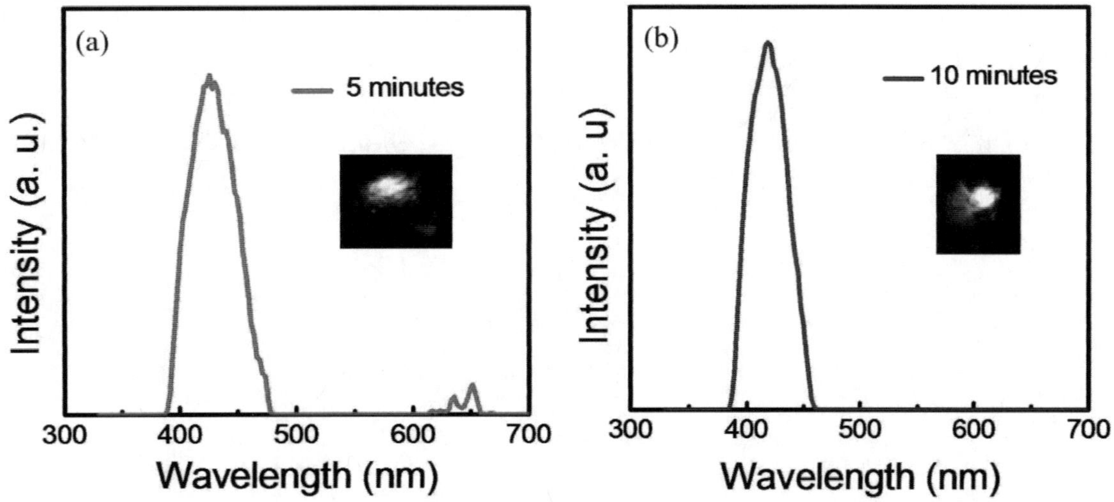

Fig. 4 Photoluminescence spectra of Si nanostructures obtain after etching of Si wafer for (a) 5 minutes and (b) 10 minutes

4. Conclusions: In summary, a facile and cost effective method of electroless chemical etching was described for the development of Si nanostructures from p-type Si wafer (100). We employed sputtering method for Au deposition on the Si surface. Crystalline properties of etched Si strongly depend on etching conditions: time, size of metal nanoparticles, and concentration of etchant solution. Si crystal plane (100) easily dissolved in etchant solution than the (111) plane. Etched Si also shows optically active nature i. e. luminescent in broad range blue to read emission corresponding to carrier confinement effect and surface defect. Thus presented work give brief overview of Si NSs formation and their properties; could be helpful for further probe and application of Si NSs for various application especially cost effective light emitting devices.

Acknowledgements: A. Kumar is also thankful to Director; CSIR-CERRI support to carry out this research work. Authors are also grateful to technical staff of CSIR-CEERI Pilani for their help during the experiment.

References:

[1]. Feynman, R.P., 2012. There's plenty of room at the bottom: An invitation to enter a new field of physics. In Handbook of Nanoscience, Engineering, and Technology, Third Edition (pp. 26-35). CRC Press.
[2]. Zhang, L. and Webster, T.J., 2009. Nanotechnology and nanomaterials: promises for improved tissue regeneration. Nano today, 4(1), pp.66-80.
[3]. Osminkina, L. A., Gonchar, K. A., Marshov, V. S., Bunkov, K. V., Petrov, D. V., Golovan, L. A.. & Timoshenko, V. Y. (2012). Optical properties of silicon nanowire arrays formed by metal-assisted chemical etching: evidences for light localization effect. Nanoscale research letters, 7(1), 524.
[4]. Patil, J.J., Smith, B.D. and Grossman, J.C., 2017. Ultra-high aspect ratio functional nanoporous silicon via nucleated catalysts. RSC Advances, 7(19), pp.11537-11542.
[5]. Huang, Z., Geyer, N., Werner, P., De Boor, J., &Gösele, U. (2011). Metal-assisted chemical etching of silicon: a review. Advanced materials, 23(2), 285-308.
[6]. Han, H., Huang, Z. and Lee, W., 2014. Metal-assisted chemical etching of silicon and nanotechnology applications. Nano today, 9(3), pp.271-304.
[7]. Toor, F., Miller, J.B., Davidson, L.M., Duan, W., Jura, M.P., Yim, J., Forziati, J. and Black, M.R., 2016. Metal assisted catalyzed etched (MACE) black Si: optics and device physics. Nanoscale, 8(34), pp.15448-15466.
[8]. Zhang, Y. F., Tang, Y. H., Wang, N., Yu, D. P., Lee, C. S., Bello, I., & Lee, S. T. (1998). Silicon nanowires prepared by laser ablation at high temperature. Applied physics letters, 72(15), 1835-1837.
[9]. Kim, J., Kim, Y. H., Choi, S. H., & Lee, W. (2011). Curved silicon nanowires with ribbon-like cross sections by metal-assisted chemical etching. Acs Nano, 5(6), 5242-5248.

[10]. Huang, Z., Shimizu, T., Senz, S., Zhang, Z., Geyer, N. and Gosele, U., 2010. Oxidation rate effect on the direction of metal-assisted chemical and electrochemical etching of silicon. The Journal of Physical Chemistry C, 114(24), pp.10683-10690.
[11]. Seo, K., Wober, M., Steinvurzel, P., Schonbrun, E., Dan, Y., Ellenbogen, T., & Crozier, K. B. (2011). Multicolored vertical silicon nanowires. Nano letters, 11(4), 1851-1856.

IOP Conf. Series: Materials Science and Engineering **594** (2019) 012002 doi:10.1088/1757-899X/594/1/012002

Structural and Optical Characteristics of Nanoscale Semiconductor Lasers for Telecommunication and Biomedical Applications: A Review

Jayprakash Vijay[1, 2], Kulwant Singh[1], Dimple Soni[2], Amit Rathi[1]
[1]Department of Electronics and Communication Engineering, Manipal University Jaipur
[2]Department of Electronics and Communication Engineering, Swami Keshvanand Institute of Technology, Management &Gramothan, Jaipur
E-Mail: jpvijay121@gmail.com
amit.rathi@jaipur.manipal.edu

Abstract. This paper presents the structural and optical characteristics of nanoscale semiconductor lasers for telecommunication and biomedical applications. Improved fabrication techniques, new materials and nano-scale heterostructures have led to improvement in the device performance. The material selection and their bandgap have an important role in the heterostructure to generate a lasing wavelength for particular applications. The bandgap modification can be done by the use of alloy semiconductor, quantum well structure, and strain layer epitaxy without changing the material itself. Semiconductor multilayers on the substrate are generally grown by using the metal-organic vapor phase epitaxy (MOVPE) and molecular beam epitaxy (MBE) process. Many researchers have provided different designs of heterostructures for the lasers. Generally, lasers are manufactured by using different semiconductor layers grown on GaAs, InP or GaSb substrate at the nanoscale. But controlling the thickness of the layer grown on the substrate at the nanoscale is the major problem in the fabrication. In a study, it has been found that for the proper functioning of semiconductor lasers it is beneficial to have light conduction and valence band masses. By using band structure engineering theory of quantum confinement and incorporation of strain on the active layer, this problem can be resolved. Red lasers are currently used in biomedical applications for treatment of superficial skin diseases like psoriasis, vitiligo etc. The manufacturing of red laser was earlier done by using nitrides material but they are harmful to skin and are expensive as well. An alternative for designing of red laser is manufacturing red lasers by using phosphides. The red laser is manufactured by using GaInP and AlGaInP ternary and quaternary compounds are widely used in the biomedical industry currently. This paper is the outcomes of the papers presented by many researchers in the field of optoelectronics.

Keywords: Double heterostructure, Laser, Quantum well, Optoelectronics.

1. Introduction

Nowadays in the world of science optoelectronic devices like LASER diodes, photodetectors, optical waveguides, directional couplers, and LED's are playing the major role in communication industries [1, 2]. The optoelectronics deals with the systems that detect and control the light and it is considered as a sub-field of photonics. Optoelectronic devices can be designed to be used for the shortwave infrared region, mid-wave infrared regions and near-infrared regions. The regions are selected on the basis of the wavelength at which the device is operating for an application [3]. Currently, for the treatment of tumor, cancer and superficial skin diseases, researchers are more concerned about the use of laser in telecommunication and biomedical applications in photo-dynamic therapy (PDT). A laser is a light source which also provides the optical amplification based on the principle of stimulated emission of electromagnetic waves. Lasers are classified on the basis of material used in active medium and broadly it is classified as gas lasers, solid lasers, liquid lasers and semiconductor lasers [4, 5]. Today one of the most important laser in optical communication is semiconductor laser [6, 7]. These lasers use semiconductors as the lasing medium and are characterized by specific advantages such as the capability of direct modulation in the gigahertz region, small size and low cost [8]. Lasers can be designed by using semiconductor heterostructures also. In several applications of semiconductors emission and absorption of light which is done in the energy bandgap by electron-hole pair combination and hence generating some photoconduction which gives importance to modification of bandgap. The modification of energy bandgap can be done by using ternary and quaternary semiconductor compounds. The laser using the heterojunction was first proposed by the Herbert Kroemer, in 1963 [9]. He suggested that the population inversion can be enhanced with the use of heterostructure. Interface of two layer having different crystalline semiconductors is the main concept of a heterojunction. These semiconductor materials have unequal bandgapsunlike to homojunction. The semiconductor materials for the heterostructures are selected on the basis of wavelength and for which application it can be used [10]. The heterostructure has various solid state devices and having different applications with solar cells, semiconductor LASER, and transistors. Depending on the band alignment between the layers of heterostructure they can be categorized as, Type-I-II-III heterostructure [11]. These are widely used to design optoelectronic devices and structures (Photodetectors, Laser diodes, Quantum well and superlattice optical and optoelectronic devices), optical components (Mirrors, Waveguides) and advanced electronic devices (Resonant tunneling devices, modulation doped FET). Figure 1 shows the band alignment for these heterostructures.

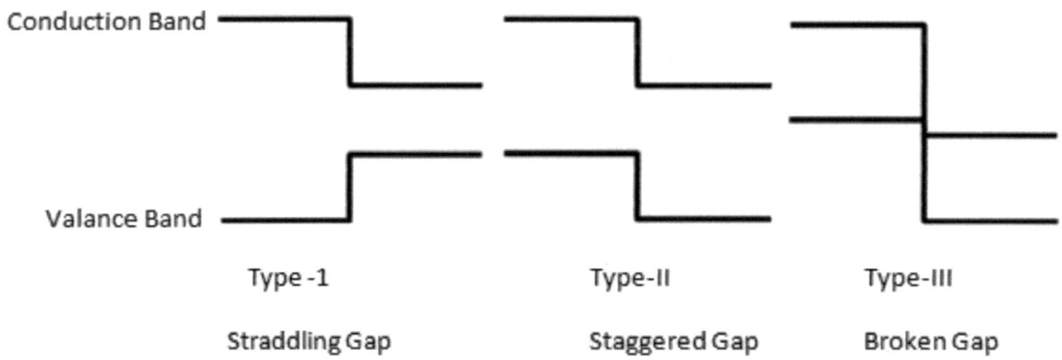

Figure 1 Band alignment types of heterostructure

A LASER is used in the field of medicine also. There are several techniques based on optical radiation which can be used for the diagnosis as well as for the treatment of the disease because of their non-invasive nature and high resolution [12, 13]. The major application in medical is fighting against cancer. The treatment of cancer can be done with different techniques. One of these techniques is photodynamic therapy which involves the use of chromophore that works as a catalyst for intermolecular energy transfer. PDT involves the injection of a photosensitizer which can collect within the tumor once exposed to the particular wavelength (630 nm-689 nm) which kills the cancer cells. Red LASER is preferred in PDT due to its short wavelength and low absorption of light by chromophore [14, 15].

2. Quantumwell LASER

In heterostructure one type is quantum well which has one thin "well" layer which is surrounded by two "barrier" layers. The layer in which these holes and electrons are confined is very thin, so we consider the electron and holes are both waves in the well. By using two techniques MBE (Molecular beam epitaxy) and MOCVD (Metal organic chemical vapor deposition) can be used for growing Quantum well structure [16]. We can use these techniques to make quantum well structure but the basic requirement is to keep Lattice constant of layers to be identical. Unidentical Lattice constants can lead to "Dangling bonds" which will create difficulty in retaining a well-defined crystal structure throughout the layers. A partial list of material used for the quantum well includes III-V group materials GaAs/GaAlAs on GaAs Type-I, GaSb/GaAlSb on GaSb Type-I, InAs/GaSb Type-II structures. II-VI group materials HgCdTe/CdTe, ZnSe/ZnMnSe [3, 10]. Quantum well lasers are laser diodes with a narrow active region which leads to quantum confinement. The determination of wavelength is just not dependent on energy bandgap of materials used but the thickness of the active region also plays an important role. The benefit of Quantum well laser is it can achieve much shorter wavelength than the conventional lasers. In Quantum well lasers the carrier motion is restricted in a particular direction normal to the well, which results in the set of discrete energy levels and density of states that can be modified. Modifications in the density of states provide us with improvement in laser characteristics such as lower threshold current, higher efficiency, and higher modulation bandwidth and dynamic spectral width. They are preferred over conventional lasers because they require fewer electrons and holes to reach the threshold. For higher quantum efficiencies in optical absorption, quantum well lasers are beneficial [17-19].

2.1 Band Structure of heterostructures

Band structure engineering involves the incorporation of strain, quantum confinement and superlattices [20]. This technology contains a conjunction of both quantum and strain confinement which is used in valence band that leads to more propitious energy dispersion in the laser as compared to the natural crystal. Recently it has been studied that lowered valence band effective mass is beneficial for semiconductor lasers. Heavy valance band mass is the main problem in semiconductor lasers [21]. Regrettably, this problem is common in III-V semiconductors heavy valence band mass and very light conduction band mass due to asymmetry. For proper functioning of the laser, these masses should be light as possible and density of states should be low. [22-23].

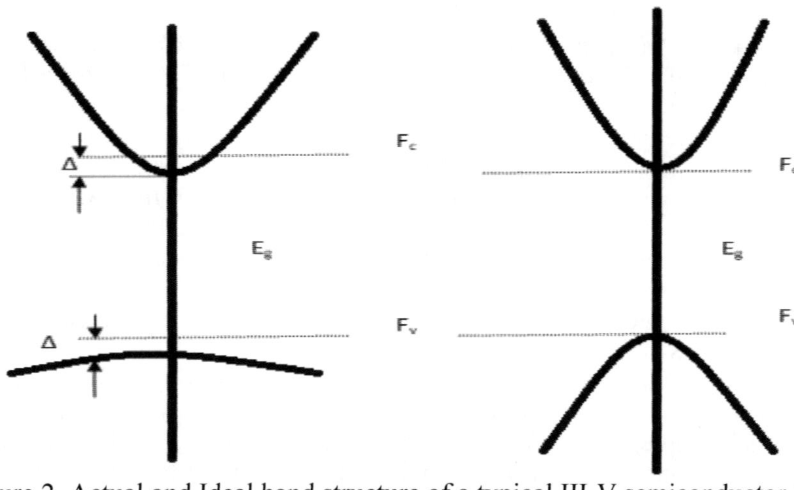

Figure 2. Actual and Ideal band structure of a typical III-V semiconductor [22]

In figure 2, the right image shows the ideal band structure for which lasing can be easily met. For this structure, both the holes and electron masses are light to achieve Bernard-Duraffourg Condition [24]. This condition is used to provide certain benefits to the laser-like threshold current, nonradiative recombination. In the laser, we have a threshold condition for gain that is given by

$$(Fc- Fv) > \hbar w \geq Eg \quad (1)$$

This equation denotes that for a better gain separation in LASER Quasi Fermi Levels (Fc-Fv) should be greater than the energy band gap. By achieving this condition valence band mass reduction takes place which leads to change in valence band structure. Hereby varying the valence band structure in order to lessen mass and population of heavy can lead to the solution of three troublesome aspects of semiconductor lasers a.) Asymmetry of conduction and valence band mass, b.) Intervalence band optical absorption and c.) The Auger recombination. Here intervalence band optical absorption is explained as a possible cause of temperature sensitivity of threshold currents in greater wavelength semiconductor lasers and in Auger recombination, two holes and one electron collide and recombine leaving only one

IOP Conf. Series: Materials Science and Engineering **594** (2019) 012002 doi:10.1088/1757-899X/594/1/012002

hole and some heat behind. Due to this Auger recombination laser threshold conditions and even nonradiative recombination is affected. So there is the need to reduce this rate of the combination. Some researchers reported that if we combine strain and quantum confinement in valence band this can provide us with more favorable energy dispersions for laser actions than in natural semiconductor crystal. This leads to band structure engineering in which laser threshold current and intervalence band absorption can be significantly diminished [17, 18, 25, 26]

2.2 Strain and electric field effects on heterostructure Lasers

In general when we talk about the stress we consider uniaxial stress i.e. force exerted along a particular direction divided by cross section of that surface. When a strain is applied, a degeneracy generated in original symmetry of semiconductor which leads to band splitting. The applied strain changes the band structure by reducing the crystal symmetry [25, 26]. Uniaxial stress results in degeneracy lifting and mixing of heavy holes and light holes bands which warps the bands and that is the reason for anti-crossing in quantum well valence band structure. Hence strain applied to change the wavefunctions and hence optical gain and bandgap energies are affected. Bandgap shifts directly affect the wavelength of ternary and quaternary compounds. It provides us extra flexibility in designing a semiconductor laser with more optimized parameters. Due to strain, change in density of state (DOS) affects the absorption efficiency of the quantum well laser [27-29]. There is a reduction in density of state for valence band which reduces the threshold current and increases gain. Strained quantum well lasers are investigated nowadays because of their potential to be used in laser printers and high power pump lasers. The Strained quantum well lasers are showing better results as compared to unstrained lasers in term of high optical gain.Chih-Sheng Chang et.al [18] has shown the effect of strain on the reliability of $In_xGa_{1-x}As/GaAs$ multiple quantum well lasers. The results clearly show that reliability improves with increment in strain in the quantum well. In order to understand the strain induced reliability improvement, the study of the propagation of defects in semiconductor heterostructure and the theory of elasticity is required [29].Different types of strain like the tensile and compressive strain that depends on lattice mismatching between substrate and bulk semiconductor are reported by the researchers [30, 34]. Jianfeng Chen [19] reported the InGaAsSb/AlGaAsSb quantum well diode structure, which is grown over GaSb by molecular beam epitaxy technique with the compressive strain of both 1% and 1.5% respectively. By observing the results it was concluded that use of InGaAsSb quantum well with compressively strain above 1% can be used for the reduction in threshold current density. For heavily strained structure threshold current density is nearly half (120A/cm^2) than that of moderately strained structure (230A/cm^2). E.V. Bogdanovet. al. [30] reported the effect of tensile strained on p-AlGaAs/GaAsP/n-AlGaAsquantum well and results are obtained for the external application of uniaxial stress. Uniaxial stress is used for wavelength tuning of laser diodes. The statement is concluded with the help of research work is done by I.V. Berman [35], who presented the effect of uniaxial stress on heavy and light holes mixing, optical gain in laser diodes and electron band. There is also an increment in gain under uniaxial compression for TE mode whereas in TM mode gain is decreased under uniaxial compression. So it was concluded that optical gain can be enhanced for tensile strained laser diode under uniaxial stress in TE mode more preferably then in TM mode.Under the effect of electric field on quantum well structure, there are variations in wavefunctions of both conduction and valence bands. Conduction bands electrons move opposite to the direction of electric field whereas valence band heavy holes move towards the direction of the electric field which changes their energy levels and hence emission wavelength is changed. The result is obtained for different values of the electric field applied and it was concluded that for high electric field optical gain is reduced but emission wavelength increases under electric field [33]. The effect of electric field on the optical gain of different semiconductor compounds with different compositions is studied in order to obtain laser operating at different wavelengths. The effect of electric field on the optical gain in $Al_{0.8}Ga_{0.2}As/GaAs_{0.8}P_{0.2}$ type-I nanoscaleheterostructure is reported [36]. The designedheterostructure is suitable for a wavelength ranging from (708nm to 826nm). Thus a wide range of tuning is available with the effect of electric field. Type-I heterostructure can be used for shorter wavelength and can be tuned by using the external electric field and strain [37-39]. Type-II heterostructure is advantageous because of their high interband transitions that lead to larger wavelength as compared to fundamental bandgap energies and widely used in optical fiber communication. It is found that GaAsSb/InGaAsheterostructure over GaAs substrate can be good material for the designing of 1.3μm wavelength the design is a 'W' type structure in which optical transitions are strongly dependent on conduction band offset. A design of type-II QW

heterostructure that is suitable for the mid-wave infrared region. MWIR photodiodes find their applications in gas monitoring, chemical sensing and medical diagnostics.H.K.Nirmal et.al [40] presented InGaAs/GaAsSb quantum well heterostructure and the behavior of optical gain under applied pressure is investigated. By applying variable pressure (2, 5 and 8 GPa) on the structure in [110] direction shows that optical gain and lasing wavelength both approaches higher value. The designed heterostructure can be tuned externally for the application of pressure and hence can be used for various applications in the SWIR region. The optical gain seems to improve up to a value of ~9000 /cm under pressure at a wavelength of 1.95μm. This shows that a high optical gain value can be obtained under a uniaxial pressure application. A.K. Singh et.al [41] presented the effect of uniaxial strain on type-II heterostructure. The results show that under x-polarization the gain is ~14500/cm and under y-polarization, the gain is ~15700/cm without strain after applying strain the envelope wavefunctions are modified valence subbands are moved to different positions which improve the gain. So, it is presented that the optical gain can be tuned by applying uniaxial strain and high gain can be achieved along [110] direction [39-41]. J.Y.T.Huang et.al [42], presented in his work that 3μm wavelength can be achieved without strain relaxation. The design is type-II 'W' quantum well structure on InP substrate along with $In_{0.8}Ga_{0.2}As/GaAs_{0.35}Sb_{0.65}$ quantum wells separated by InP barriers layers. In his work, it is reported that the higher band offset of AlAsSb over GaInP is preferred for improved carrier confinement. So it is concluded that the tuning of wavelength is possible with the help of external strain without changing the material itself.
Heterostructures can be categorized on the basis of confinement structures also such as simple separate confinement heterostructure (SCH) and graded index separate confinement structure (GRIN-SCH). P. A. Alvi et.al [43], intensive study of a graded index (GRIN) SCH type heterostructure is done, which can be used for the generation of 1.55μm wavelength, which is mostly used silica made optical fiber wavelength. Here both simple and GRIN structure is examined and results show that the designed structure has the high optical gain at 1.55 μm and 1.38 μm in TE and TM polarizations respectively. The conclusions are made that simple structure is better than GRIN structure.

3. Quantum dot Laser

Brief research and study are also made on quantum dot lasers which are also important in optical communication for new technology and for the requirement of high optical gain. Firstly the concept of quantum dots for semiconductor lasers was proposed by Arakawa and Sakaki in 1982 with a prediction for lesser threshold current density [44]. They are preferred as they are less temperature dependent compared to conventional semiconductor lasers. The advantages of Quantum dot lasers can be summarised as temperature independent, emission wavelength depending on energy levels of dots rather than bandgap energy, has the maximum material gain at least 2-3 times greater than quantum well lasers. Quantum dot lasers make use of quantum dots (electrons are confined in length according to the desired wavelength) as their active region for the transition of electrons. Quantum dots are used due to its small size (~2-10 nm or 10-50 atoms in diameter) and the emission frequency is sensitive to dots size and composition. In quantum dot laser freedom of electron, propagation is confined in all degrees of freedom as compared to quantum well lasers in which degree of freedom is limited to one direction. We can change the bandgap energy of quantum dot laser just by changing the geometry of the surface HuiziDiwu et.al [45], written a review paper in which introduction of quantum dot lasers, history, and physics of quantum dot lasers, types of quantum dot lasers, their fabrication techniques and market need of quantum dot laser is explained. Quantum confinement of quantum dot laser is better than quantum well laser hence its tunability is better than quantum well lasers [46-48]. Thomas Frostet. al [49], reported a nitride-based red laser with the help of InGaN/GaN quantum dots laser. The technique used for the growth of GaN, AlGaN, and InGaN on GaN substrate is plasma-assisted molecular beam epitaxy (PA-MBE). Here light-current characteristics are presented with a dot density of $3*10^{10}/cm^2$ emitting light at a wavelength of 635nm with cavity length 5μm and internal quantum efficiency to be 36% lesser than the quantum well laser. Laser arrays are complicated to design so a single band laser with high power laser operating at 635nm is designed to overcome the complications of array lasers. The structure consists of a single quantum well of GaInP with waveguide barriers of AlGaInP and cladding layers of AlInP. This high power 635nm laser finds its application in digital video disk and in data storage. The thinner quantum wells are preferred for high positive gain as compared to a thicker quantum well. The fabrication of quantum dot lasers involves commercial techniques like metal organic vapor phase epitaxy (MOVPE), Molecular beam epitaxy (MBE) and monolayer fluctuations [50-52].

4. Biomedical Lasers

Currently, from the research of laser in biomedical, it comes to know that red lasers have wide applications in medicine also. The formation of the red laser diode with nitride is expensive as well as harmful. So, a study about other alternatives for this red laser is made, and it is proposed that phosphides can be used for the generation of red wavelength [53, 54]. The red lasers are widely used in photodynamic therapy (PDT) for the treatment of damaged human tissues. For PDT application a large beam divergence is required. For the biomedical application of laser the power is a limited factor and depends on the type of diseases. Diode lasers operating in the range of 630 nm to 740nm are suitable for PDT and dermatological applications. PDT makes use of photosensitizer drug injected with a specific wavelength of light to be absorbed by the cell [55-59].

FarhadH.Mustafa [13] reported the effect of laser wavelengths on PDT for the treatment of superficial skin diseases. He has compared illumination intensity of several wavelengths within the range of 308nm, 365nm, 405nm, 532nm and 635 nm in a skin sample with Advanced System Analysis Program (ASAP) software for the calculation of transport intensity of laser in the skin. In this work study of dose absorption by photosensitizer and melanin is done. The ability of short and long wavelength is studied to show power uptake effect in chromophore cell in human skin. A comparison between UV laser and red laser is for dose absorption is made. The results obtained are the peak power uptake by UV laser is 501mW/mm-3 and by the red laser, it is 71.5mW/mm-3 which is 7 times less and for the same skin depth (Z = 0.019mm). By this, we conclude that UV laser transmits the minimum amount of laser as it is highly absorbed by the chromophore. This is the reason red laser is preferred over UV laser for treatment in PDT as there is less interaction by Red laser.

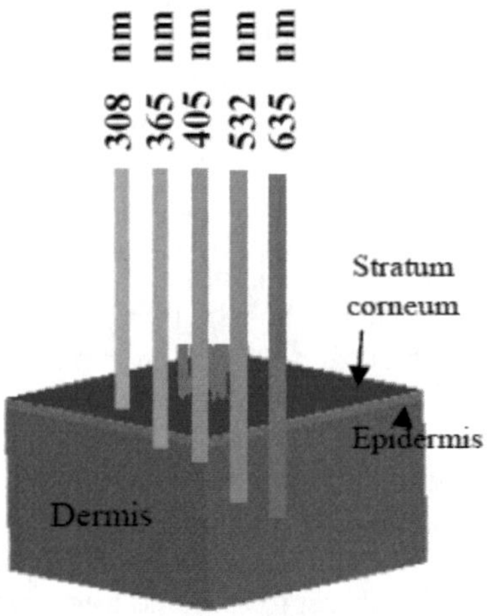

Figure 3. Stratum, corneum, epidermis and dermis layers of skin with 5 different laser wavelengths [13]

Table1. List of photosensitizers with activating wavelength [60]

Wavelength (nm)	Photosensitizer (Company)
630	Photofrin II (QLT)
630-635	ALA (DUSA, ESC)
652	mTHPC/Foscan (Scotia Quanta Nova)
665	SnET2 (Miravant)
688	HBP(QLT)

L. Brancaleonet.al [61] discussed the characteristics of light sources of PDT, including dye lasers pumped by argon, frequency doubled YAG lasers and non-laser sources such as tungsten filament, metal halide, xenon arc, and fluorescent lamps. Argon lasers are mostly used devices for PDT treatment. The argon pumped dye lasers are coupled with optical fiber to be used in the treatment of lung cancer, oral precancer, bladder cancer and for the treatment of superficial skin disease. The metal vapor pumped dye lasers [62] are also used in PDT but unlike argon laser, they are normally pulsed at a pulse width ranging from 10ns to 50ns. These lasers are able to deliver at irradiance up to several hundred mW/cm . Nowadays PDT is not limited to lasers, it makes use of high power lamps for filtered output.

They are preferred over lasers because they provide large spectral range as compared to lasers. They have a drawback that their output power can't be easily coupled to small optical fibers and hence is limited in the treatment of skin lesion only and not been used for endoscopic PDT. Tungsten filament Quartz Halogen lamps were first introduced by Pottier and Kennedy [60]. These lamps are used as they can deliver on a wide spectral range of (350-850 nm) with output power 250mW/cm2. Another category of lamps used is Xenon Arc lamps providing a higher spectral region for emission (300-1200 nm) along with high output power up to 8 W for direct exposure and up to 1W using liquid light guide.
PDT can be used for the treatment of tumors [63, 64]. The length of diffuser for the treatment of a tumor depends on the size of the tumor to be treated. The method of light delivering depends on the geometry of the tumor and the extent of the tumor. Sometimes surface illumination is not sufficient for the treatment, so interstitial illumination is required for that purpose balloon catheters are designed [65]. Balloon catheters are the latest development in the field of PDT. In this technique, a transparent balloon with cylindrical fiber positioned at the center is inflated in the patient esophagus. The diameter of balloon inflated is limited to 25mm only as large diameter can lead to serious tissue damage which further to blood flow. Many more advancements and research work are going on in PDT for other treatment with the help of various photosensitizer.
P. Babilas et.al [66] presented a comparison between in vitro and in vivo for photodynamic therapy with two different light sources. To understand the effectiveness of LED over incoherent lamp this comparison trial is done by using 5-aminolaevulinic acid (ALA) for the treatment of actinic keratosis (AK). For the treatment after incubation with ALA irradiation by LED system (40 Jcm^{-2}, 80mWcm-2) on one side and on the other side irradiation is done by the incoherent lamp (100 J/cm^2, 160 mW/cm^2). The LED system used is operating on wavelength 633 nm and incoherent lamps range between 580nm-700nm respectively. This treatment was done on 40 patients for 24h the result shows the very minor difference between ALA-PDT using LED system and incoherent lamps. These light sources can be implanted also for treatment of various cancers and tumors. The results obtained provide us with the idea of implanting of multiple sources to get optimal energy for localized PDT. In PDT penetration depth needs to be measured to understand the effect of photosensitizer on human tissues. S. Stolik et al. [64] reported the results for the penetration depth of more than 10 tissues at different wavelengths.

Table 2. Penetration depths of different tissues with wavelength [64]

Tissue	Wavelength (nm)	Δ (mm)
Lung	630	0.81±0.06
Lung Carcinoma	630	1.68±0.15
Uterus	630	2.14±0.18
Brain	675	1.38±013
Liver	630	1.20±0.13

Earlier PDT used for cancer treatment involves injection by the light source but now Photosensitizers can be directly generated inside the tumor with the help of pre photosensitizer. So PDT is a specific method for destroying malignant, premalignant and other benign tissues while sparing the surrounding healthy cells. The IPDT (Interstitial Photodynamic therapy) is used for the treatment of deep-seated malignant tumors and brain tumor treatments. It is highly compatible with currently available standard and experimental treatment options for malignant glioma [67, 68].

5. Conclusion

In this paper, we came across various structural parameters and optical characteristics of semiconductor lasers. The requirement of the laser is the high optical gain and controllable performance. The review result also shows the effect of quantum well thickness, the effect of external strain and temperature on the optical gain for type-I and type-II heterostructure. The red laser is widely used in biomedical for treatment of various diseases. Currently, 635 nm red laser fabricated using the phosphides is widely used in PDT for the treatment of superficial skin diseases, because these lasers are less absorbing and can be precisely focussed to the infected tissue. A high optical gain is required in order to reduce the absorption of drugs in PDT.

6. References

[1] Shun Lien Chuang, "Physics of Optoelectronic Devices", ISBN 0-471-10939-8, Wiley Interscience Publication, March 1995.
[2] Chin-Lin Chen "Elements of Optoelectronics and Fiber Optics'', School of Electrical and Computer Engineering, Purdue University, Irwin, Chicago, Bogota, Boston. 1996.
[3] David A.B. Miller, "Optical Physics of Quantum Well", Research Gate publications, Vol.-38, no. 8, pp. 25 July 2014.
[4] M. J. Weber, "Handbook of Lasers", CRC Press, 2000.
[5] O. Svelto, "Principles of Lasers," Plenum Press, NY 1998.
[6] Coldren, L. A.; Corzine, S. W. "Diode Lasers and Photonic Integrated Circuits", University of California, Santa Barbara, California, John Wiley & Sons, Inc., New York, Toronto, Singapore; 1995
[7] Chow, W. W.; Koch, S. W. (2011). Semiconductor-Laser fundamentals. Springer. ISBN 978-3540641667.
[8] Hader, J.; Moloney, J. V.; Koch, S. W.; Chow, W. W. (2003). "Microscopic modeling of gain and luminescence in semiconductors". IEEE J. Sel. Top. Quant. Electron. 9 (3): 688–697. DOI:10.1109/JSTQE.2003.818342.
[9] Kroemer, H. (1963). "A proposed class of hetero-junction injection lasers". Proceedings of the IEEE. 51 (12): 1782. DOI:10.1109/PROC.1963.2706 .
[10] W.R.Frensley and N. G. Einspruch, "Heterostructures and Quantum Devices", eBook ISBN 978-1-48329-517-6, Academic Press, May 15, 1998.
[11] Franciosi A.; Van de Walle C.G: Heterojunction band offset engineering, Surface Science Reports, Volume 25, Number 1, October 1996, pp. 1–140.
[12] T.Vo-Dinh, "Biomedical Photonics Handbook", ISBN 0-8493-1116-0, Boca Raton: CRC Press, February 2003.
[13] FarhadH.Mustafa, Mohamad S. Jaafar, AsaadH.Ismail, Hend A.A. Houessin, "The effect of laser wavelength in photodynamic therapy and phototherapy for superficial skin diseases", IEEE International Conference on Imaging Systems and Techniques (IST), pp.232-236, July 2011.
[14] Rishi P, Agarwal V. Current Role of Photodynamic Therapy in Ophthalmic Practice, Sci J Med & Vis Res Foun 2015; XXXIII:97–99.
[15] Josefsen LB, Boyle RW. Photodynamic Therapy and the Development of Metal-Based Photosensitisers. Metal-Based Drugs. 2008; 2008:276109. DOI:10.1155/2008/276109
[16] A.Y.Cho, "Advances in molecular beam epitaxy (MBE)" Journal of Crystal Growth, Volume 111, Issues 1–4, 2 May 1991, Pages 1-13.
[17] Baile Chen, W.Y.Jiang, A.L.Holmes Jr., Design of strain-compensated InGaAs/GaAsSb type-II quantum well structures for mid-infrared photodiodes, Opt. Quant. Electron.44(2012)103–109.
[18] Chih-Sheng Chang, Shun Lien Chuang, " Modelling of strained quantum well lasers with spin-orbit coupling", IEEE Journal of Selected Topics in quantum electronics, Vol.-1, no.-2, pp. 218-229, June 1995.
[19] Jianfeng Chen, Dmitry Donetsky, Leon Shterengas, Mikhail V. Kisin, Gela Kipshidze, "Effect of Quantum Well Compressive Strain Above 1% On Differential Gain and Threshold Current Density in Type-I GaSb-Based Diode Lasers", IEEE Journal of Quantum Electronics Vol.- 44, no.-12, pp. 1204-1210, December 2008.
[20] E. Yablonovitch and E. O.Kane, "Band Structure Engineering of Semiconductor Lasers for Optical Communications," Journal of Lightwave Technology, Vol.- 6, no.- 8, pp. 1292-1299, August 1988.
[21] R. Adams, "Band structure engineering for low threshold high-efficiency semiconductor lasers," Electron Lett., Vol.- 22, no.- 5, pp. 249-250, February 1986.
[22] I.Vurgaftman, J. R. Meyer and L.R. Ram-Mohan, "Band parameters for III–V compound semiconductors and their alloys", Journal of Applied Physics, Vol.- 89, no.-11, pp. 5815-5863, June 2001.
[23] G.C. Osbourn, R.M. Biefeld, and P.L. Gourley, "A GaAsP/GaP strained layer superlattice," Appl. Phys. Lett. , Vol.-41, no. - 2, pp. 172, August 1982.
[24] M.G.A. Bernard and G.Duraffourg, "Laser conditions in semiconductor," Phys. Stat. Solidi, Vol.-1,no.-7, pp. 699-703, January 1961.
[25] E. Yablonovitch and E. O.Kane, "Reduction of lasing threshold current density by the lowering of valence band effective mass," J.Lightwave Technol., Vol.-4, no.-5, pp. 504-506, May 1986.
[26] Y. Sun, S. E. Thompson, and T. Nishida, "Physics of strain effects in semiconductors and metal-oxide-semiconductor-field-effect transistors", Journal of Applied Physics, Vol.-101, no.-10, pp. 104503(1)-104503(21), March 2007.
[27] Yongke Sun, Scott E. Thompson and Toshikazu Nishida, "Strain effects in Semiconductors" ISBN 978-1-4419-0552-9, Springer, 2010.
[28] G.C.Osbourn, InxGa1-xAs–InyGa1-yAs strained-layer superlattices: a proposal for useful, new opto electronicmaterials", Phys.Rev.B27(1983)5126–5128.

[29] H. Yoon, Y. C. Chen, L. Davis, H. C. Sun, K. Zhang, J. Singh, and P. K. Bhattacharya, "Reliability of Strained Quantum Well Lasers", Proceedings of IEEE/Cornell Conference on Advanced Concepts in High-Speed Semiconductor Devices and Circuits, Vol.-2, no.- 11, pp. 348-356, August 1993.

[30] E.V. Bogdanov, N. YaMinina, J.W. Tomm, H. Kissel, "Effect of uniaxial stress on electroluminescence valence band modification optical gain and polarization modes in tensile strained p-AlGaAs/GaAsP/n-AlGaAs laser diode structures: numerical calculations and experimental results", Journal of Applied Physics Vol.-112, no.-9,pp. 09311(1) -09311(9), October 2012.

[31] Hidenao Tanaka, Jun-Ichi Shimada, Yoshio Suzuki, "Highly efficient TE/TM mode switching of GaAsP/AlGaAs strained quantum-well laser diodes", Applied physics letters Vol.-64, no.-2, pp. 158-160, June 1998.

[32] Baile Chen, W.Y. Jiang, A.L. Holmes Jr., "Design of strain compensated InGaAs/ GaAsSb type-II quantum well structures for mid-infrared photodiodes", Opt. Quant Electron, Vol.- 44,no.-3-5, pp. 103-109, June 2012.

[33] Baile Chen, A.L. Holmes Jr, " Modelling of the type-II InGaAs/GaAsSb quantum well designs for mid infrared laser diodes by k.p. method", SPIE Laser Technology for Defense and Security VIII, Vol.-8381, no.- 23, pp. F 1- F 7, 2012.

[34] Sang-Wan Ryu, P.D. Dapkus, "Optical characterization and determination of conduction band offset of type-II GaAsSb/InGaAs QW", Semicond. Sci. Technol. Vol.-19, no.-12, pp. 1369–1372, October 2004.

[35] I. V. Berman, E. V. Bogdanov, H. Kissel, N. Ya. Minina, S. S. Shirokov, and A. E. Yunovich, Phys. Status Solidi B Vol.-246, pp. 522, 2009.

[36] Md. Riyaj, A. K. Singh, Sandhya K., Amit Rathi, P. A. Alvi, Garima Bhardwaj, "Wavefunctions and optical gain in Al0.8Ga0.2 As/GaAs0.8 P0.2 type-I QW-heterostructure under the external electric field", International Conference on Computer, Communications and Electronics (Comptelix), pp.59-62, July 2017.

[37] Md. Riyaj, A. K. Singh, Sandhya K., Amit Rathi, P. A. Alvi, "Optical properties of Type-I GaAsP/AlGaAsnano-heterostructure under external uniaxial strain", AIP Conference Proceedings Vol.-1832, no.-1, pp. 120022.1-120022.3, 2017.

[38] H. K. Nirmal, S. G. Anjum, PyareLal, Amit Rathi, S. Dalela, M. J. Siddiqui, P. A. Alvi "Field effective band alignment and optical gain in type-I Al0.45Ga 0.55 As/GaAs 0.84 P 0.16 nano-heterostructures", International Journal of light and electron optics: Optik Vol.-127, no.-18, pp. 7274-7282, May 2016.

[39] P.A. Alvi, "Strain-induced non-linear optical properties of straddling-type indium gallium aluminum arsenic/indium phosphide nanoscale-heterostructures", Material Science in Semiconductor Processing, Elsevier, Vol.-31, no.-10, pp.106–115, March 2015.

[40] H.K.Nirmal, Nisha Yadav, S.Dalela, Amit Rathi, M.J.Siddiqui, P.A.Alvi, " Tunability of optical gain (SWIR region) in type-II In0.70Ga0.30As/GaAs0.40Sb0.60 nano-heterostructure under high pressure", Physica E: Low-dimensional Systems and nanostructures, Elsevier Vol.- 80, no.-10, pp. 36-42, June 2016.

[41] A.K.Singh, Md.Riyaz, S.G.Anjum, Nisha Yadav, Amit Rathi, M.J.Siddiqui, P.A.Alvi, "Anisotropy and optical gain improvement in type-II In0.30Ga0.70As/GaAs0.4Sb0.6 nano-scale heterostructure under external uniaxial strain", Superlattices and Microstructures Vol.- 98, no.-C, pp. 406-415, August 2016.

[42] J.Y.T.Huang, L.J.Mawst, T.F.Kuech, X.Song, S.E. Babcock, C.S.Kim, "Design and characterization of strained InGaAs/GaAsSb type-II 'W' quantum wells on InP substrates for mid infra red emission", Journal of Physics D: Applied Physics Vol.-42, no.-2, pp. 025108.1-025108.9, December 2009.

[43] P.A. Alvi, PyareLal, S. Dalela, M.J. Siddiqui, "An extensive study on simple and GRIN SCH-based In0.71Ga0.21Al0.08As/InP lasing heterostructures", Physica Scripta, Vol.- 85, no.-3, pp. 035402(1)-035402(9), 2012.

[44] Arakawa, Y.; Sakaki, H. (1982). "Multidimensional quantum well laser and temperature dependence of its threshold current". Applied Physics Letters. 40: 939

[45] HuiziDiwu and BetulArda, "Quantum Dot Lasers", ECE-580 Term Project, 2012.

[46] T. Kita, N. Yamamoto, T. Kawanishi, and H. Yamada, "Ultra-compact wavelength-tunable quantum-dot laser with silicon-photonics double ring filter," Appl. Phys. Express 8(6), 062701 (2015).

[47] J. Norman, M. J. Kennedy, J. Selvidge, Q. Li, Y. Wan, A. Y. Liu, P. G. Callahan, M. P. Echlin, T. M. Pollock, K. M. Lau, A. C. Gossard, and J. E. Bowers, "Electrically pumped continuous wave quantum dot lasers epitaxially grown on patterned, on-axis (001) Si," Opt. Express 25(4), 3927–3934 (2017)

[48] S. Banyoudeh, O. Eyal, A. Abdollahinia, F. Schnabel, V. Sichkovskyi, J. P. Reithmaier, and G. Eisenstein, "High-bandwidth temperature-stable 1.55-μm quantum dot lasers," Proc. SPIE 10123, 1012306 (2017).

[49] Thomas Frost, Animesh Banerjee, Pallab Bhattacharya, " Red-emitting InGaN/GaN Quantum Dot Laser", 71st Device Research conference IEEE, Vol.-13, no.-10,pp. 215-216, October 2013.

[50] Peter M. Smewton, Gareth M.Lewis, Peter Blood, Weng W. Chow, "Optimization of 635-nm tensile strained GaInP Laser Diodes", IEEE Journal of Selected Topics In Quantum Electronics, Vol.- 9, no.- 5, pp.1246-1251, October 2003.

[51] Chang Zoo Kim, Je Hyuk Choi, Keun-Man Song, Chan Soo Shin, and ChulGiKo, "Optical Investigation of GaInP-AlGaInP Quantum-well Layers for High-power Red Laser Diodes", Journal of the Korean physical society, Vol.- 62, no.- 9, pp. 1301-1306, May 2013.

[52] Antti Tukiainen , JariLikonen , LauriToikkanen , TomiLeinonen, "Unintentional boron contamination of MBE-grown GaInP/AlGaInP quantum wells", Journal of Crystal Growth, Vol.- 425, pp. 60-63, September 2015.

[53] L. Brancaleon and H. Moseley, "Laser and Non-laser Light Sources for Photodynamic Therapy", Laser Medical Science, Vol.-17, no.-3, pp. 173-186, April 2002.

[54] Ivan Charamisinaa, Germunu S. Happawana, Gary A.Evans, Jay B. Kirk, David P. Boar, Arye Rosen, R.Alexander, " High- Power Semiconductor Red Laser Arrays for use in Photodynamic Therapy", IEEE Journal Of Selected Topics in Quantum Electronics, Vol.- 11, no.- 4, pp. 881-891, December 2005.

[55] T.S. Mang, "Lasers and light sources for PDT: Past, present and future", Photodiagnosis and Photodynamic Therapy, Vol.-1, no.-1, pp. 43-48, May 2004.

[56] B. Maiya, "Photodynamic Therapy (PDT)", Resonance Springer, Vol.-5, no.-4, pp. 13-21, August 2000.

[57] Douglass Mead, "High Power Red Diode Laser Systems for Medical Applications", IEEE Applied Optronics Corporation, Vol.-49, no.-4, pp. 394-395, August 2002.

[58] D Cerniglia, A Pantano, C Mineo, "Influence of laser beam profile on the generation of ultrasonic waves", Journal Applied Physics A, Vol.- 105, no.-4, pp. 959-967, December 2011.

[59] Douglass Mead, "High Power Red Diode Laser Systems for Medical Applications", IEEE Applied Optronics Corporation, Vol.-49, no.-4, pp. 394-395, August 2002.

[60] Kennedy JC, Marcus SL, Pottier RH," Photodynamic therapy (PDT) and photodiagnosis (PD) using endogenous photosensitization induced by 5-aminolevulinic acid (ALA): mechanisms and clinical results", J Clin Laser Med Surg. 1996 Oct;14(5):289-304.

[61] L. Brancaleon and H. Moseley, "Laser and Non-laser Light Sources for Photodynamic Therapy", Laser Medical Science, Vol.-17, no.-3, pp. 173-186, April 2002.

[62] Piper J.A. (1996) Frequency Extension of Metal Vapour Lasers. In: Little C.E., Sabotinov N.V. (eds) Pulsed Metal Vapour Lasers. NATO ASI Series (Series 1: Disarmament Technologies), vol 5. Springer Dordrecht.

[63] P.Babilas and R.M. Szeimies, "Photodynamic Therapy in Laser and IPL Technology in Dermatology and aesthetic medicine", ISBN 978-3-642-03438-1, Springer-Verlag Berlin Heidelberg, 2011.

[64] S. Stolik, J.A. Delgado, A. Pe´rez, L. Anasagasti, "Measurement of the penetration depths of red and near infrared light in human ''ex vivo'' tissues", Journal of Photochemistry and Photobiology, Vol.-57, no.- 2-3, pp. 90-93, September 2000.

[65] Roif H. Eichenauer, GereonHuttmann, Stephan Wormert, Norbert Koop, Wolfgang Beyer, Dieter Jocham, "A new balloon catheter system used for PDT in the human urinary bladder, accuracy of light distribution", SPIE Vol. 3247 • 0277-786X

[66] P. Babilas, E. Kohl, and T.Maisch, "In vitro and in vivo comparison of two different light sources for topical photodynamic therapy", British Journal Dermatology, Vol.-154, no.-4, pp.712-718, April 2006.

[67] L. Pacheco, S. Stolik, J.De la Rosa, A.De. la Cadena, " Laser Diode System for Interstitial Photodynamic Therapy", IEEE Latin America Transactions, Vol.-12, no.- 4, pp. 574-579, June 2014.

[68] A. Johansson, FW Kreth, W. Stummer, H.Stepp, "Interstitial Photodynamic Therapy of Brain Tumors", IEEE Journal of selected topics in quantum electronics, Vol.-16, no.- 4, pp. 841-853, August 2010.

IOP Conf. Series: Materials Science and Engineering **594** (2019) 012003 doi:10.1088/1757-899X/594/1/012003

Simulating 1.55 μm Optical Gain in Type-II InAlAs/InGaAs/GaAsSb Nanoscale Heterostructure

A. M. Khan[1], Garima Bhardwaj[2], M. Abu-Samak[3], S. H. Saeed[1], P. A. Alvi[4*]

[1]Department of Electronics & Communication Engineering, Integral University, Lucknow- 226026, UP (INDIA)
[2]Department of Electronics, Banasthali Vidyapith, Banasthali-304022, Rajasthan (INDIA)
[3]Department of Physics, Al-Hussein Bin Talal University, P.O. Box 20, Ma'an, JORDAN
[4]Department of Physics, Banasthali Vidyapith, Banasthali-304022, Rajasthan (INDIA)
E-mail: *drpaalvi@gmail.com

Abstract. This particular work has been designed to give an idea regarding W shaped nano based type II heterostructure and for the above particular design simulation concept have also been carried out in a theoretical way to obtain optical gain for the given material system InAlAs/InGaAs/GaAsSb, which is used to be operated at 1.55 μm near IR. The designed heterostructure contains two quantum wells (width ~ 2 nm) of InGaAs material. To solve the carrier's localities within the heterostructure, k.p method has been adopted. As far as calculations are concerned we have opted a defined phenomena to measure the significant characteristics with respective to optical nature like dispersion, gain and many other properties have also been found in order to get good results that would be helpful to emit radiations of the order of the working wavelength and to employ great communication systems applications that can be optical fiber based

Index-Terms: Dispersion, Gain, InGaAs, InAlAs, GaAsSb

1. Introduction

Since last two decades, the InGaAlAs (quaternary compound semiconductor) material based type-I heterostructure grown on InP substrate has been very popular due to emission of radiations of ~ 1.55 μm wavelength. The defined wavelength is the main catch of this particular work due to its potential applications in optical fiber communications because of very low attenuation [1-3]. The study regarding the formed heterostructure states that the charge carriers in active region are confined and the study also gave an idea regarding photons in the material system which are also confined in the above said region containing a quantum well along with an impression that this particular statement with respect to electrons, holes and photons that they all have a good refractive index in the defined active region. So this leads to the proper operation of laser diodes at room temperature as there is a dominant nature and good efficiency due to double confinements, as discussed above. The above phenomenon of double

IOP Conf. Series: Materials Science and Engineering **594** (2019) 012003 doi:10.1088/1757-899X/594/1/012003

confinements also shows that other parameters like stability with respect to gain and tunable wavelength are having great performance at high temperature also. But the substrate of GaAs which is of great concern for this work finds a suitable way for such type-II material system to emit identical radiations of wavelength that were also being radiated by type –I material system in a positive manner. So we can say that this particular work corresponds to InAlAs/InGaAs/GaAsSbBi heterostructure, the proposed work is also compared with the same material system as mentioned above [4].

2. Structure and Theory

The idea of two QWs has shown a remarkable behavior as it is used for W shaped heterostructure which corresponds to good material, the barrier and claddings of InGaAs, GaAsSb and of InAlAs, respectively

Table 1: nature of material

S.no	Parameters	Width (nm)
1.	Quantum Well	2
2.	Barrier	6
3.	Claddings	10

So in a symmetrical way the above heterostructure of total size 30nm is grown on GaAs substrate in order to get required results as well as to match lattice parameters for removing redundant strain production that would lead to bad device performance. In this paper 6 × 6 Luttinger-Kohn Hamiltonian method is used to determine the confinement of charge carriers which are allied to the standard quantum well for a given material system. By considering estimated effective mass for proper calculations [5-6], the material in an active region shows a vital performance in order to get stability, good strain and many other advantageous can be carried out with the help of this above said method. Therefore the optical gain of the entire material system can be measured or calculated with respect to the symmetrical heterostructures containing QWs that corresponds to the cladding on both sides, which is given by the equation [5]

$$G(\hbar\omega) = \frac{2.\pi e^2}{nc\epsilon\omega L m^2} \sum_{\sigma=U,L} \sum_{n,m} \int \left| \left(\hat{e}.\, M_{nm}^{\eta\sigma}(k_t) \right) \right|^2 \times \frac{\left(f_n^c(k_t) - f_{\sigma m}^v(k_t) \right) \left(\frac{\gamma}{\pi} \right)}{\left(E_{\eta,\sigma nm}^{c,v}(k_t) - \omega\hbar \right)^2 + \gamma^2} \frac{k_t dk_t}{2\pi}$$

3. Results and Discussion

The electronic wave functions and hole wave functions are carried out to determine localities between the charge carriers in order to confirm the nature of defined type-II heterostructure. Also according to the readings and calculations the dipole transition elements can be carries out or determined after having above defined parameters. By considering the wave functions of carriers on nano scale, it can be seen that the electrons are related to the conduction bands of the quantum well only whereas holes find its place in barrier region only, figure 1 (a & b). therefore the transition of elements according to the wave functions occurs from well region associated with InGaAs to barrier region which is associated with GaAsSb, and this mentioned property of transition is not there for type –I material system (InGaAlAs/InP) Now the optical gain comes to the picture as it is most significant and basis requirements in order to understand the heterostructures that can be applied to make other opto- electronic devices which leads to good performance in near future for laser based devices. However the recombination process of carriers is responsible for great emission of radiation in as stimulated way by considering type-II InAlAs/InGaAs/GaAsSb nanoscale heterostructure. The simulated work also shows the clear behavior of optical gain with a photonic energy of 0.80eV which is operated at the same wavelength 1.55 μm, figure 2. The figure 3, corresponds to the comparable work of type-II InAlAs/InGaAs/GaAsSbBi nanoscale heterostructure which shows the optical gain phenomena associated with Indium phosphide substrate. It can also be seen from figure 3, that the Bi concentration is having a wavelength from 2 μm to 3 μm. So the low gain can be increased by the process of carrier injection in mid infrared region and the other region like NIR and SWIR used to be operated for InAlAs/InGaAs/GaAsSb and InAlAs/InGaAs/GaAsSbBi heterostructure, respectively

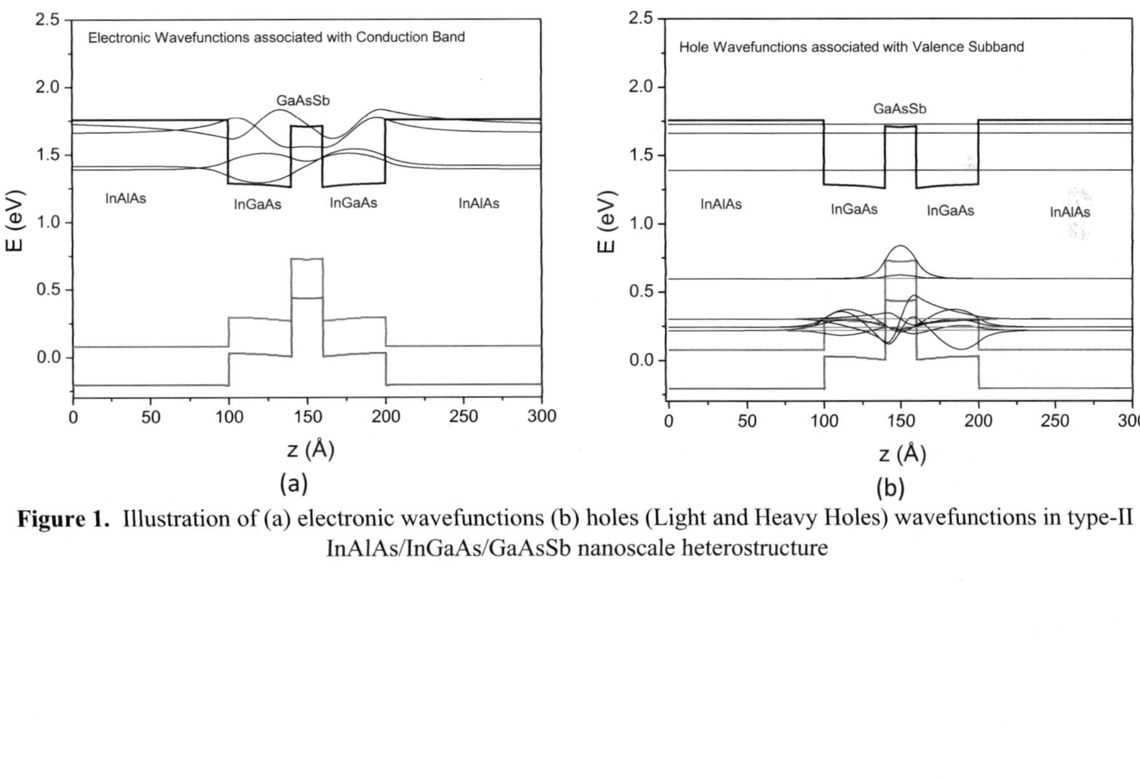

Figure 1. Illustration of (a) electronic wavefunctions (b) holes (Light and Heavy Holes) wavefunctions in type-II InAlAs/InGaAs/GaAsSb nanoscale heterostructure

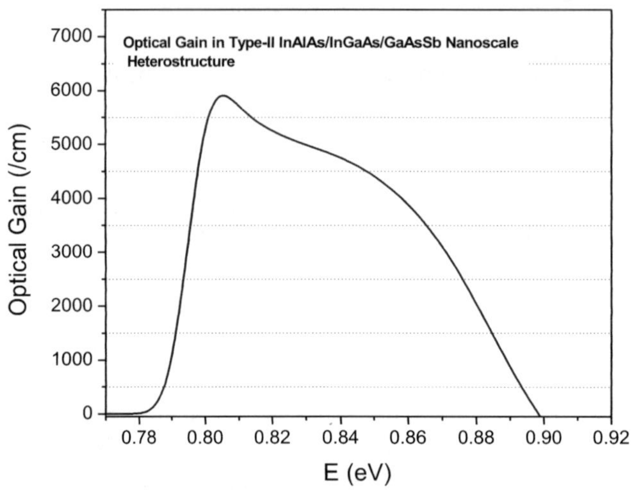

Figure 2. Optical gain for type-II InAlAs/InGaAs/GaAsSb nanoscale heterostructure

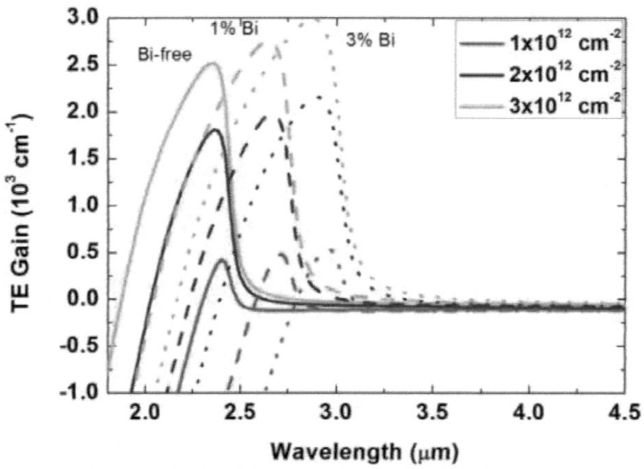

Figure 3. Optical gain for type-II InAlAs/InGaAs/GaAsSbBi nanoscale heterostructure [4].

Conclusion

So from the above proposed work of nano based it is clear that the type –II heterostructure is operated at 1.55 μm and all the results have been simulated accordingly for great performance using well known K-P method. Also the results show that the material of such measurements, wavelength , strain and transition of charge carriers from well to barrier region is having a great impact on optical fiber communication systems

Acknowledgement

P. A. Alvi is very thankful to the DST Government of India, New-Delhi for providing computational facilities to Banasthali Vidyapith under CURIE programme.

IOP Conf. Series: Materials Science and Engineering **594** (2019) 012003 doi:10.1088/1757-899X/594/1/012003

References

[1]. P. A. Alvi, Pyare Lal, S. Dalela, M. J. Siddiqui, "An Extensive Study on Simple and GRIN SCH based $In_{0.71}Ga_{0.21}Al_{0.08}As$/InP Lasing heterostructure", Physica Scripta, 85, 035402 (2012).

[2]. P. A. Alvi, Pyare Lal, Rashmi Yadav, Shobhna Dixit, S. Dalela, "Modal gain characteristics of GRIN-InGaAlAs/InP lasing nano-heterostructures" Superlattices and Microstructures, Vol. 61, pp. 1-12 (2013).

[3]. P. A. Alvi, "Strain-induced non-linear optical properties of straddling-type indium gallium aluminum arsenic/indium phosphide nanoscale-heterostructures", Materials Science in Semiconductor Processing, Vo. 31 (2015) pp. 106-115.

[4]. Baile Chen, "Active Region Design and Gain Characteristics of InP-Based Dilute Bismide Type-II Quantum Wells for Mid-IR Lasers", IEEE Transactions on Electron Devices, Vol. 64, NO. 4, 2017.

[5]. S. L. Chuang, Physics of optoelectronic devices (Wiley, New York, 1995).

[6]. H. K. Nirmal, Nisha Yadav, F. Rahman, P. A. Alvi, "Optimization of High Optical Gain in Type-II $In_{0.70}Ga_{0.30}As/GaAs_{0.40}Sb_{0.60}$ Lasing Nano-heterostructure for SWIR Applications" Superlattices Microstruct., 88, 154-160 (2015).

IOP Conf. Series: Materials Science and Engineering **594** (2019) 012004 doi:10.1088/1757-899X/594/1/012004

Effect of Fe Doping on Magnetic Behavior of SnO_2 Nanoparticles for Spintronics Applications

Mayuri Sharma[1], S. N. Doila[2], Shalendra Kumar[3], P. A. Alvi[1*]

[1]Department of Physics, Banasthali Vidyapith, Banasthali-304022, Rajasthan (INDIA)
[2]Department of Physics, University of Rajasthan, Jaipur-302004, Rajasthan (INDIA)
[3]Electronic Materials & Nanomagnetism Lab, Department of Applied Physics, Amity School of Applied Sciences, Amity University Haryana, Gurgaon-122413 (INDIA)
E-mail: [*] drpaalvi@gmail.com

Abstract. Doping of magnetic ion (Transition metal) in oxide based semiconducting material at nanoscale level serves a variety of spintronics based applications. In order to understand the influence of doping concentration of Fe ions on the magnetic behavior of semiconducting oxides, we have synthesized Fe doped tin oxide nanoparticles with 1 and 2% doping concentration via chemical co-precipitation method and then magnetic measurements have been taken through Superconducting quantum interference device (SQUID) at 5K and 300K temperatures for both the samples. Various magnetic quantities such as coercivity (H_c), saturation magnetization (M_s), remanent magnetization (M_r) have been extracted from the obtained hysteresis loops and found reduction in their values with increasing doping concentration which might corresponds to the tuning of oxygen defects created due to the substitution of dopant ions in the host lattice. In this way, our study can contribute to fulfill the necessity of ferromagnetism based applications such as data storage devices, gas sensors, communication, quantum computation etc.

Keywords. Dilute magnetic semiconductors, Transition metals, Spintronics, SQUID, Ferromagnetism, etc.

1. Introduction

Spintronics technology is one of the current areas of curiosity and has attracted enormous efforts of the researchers to enhance the efficiency of electronic devices for high speed information processing and in supplement to enrich the spin degree of freedom in order to enlarge data storage capacity [1]. To accomplish this multi-functional requirement, a new class of materials has come into sight, usually known as dilute magnetic semiconductors (DMS) which offers immense technical opportunities due to their possible utilizations in spin transport electronic devices [2-7]. Practically, DMS are alloys of semiconducting crystal lattice consist of small (dilute) substitution of magnetic dopants, such as $Pb_{1-x}Eu_xTe$, $Ga_{1-x}Mn_xAs$, $Zn_{1-x}Fe_xSe$ etc. The nature of DMS makes them suitable to tune their electrical, structural, magnetic and optical properties simply by altering the composition(x).

A lot of work has been performed in this field in order to understand the actual motivation present behind the performance of DMS materials. Few literatures reported that pristine SnO_2 is diamagnetic, while few other reveals that annealing treatment plays an important role in magnetic polarization means sometimes Ni-doped tin oxide films exhibits paramagnetic behavior and achieve ferromagnetism after annealing treatment in nitrogen ambient atmosphere. Also, significant increment in magnetic moment has been noticed with rise in annealing time [8, 9]. Some of the metal oxide nanostructured materials like CeO_2, TiO_2 and ZnO behaves ferromagnetically even without dopant magnetic ions which also supports the importance of surface defects like oxygen (ligand) holes instead of the magnetic doping ions of TM elements [10]. One of the literatures reported magnetic study of Fe doped tin oxide prepared through simple chemical co-precipitation technique and claimed that anti-ferromagnetic alignment decreases with increment in doping concentration [11]. Hence, occurrence of room temperature ferromagnetism in DMS based materials, which is a key requirement for spintronics based applications, is still an unclear issue, whether it is intrinsic property of material or doping is responsible for it.

2. Experimental Details

$Sn_{1-x}Fe_xO_2$ (x=0.01 and 0.02) samples have been successfully synthesized using chemical co-precipitation method. The chemicals used for synthesizing of samples were of analytical grade and no further purification is required. Synthesis includes tin (IV) chloride pentahydrate and iron (III) chloride (AR grade). The chemicals were weighed accurately as per the proportions required by using microbalance. Prepare a clear solution of $SnCl_4.5H_2O$ and $FeCl_2.6H_2O$ by dissolving them into deionized water with the help of magnetic stirrer. Now add 0.1M NH_4OH solution drop by drop into the prepared solution of tin (IV) chloride and iron (III) chloride with constant stirring until its pH value raised upto 7 and then aged the solution for half an hour. The obtained white precipitate was filtered with the filter paper and washed it 2-3 times with deionized water in order to remove ionic impurities. Afterwards the wet colloidal solution was kept at 40°C in air to remove moisture and the dried powder was collected. The prepared samples have been characterized through SQUID in order to investigate the magnetic behavior of Fe doped SnO_2 having two different doping concentration.

3. Result and Discussion

Fig 1(a) and (b) represents magnetization (M) vs. applied magnetic field (H) curves for 1% and 2% Fe doped SnO_2 nanoparticles recorded at 5K and 300K temperatures.

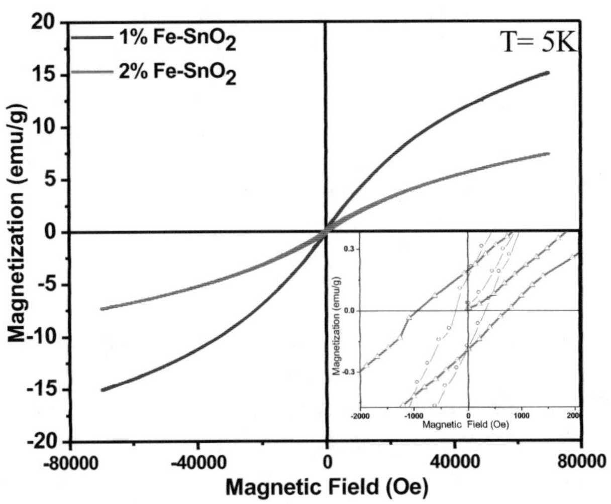

(a)

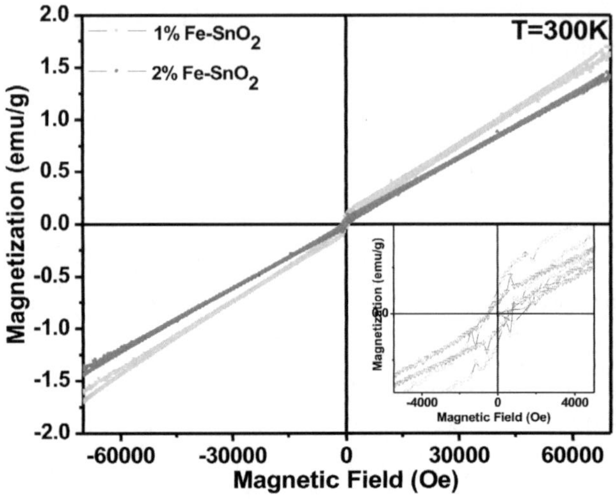

(b)

Fig.1 Curves of magnetization (M) vs. applied magnetic field (H) (a) at 5K and (b) at 300K temperatures for 1% and 2% Fe-SnO_2 nanoparticles.

At 5K, saturation magnetization (M_s) reduces from 15emu/g to 7.45emu/g while the width of the hysteresis behavior enhances with increase in the concentration of dopant ions from 1% to 2% with the coercivity (H_c) of 239.85 Oe and 922.572 Oe respectively. These observations might correspond to the charge imbalance of dopant ions with the host ion due to the substitution of Fe^{3+} ion at the Sn^{4+} sites which creates oxygen vacancies as a result of formation of Fe^{3+}-Vo^{2-}-Fe^{3+} clusters in the host lattice sites. Thus, when an electron caught in these oxygen vacancies will interact with two neighboring irons through F-centers and leads to ferromagnetic ordering. While at higher temperatures i.e. at 300K (room temperature), the magnetic response gets drastically modified. Both the samples show ferromagnetic response at lower magnetic fields which becomes nearly paramagnetic at higher applied fields.

4. Conclusion

$Sn_{1-x}Fe_xO_2$ (x=0.01 and 0.02) nanoparticles were prepared using chemical co-precipitation method and then SQUID measurements have been taken to study the influence of dopant ion on the magnetic properties of the host SnO_2 nanoparticles. A considerable ferromagnetism has been observed at low temperature (5K) with wider hysteresis loops which could be attributed to the presence of dopant ions and the defects produced at the grain boundaries while at room temperature (300K), small ferromagnetism has been attained with increasing magnetization. Thus, Fe doped SnO_2 nanoparticles could be considered as a promising candidate from the practical applications point of view especially in data storage devices.

Acknowledgement

P. A. Alvi and Mayuri Sharma are very thankful to the DST Government of India, New-Delhi for providing facilities to Banasthali Vidyapith under CURIE programme and Dr. Shalendra Kumar is grateful to DST, Government of India, New Delhi (YSS/2015/001262) for financial assistantship.

References

[1] Ashutosh Tiwari, V. M. Bhosle, S. Ramachandran, N. Sudhakar, J. Narayan, S. Budak, and A. Gupta, "Ferromagnetism in Co doped CeO_2: Observation of a giant magnetic moment with a high Curie temperature", Applied Physics Letters 88, no. 14 (2006): 142511.

[2] J.K. Furdyna, "Diluted magnetic semiconductors", Journal of Applied Physics 64.4 (1988): R29-R64.

[3] H. Ohno, "Making nonmagnetic semiconductors ferromagnetic", science 281.5379 (1998): 951-956.

[4] A.H. MacDonald, P. Schiffer, and N. Samarth, "Ferromagnetic semiconductors: moving beyond (Ga, Mn) As", Nature Materials 4.3 (2005): 195-202.

[5] T. Dietl, "A ten-year perspective on dilute magnetic semiconductors and oxides", Nature Materials 9.12 (2010): 965-974.

[6] Mayuri Sharma, Shalendra Kumar, and P. A. Alvi, "Electronic structure and magnetic properties of Ni-doped SnO_2 thin films",AIP Conference Proceedings, 1953, 120010 (2018).

[7] Pradeep Kumar, Kavita Kumari, Mayuri Sharma, P. A. Alvi, and Shalendra Kumar, "Structural, magnetic and surface morphological study of Ni doped SnO2 thin films", AIP Conference Proceedings 2006, 030039 (2018).

[8] N.H. Hong, A. Ruyter, W. Prellier, J. Sakai and N. T. Huong, J. Phys. Cond. Mat., 17 (2005) 6533.

[9] P.I. Archer, P.V. Radovanovic, S.M. Heald and D. R. Gamelin, J. Am. Chem. Soc., 127 (2005) 14479.

[10] P. Sharma, A. Gupta, K.V. Rao, F.J. Owens, R. Sharma, R. Ahuja, J.M. Osorio Guillen, B. Johansson, and G. A. Gehring, Nat. Mater. 2 (2003) 673.

[11] Adhikari, R., A. K. Das, D. Karmakar, TV Chandrasekhar Rao, and J. Ghatak. "Structure and magnetism of Fe-doped SnO_2 nanoparticles." Physical Review B 78, no. 2 (2008): 024404.

IOP Conf. Series: Materials Science and Engineering **594** (2019) 012005 doi:10.1088/1757-899X/594/1/012005

Bandstructure Effects in the Electronic Transport of Silicon Nanowire in [100] and [110] Transport Orientations

Manoj Kumar[1,*], Vijay Kumar Lamba[2]
[1]BRCM College of Engineering and Technology, Bahal, Bhiwani (India)
[2]Global College of Engineering & Technology, Kahanpur Khui, Punjab (India)
E-mail:* manoj10276@gmail.com

Abstract. An electronic transport study about band structure effects of silicon nanowire (Si-NW) in [100] and [110] transport orientations has been done. A dispersion calculation has also been done for a $sp^3d^5s^*$ hybridized atomistic model along with a Poisson solver. A ballistic model has been used for the evaluation of I-V characteristics. For [100] and [110] transport-oriented nanowires, the total gate capacitance was decreased by 30%. The carrier velocity was affected by both [100] and [110] transport-oriented nanowires. The velocities depend on degeneracy and effective mass of the dispersion. In our study, we have used 3nm thick nanowire oriented in [100] and [110] directions. The [100] oriented nanowire gives better ON-current performance as compared to [110] oriented nanowire. We have studied the valley splitting effects which can lift the degeneracies of 3nm wires. The effective mass which changes with various transport orientations and quantization, was increased with quantization for [100] while it was decreased for [110] transport orientated nanowire.

Index-Terms- silicon nanowire, tight binding, band structure, injection velocity, nano FETs, effective mass, quantum capacitance.

1. Introduction

As the sizes of transistor required being of nanoscale, the development of CMOS finds an alternative place in the designing of different structures and devices. The CMOS transistors that were developed from planar technology nowadays were used as nanodevices. The use of nanowire devices has made a good attention due to its possibility of enhanced electrostatic control. Different experimental results show that the nanowire transistors below 3nm size have been designed [1]. With the device's dimension going down, the lattice structure, bonding between the atoms were affected. Therefore, a good molecular modeling was required so that the nanowire device should provide an accurate electrical characteristic.

In our work, we had used the atomistic model ($sp^3d^5s^*$) along with a 2D Poission Solver to calculate the electronic structure of the silicon nanowire. The band structure parameters which affected the properties of silicon nanowires were also identified [2]. To calculate the transport properties of silicon nanowire, a basic semi-classical ballistic model, shown in figure 1 was utilized [3]. In our work, we have taken silicon nanowire of 3 nm width, which were oriented in [100] and [110] transport direction to study and compare the transport characteristics. 2D Poission solver shown in figure 1 has been used to calculate the electrostatic potential variation in the lattice structure.

While examining the silicon nanowire transport characteristics, the potential variation in the cross section of nanowires played a criticizing role. This factor causes just little shifting of the energy levels, yet the placement of the charge was affected by it. The capacitance of the device was decreased by 30% due to charge placement along with density of states in silicon nanowires and this reduction was same for both [100] and [110] transport directions. The device performance was controlled by the injection velocity and the orientation of the nanowire. The transport mass controls the orientation velocity and orientation controls the degeneracy. The [110] oriented devices with a lowest mass and highest carrier injection velocity were best suitable for charge transportation in small nanowires than [100] orientated devices with a little higher mass [4]. The [100] and [110] transport directions and structural quantization works on the transport masses and degeneracy [5].

1.1. Need of atomistic modelling:
The atomistic effects were not usually captured by effective mass approximation (EMA) method. The atomistic description of the devices actually captures the valley projections and extracts dispersion of the oriented nanowires. The problem on correct band structure and effective masses of nanowires has been discussed in references [6-7]. The main features on the electronic structure have also been discussed [8]. In the electronic structure calculation techniques, the mass variation in nano structure has been described [9]. The mass variation in nano structure resulted in various threshold voltages. The effective mass approximation method could be used especially for [100] oriented nanowire for the calculation of conduction band only. As already discussed, the EMA method was not always valid, so the more useful method for nanowire of few nanometers was the atomistic simulation method.

In our work, we have used Tight Binding (TB) model for which we have described its simulation approach. We have described its simulation approach. We have described the behavior of band structure of the lattice for 3nm cross-section square nanowires of [100] and [110] orientations. Nanowire's performance in terms of injection velocity, quantum capacitance, drive current capabilities and gate capacitance for [100] and [110] oriented nanowires has been described.

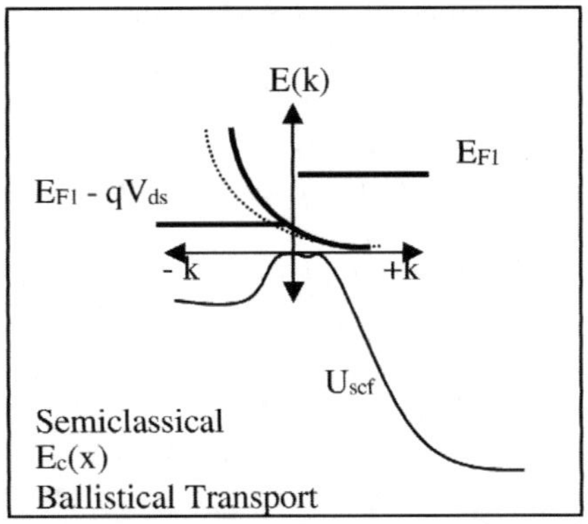

Figure 1(a): semi classical ballistic model

2. Methodology

2.1. Why we have selected Tight Binding (TB) model:
As we know that the MOS technology has come to saturation point, the requirements for new technology at nanoscale was to find new material and new devices. The quantum mechanics of the atomic composition and atomic structure had to be examined. For that an atomic structure model which may meet the requirement of nanoscale device was required. In a typical semiconductor device, the stability of bands allows the requirement of a reduced model. The hetero-structure model requires a nearest neighbor model. To simulate larger structures containing a large no. of atoms, there was a need of a reduced model [10]. To accurately model band gaps within a few mV, there occurs a requirement of an empirical band structure model. To fulfill all the above requirements, an empirical Tight Binding (TB) model was the best choice.

2.2. The sp3d5s TB Model:*

In our work, we have used $sp^3d^5s^*$ nearest neighbor model made up of orthogonal localized orbital. This model was best suited model for electronic structure of nanometer scaled nanostructures. This model can be used to treat potential variations and materials at nanoscale. The main feature of this method was that by using genetic algorithm, the band edges of silicon band structure can be reproduced by extracted parameters. The explanation of this model was given in reference [11]. Energy bands drawn from this model for the nanowires were up to the requirements as compared with the theoretical calculations using ab-initio method. For charge self-consistency, the electrostatic potential was used in the Hamiltonian equation for shifting the bands.

2.3. Verification of Model:

As the accuracy of the results were required and since the accuracy will depend on the verification of the model we were using. The verification of model has been done by using the calibration parameters to explain the experimental data. Some samples were used e.g. under high bias voltage with charge self-consistency, resonant tunnel diode was tested for transportation of charge [12-13]. The experimental data for valley splitting, band gap of nanowires and electronic structures of silicon nanowires compared with the results given by the Tight Binding (TB) model [14-17]. Also, the theoretical works were compared with results given by this model [18]. Especially this model validates thc calculation of experimental results of band gaps of the oriented nanowires in Brillouin zone of silicon nanowire. The model also shows that anisotropy at high energy levels affects the band edges and atomic masses of nanowires.

2.4. The simulation approach:

We have used a square nanowire of 3nm x 3nm dimension. Our model device was simulated for [100] and [110] transport orientations. The simulation process was done using these following steps:
First of all, by using atomistic Tight Binding (TB) model, we calculated the band structure of the nanowire. All the edges in lattice structure (Zincblende lattice) was shown by sp3 hybridization ($sp^3d^5s^*$) in Hamiltonian wire. Only conduction band was considered for study. The accuracy of the results was not affected, and this model was perfect for computational efficiency. The sp^3 hybridization scheme was used in which the atoms were persisted on the surface of nanowires. This technique was very useful for removing those bonds which could make surface state with eigen – energies in band gaps of the device. The channel atoms were available for atomistic calculations. The model uses dispersion state energy for computation. We have studied the band structure effects such as effective mass variation, valley splitting of the [100] and [110] transport-oriented nanowire.

Secondly, we have used a semi-classical ballistic model for computation of transport characteristics and to fill the dispersion state. This model makes use of drain Fermi level to fill negative going states and source Fermi level to fill the positive going state. Then the charge distribution takes place in all the orbitals of the shells using their wave functions when the dispersion state occupancy was computed.

Thirdly, the electrostatic potential was obtained by 2D Poisson equation with the help of charge distribution. 2D Poisson equation was solved in 2D plane. Also, in 2D plane, atomic locations were collapsed. The Poisson domain contains the nanowire core, dielectric and metal. The diagonal elements of Hamiltonian matrix were added with the electrostatic potential to re-calculate the band structure for achieving self consistency. The dielectric present in Poisson domain was believed to be of SiO_2 of 1.1nm thick layer, was not included in the Hamiltonian. Due to the potential difference, any effects produced by the transport orientations were not considered. It has also been studied that the carrier injection was very essential for the transport properties of nanowires.

Our study of transport model explains that the ballistic transport characteristics will be affected by the band structure of nanowires. 3D quantum non-equilibrium green function (NEGF) simulation also provides the same results. The results of our study will be valid for other nanowire dimensions, as our results shows that the electronic properties of nanowires were dependent upon the quantization size, not on quantization shape.

3. Results and Discussions

3.1. How nanowire dispersion and charge distribution was affected by potential variation:

3.1.1. Dispersion characterization in [100] oriented nanowires: Figure 1a shows dispersion in [100] oriented nanowire. It shows four-fold valleys at Γ which was at a distance of $k_x = 0$ and was projected from four silicon ellipsoids. These ellipsoids reside in a quantized plane. From figure 1a, two more valleys can be seen which were available away from Γ. The position of two valleys was opposite to each other. One was in positive side of k_x axis and another resides in the negative side.

Out of four valleys, first two were seen lower in energy level due to their light transport mass (m_x=m_t=0.19 m_0) and heavy quantization mass (m_y=m_l=0.89m_0), where m_x and m_y were nanowire masses.

3.1.2. Potential variations change the dispersion of [100] oriented nanowire: Our results shows how potential variations can change the dispersion of the oriented nanowires. It was also shown how shape of lattice was changed when it was filled up with charge. Figure 1 shows the effects of low and high gate voltages on 3nm square [100] oriented nanowires. We have used drain voltage, V_D= 0.5V. Lattice was approximately charge less and dispersion was stable, when gate voltage applied was low. Charge filling in lattice occurs when high gate voltage was applied. Charge distribution also comes into a shape of inversion layer. When inversion layer becomes high, our results show that the wave function was displaced by 0.5nm apart from Si/Sio_2 interface. The dispersion was considered to be a material parameter and it was independent of charge filling of lattice.

Due to the charge filling in lattice structure, there occurs a change in dispersion of 3nm nanowire length. Here a small change in dispersion occurs but as wave function was associated with it, so it increases the charge distribution. As a result, the transport characteristics and the capacitance of device were affected.

3.1.3. Orientation variances in the velocity: Our simulation result shows that the lighter mass (0.16m_0) produces highest velocity in [110] oriented nanowire. The velocity increases when higher k-states got filled. When Fermi level entered the conduction band as shown in figure 2, the carrier velocity in the [100] oriented nanowire becomes equal to the velocity of [110] oriented nanowire. The heavier mass (0.27m_0) in [100] oriented nanowire Γ valley was taken over by the lighter masses (0.16m_0) in [110] oriented nanowire's Γ valley. When heaviest four-fold deviate off- Γ valleys of the device [110] start to populate with the heavy two fold deviate off-Γ valley of the device [100], a factor known as carrier velocity gets looks alike. Our study analysis told that [100] oriented nanowire and [110] oriented nanowire has different properties.

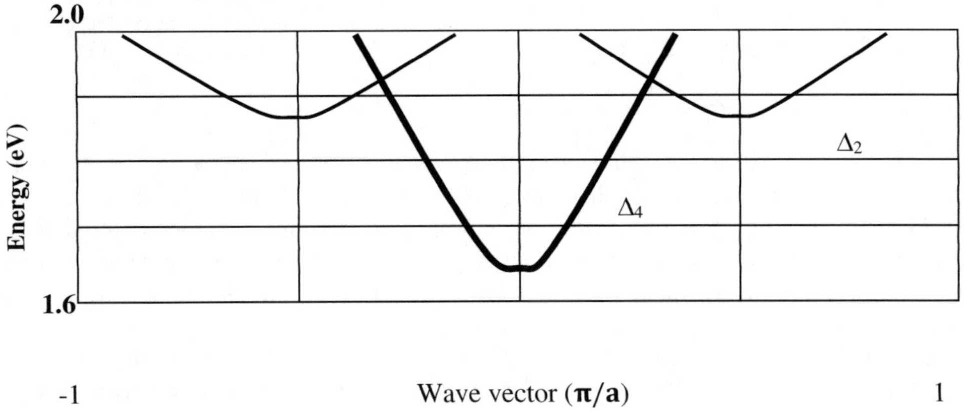

(Conduction bands)

(2a)

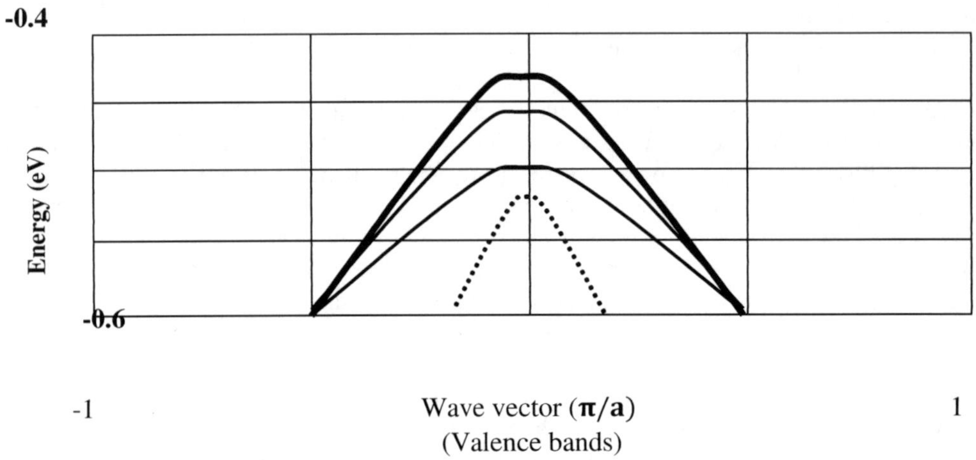

(2b)

Figure 2: Si 5x5nm nanowire band structure estimated from the empirical tight binding sp³d⁵s* model (2a) Conduction and (2b) valence band

3.2. Performance analysis of device for nanowires in different orientations:
We have considered the [100] oriented nanowires comparison at same Fermi level position. For the comparison of performance of different oriented nanowires, we have used self consistent model. The performance comparison has been shown in figure 4. The comparisons for all devices have been done at same off current (I-off).

3.2.1. C_s controlling factors: C_s was referred as differentiation of charge present in nanodevice w.r.t. surface voltage (ψS). This charge was integral of convolution of density of states (g_{1D}) and Fermi function (f (E_G - E)).

$$C_{\&} = \frac{Q^{*} \pm = C}{C_{+}} \; C_{+} (\int q g_{12} f 4 \; \frac{5_6 7 5_8 7 9_:}{;<=} > dE \qquad (1)a$$

where q was charge of electron, E_f was Fermi level, ψs was the surface potential, E_c was conduction band energy, and ε_i was separation of ith sub-band above E_c. Moving to coordination, the conditions above outcomes was:

$$C_{\&} = \frac{)^{B}}{C} D \frac{EF}{GB} H^{B} Jk_L T \sqrt{\pi} \frac{C}{O,P} (\Im_{7I} 4^{\frac{5_6 7 5_8 7 \mathcal{E}_:}{;<=}} >) \qquad (1)b$$

$$= q^{E} \frac{D^{EF} H^{B}}{CG^{B}} Jk_L T \; {}_{P}\Im_{7_{\overline{B}}}^{\,S} \, 4^{\frac{5_6 7 5_8 7 \mathcal{E}_:}{;<=}} > (1 - \frac{(9_:}{C_+}) \qquad (1)c$$

$$= C^{V} (1 - \frac{(9_:}{C_+}) \qquad (1)d$$

In Eqn. 1d, C_Q, represents quantum capacitance. At Fermi level, it was a measure of the thickness of Density of states. From equation, we see that the changes in $\mathcal{E}_i$ correspond to decrease in C_s from C_Q. Ideally $\mathcal{E}_i$ should be constant at high inversion conditions which mean that conduction band enrgy level, E_c and quantization level shifts by same amount. At this condition, sub bands level reaches at Si/SiO_2 interface. Due to this, wave function comes closer to interface. As charge was accumulated in the device, $\mathcal{E}_i$ floats up. It corresponds to a differential term in equation 1d and wave function shift from interface. Besides shifting of wave function, C_Q being small becomes decreasing factor of C_s as shown in equation 1d. C_Q in both [100] and [110] nanowires, being measured at Fermi level, was a function of gate bias, V_G. We found a large significance of the position of charge distribution on the performance of a device in the fact that it degrades C_s from C_Q by four-times.

3.2.2. Variance in C_Q between [100] and [110] oriented nanowires: C_Q was not constant even it shows a large transition at large gate voltages when Fermi level was thrown into the sub bands for both [100] and [110] oriented nanowire cases. This was right because dispersion causes difference in C_Q and also at the Fermi level, C_Q was a source of density of states. When we compare the C_Q for different oriented wires, we found that [110] oriented nanowire has more C_Q because of high degeneracy of its valleys (D=6) and higher masses ($0.47m_0$) for all gate bias ranges while [100] oriented nanowire has less C_Q because of low degeneracy and light mass (less density of state). The [100] oriented nanowire gets populated when a high gate bias was applied as shown in figure 3. As a result, a continuous increase in $C_{Q\ takes}$ place. At a lower gate voltage, the Fermi level in the [110] nanowire, the smaller valley degeneracy's and lower mass reaches the upper valleys faster as compared to [100] nanowire. As a result, the C_Q of [110] nanowires crosses the C_Q of [100] nanowire at gate voltages around 0.4V (V_G=0.4V).

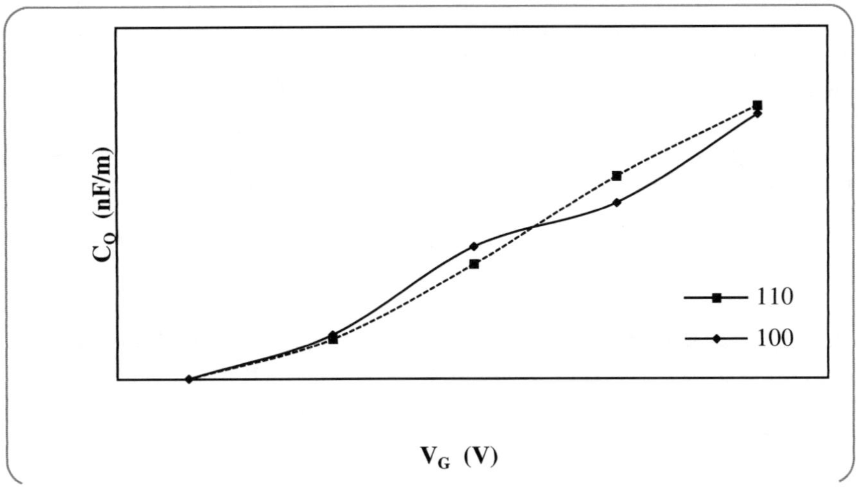

Figure 3(a): C_Q vs. V_G plot for two devices

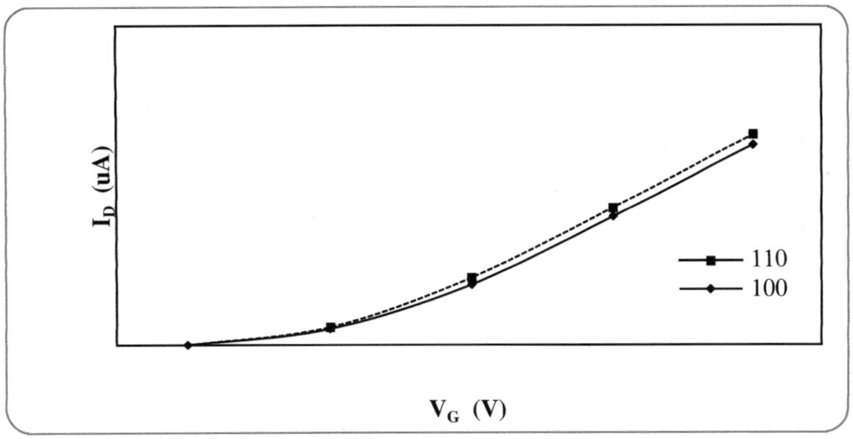

Figure 3(b): I_{DS} vs. V_G plot for the two wires having same value of I.

IOP Conf. Series: Materials Science and Engineering **594** (2019) 012005 doi:10.1088/1757-899X/594/1/012005

3.2.3. I-V characteristic was affected by velocity differences: Velocity was totally relied on I_{DS}. A comparisons between [100] and [110] nanowires for current driving capabilities has been done at same I_{off}. In terms of ON-current capabilities, [110] nanowire performs better it gives 5% better ON-current capability than [100] nanowire because of its lower mass. The results generated by 3nm nanowires must be taken into consideration as band structure of nanowires depends on their quantization.

3.3. Mass variation, valley splitting was affected by quantization:
The device performance was controlled by mass variation and valley splitting that lifts degeneracy's and these two parameters were affected by quantization. So we can say that device performance was controlled by quantization. In our study, we have examined the effects of quantization.

3.3.1. Weak valley splitting in [100] quantized nanowire: Energy Vs wave vector (E-k diagram) graph of a 3nm nanowire in [100] and [110] orientation has been shown in figure 2a-2b. A little valley splitting under quantization has been examined. For [100] oriented nanowire, it was 10meV. Even the room temperature value (26meV) was greater than this value, so the transport properties will not be affected at room temperature.

3.3.2. Strong valley splitting in [110] quantized nanowire: The valley splitting was larger in [110] oriented nanowire. Figure 4c, shows the splitting of 76meV at Γ valley and 14meV off- Γ valley. The larger nanowires greater than 5nm were not affected. The value of splitting may attain 200meV at Γ valley in 1.5nm narrow nanowires. The splitting values off- Γ valley were not being affected so much. Very little splitting of a few tenths of meV was seen in this case.

3.3.3. With increase in quantization the Γ valley mass of [110] oriented nanowire decreases: The [110] nanowire Γ valley mass decreases as compared with the rest of the valleys, when quantization, was increased. Figure 4b shows the mass reduction of 32% when thickness of nanowire reduces from 7.1nm to 1.5nm.

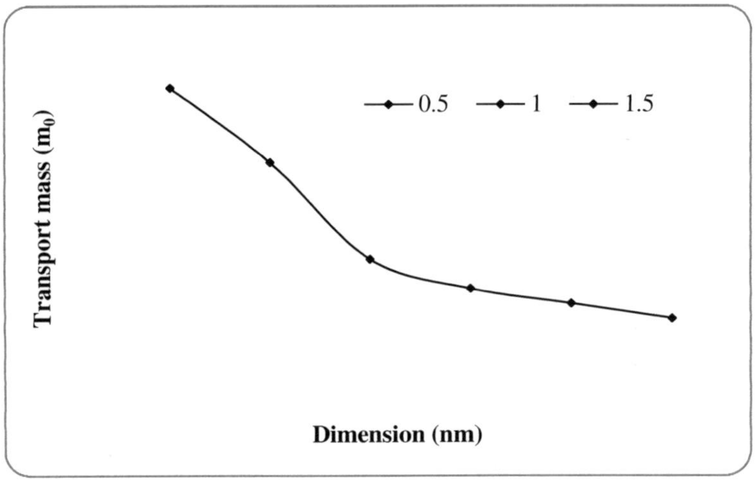

Figure 4a: nanowire thickness vs. transport mass at conduction band minima

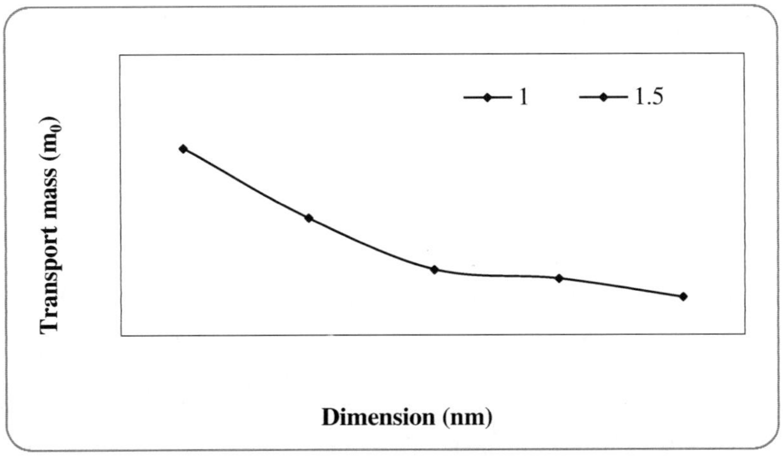

Figure 4b: Conduction electron effective masses versus wire thickness at the Δ2 valley.

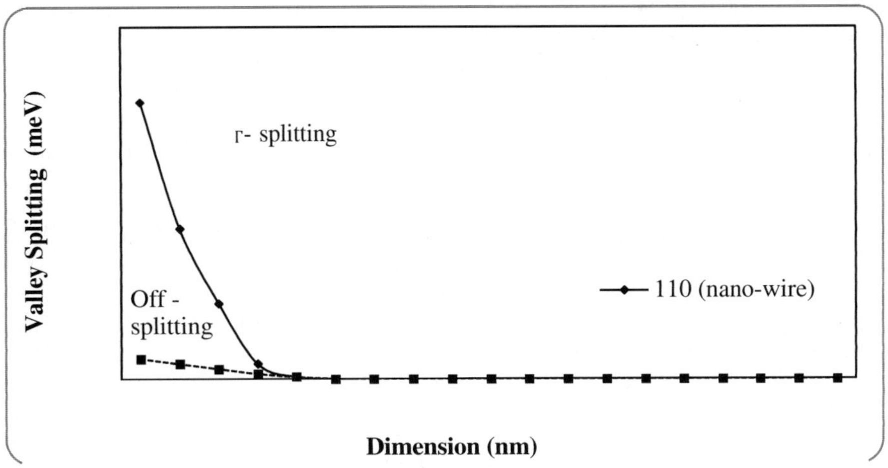

Figure 4c: valley splitting in [110] nanowire

4. Conclusion

A sp3d5s* hybridized atomistic model along with a Poisson solver was used to examine transport properties of [100] and [110] transport oriented nanowires. The I-V characteristics of the nanowires were calculated using a semi classical ballistic model. The dispersion of the nanowires which could shift the degeneracy's and shift the sub bands, was changed when gate bias was changed. These changes couldn't change the density of states and velocity of the nanowires but the nanowire's capacitance could be decreased by 30% by the charge distribution and density of states. The quantum capacitance was found to be of same magnitude in both [100] and [110] oriented 3nm nanowires depended on gate voltage. It was found largest for [110] oriented nanowires because of higher degeneracy and higher mass for all gate voltages. The gate capacitance was same in both [100] and [110] oriented nanowires and the inversion charge was also same. The [110] oriented nanowires have lower injection velocity because of their bigger masses whereas [100] nanowire have maximum injection velocity due to their lighter masses. The current capabilities of the nanowires were reflected by the injection velocities. The ON-current capabilities represented the best performance of the device. The [100] oriented nanowire gave the best performance while [110] nonowire gave worst performance as compared to [100] nanowire. The effective mass of nanowires was a function of nanowire's cross-section. When the dimension of nanowire was decreased from 7.1 to 1.5nm, [100] oriented nanowire mass became almost double. Valley splitting was dependent on quantization. With increase in quantization, the [110] nanowire valley mass was decreased as compared with rest of valleys. The [110] oriented nanowire of dimension below 3nm were very sensitive to the splitting valley.

References

[1] Xiang, J., Lu, W., Hu, Y., Wu, Y., Yan, H. and Lieber, C.M., (2006), "Ge/Si Nanowire Heterostructures as High-Performance Field-Effect Transistors," Nature, 441(7092), pp. 489.

[2] Boykin, T. B., Klimeck, G. and Oyafuso, F., (2004), "Valence Band Effective-Mass Expressions in the $Sp^3d^5s^*$ Empirical Tight-Binding Model Applied to a Si and Ge Parametrization," Physical Review B, 69(11), pp. 115201-115210.

[3] Rahman, A., Guo, J., Datta, S. and Lundstrom, M. S., (2003), "Theory of Ballistic Nanotransistors," IEEE Transactions on Electron Devices, 50(9), pp. 1853-1864.

[4] Luisier, M., Schenk, A. and Fichtner, W., (2007), "Full-band atomistic study of source-to-drain tunneling in Si nanowire transistors," Proc. Simulation of Semiconductor Processes and Devices, T. Grasser et al., eds., Springer, Vienna, Austria, pp. 221-224.

[5] Boykin, T. B., Klimeck, G., Friesen, M., Coppersmith, S. N., von Allmen, P., Oyafuso F., and Lee, S., (2004), "Valley splitting in low-density quantum confined heterostructures studied using tight-binding models," Phys. Rev. B, 70(16), pp. 165325.

[6] Luisier, M., Schenk, A., Fichtner, W. and Klimeck, G., (2006), "Atomistic simulation of nanowires in the $sp^3d^5s^*$ tight-binding formalism: From boundary conditions to strain calculations," Physical Review B, 74(20), pp. 205323.

[7] Gnani, E., Reggiani, S., Gnudi, A., Parruccini, P., Colle, R., Rudan, M. and Baccarani, G., (2007), "Band-structure effects in ultrascaled silicon nanowires," IEEE Transactions on Electron Devices, 54(9), pp. 2243-2254.

[8] Wang, J., Rahman, A., Ghosh, A., Klimeck, G. and Lundstrom, M., (2005), "On the validity of the parabolic effective-mass approximation for the IV calculation of silicon nanowire transistors," IEEE Transactions on Electron Devices, 52(7), pp. 1589-1595.

[9] Neophytou, N., Paul, A., Lundstrom, M. S. and Klimeck, G., (2008), "Simulations of nanowire transistors: Atomistic vs. effective mass models," Journal of Computational Electronics, 7(3), pp. 363- 366.

[10] Klimeck, G., Ahmed, S. S., Bae, H., Kharche, N., Clark, S., Haley, B., Lee, S., Naumov, M., Ryu, H., Saied, F. and Prada, M., (2007), "Atomistic simulation of realistically sized nanodevices using NEMO 3-D—Part I: Models and benchmarks," IEEE Transactions on Electron Devices, 54(9), pp. 2079- 2089.

[11] Klimeck, G., Oyafuso, F., Boykin, T. B., Bowen, R. C. and von Allmen, P., (2002), "Computer Modeling in Engineering and Science (CMES)," Proc. Simulation of Semiconductor Processes and Devices, T. Grasser et al., eds., SpringerWien, NewYork, 3(5), pp. 601-642.

[12] Bowen, R. C., Klimeck, G., Lake, R., Frensley, W. R. and Moise, T., (1997), "Quantitative resonant tunnelling diode simulation," J. Appl. Phys, 81, pp. 3207-13.

[13] Bowen, C., Fernando, C. L., Klimeck, G., Chatterjee, A., Blanks, D., Lake, R., Hu, J., Davis, J., Kulkarni, M., Hattangady, S. and Chen, I. C., (1997), "Physical oxide thickness extraction and verification using quantum mechanical simulation," In Electron Devices Meeting (IEDM'97), Technical Digest, pp. 869-872.

[14] Ma, D. D. D., Lee, C. S., Au, F. C. K., Tong, S. Y. and Lee, S. T., (2003), "Small-diameter silicon nanowire surfaces," Science, 299(5614), pp. 1874-1877.

[15] Zhao, X., Wei, C. M., Yang, L. and Chou, M. Y., (2004), "Quantum confinement and electronic properties of silicon nanowires," Physical review letters, 92(23), p. 236805.

[16] Kharche, N., Prada, M., Boykin, T. B. and Klimeck, G., (2007), "Valley splitting in strained silicon quantum wells modeled with 2 miscuts, step disorder, and alloy disorder," Applied Physics Letters, 90(9), p. 092109.

[17] Liang, G., Xiang, J., Kharche, N., Klimeck, G., Lieber, C. M. and Lundstrom, M., (2007), "Performance analysis of a Ge/Si core/shell nanowire field-effect transistor," Nano letters, 7(3), pp. 642-646.

[18] Lee, S., Oyafuso, F., Von Allmen, P. and Klimeck, G., (2004), "Boundary conditions for the electronic structure of finite-extent embedded semiconductor nanostructures," Physical Review B, 69(4), p. 045316.

IOP Conf. Series: Materials Science and Engineering **594** (2019) 012006 doi:10.1088/1757-899X/594/1/012006

Polarization Emancipated Ultrathin Metamaterial Absorber

Shwetank yadav, Garima tiwari
Department of Electronics and Communication Engineering, Jabalpur Engineering College, Jabalpur, India
E-mail: *shwetankyadav5101991@gmail.com

Abstract: Through this paper, we propose quad-band electromagnetic wave metamaterial absorber structure consist of square rings resonators arranged in two-dimensional periodic array with a metallic ground plane. Under normal incidence this structure manifests four sharp peak absorptions at 3.75, 5.05, 7.35 and 8.96 GHz with pinnacle absorptivity of94.28, 99.67, 99.98 and 99.90% respectively. This structure with four-fold rotational symmetry in design provides nearly 100% absorption for all angle of polarization. Moreover, this outcome is polarization Emancipated, which manifest aloft absorption (more than 90%) for oblique incident angles till 45°. Corresponding to the highest frequency the absorber is ultra-thin and uses commercially available FR-4 as substrate. The designed quad-band absorber can be employed in innumerable feasible execution like thermal emitters, sensors, receivers, shielding, optical organization, radar signal absorption, forbid drugs and many communication utilizations. It can be easily manufactured with simple PCB manufacturing techniques.

Index-Terms: metamaterial; microwave absorber; quad-band; polarization-insensitive.

1. Introduction

Metamaterials (MTMs) [1]are artificial complex periodic structure of unit cell through which it derive its electromagnetic properties. It exhibits manifest which are eluded in natural materials. Due to their peculiar properties, a significant research interests has been drawn and has been used to design antenna[2], perfect lens[3], cloaking [4] etc. Owing the advantages such as near unity absorption, wide incident angle absorption, ultrathin thickness, easy to fabricate, polarization insensitivity and so on, recently, researchers are paying more attention towards the applicability of MTMs as electromagnetic wave absorbers.

Usually, a meta-material absorber structures is manufactured by placing periodic arrays surfaces which is selected according to frequency as top layer are escort by dielectric substrate[5] in the middle and an absolute metal plate in the end. Actually, a sub wavelength unit, is assembled, consist of simulated electric and magnetic field components of the incident radio frequency waves. Magnetic field component affects dielectric substrate while electric component affects top layer of FSS material. This allows the procurement of tailored mediums with random but fruitful material parameters. Ultimately current flows at lower and upper metal is of conductive behavior & displacements current through dielectric. By achieving resonant frequency, permeability and permittivity of the racially balanced output can be equated to one another. As the input impedance of the structure matches to the free space impedance of air, reflection from the absorber drastically decreases.

Application of electromagnetic wave absorber can be used as thermal detectors[6], radar application[7],stealth technologies[8], electromagnetic wave absorption[9], IR concealments[10], wireless communication [11], EMI shielding [12], satellite[13], solar cell [14], spectroscopic detection and also in defense applications[15]. Meta-material absorber resulting inherent advantages such as fragile, more efficiency and very small thickness[16] Usual absorber being of large size and fragile should be replaced. The fore most metamaterial absorber has narrow bandwidth and is polarization dependent, which is practically useless. After lots efforts and research, the bandwidth of metamaterial absorbers can be enhanced. Multiband metamaterial absorption has been evolve using number of resonant single cells and making array of them through a co-plane organization.

By this paper, we introduce a quad band metamaterial absorbing structure arranged in 2-D array of the single cell structure and operating in the microwave region with simple and highly symmetric geometry unit cell structure based on square shape split ring resonators. The imitate results shows that near unity absorption (>90%) at four distinct frequencies. Also, to give an explanation of the mechanism of the electric and magnetic resonances of the proposed structure the electric, magnetic field and surface current density plotted for peak absorption frequencies and hence, Mechanism of loss is investigated and

IOP Conf. Series: Materials Science and Engineering **594** (2019) 012006 doi:10.1088/1757-899X/594/1/012006

discussed. With having four-fold symmetric structure, it shows polarization-immune response under normal occurrence. For up-to 45^0oblique incidence of wave's simulation shows that structure still has higher absorption (over 90%).

2. Configuration of Unit Structure

A wedge three layer structured is given as fig. 1, consisting of 3 layers – top and lower layer are metal while the middle dielectric portion is sandwiched among them. Lower layer does not allow any waves to go through it. Flame Retardent-4 dielectric having comparative $\epsilon r = 4.2$ and $\tan(\delta) = 0.02$ is contemplate for middle layer with thickness of 1 mm~ lambda/40 compared to highest frequency which is used as substrate. For both top and lowest layers Copper ability to current flow $\sigma = 5.8*107$ S/m with 0.035 mm as thickness is used. 3 quadrilateral loops incorporate with Upper layer which has 2-d array – exterior most loop, middle loop and inner most loop. The outer most loops have four splits whereas inner two loop are complete square-loop. The inner most one being cross couple to impart more distance to surface current flow.

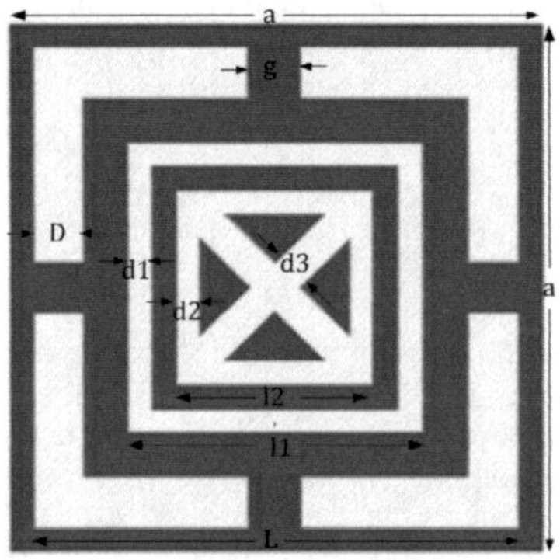

Fig.1. Front-view of unit cell geometry of the proposed structure(Dim-ension of the unit cell: a=23mm, L=21mm, l1=12.7mm, l2=8.4mm, D=2.2mm, d1=d2=1mm, d3=1.5mm, g=2.2mm)

Last layer of copper which reflects incident waves is used to reflect the power received $|S_{21}(\omega)|^2$. Hence by properly crafting the unit cell structure parameter reflect powers $|S_{11}(\omega)|^2$ reduces. Therefore, with reducing reflected max value $A(\omega)$ absorptivity can be reached.

$$\text{Absorptivity}=1 -\{ |S_{11}(\omega)|^2 - |S_{21}(\omega)|^2 \} = 1 - |S_{11}(\omega)|^2 (1)$$

3. Simulation Analysis and Results

Simulation of the unit cell structure Ansys HFSS software is used, to manifest four definite values of return loss $|S_{11}(\omega)|$ at 3.75, 5.05, 7.35 and 8.96 GHz, leading to absorptivity peaks of 94.28%, 97.27%, 99.67% and 99.90% respectively as shown in Fig. 2.

3.1 Absorption mechanism

For attaining impedance matching with free space by scheming both electric and magnetic resonances in a metamaterial the functional permittivity and permeability of the metamaterial can be tailored. Hence, the interface reflection is blocked, and incident energy has a chance to be absorbed inside the metamaterial absorber.

$$Z(\omega) = \sqrt{\frac{\mu(\omega)}{\epsilon(\omega)}} \qquad (2)$$

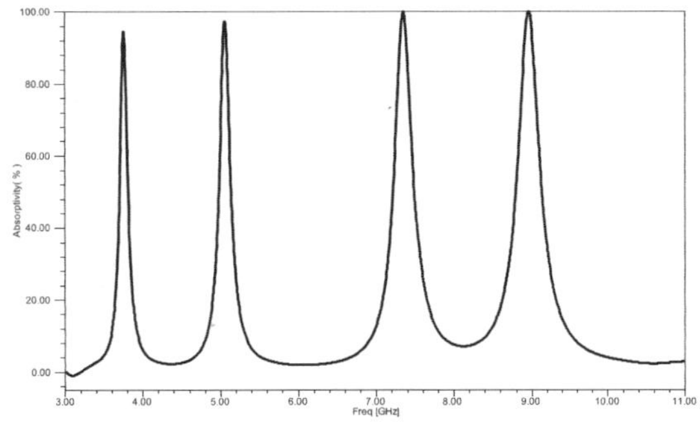

Fig.2. Imitated absorption spectra of the four-band absorber under normal incidence of wave.

Complements free space impedance η_0 (377 Ohm) and therefore reduce the reflection.

$$Z(\omega) = \sqrt{\frac{\mu_0\mu_{eff}}{\epsilon_0\epsilon_{eff}}} = \eta_0\sqrt{\frac{\mu_{eff}}{\epsilon_{eff}}} = \eta_0\sqrt{\frac{\mu'+j\mu}{\epsilon'+j\epsilon}} \quad (3)$$

Hence, at absorption frequencies

$$\epsilon' \approx \mu' \text{ and } \quad \epsilon \approx \mu \quad (4)$$

normalized input impedance Z(w) of the introduced structure shown in fig.3 which matches compact with the free space impedance η0 and perceived from here that the real parts of normalized input impedance Z(w)/377Ω (377 OF ATMOSPHERE) are close to 100% For all 4 absorptivity peaks, while the imaginary portion as given, of normalized input impedance Z(w)/377Ω are near to 0%and hence it nourish the absorption circumstances.

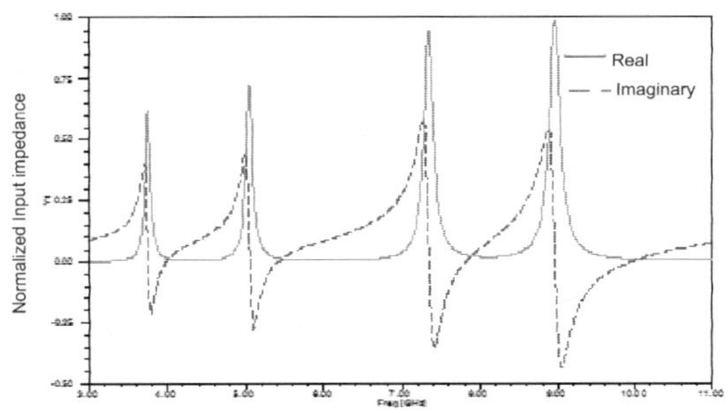

Fig.3. Normalized input impedance Z(w) of the introduced structure

3.2 Polarization Insensitivity

The electrics-filed, magnetic-field with currents distributions of the top and lowest surface have been shown in Fig. 4, Fig. 5 and Fig. 6 at all the four different absorption frequencies in order to better understand the absorption phenomenon of the structure. Fig. 4 shows the as middle loop, exterior split loop, interior cross connected loop and also too outer split ring with all the apex absorption frequencies 3.7, 5.05, 7.30 and 8.9 GHz respectively by electric and magnetic field distributions. Electric and magnetic innervations is described by too concentrated electromagnetic fields at the distinct segments of the structure.

The current density shows in Fig. 5 and Fig. 6 and is pinched into the central square loop at 3.7 GHz. However, interior coupled square gives more current at 7.30 GHz. Path for the flow of current higher due to coupled path present in interior arrangement, yet due to combination which is parallel of cross and loops acting inductance has been diminished, resulted in growing of absorption frequency. Improbable single square middle loop (without splits) providing resonance at solo frequency, outer square loops with splits provide resonances at two different frequencies. Surface current is mainly dispersed in outer most split loop at the frequency 5.05 GHz. Also, near-unity absorption has been realized at 8.9 GHz due to the split in the outer loop. In contrast to the other 2 diminished last loops, outer loop extent further and it is likely to have little resonance (hence absorption) frequency. Agile inductance gets demonized due to splitting of loops. This decreases total inductance and add to the absorption frequency. More observed from the Fig. 5 and Fig. 6, at the top and bottom layers constitute the anti-parallel directions of surface currents and thus a circulating loop current which is combination of conduction current (top and bottom layer copper) and displacement current (in-between top and bottom layer through substrate). Circulating loop current is perpendicular to the direction of incident magnetic field leads to creation of magnetic excitation. On the contrary, creation of high induced electric field in the direction of incident electric field shows the electric excitation within the structure at top and bottom copper. Therefore, strong electromagnetic absorption occurs due to both electric and magnetic resonances at highest absorption frequencies.

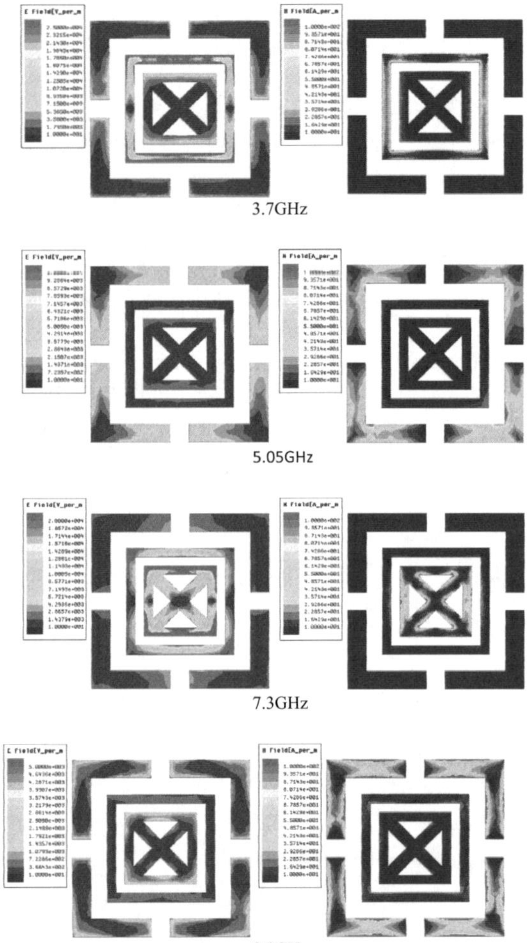

(a) E-Field (b) H-Field

Fig.4. (a) Electric field distribution and (b) magnetic field distribution with

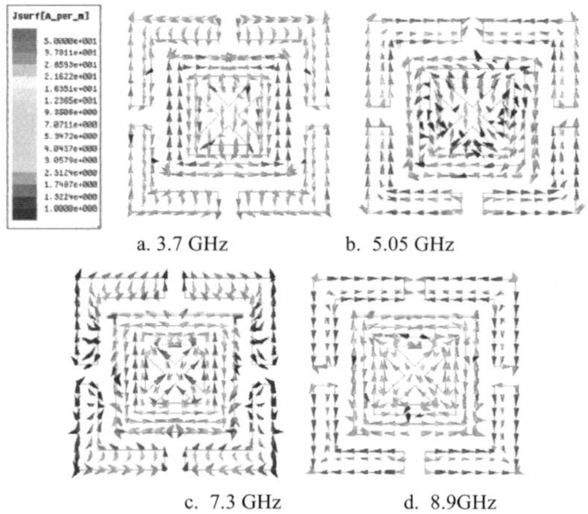

Fig.5. Surface current distributions attop surface at all the four

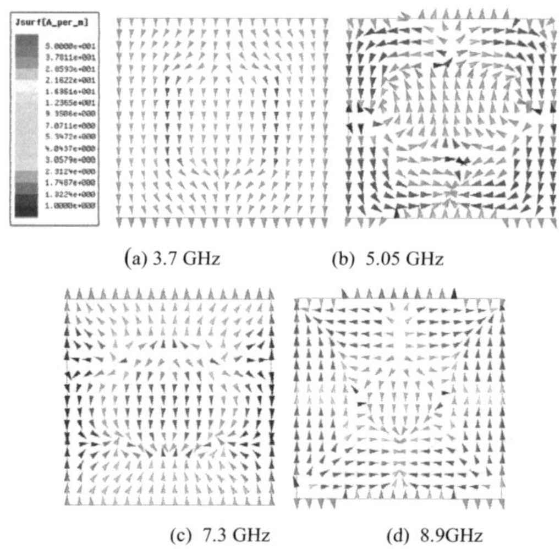

Fig. 6. Surface current distributions with top surface at all the four frequency of absorption.

Polarization-immunity because of 4-fold isometric unit cell structure in suggested structure has been established. The structure response with the divergence of phi-angle of incidence as shown in Fig. 7. Absorptivity is constant even till 60^0 of angle of incidence

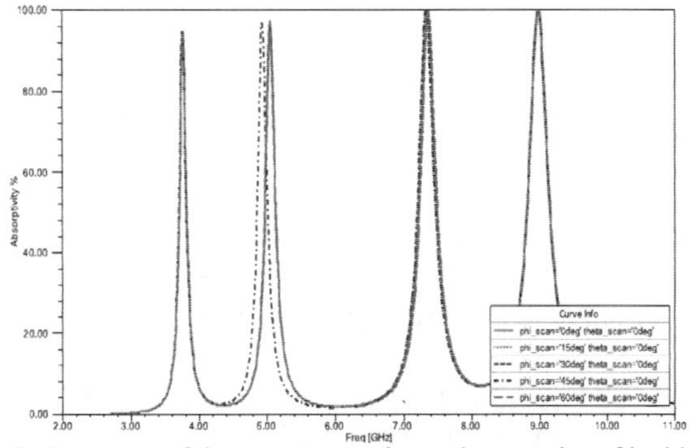

Fig. 7. Response of the structure under various angles of incidence (Phi-Angle variation).

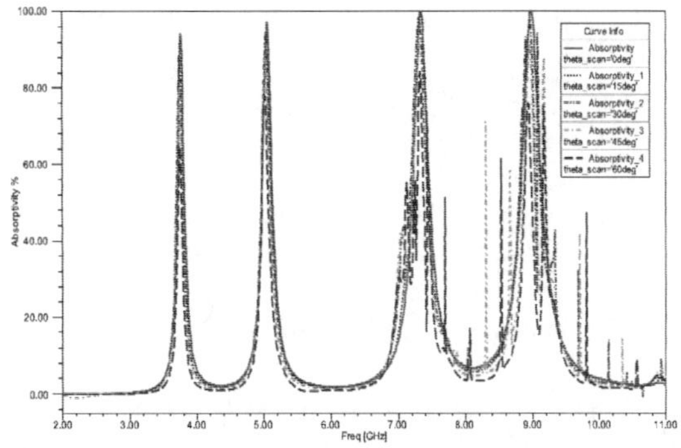

Fig. 8. Response of the structure under various oblique angles of incidence (Theta - Angle variation).

Theta-angle of incidence of different as shown in Fig. 8. The structure is inspected till 60^0 angle of incidence (with variation of theta), it has high absorption but further on absorptivity moderately diminished.

4. Conclusion

A fragile four band square loop metamaterial absorber has been displayed. To know what absorption actually the electric and magnetic field and their current distribution at various frequencies has been displayed. Also, a 4fold symmetric structure is used to make the structure polarization emancipated. A good absorption result nearly unity up to 45° angle of incidence can be achieved and will be used for various applications.

References

[1] D. R. Smith *et al.*, "Composite Medium with Simultaneously Negative Permeability and Permittivity," *Phys. Rev. Lett.*, vol. 84, no. 18, pp. 4184–4187, 2000.

[2] S. Enoch, G. Tayeb, P. Sabouroux, N. Guérin, and P. Vincent, "A Metamaterial for Directive Emission," *Phys. Rev. Lett.*, vol. 89, no. 21, p. 213902, 2002.

[3] X. Z. Nicholas Fang, Hyesog Lee, Cheng Sun, "Sub–Diffraction-Limited Optical Imaging with a Silver Superlens Nicholas," *Science (80)*, vol. 308, no. 2005, pp. 534–537, 2005.

[4] A. Rajput and K. V. Srivastava, "Design of a two-dimensional metamaterial cloak with minimum scattering using a quadratic transformation function," *J. Appl. Phys.*, vol. 116, no. 12, 2014.

[5] S. Ghosh, S. Bhattacharyya, D. Chaurasiya, and K. Srivastava, "An Ultra-wideband Ultra-thin Metamaterial Absorber Based on Circular Split Rings," *IEEE Antennas Wirel. Propag. Lett.*, vol. 1225, no. c, pp. 1–1, 2015.

[6] C. Shemelya *et al.*, "Stable high temperature metamaterial emitters for thermophotovoltaic applications," *Appl. Phys. Lett.*, vol. 104, no. 20, pp. 10–14, 2014.

[7] J. Su, Y. Lu, Z. Li, R. Zhang, and Y. L. Yang, "A Wideband and Polarization-Independent Metasurface Based on Phase Optimization for Monostatic and Bistatic Radar Cross Section Reduction," *Opt. Express*, vol. 2016, 2016.

[8] H. Tao, N. I. Landy, C. M. Bingham, X. Zhang, R. D. Averitt, and W. J. Padilla, "A metamaterial absorber for

the terahertz regime: design, fabrication and characterization.," *Opt. Express*, vol. 16, no. 10, pp. 7181–7188, 2008.

[9] C. M. Watts, X. Liu, and W. J. Padilla, "Metamaterial electromagnetic wave absorbers," *Adv. Mater.*, vol. 24, no. 23, 2012.

[10] S. Devkinandan Chaurasiya, Saptarshi Ghosh, "An Ultrathin Quad-Band Polarization-Insensitive Wide- Angle Metamaterial Absorber," *Microw. Opt. Technol. Lett.*, vol. 57, no. 12, pp. 697–702, 2015.

[11] A. A. Dastranj, H. Abiri, and A. R. Mallahzadeh, "Cosecant-squared pattern synthesis method for broadband-shaped reflector antennas," *IET Microwaves, Antennas Propag.*, vol. 8, no. October 2013, pp. 328–336, 2014.

[12] D. Chaurasiya, S. Ghosh, S. Bhattacharyya, and K. V. Srivastava, "Dual-band Polarization-Insensitive Metamaterial Absorber with Bandwidth-Enhancement at Ku-band for EMI / EMC Application," *IEEE Int. Microw. RF Conf.*, pp. 96–99, 2014.

[13] Q. Wu *et al.*, "Design Synthesis of Metasurfaces for Broadband Hybrid-Mode Horn Antennas With Enhanced Radiation Pattern and Polarization Characteristics," *IEEE Trans. Antennas Propagation,* vol. 60, no. 8, pp. 3594–3604, 2012.

[14] F. Dincer, O. Akgol, and M. Karaaslan, "Polarization Angle Independent Perfect Metamaterial Absorbers for Solar Cell Applications in the Microwave , Infrared , and Visible Regime," *Prog. Electromagn. Res.*, vol. 144, no. November 2013, pp. 93–101, 2014.

[15] J. Chen, G. Wang, Z. Chen, M. Liu, and X. Hu, "Multi-band Microwave Metamaterial Perfect Absorber Based on Mie Resonance Theory," *PIERS Proceedings,Guangzhou*, no. Aug. 25–28, pp. 541–547, 2014.

[16] P. Saville, "Review of Radar Absorbing Materials Defence R & D Canada – Atlantic," *Def. Res. Dev. Canada*, no. January, 2005.

[17] N. I. Landy, S. Sajuyigbe, J. J. Mock, D. R. Smith, and W. J. Padilla, "Perfect metamaterial absorber," *Phys. Rev. Lett.*, vol. 100, no. 20, pp. 1–4, 2008.

[18] H.-M. Lee and H. Lee, "a Dual-Band Metamaterial Absorber Based With Resonant-Magnetic Structures," *Prog. Electromagn. Res. Lett.*, vol. 33, no. May, pp. 1–12, 2012.

[19] S. Bhattacharyya and K. Vaibhav Srivastava, "Triple band polarization-independent ultra-thin metamaterial absorber using electric field-driven LC resonator," *J. Appl. Phys.*, vol. 115, no. 6, 2014.

[20] Y. Bai, L. Zhao, D. Ju, Y. Jiang, and L. Liu, "Wide-angle, polarization-independent and dual-band infrared perfect absorber based on L-shaped metamaterial," *Opt. Express*, vol. 23, no. 7, p. 8670, 2015.

[21] and K. V. S. Praneeth Munaga, Saptarshi Ghosh, Somak Bhattacharyya, "A Fractal-Based Compact Broadband Polarization Insensitive Metamaterial Absorber Using Lumped Resistors ," *Microw. Opt. Technol. Lett.*, vol. 58, no. 2, 2016.

[22] D. C. Donghao Zheng, Yongzhi Cheng, "Four-Band Polarization-Insensitive Metama- Terial Absorber Based On Flower-Shaped Structures," vol. 142, no. May, pp. 221–229, 2013.

IOP Conf. Series: Materials Science and Engineering **594** (2019) 012007 doi:10.1088/1757-899X/594/1/012007

Novel Power Coefficient for extracting the maximum power in Wind power based Doubly fed Induction generator (DFIG) using vector control

Arunesh kumar Singh[1,*], Abhinav Saxena[1], Prashant[2], MdWaris[2], Arun Singh Rana[3]

[1]Department of Electrical Engineering , Jamia Millia Islamia, India
[2]Department of Electrical Engineering , JSSATE Noida, India
[3]Department of Electrical Engineering , GITM,India

Email: *abhinaviitroorkee@gmail.com

Abstract. This Paper shows behaviour of the Doubly fed Induction generator(DFIG) for extracting the maximum power using vector controlled mechanism. This paper present the controlling of rotor side converter(RSC) and grid side converter(GSC) through conventional method of gate triggering by using d-q theory The paper includes the simulation model of DFIG and further the maximum power extraction from wind is estimated by using power coefficient which is completely dependent on pitch angle, tip to speed ratio(TSR). The novelty in power coefficient has design to harness the maximum power. Firstly the mathematical model of DFIG is developed with incorporated wind turbine as mechanical input and then vector controlled method is used. DFIG is broadly used in wind power applications which are more robust, reliable and very efficient system. The complete system is implemented using MATLAB software tool.

Keywords—Doubly fed induction machine, vector controlled ,TSR, Wind

1. Introduction

The conventional sources of electrical energy generation are reducing at a faster rate and environmental issues are also the cause to worry about the future[1]. This situation is forcing us to think about depletion of harnessing of electrical energy generation[2]. But the demand of energy consumption increases day by day as the population density increases exponentially at a global level so it become necessary to switch to renewable sources of energy from non renewable source of energy[3].

Nowdays energy generation can also be done at small scale level using different non conventionally energy like solar power plant,wind power plant, small hydropower plant(preferred to be less than 25MW) because the infrastructure cost and operating is considered to very small as compared to large scale power generation. The hydropower plants are only installed where hydro power plant is abundant[4]. On other hand solar & wind power generation can be installed irrespective of the location because wind & solar both are freely available. So wing & solar power is generally preferred to harness energy[5]. In this Doubly fed Induction machine(DFIM) is used in both generating and motoring operation using wind power including wind turbine as a mechanical input for

DFIM[6] but the speed measurement was the biggest issue because the conventionally speed was measured through the speed sensor which shows error of 7-8% from its actual value due to the mechanical vibrations Nowdays DFIM play as significant role in wind power generation so as to recover from slip power losses to increases the efficiency, reliability of the machine so that maximum of the output can be generated[8]. DFIM in proper hostile environment, hazardous places, noiseless environment where the speed can be measured at the short accessible remote location[10]. Now the main objective to generate or absorbing the maximum power under generating or motoring mode has to be accomplished[12].For the maximum power extraction we need to consider the effect of wind turbine parameters like pitch angle ,tip speed ratio(TSR),power coefficient with latest control equations to have maximum power coefficient which is directly proportional to generated power. Basically, the energy consumption per unit time is measured by DFIM under generating and motoring mode given by

$$E = Kw^3$$

After that energy consumption is calculated for sensor speed and sensor less speed and comparison will be analyse to check the performance under both cases and to show that why sensorless is preferred to overcome its limitation[13]

2. DFIM Dynamics

2.1 Basic introduction to DFIM

Doubly fed Induction are the typical conventional induction machine which can operate as motoring as well as generating mode where the powers can be absorbed or delivered from same grid[14] but to make the system more efficient and reliable two com rotor to stator so that rotor speed is somehow controlled by rotor side converters further grid voltage is directly voltage side converter and chock coil is connected between GSC and main +grid to avoid the problem of harmonics in the supply[15]. Depends upon the operating mode of DFIM, reactive and active power behaves differently under generating mode machine absorbs reactive and active power both from the grid in that case grid side converter act as rectifier while rotor side converter works as inverter while under motoring mode machine will deliver both active and reactive power to the grid both in that case grid side converter works as inverter while rotor side converter works as rectifier[16]. Firstly, the Mathematical model of DFIG has been developed using different reference frame[17]. Secondly; the rotor currents control and voltages enable the Doubly fed induction machine to be connected with the grid under synchronization while the rotor speed varies. Nowdays variable speed operating machine is preferred to fixed speed specially in case of lower speed operation of DFIM. Thirdly converter cost is also needed to consider because certain fraction of slip power losses (25-30%) is fed back to the grid to increases the efficiency of the DFIM [18] which makes the system more immune towards disturbance. Now the speed measurement using rotor shaft is avoided to avoid the problems of mechanical vibrations, more noise, more fluctuations in the grid voltage and current and the major danger to workers who are working under the hazardous environment so we prefer sensor less scheme to measure the rotor speed to avoid the above mentioned problems by measuring the speed using rotor current signal in different reference frame i.e synchronous rotor, stationary reference frame, further in continuation rotor position is measured by taking the angle between rotor currents in synchronous and rotor reference frame[19].

2.2. Power flow in DFIG

Fig 1 represent the structure of DFIG with power flow. Generally, we know that output of electrical power of rotor P_r is completely depend on slip value under the generating mode slip is negative so power will flow in the delivering state from machine to the grid directly from the stator while both reactive and active power will be absorbed through both grid side and rotor side converter from the same grid. Machine will absorb both powers in that case GSC acting as rectifier while RSC working like inverter[20] while delivering power from grid to the machine via GSC &RSC while voltage across dc link this will help to maintain constant voltage across grid terminals. Now the rotor power P_r is function of stator power Ps under the steady state both the powers are equivalent to each other[21]. DC link will absorb the power to maintain the voltage constant at the grid terminal under super synchronous operation while reverse phenomenon will for motoring mode machine will absorb power from grid directly to the winding of the stator and rotor will deliver same active & reactive power at slip frequency to the grid indirectly through RSC & GSC in that case grid side converter works as inverter while rotor side converter works as rectifier[22]. Phase sequence of generated voltage on the rotor changes under sub synchronous in addition to super synchronous mode of operation it will be change from positive under subsynchronous to the negative under super synchronous mode. The chock coil was a filter to remove the harmonics from the generated AC voltage. Simulation of 2KW DFIM is designed

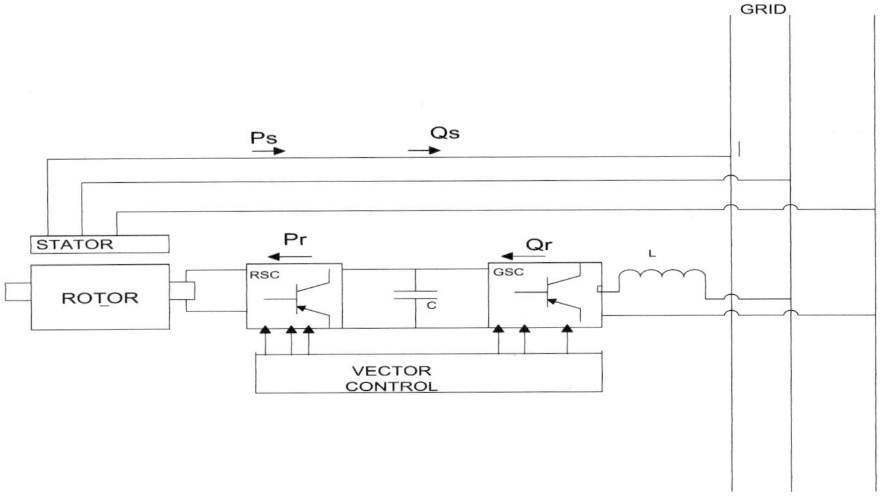

Fig.1 Power flow in DFIG

2.3 Modelling of DFIG:

The modelling of DFIG can be understood by considering the voltage source at both end (stator as well as rotor) as shown in the Equivalent circuit Diagrams of DFIG. The Doubly fed induction generator can be represented by adding a secondary voltage source V_2 to its secondary winding as shown in Fig.2 where X_r and R_r represent rotor reactance and rotor resistance referred to the stator side while R_l and X_l represent the stator Resistance and rotor resistance

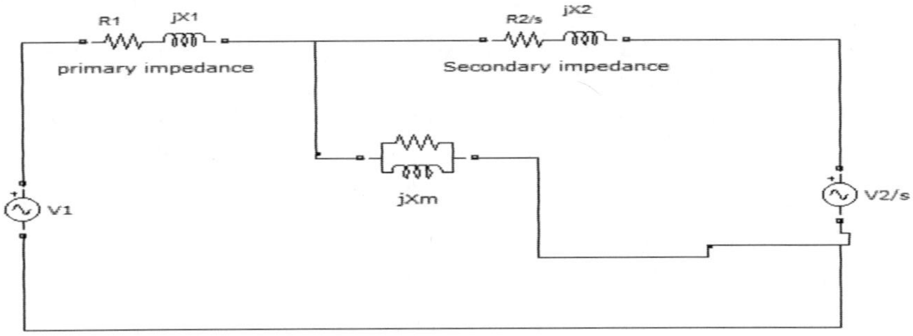

Fig.2 Equivalent Circuit Diagram of DFIG

Let $V_{qs}, V_{ds}, V_{dr}, V_{qr}, I_{qs}, I_{ds}, I_{qr}, I_{dr}$ are the corresponding stator and rotor voltages and currents in direct and quadrature axis coordinates their relationship are given below[12]

$$V_{qs} = (R_s + pL_s)I_{qs} + \omega L_s I_{ds} + pL_m I_{qr} + \omega L_m I_{dr} \qquad (1)$$

$$V_{ds} = -\omega L_s I_{qs} + (R_s + pL_s)I_{ds} - \omega L_m I_{qr} \quad pL_m I_{dr} \qquad (2)$$

$$V_{qr} = L_m pI_{qs} + (\omega - \omega_r)L_m I_{ds} + (R_r + pL_r)\, I_{qr} + (\omega \quad \omega_r)L_r I_{dr} \qquad (3)$$

$$V_{dr} = -(\omega - \omega_r)L_m I_{qs} + L_m pI_{ds} - (\omega - \omega_r)L_r I_{qr} + (R_r + pL_r)I_{dr} \qquad (4)$$

Electromagnetic torque is given by

$$T_e = 3(I_{qs} I_{dr} - I_{ds} I_{qr})pL_m / 2 \qquad (5)$$

Speed can be calculated by given relation

$$\omega_r = (T_e - T_L)/(B + pJ) \qquad (6)$$

Where $p = \frac{d}{dt}$

T_L = load torque

ω = reference frame speed

for rotor reference frame $\omega = \omega_r$

for synchronous reference frame $\omega = \omega_s$

for stationary reference frame $\omega = 0$

3. Wind Turbine Dynamics:

We know that load torque is applied as mechanical input to the Doubly fed Induction Machine[23]. This load torque or mechanical torque(Tm) is completely drive by wind turbine which depends on the wind power coefficient (C_p), wind speed(V_w), density of the air(ṗ),swept circular area(S).

$$T_m = 0.5S\rho C_p(\lambda, \beta)V_w^3 / w_t \qquad (7)$$

Performance parameter(C_p) of the wind turbine[8] depends on the pitch angle(β) and TSR(λ) , swept circular area $S = \pi R^2$

TSR(tip to speed ratio) is the ratio of the linear speed of the blades to the wind turbine

$$TSR = = \omega_{turbine} R / V_\omega \qquad (8)$$

Where R is radius of blades, $\omega_{turbine}$ is the linear speed of the wind turbine. The characteristic between C_p and λ has been drawn for variation in pitch angle(β). For the any value of the wind speed, output power(P) as mechanical input is given by

$$P = 0.5S\rho C_p(\lambda, \beta)V_w^3 \qquad (9)$$

In the above equation(9) Wind power is directly dependent on (C_p) with some constant. if maximum power is to be drawn from the wind then power coefficient(C_p) should be on maximum at particular values of pitch angle and TSR. In the given characteristics (Cp-λ) as shown in the figure Fig.3 maximum power coefficient is obtained at pitch angle (β) of 0 degree with the different wind speed variation of 6,8.2,8.4,9,12 with the turbine speed of 0.2,0.4,0.6,0.8,1 and it is observed that maximum power coefficient is obtained at an angle of 0 degree. A Single graph depict drawn from Fig3. in which Characteristic between Power Coefficient (C_p) and TSR (λ) is obtained for different values of pitch angle and at pitch angle of 0 degree peak value of C_p is obtained [24].

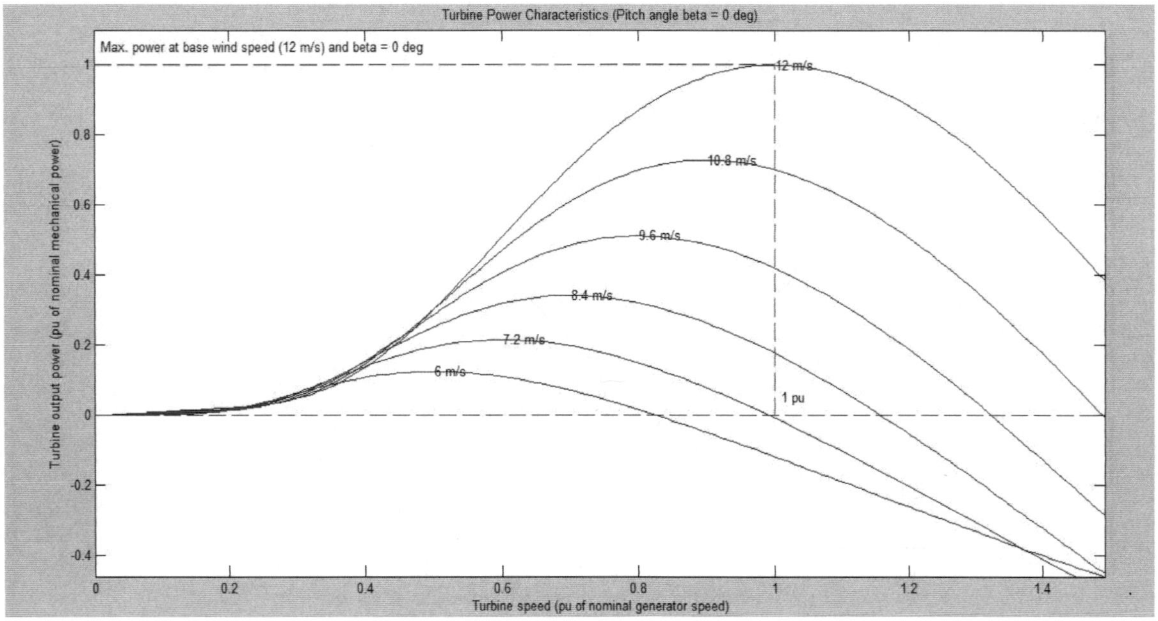

Fig.3 Power Coefficient Curve

4. Vector control of DFIG:

The equations which are used for vector controlled of DFIM is done in both RSC and GSC which is given inFig.4 and Fig.5 for the 415V and 15 KVA DFIG.

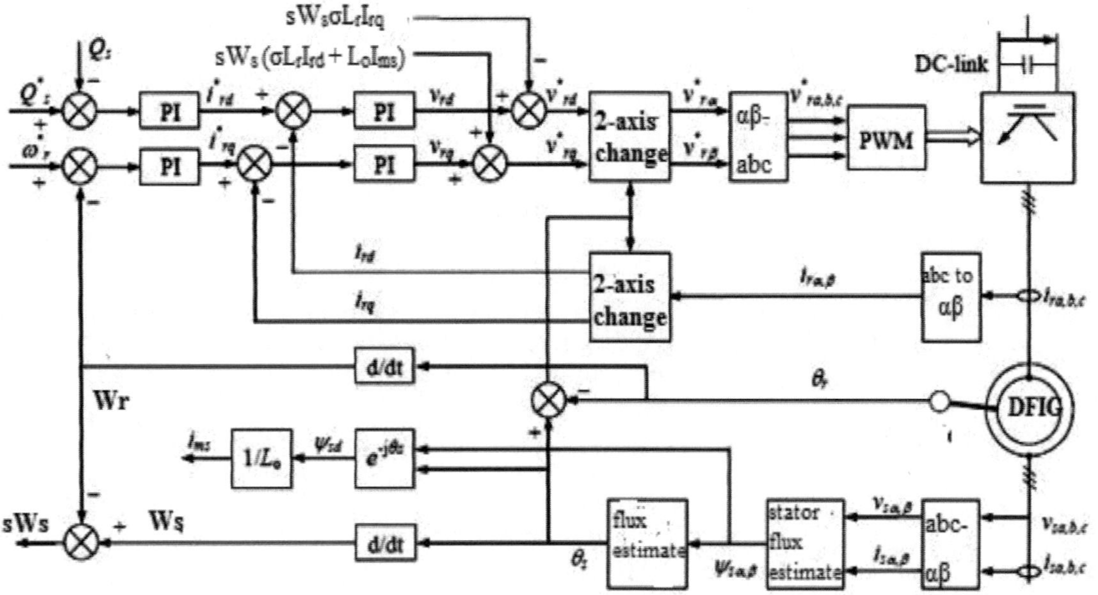

Fig.4 Vector Control of Rotor Side Converter

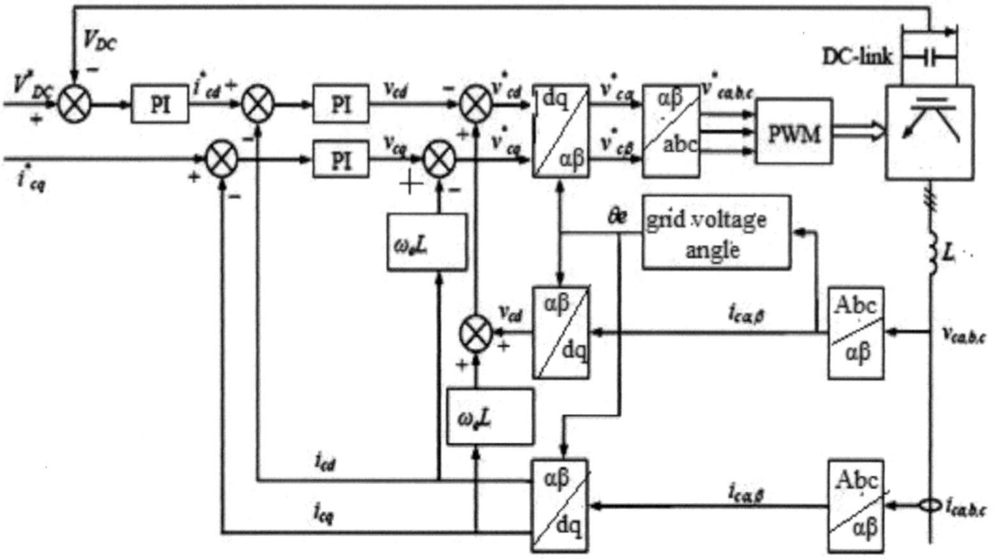

Fig.5 Vector Control of Grid Side Converter

4.1 Grid Side Converter (GSC)

Following equation for GSC are used for fig.5 as shown and for generating pulses in GSC a complete control mechanism using PI controller are represented from eq.(10) to eq.(14)

$$i^*_{cd} = (k_1 + \frac{k_2}{s})(V^*_{dc} \quad V_{dc}) \tag{10}$$

$$V_{cd} = (k_3 + \frac{k_4}{s})(i^*_{cd} \quad i_{cd}) \tag{11}$$

$$V_{cq} = (k_5 + \frac{k_6}{s})(i^*_{cq} \quad i_{cq}) \tag{12}$$

$$V_{cd} - w_e(L\, I_{cq}) = V^*_{cd} \tag{13}$$

$$V_{cq} - w_e(L\, I_{cd}) = V^*_{cq} \tag{14}$$

GSC is used to maintain the constant voltage across the grid by allow the dc link to absorb reactive power

4.2 Rotor Side Converter (RSC)

It is used to control the speed of the rotor in such a way to absorb more reactive power from the grid as shown in the Fig.4 through generate the different reference voltages and current by passing error through PI Controller represented from eq.(15) to eq.(20)

$$i^*_{rd} = (k_1 + \frac{k_2}{s})(Q^*_s \quad Q_s) \tag{15}$$

$$i^*_{rq} = (k_3 + \frac{k_4}{s})(w^*_r \quad w_r) \tag{16}$$

$$V_{rd} = (k_5 + \frac{k_6}{s})(i^*_{rd} \quad i_{rd}) \tag{17}$$

$$V_{rq} = (k_7 + \frac{k_8}{s})(i^*_{rq} \quad i_{rq}) \tag{18}$$

IOP Conf. Series: Materials Science and Engineering **594** (2019) 012007 doi:10.1088/1757-899X/594/1/012007

$$V_{rd} - sw_s(\sigma L_r I_{rq}) = V_{rd}^* \tag{19}$$

$$V_{rq} + sw_s(\sigma L_r I_{rd} + L_o \mathrm{I}_{ms}) = V_{rd}^* \tag{20}$$

Table.1 Machine Parameters

Parameter	Value
resistance of stator	0.023 ohm
inductance of stator	0.18 henry
resistance of rotor	0.016 ohm
inductance of rotor	0.16 henry
Stator magnetizing inductance	0.16 henry
Frequency	50 Hz
Pole pair	2
Wind Speed	12 m/s
DC link Voltage	800 V
Filter Specification	L=24μH, R=10Ω
Rotor side converter	K1=3.6,k2=8.9,k3=7.8,k4=11.3,k5=15.26,k6=17.56,k7=18.5k8 =21.25
Grid side converter	K1=25.6,k2=18.6,k3=27.8,k4=10.3,k5=32.86,k6=19.78

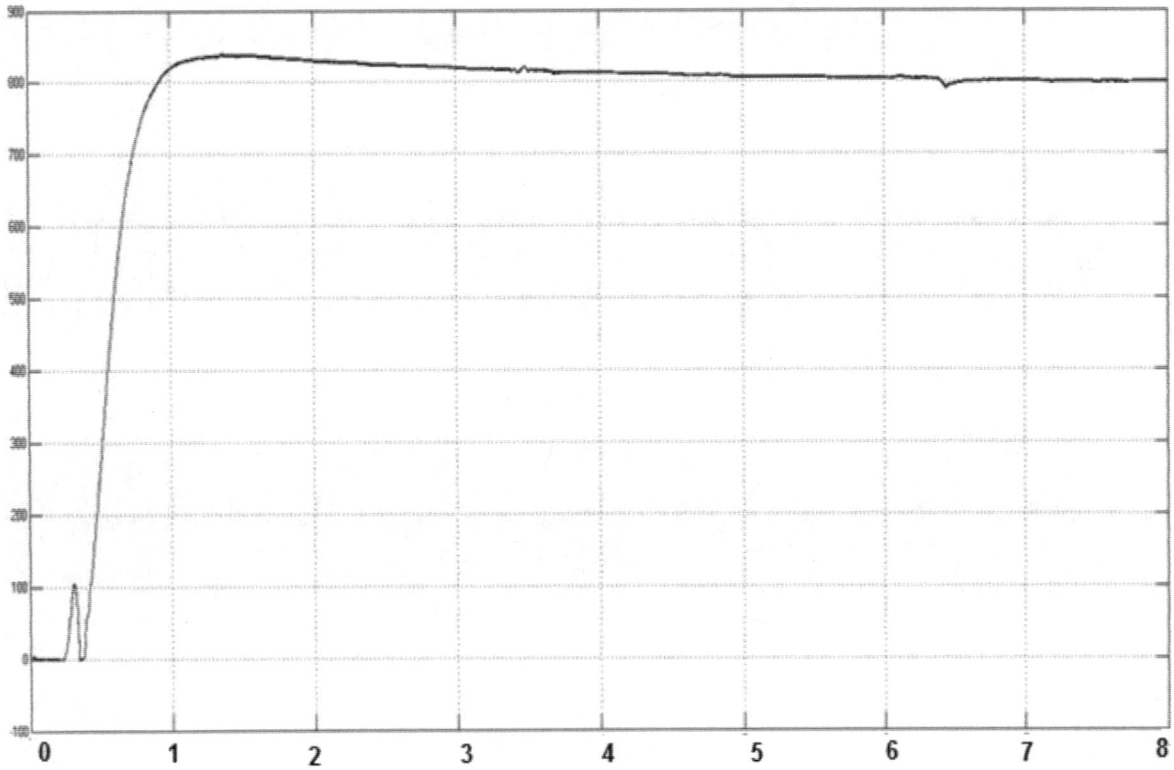

Fig.6 DC Link Voltage Waveform

IOP Conf. Series: Materials Science and Engineering **594** (2019) 012007 doi:10.1088/1757-899X/594/1/012007

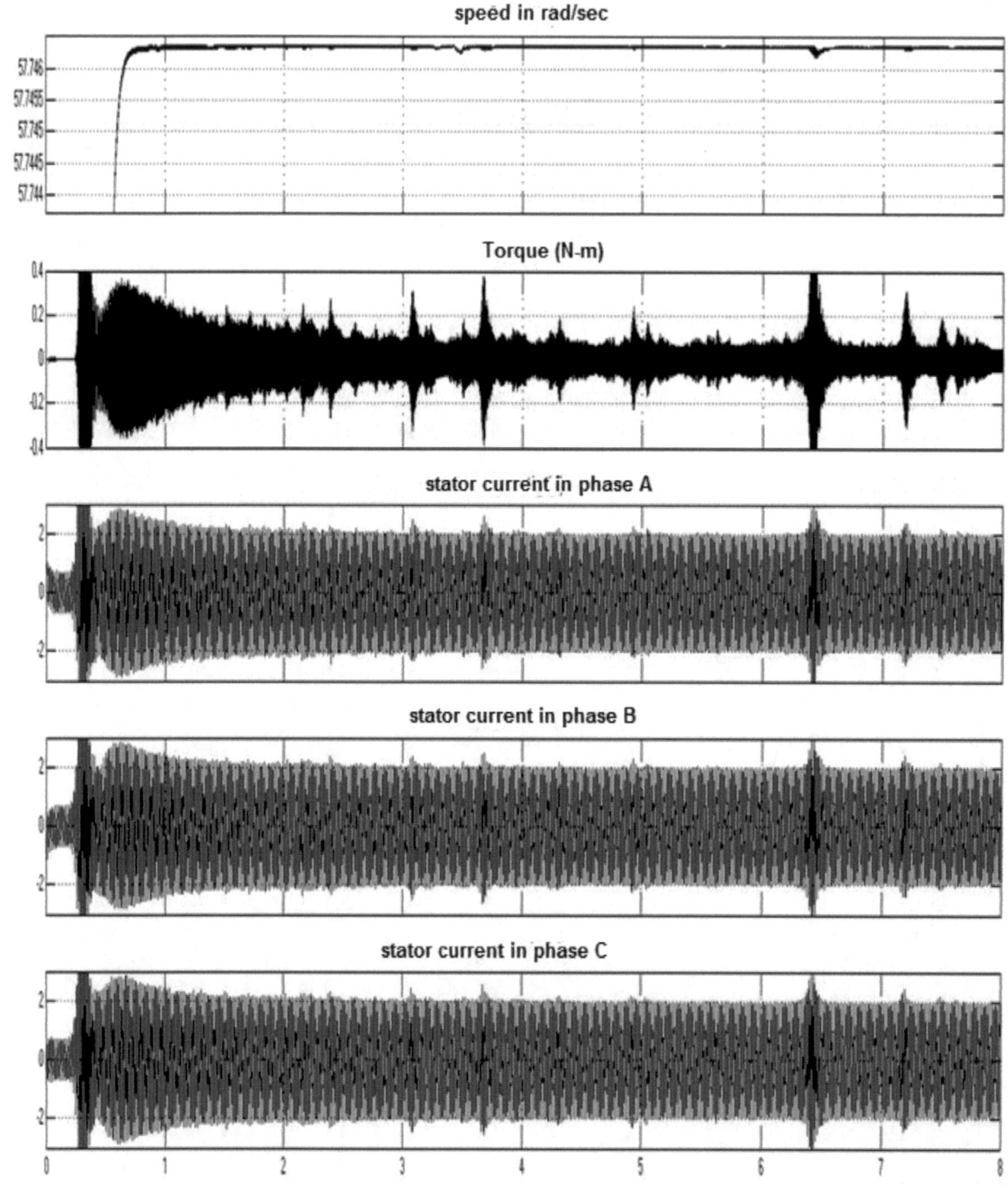

Fig.7 Graphical Representation of the Different Performance Parameter (Speed,Torque,3-Phase Current) of DFIG

IOP Conf. Series: Materials Science and Engineering **594** (2019) 012007 doi:10.1088/1757-899X/594/1/012007

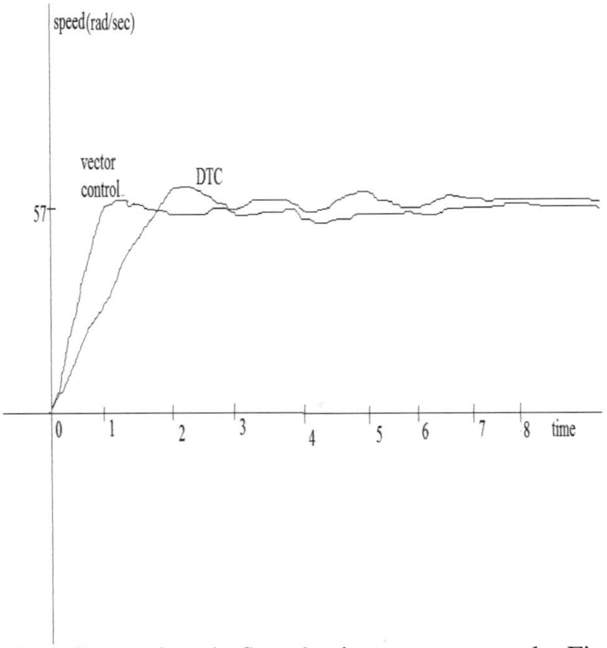

Fig.8 Comparison in Speed using vector control & DTC

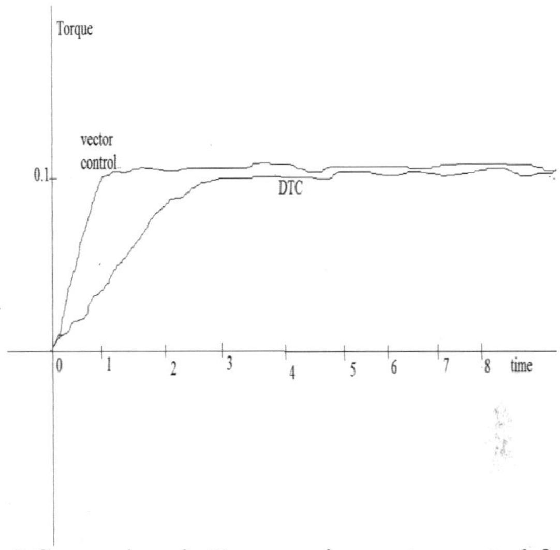

Fig.9 Comparison in Torque using vector control & DTC

5. Result and Analysis

From the different graph obtained of the Fig.7 for speed, torque, current by using the vector control for harnessing the maximum energy From wind turbine fed to the DFIG by varying the pitch angle and its maximum value obtained at $\beta=0$ degree at the wind speed of 12m/s to obtained the maximum power coefficient which directly depends on the power to harness in the maximum way and Vector control for GSC & RSC is also performed to harness the maximum energy from wind and performance parameter must be settled

6. Conclusion

This paper presents the maximum extraction or delivering of power .it is clearly shown that after designing the model of DFIG in different reference frame specially synchronous reference frame under vector control technique. Maximum power is obtained when power coefficient is maximum for pitch angle equal to 0 degree under different value of the tip speed ratio. This shows that power can be drawn maximally from wind turbine by controlling power coefficient. By choosing the suitable value of proportional constant of PID controller for vector control of RSC and GSC voltage across dc link voltage is maintained constant. It is also shown that performance parameter like speed, torque shows the better performance when system operated in vector control in comparison to Direct torque control method because it takes lesser settling time for attaining the final value

References

IOP Conf. Series: Materials Science and Engineering **594** (2019) 012007 doi:10.1088/1757-899X/594/1/012007

[1] G. SaravanaIlangoC. NagamaniM. A. AshaRaniM. Jaya BharataReddy Sensor-Less Estimation of Rotor Position in a Doubly Fed Induction Machine',Frontier energy Springer,171-206,March 2017
[2]:NguyenPhungQuang,Jörg-Andreas Dittrich'Nonlinear Control Structure for Wind Power Plants with DFIM',Frontier Energy Springer,327-343,May 2015
[3] F. RüncosR. CarlsonN.SadowskiP.Kuo-Peng,'Performance Analysis of Doubly fed Twin Stator Cage Induction Generator'Frontier energy Springer,361-373,Springer 2006
[4] MithunVyas,MohitSingh,SuryaSantoso'Grid Integration of Wind Power Systems: Modeling of Wind Power Plants'153-209,January 2014, Frontier Energy Springer
[5] SadeghGhaniVarzaneh,MehrdadAbedi,Govork B. Gharehpetian,'Enhancement of output power smoothing and Transcient Stability of DFIG-Based Wind energy Conversion system using Optimized Controllers Parameters',Electric power system 726-738,Taylor francis,2017
[6] Abdel GhaniAissaoui,MohamedAbid, Ahmed Thour,'Robust Fuzzy gains Scheduling of RST 3380, 617-626(2016)
[7] Lihui Yang, Zhao Xu, Zhao Yang Dong,' Advanced Control strategy of DFIG Wind Turbines for Power system Fault Ride Through', IEEE Transaction on Power system,Vol 27 No.2,May 2012
[8] M. Bezza, B.EL.Moussaoui, A.Fakkar,' Sensor less MPPT fuzzy controller for DFIG wind turbine',Science direct,339-348, Elseveir 2016
[9] Jun Yao, Yong Liao,' An improved Control Strategy of Limiting the DC-Link Voltage Fluctuation for Doubly fed Induction Wind generator', IEEE Transaction on Power electronics,Vol.23,No. 3, May 2008
[10] A k singh, A saxena, A pawar, S singh, "Analysis of Intelligence Techniques on Sensor less Speed Control of Doubly fed - Induction Machine(DFIM)", IEEE conference(INDIACOM 2016), 16-18th mar 2016, ISSN 0973-7529; ISBN 978-93-80544-20-5

[11] A saxena, D Singhal, D sharma, P gupta, S singh "Comparative Analysis of Sensor-Less Speed Control of Three Phase Induction Motor", IEEE conference(INDIACOM 2016), 16-18th mar 2016, ISSN 0973-7529; ISBN 978-93-80544-20-5

[12] W. Kartikeye,'An implicit sensor less control of DFIG, Elseveir 2012 vol.9, pg325-332

[13]NavneetKumar,Thanga Raj Chelliah,'Impact of Unbalance Grid Voltage on the Performance of Doubly fed Induction Machine at Motoring Mode',1291-1306,Taylor Francis,2016

[14] .Babouri,D.Aouzellag,K.Ghedamsi,'Introduction of Doubly fed Induction Machine in an Electric Vehicle',1076-1084, Elseveir,2013

[15]JinxinOuyang, XiauofuXionegǁ, Researchon the short-circuit capacity current of doubly fed induction generator under the varaiable non-deep voltagedropǁ,Elsevier2014

[16]Ranjan k. beheraandwenzhonggao,a Novel Controller Design for Grid-side Converter of Doubly Fed Induction Generator for Wind Power Interface: An Experimental Investigation,TaylorFrancis,ISSN: 1532-5008, Electric Power Components and Systems, 38:1531–1545, 2010

[17]Shuhui li and timothy a. haskew,'Characteristic Study of Vector-controlled Doubly-fed Induction Generator in Stator-flux-oriented Frame',Taylor&Francis,ISSN: 1532-5008,36:990–1015, 2008

[18] S Nouali , A Ouali,'Sensorless Control stragedy of WECS based on Cascaded DFIG using ANN',IJEP,ISSN:1990-7958,2011

[19] Wang Xuetal .Modeling and Simulation of PWM Rectifier Based on the Space Vector Modulation[J].Electric Drive,2009,32(2):27-31.

[20] S.T.Wange, J.Wangs, etal,— MRAS method of stator resistance using DTC control Induction motor drive,ǁ IEEEConferenceofIPEC,pp.213-220,2006.

[21] M H Khan,ArshiaAzam,'Sensorless control of induction motor drive for low speed range using IFLC based MRAS,'IJAREEIE,ISSN:2278-8875,2013

[22]K.S.T Szabat, T. Orlowskar-Kowalskar,—Optimization based fuzzy-logicsetsoof the speedcontroller withnonelastic pointsǁ,IEEETrans.onIndustry Applic.,vol.41,No.4, pp.1139-1145,2003

IOP Conf. Series: Materials Science and Engineering **594** (2019) 012007 doi:10.1088/1757-899X/594/1/012007

[23] Tamai S,D Sugimoto H and Yano Side M, Speed sensorless vector control of Induction motor with model reference adaptive system, Conf. Rec. IEEE/IAS annual meeting, 1987.

[24] Schauder Z C, Adaptive speed estimation and vector control identification of DFIM without considering rotational transducers, Conf. rec. IEEE/IAS annual meeting, 1999.

IOP Conf. Series: Materials Science and Engineering **594** (2019) 012008 doi:10.1088/1757-899X/594/1/012008

Feasibility Analysis of 100 kW Solar Plant for Chandigarh, India

Amit Rai[1,*], Ashish Shrivastava[2], Kartick C. Jana[1], Sandeep Tripathi[1] and Amit Agrawal[3]
Indian Institute of Technology (ISM) Dhanbad[1]
Manipal University Jaipur[2],
ABES Engineering College Ghaziabad[3]
E-mail: *rai.amit21@gmail.com

Abstract. The electricity demand is increasing worldwide with a fast pace. Accomplishment of this growing demand with renewable energy sources is gaining importance globally due to it's negligible pollutant emission. In this work, solar power potential is analysed for Chandigarh. A 100-kW photovoltaic plant is analysed for selected area with different PV panel combination for finding optimal solution of power generation. This selected area is having a good solar radiation reception potential of 5.07 kWh/m^2/day annually at annual 25.4°C temperature. The plant is designed with fixed axis orientation and has tilt of 31°. The plant has potential to export 156120 kWh of electricity to the grid with performance ratio of 76%.

1. Introduction

The growing demand of energy can be full field with renewable or non-renewable energy generation systems. The non- renewable generation systems have a major problem of greenhouse gas emission. However, the renewable energy sources are environmentally friendly and sustainable way to accommodate this supply this energy [1]. The geographical position of India has enormous solar based plant potential, as it receives solar radiation from 4 to 7 kWh [2]. The National Solar Mission of India has a massive target of 20,000 MW power generations through solar power by 2020 [3].

The solar panels have zero carbon emission because they use semiconductor device to generate electricity. Recent developments in manufacturing of these panels have reduced the production cost significantly and have provided, solar a competitive platform in energy generation systems. However, power output from solar depends on the varying climatic conditions which creates shading and also reduces the radiation [4]. To solve this problem of power fluctuation various dc-dc converters [5-6], maximum power point tracking algorithms are proposed. Different researchers in their work have analysed the PV plant feasibility in terms of cost and energy production for different areas of world [7-11].

In this work feasibility analysis of 100 kW solar power plant is analysed for Chandigarh, India. The technical performance is estimated on the basis of capacity factor, annual energy production, and energy production and performance ratio. Moreover, financial analysis is also presented in terms of Levelized COE, payback period and net present cost.

IOP Conf. Series: Materials Science and Engineering **594** (2019) 012008 doi:10.1088/1757-899X/594/1/012008

2. Another System Description

The location of the plant is 30.75°N latitude and 76.5°E longitude. At this location, the value of GHI is 5.07kWh/m²/day with the annual mean temperature of 25.4°C. The selected solar panel is Trina Solar TSM which has 16.76 % efficiency and other parameters are shown in table.1. The selected inverter is of SMA American with maximum ac power capacity of 3800 Wac and other parameters are shown in table.2. To generate 100 kW of power total system will need 306 solar panels in which 34 strings of 9 panels each string, shown in table.3.

Table. 1. Solar panel Specification

Panel	Value
Nominal efficiency	16.76%
Max. Power	325.2 Wdc
Voc	46 Vdc
Isc	9.2 Adc
Module Length	1.94 meter

Table.2 Inverter Specification

Inverter	Value
Max. AC Power	3800 Wac
Max. DC Power	3928.11 Wsc
Max. DC Voltage	600 Vdc
Nominal DC Voltage	398.4 Vdc

Table.3. System Configuration at reference condition

Configuration	Value
Nameplate Capacity	99.5 kWdc
Number of Modules	306
Modules/String	9
String in Parallel	34
Total Module Area	593.6 m2
String Voc	414 V

3. Methodology

The performance of PV plant is evaluated through System Advisor Model (SAM), a software for performance evaluation, developed by National Renewable Energy Laboratory, USA. Analysis of this plant includes:

A. Technical feasibility of the plant in terms of capacity, production and energy yield
B. The financial feasibility of the plant in terms of payback period, COE and net capital cost
C. The carbon emission reduction potential of the plant.

IOP Conf. Series: Materials Science and Engineering **594** (2019) 012008 doi:10.1088/1757-899X/594/1/012008

4. Result and Discussion

4.1 Power Production and Capacity

The power generated from PV panel is directly proportional to the solar irradiance incident on it. However, shading and seasonal variation play a crucial role in the DC power output of the plant, and ac power supplied to the grid. In the month of April irradiance and power production both is maximum as there is a clear sky condition is in this month with irradiance of 115731 kWh and dc output is 15771.9 kWh. The minimum power and irradiance is in the month of January, shown in fig.1. The annual energy production of first year will be 156120 kWh and after this 0.5% degradation in power is taken in simulation to match practical limitation of the plant. The capacity factor of the plant is 17.90% and performance ratio is 76 %, which shows the feasibility of the plant is in acceptable limit and location of the PV plant is viable for the production, shown in table 4.

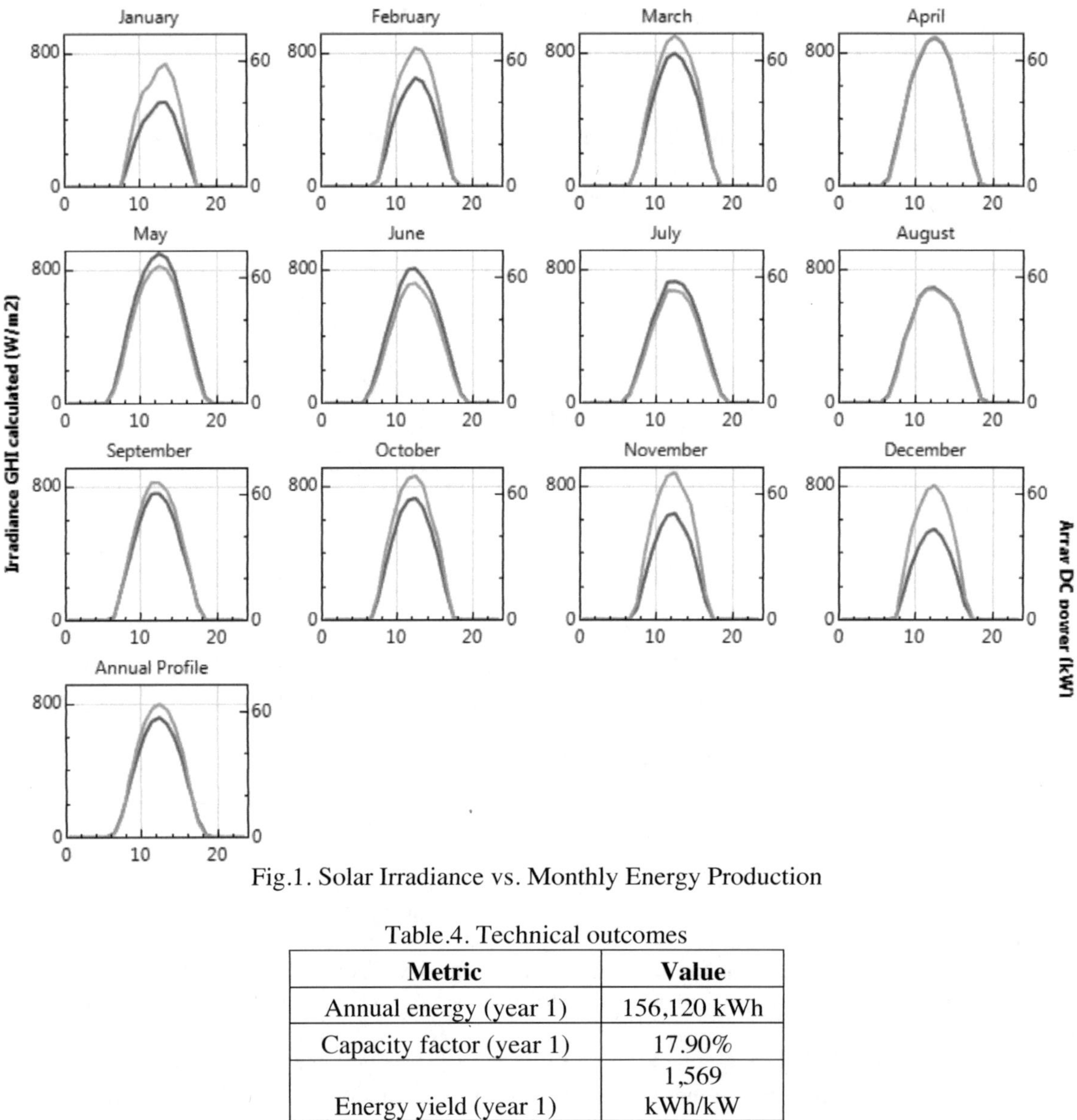

Fig.1. Solar Irradiance vs. Monthly Energy Production

Table.4. Technical outcomes

Metric	Value
Annual energy (year 1)	156,120 kWh
Capacity factor (year 1)	17.90%
Energy yield (year 1)	1,569 kWh/kW
Performance ratio (year 1)	0.76

IOP Conf. Series: Materials Science and Engineering **594** (2019) 012008 doi:10.1088/1757-899X/594/1/012008

4.2 Financial Viability of the Plant

The financial analysis is a key factor for decision making in any plant. In this section, for analysis various inputs and parameters are specified, like project life time is 25 years, loan term is 20 years with the rate of 2%, inflation rate is considered 1%. With this input parameter as shown in table.5, the COE of the solar plant is coming 2.57 cent per kWh which is a comparable COE. The payback period is 16.9 years, which shows with 100% loan, the project is financially viable.

Table.5. Financial Parameters and Outcome

LOAN PARAMETERS	
Debt fraction (% of total installed cost)	100.00
Loan term (years)	20
Loan rate (%/year)	2.00
ANALYSIS PARAMETERS	
Analysis period (years)	25
Inflation rate (%/year)	1.00
Real discount rate (%/year)	5.50
Nominal discount rate (%/year)	6.56
TAX AND INSURANCE RATES	
Federal income tax rate (%)	7.00
State income tax rate (%)	2.00
Sales tax (% of direct costs)	0.00
Insurance rate (%/year)	1.00
Property Tax (%/year)	
Assessed percentage (% of total installed cost)	100.00
Assessed value ($)	99,523
Metric	**Value**
Levelized COE (nominal)	2.57 ¢/kWh
Levelized COE (real)	2.33 ¢/kWh
Electricity bill without system (year 1)	$1,091
Net savings with system (year 1)	$5,043
Net present value	$17,786
Payback period	16.9 years
Net capital cost	$99,523
Debt	$99,523

5. Conclusion

The performance 100 kW PV plant is analysed for Chandigarh, India. The annual production and performance ratio of 156120kW and 76% respectively shows that the project is comparable to established PV plants in similar climatic condition. The COE of the plant is low and payback period is

IOP Conf. Series: Materials Science and Engineering **594** (2019) 012008 doi:10.1088/1757-899X/594/1/012008

ensuring the return on investment. With above outcome, it can be concluded that this project is suitable and will reduce the dependency on the conventional non-renewable plants.

References

[1] Goura, R., 2015. Analyzing the on-field performance of a 1-641 megawatt-grid –tied PV system in South India. Int. J. Sustainable Energy 34, 1–9.
[2] Sudhakar, K., Srivastava, T., Satpathy, G., Premalatha, M., 2013. Modelling and estimation of photosynthetically active incident radiation based on global irradiance in Indian latitudes. Int. J. Energy Environ. Eng. 4 (21), 2–8.
[3] Ministry of New and Renewable Energy, Solar energy in India. February 12, 2014.
[4] Makrides, G., Zinsser, B., Norton, M., Georghiou, G.E., Schubert, M., Werner, J.H., 2010. Potential of photovoltaic systems in Countries with high solar irradiation. Renewable Sustainable Energy Rev. 14, 754–762.
[5] Amit Agrawal, Ashish Shrivastava, Kartick C. Jana, "Uniform Model and Analysis of PWM DC-DC Converter for Discontinuous Conduction Mode", IETE journal of Research, Vol. 64, issue-4, 2018, pp. 569-581.
[6] Bhim Singh and Ashish Shrivastava, "Buck converter based power supply design for low power light emitting diode lamp lighting," IET Power Electronics, vol. 7, no. 4, pp. 946-956, 2014.
[7] Ayompe, L.M., Duffy, A., McCormack, S.J., Conlon, M., 2011. Measured performance of a 1.72 kW rooftop grid connected photovoltaic system in Ireland. Energy Convers. Manage. 52, 816–825.
[8] Besarati, S.M., Padilla, R.V., Goswami, D.Y., Stefanakos, E., 2013. The potential of harnessing solar radiation in Iran: generating solar maps and viability study of PV power plants. Renew. Energy 53, 193–199.
[9] Elhodeiby, A.S., Metwally, H.M.B., Farahat, M.A., 2011. Performance analysis of 3.6 kW Rooftop grid connected photovoltaic system Egypt. In: International Conference on Energy Systems and Technologies, Cairo, Egypt, CEST 2011, 11–14 March.
[10] IEA (International Energy agency) Country report on analytical monitoring of photovoltaic systems IEA PVPS Task13, 2014
[11] Labed, S., Lorenzo, E., 2004. The impact of solar radiation variability and data discrepancies on the design of PV systems. Renew. Energy 29, 1007–1022.

IOP Conf. Series: Materials Science and Engineering **594** (2019) 012009 doi:10.1088/1757-899X/594/1/012009

Response of monopolar HVDC system in presence of single double-tuned dc filter

Soma Deb and H K Verma

Department of Electrical and Electronics Engineering
School of Engineering and Technology
Sharda University, Greater Noida, India

E-mail: soma.deb@sharda.ac.in

Abstract. Study of the response of an HVDC system to various conditions, types and locations of fault on the system is basic to developing suitable protection techniques and algorithms. This paper presents a study of the behaviour of monopolar HVDC system in the presence of a single double-tuned dc filter. CIGRE Benchmark model for monopolar HVDC system has been simulated on PSCAD/EMTDC for the study. Five different fault locations, viz. three dc line faults (two faults near the line ends and the third fault at the middle point of the line) and two ac side faults (one near the rectifier and the other near the inverter), and five different values of fault resistance (ranging from 0 to 2000 ohms) have been considered. It has been observed that incremental current and incremental power in the HVDC line can discriminate only mid-line dc faults from ac fault. Also, incremental current can discriminate between rectifier-side ac fault and inverter-side ac fault for all values of fault resistance, whereas incremental power can do so only for lower values of fault resistance (0 ohm and 10 ohms). Transient energy can discriminate dc line faults at all locations from ac faults with all values of fault resistance. However, transient energy can discriminate rectifier-side ac fault from inverter-side ac fault only for low values of fault resistance (0 ohm and 10 ohms) and not for high values. This study can be very effectively used in developing a protection algorithm/technique. The study can be further extended to analyze the performance of HVDC system in the presence of any type of the dc filter and for any length of the line.

1. Introduction

HVDC system is widely used these days for transporting bulk power over long distances. With rapid developments in the technology of line-commutated converter based high voltage direct current (HVDC) system, the reliability and sensitivity of the associated protection system are becoming increasingly important. For development of suitable protection techniques and algorithms, study of the response of the given HVDC system to various conditions, types and locations of fault is essential so as to determine the system variable to be used as the basis of protection. A survey of the literature on the development of protection techniques and algorithms for HVDC systems reveals that the following system variables have been used or recommended for protection of monopolar HVDC system by various researchers:

i Current [1-5]
ii Voltage [6]
iii Power [1]

IOP Conf. Series: Materials Science and Engineering **594** (2019) 012009 doi:10.1088/1757-899X/594/1/012009

iv Reactive Energy [7]
v Transient Energy [1,8]

Study of the response of an HVDC system essentially requires its simulation under healthy and various fault conditions. This paper is aimed at evaluating the response of a monopolar HVDC system and choosing system variables suitable for designing a protection scheme. Soma Deb and Verma [1] reported a detailed performance evaluation of CIGRE benchmark model, wherein they concluded that change in current, power and transient energy are the best suited system variables for protection, while voltage, impedance etc. generally do not give the desired selectivity of the protection scheme. As such, this paper also evaluates the system performance only in respect of the change in current, power and transient energy.

In any HVDC system, dc filters are a major and important component and are used to serve two purposes, providing reactive power and reducing harmonic currents. DC filters are installed on the line side of the HVDC smoothing reactor and shunt with the dc line. The DC filters used, employ different configurations of reactors, capacitors and resistors to achieve single, double or triple tuning.

Commonly used types are cascaded double tuned filter [9], parallel double-tuned filter [8], triple tuned filter [10] and parallel triple tuned filter [4]. Type of dc filter installed affects the response of HVDC system to faults. Reference [1] reported the system response in the presence of cascaded double tuned dc filter while, this paper considers only one single double tuned dc filter. Parameters of the simulated HVDC System along with parameters of the dc filter are given in Section 2. Simulation methodology and the results of the simulation study are given in Section 3 and Section 4 respectively, followed by discussion and conclusion.

2. Simulated HVDC System

HVDC system considered in this paper for simulation study is based on CIGRE benchmark model for monopolar HVDC system, which has been simulated using PSCAD/EMTDC [11].

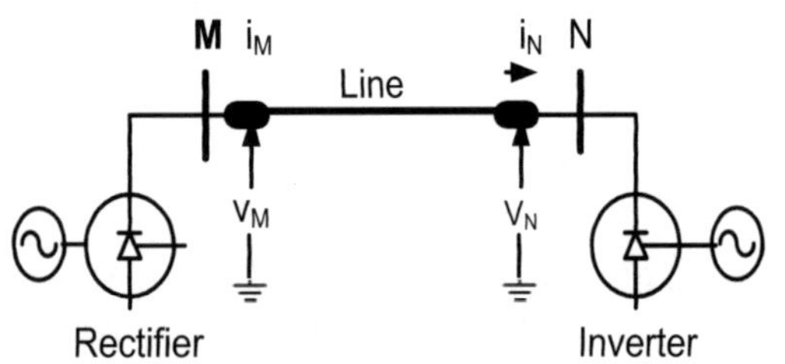

Figure 1. Simplified Diagram of HVDC System.

Table 1. Parameters of Simulated System

Parameters	Values
AC Source 1	345 kV, 50 Hz
AC Source 2	230 kV, 50 Hz
Rated Power	1000 MW
Rated Voltage	500 kV
Rated Current	2 kA
DC line length	800 km

Figure 1 above shows a simplified diagram of monopolar LCC-HVDC system with two-end measurements. Here, v_M, i_M represent the voltage and current measurements at sending end M, and v_N, i_N represent at receiving end N. The assumed directions of currents and voltages are as shown.Table 1 above shows the parameters of the simulated HVDC system.

Configuration of a single double-tuned dc filter is shown in figure 2. It comprises of a series resonant circuit in series with a parallel resonant circuit. Filter parameters of the dc filter is given in Table 2.

IOP Conf. Series: Materials Science and Engineering **594** (2019) 012009 doi:10.1088/1757-899X/594/1/012009

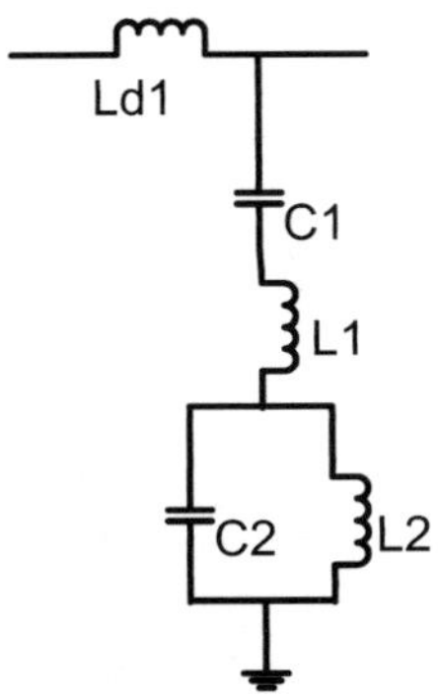

Figure 2. Single double-tuned dc filter.

Table 2. Parameters of DC Filter

Parameter	Value
C1	2 μF
L1	11.71 mH
C2	9.074 μF
L2	5.874 mH
Ld1	0.29 H

3. Simulation Methodology

Simulation study has been carried for faults occurring at five different locations. Two three phase faults – one at the rectifier side (RF) and the other at the inverter side (IF), and three internal dc faults- sending end (DF_1), mid-point (DF_2) and receiving end (DF_3) have been created. Every time the fault is created at 0.5 s. The data has been collected at a rate of 2 kHz.

4. Simulation Results

In this simulation study, system response has been obtained in terms of change in current (incremental current), change in power (incremental power) and transient energy. Five different values of fault resistance, viz, 0 ohm, 10 ohms, 100 ohms, 1000 ohms and 2000 ohms, are considered.

4.1. System Response in terms of incremental current

Variation of the incremental current with time for three internal fault locations (DF_1, DF_2 and DF_3) is shown in figures 3 to 5 respectively, and for external fault locations (RF and IF) in figures 6 and 7 respectively.

4.1.1. Incremental-Current Response for Internal Faults (DF_1, $DF_2$and DF_3)

Response curves in terms of change in current for rectifier side dc fault DF_1 are shown in figure 3. Separate curves are shown for five different values of fault resistance. In each case, a pair of curves is plotted, one for the sending end M and the other for the receiving end N. It may be noted that the scale is progressively amplified as the values of the incremental current are decreasing with the increase in the fault resistance. It is observed that on solid fault, the incremental current at the fault end (ΔI_M) has huge oscillations and changes its polarity between positive and negative values for a few cycles of the fault inception. The incremental current at the other end (ΔI_N) is almost zero with no oscillations. As the fault resistance increases, the magnitude and oscillations in the incremental current at the sending end reduce while it remains positive throughout.

Response curves in terms of change in current for mid-line dc fault DF_2 are shown in a similar style in figure 4. It may be noted that for mid-line fault, the oscillations in the sending end current are less severe than for fault at the rectifier end DF_1. Moreover, the incremental current at the other end is also more than that for DF_1

Figure 5 shows the system response in terms of change in current for dc fault near the inverter end (DF_3). It is observed that the oscillations are more severe in the incremental current at the fault end, i.e. ΔI_N and very little oscillations in the incremental current at the other end, i.e. ΔI_M

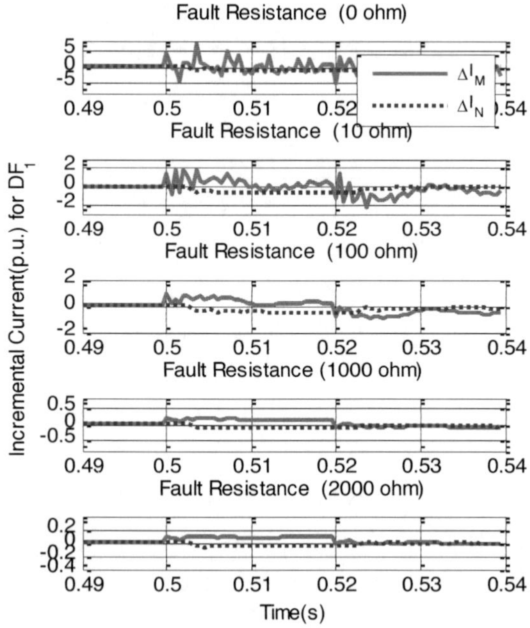

Figure 3. Response curves in terms of change in current for rectifier-side dc fault (DF_1).

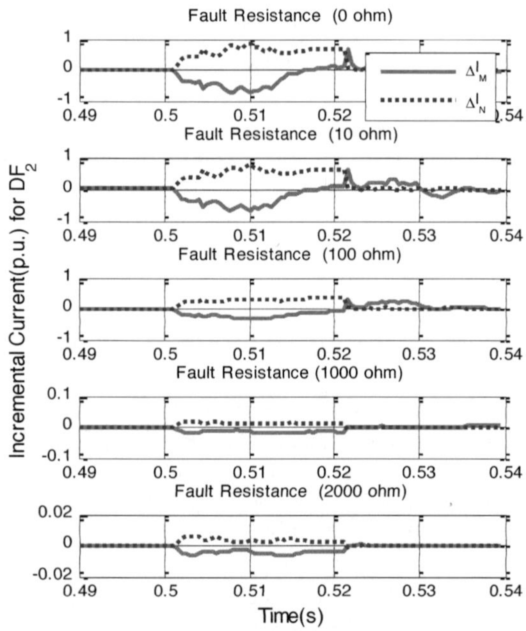

Figure 4. Response curves in terms of change in current for mid-line dc fault (DF_2).

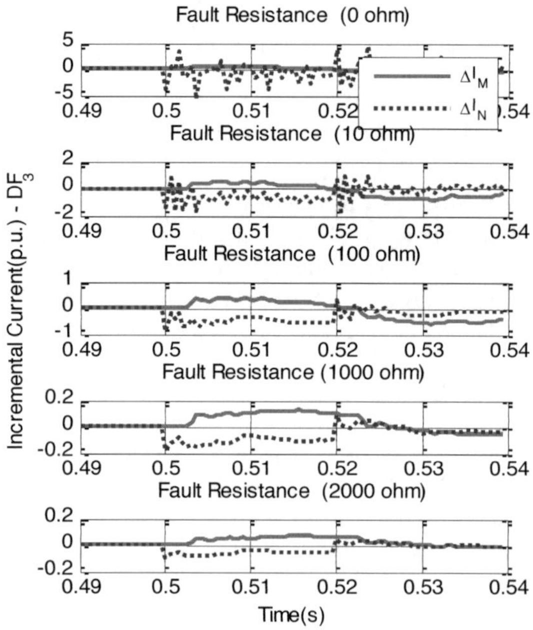

Figure 5. Response curves in terms of change in current for inverter-side dc fault (DF_3).

4.1.2. Incremental-Current Response for External Faults (RF and IF)

Response curves in terms of change in current for rectifier side ac fault (RF) are shown in figure 6. It can be noted that both ΔI_M and ΔI_N are negative and free from oscillations. Moreover, their values reduce with increase in the fault resistance, as would be expected.

Response curves in terms of change in current for inverter side ac fault (IF) are shown in figure 7. It is noted that both ΔI_M and ΔI_N are positive (at least for a couple of cycles) and have no oscillations. As expected, their values reduce as the fault resistance is increased.

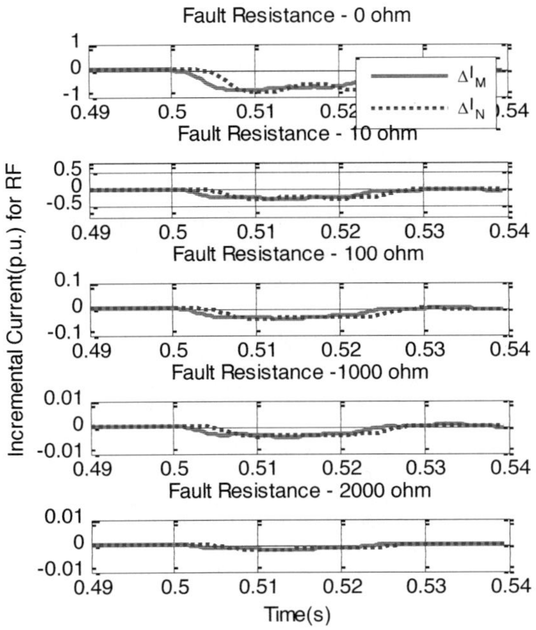

Figure 6. Response curves in terms of change in current for rectifier-side ac fault (RF).

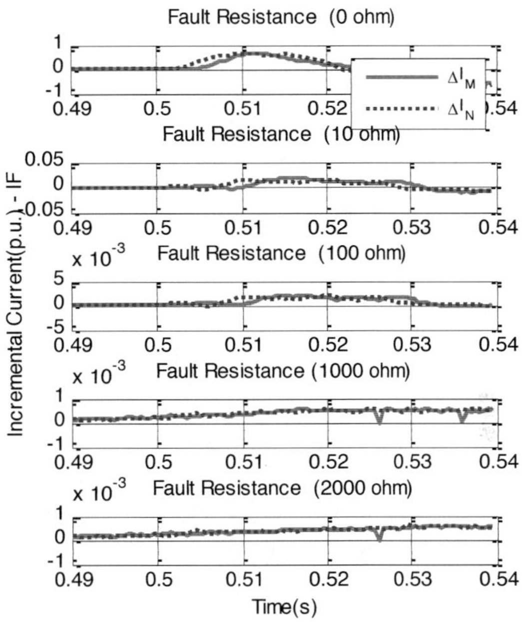

Figure 7. Response curves in terms of change in current for inverter-side ac fault (IF).

4.2. System Response in terms of incremental power

Variation of incremental power for the same five locations of fault is shown in figures 8 to 12 respectively.

4.2.1. Incremental-Power Response for Internal Faults (DF_1, DF_2 and DF_3)

Figure 8 shows the response curves in terms of change in power for DF_1. As shown in the figure, for solid fault resistance and lower values of fault resistance, the incremental power at the fault end oscillates between positive and negative values, whereas the incremental power at the end is largely free from such oscillations.

Response curves in terms of change in power for rectifier side dc fault DF_2 are shown in figure 9. The value of change in power (ΔP_M) is negative and that of ΔP_N is positive for this fault location. As expected, the magnitude of each incremental power reduces with increase in fault resistance.

Response curves shown in figure 10 show that for DF_3, the incremental power at the fault end (which is the end N) has oscillations, like the oscillations in incremental power at the fault end M for DF_1.

IOP Conf. Series: Materials Science and Engineering **594** (2019) 012009 doi:10.1088/1757-899X/594/1/012009

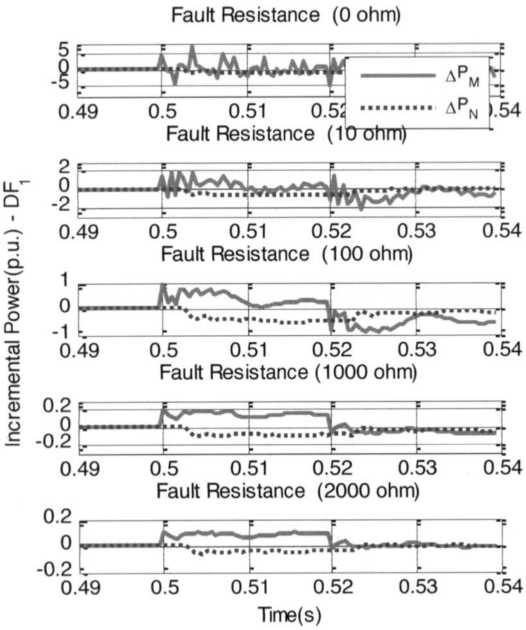

Figure 8. Response curves in terms of change in power for rectifier-side dc fault (DF_1).

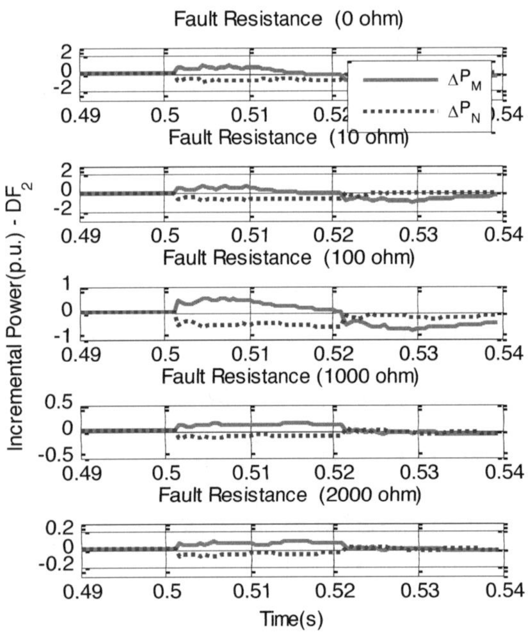

Figure 9. Response curves in terms of change in power for mid-line dc fault (DF_2).

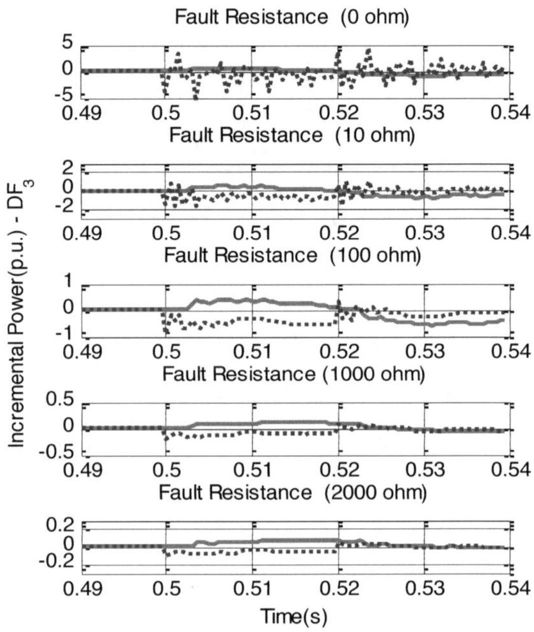

Figure 10. Response curves in terms of change in power for inverter-side dc fault (DF_3).

IOP Conf. Series: Materials Science and Engineering **594** (2019) 012009 doi:10.1088/1757-899X/594/1/012009

4.2.2. Incremental-Power Response for External faults (RF and IF)

Change in power at both the ends ΔP_M and ΔP_N are positive for rectifier side ac fault RF as shown in figure 11. It may be observed that in this case too, the magnitudes reduce as the fault resistance increases.

As shown in figure 12, change in power ΔP_M and ΔP_N are both negative for inverter-side ac fault IF for lower values of fault resistance (0 and 10 ohms), whereas these values are positive, though the magnitudes are small, for higher values of fault resistance (100 ohms and above).

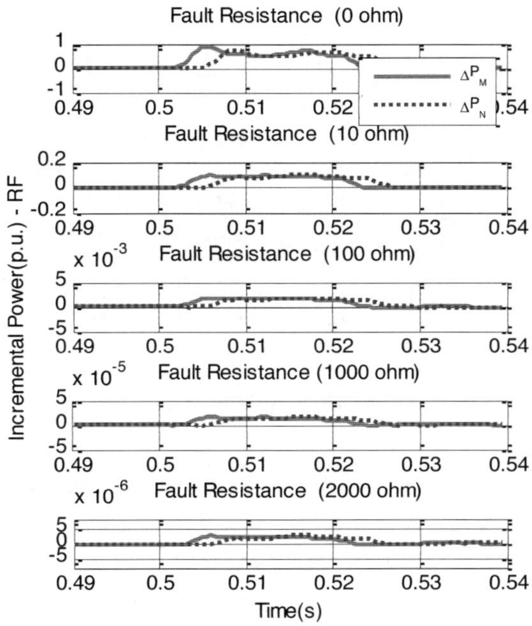

Figure 11. Response curves in terms of change in power for rectifier-side ac fault (RF).

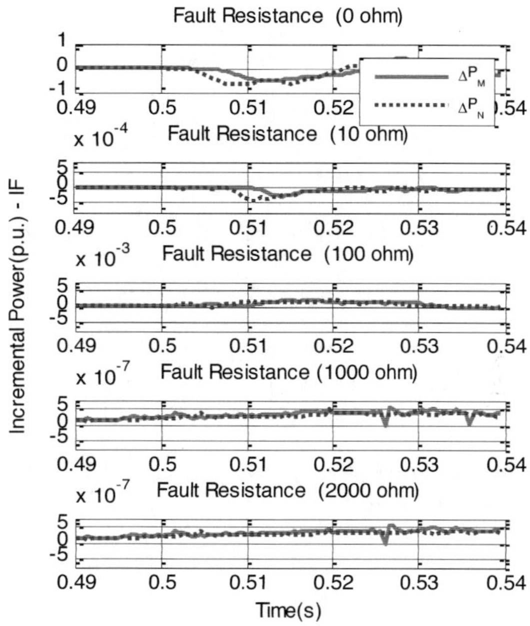

Figure 12. Response curves in terms of change in power for inverter-side ac fault (IF).

4.3. System Response in terms of transient energy

Variation of transient energy with time for the same five locations of fault is shown in figures,13 to 17 respectively.

4.3.1. Response Curves for Internal Faults (DF_1, DF_2 and DF_3)

The response curves in terms of transient energy at the two ends for the three locations of internal fault (dc fault), viz, DF_1, DF_2 and DF_3, are shown in figures 13, 14 and 15 respectively. In each case, the transient energy ΔE_M at rectifier side M remains negative whereas that for inverter side N remains positive. Their magnitudes reduce with increase in fault resistance, as expected.

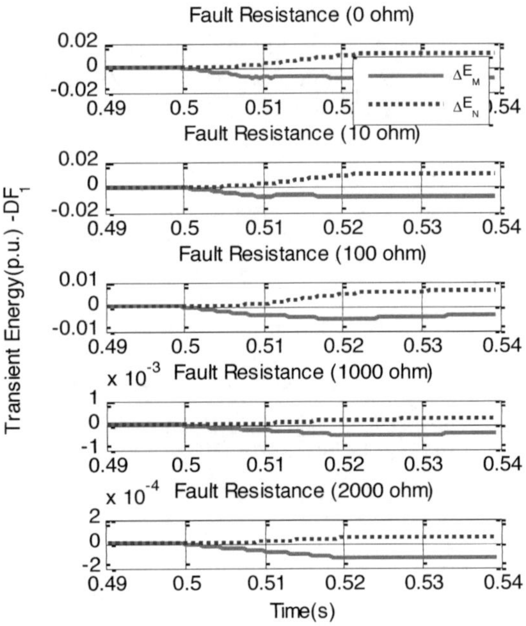

Figure 13. Response curves in terms of transient energy for rectifier-side dc fault (DF_1).

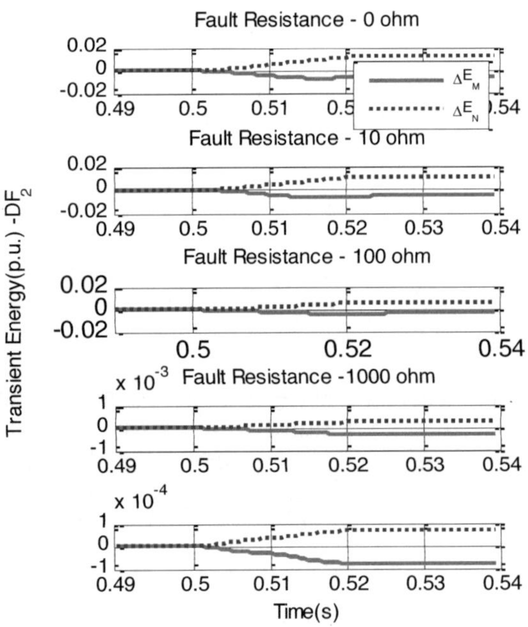

Figure 14. Response curves in terms of transient energy for mid-line dc fault (DF_2).

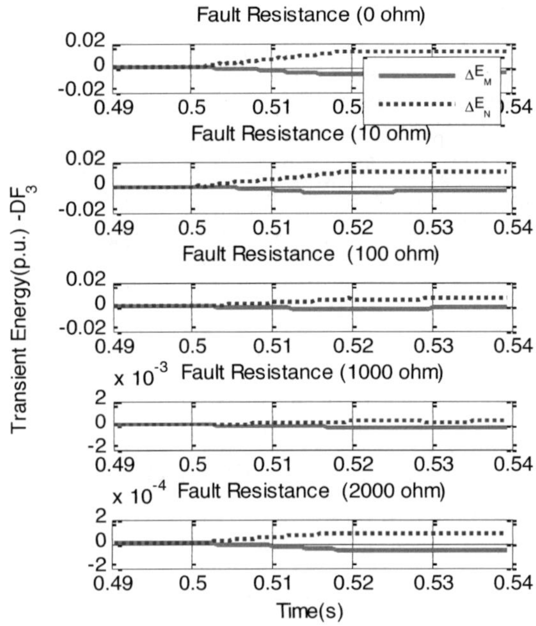

Figure 15. Response curves in terms of transient energy for inverter-side dc fault (DF_3).

IOP Conf. Series: Materials Science and Engineering **594** (2019) 012009 doi:10.1088/1757-899X/594/1/012009

4.3.2. Response Curves for External Faults (Rf and IF)

Response curves in terms of transient energy for rectifier side ac fault RF are shown in figure 16, and those for inverter side ac fault IF in figure 17. It is observed from Figure 16, that the transient energy at each end is positive for RF; their values reduce with increase in fault resistance. The transient energy curves given in Figure 17 reveal a different type of response for IF. The transient energy at both the ends is negative for low values of fault resistance (0 ohm and 10 ohm) and it is positive for both the ends for high values of fault resistance, (1000 and 2000 ohms). For a middle value of fault resistance, ΔE_M changes from initial positive value to some negative value whereas ΔE_N remains negative throughout.

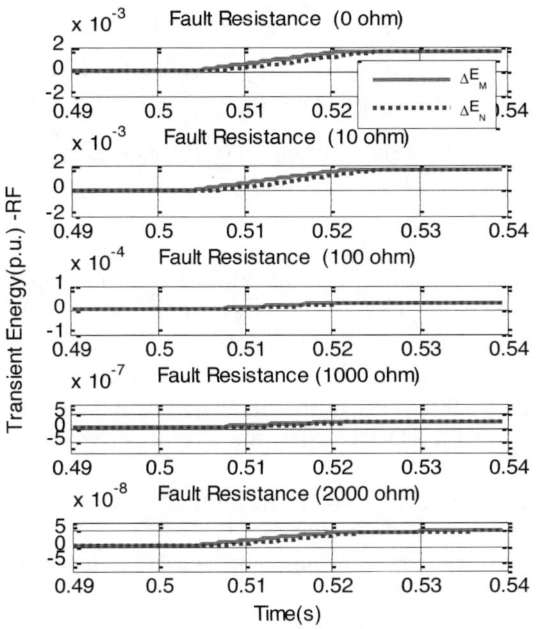

Figure 16. Response curves in terms of transient energy for rectifier-side ac fault (RF).

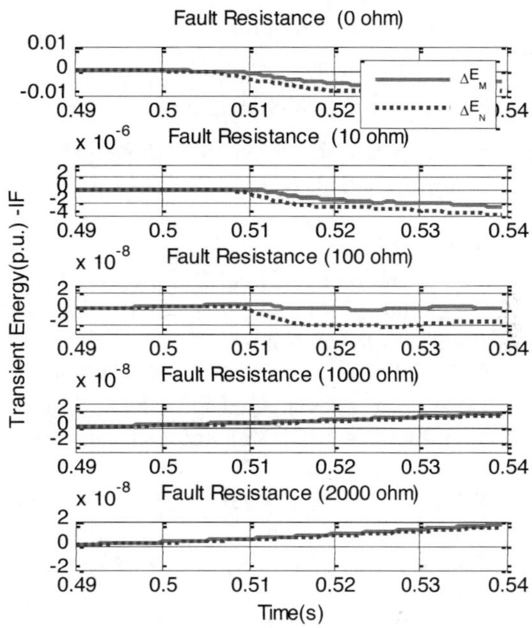

Figure 17. Response curves in terms of transient energy for inverter-side ac fault (IF).

5. Discussion and Conclusion

Monopolar HVDC system in presence of single double tuned dc filter has been simulated in this paper to analyze system response to dc and ac faults: three locations on dc line and two locations on ac side. Five different values of fault resistance (0 ohm, 10 ohms, 100 ohms, 1000 ohms and 2000 ohms) have been simulated. Incremental current, incremental power and transient energy are the system variables selected for the simulation study.

A critical analysis of the simulation results leads to the following significant observations:

(a) For line end dc faults, incremental current and incremental power at the fault end have oscillations whereas transient energy is free from any such oscillations.

(b) For mid-line dc faults, incremental current, incremental power and transient energy, all the three are free from oscillations.

(c) For external fault at the rectifier end, incremental current at both the ends remain negative for a minimum of two ac cycles whereas incremental power and transient energy remains positive. Similar tends are observed for all values of fault resistances.

(d) For external fault at the inverter end, incremental current and incremental power at both the ends remain positive for a minimum of two ac cycles for all values of fault resistances. However, transient energy at both the ends is negative only for lower values of fault resistances (0 ohm and 10 ohms). For 100 ohm fault resistance, transient energy at rectifier end is positive on fault inception and tends negative after a quarter of ac power cycle, whereas transient energy at inverter end is negative. For higher values of fault resistances (1000 ohms and 2000 ohms), transient energy at both the ends are positive.

Following proposals are made for extending the reported work in future:

i. On similar lines, system variables can be analyzed for any other dc filter.

ii. The actual system for which the relaying is to be designed can also be simulated.

iii. The line length taken in the present study is 800 km. However, the lines in several systems are much longer which may have an effect on reducing the magnitudes of the signals.

iv. Protection algorithm can be developed using the results of the simulation study such as reported or proposed above.

References

[1] Deb S and Verma H K 2017 Selection of fault discriminators for unit protection of monopolar hvdc transmission system *Internatuional Conference on Condition Assessment Techniques in Electrical Systems* 20-5.

[2] Naidoo D and Ijumba N M 2004 HVDC line systems *Proc. IEEE Powercon Conf. 1327-32.*

[3] Li A, Cai Z, Sun Q, Li X, Ren D and Yang Z 2009 Study on the dynamic performance characteristics of HVDC control and protections for the HVDC line fault *Proc. Power Energy Soc. Gen. Meeting 1-5.*

[4] Kong F, Hao Z and Zhang B 2017 Improved differential current protection scheme for CSC-HVDC transmission lines *IET Generation, Transmission & Distribution* **11** 978-86.

[5] Gao S, Liu Q and Song G 2017 Current differential protection principle of HVDC transmission system *IET Generation, Transmission & Distribution* **11** 1286-92.

[6] Wu J, Li H, Wang G, Liang Y 2017 An improved travelling wave protection scheme for LCC-HVDC transmission lines *IEEE Transaction on Power Delivery* **32** 106-116.

[7] Luo S, Dong X, Shi S and Wang B 2016 A directional protection scheme For HVDC transmission lines based on reactive energy *IEEE Tran. on Power Del.* **31** 559-66.

[8] Xiaodong Z, Tai N, Guangliang Y and Ding H 2012 A transient protection scheme for HVDC transmission line *IEEE Trans. Power Del.* **27**, 718-24.

[9] Zheng X, Tai N, Thorp J S and Yang G 2012 A transient harmonic current protection scheme for HVDC transmission line *IEEE Trans. Power Del* **27** 2278-85.

[10] Guo J, Wang G, Liang Y and Zeng D 2015 Global-sensitivity-based theoretical analysis and fast prediction of travelling waves with respect to fault resistance on HVDC transmission lines *IEEE Trans. Power Del* **30** 2007-16.

[11] Faruque M O, Zhang Y and Dinavahi 2006 Detailed Modeling of CIGRE HVDC benchmark system using PSCAD/EMTDC and PSB/SIMULINK *IEEE Trans. Power Del* **21** 378-87.

IOP Conf. Series: Materials Science and Engineering **594** (2019) 012010 doi:10.1088/1757-899X/594/1/012010

Economic Viability Solar PV Power Plant in Distribution System

Shiva Pujan Jaiswal[1], Vivek Shrivastava[2], Shashank Singh[1]

[1]EEE Department, SET, Sharda University
[2]EE Department, NIT, Delhi
E-mail: shivajaiswal@gmail.com

Abstract. Silicon is definitely, the most common substance used for manufacturing solar cells. Distributed Generation (DG) based on Solar Photo Voltaic (PV) cells enhance the voltage stability and economical profit as well as reduce the carbon emission. This manuscript optimized the location and size of PV base DG for the 13-Bus distribution network, using the nature base Particle Swarm Optimization (PSO) algorithm. This manuscript also gives an overview of the process of manufacturing of PV cell. The core intention of this manuscript is to demonstrate the economical benefits of optimal PV based generation unit. Results prove that the PV base power plant improves the technical performance as well as economical performance of the system.

1. Introduction

Now it is possible to generate solar energy competitively with the conventional source of electricity. Planners have recognized solar energy contribution to mitigating the difference between the energy demand and supply of power in India since long and overall target fixed is 17.5 GW by the year 2022 [1]. This requires mass production of commercially viable and cheaper solar panels. Silicon wafer is the most important and vital component in solar panel. The production cost of the solar cell is guided and influenced to a great extent by the cost of production of silicon wafers. Technology improvement in the process of getting raw silicon material, converting into wafers and technological advancement of tools and equipment, is the need of the hour to achieve the target. The refinement in processes of separation of pure silicon can improve and enhance mass production. Solar PV cells are categorized as organic photovoltaic and concentration photovoltaic[2]. Process of manufacturing both types of photovoltaic cells challenges the scientific community and investors to produce these in huge quantity. A photovoltaic cell is a tool which converts light into electricity. Photons of sunlight fall on the outer surface of the solar panel then electron are excited after gaining energy from a photon of sunlight. Once an excited electron breaks the valence bond and reaches the conduction band, then it is free to move[4]. Due to movements of electrons, current flow through the PV cells, i.e. DC electricity generated[4]. In this VP based Distributed Generation (DG) is sited at an optimal location to enhance the economics of utility. For optimization of place and size, the nature base algorithm is selected. To obtain the maximum benefit, an objective function is proposed with three objectives, namely minimization of cost, loss and voltage instability. The whole manuscript is prearranged in the six segments. Section 2 presents a summary of the materials and methods used for the fabrication of commercial solar cells. Section 3 shows the ability of inverters to control voltage. Section 4 develops a methodology to estimate the benefits of PV systems. Section 5 evaluates the proposed technique, and section 6 demonstrates the results of the proposed method.

2. Methods and materials for the manufacturing of solar cells

Crystal Solar Approach simplifies the process of manufacturing silicon wafers for solar cells and eliminates the use of expensive technology and equipment. A Korean company has invented the first commercially viable solar panel by using new technology for producing silicon wafers[5]. Silicon wafer is the most expensive and important part of a solar cell (one third to half the cost of a solar panel). The technology help to make wafers that are less than one third the thickness of the conventional wafer[6]. The wastage of silica during the manufacturing process is reduced drastically as compared to that of conventional approaches, and requirement of equipment needed to make the wafers is also reduced. There are substantial saving wafer costs to almost half[7].

Given the high cost of manufacturing wafers, investors used to invest resources in thin-film solar panels, which are alternatives to silicon solar panels — the normal way to manufacture silicon wafers, which requires highly purified silicon (called poly-silicon)[8]. In the earlier conventional process, pure silicon was derived from a gas that contains silicon and other elements. Now a company has developed a process to create thin crystalline silicon wafers directly from that gas. In this process, the need to first make poly-silicon, then melt and sawing into wafers is not required. This version of a process is normally used in the chip industry. There is a further scope of improvement in this technology to bring down costs lower, which can be investigated to improve machine technology to produce in numbers with quality.

Solar cells are typically made of the semiconductor and named as per the material used to manufacture it. The semiconducting materials should have a suitable quality to absorb sunlight. Some cells are designed in such a way that these can handle sunlight reaching the Earth's surface, while some others are designed for use in space arena and other application[9]. Solar cells should be made of only one single layer or multiple layers (multi-junction) to take advantage of various absorption qualities and electrons separation technique.

Solar cells classified into first generation cells, second generation cells and third generation cells. These are described as under:

a) The first generation cells are also called conventional wafer cells. These are made of crystalline silicon. This crystalline silicon was the earliest commercially usable PV technology.

b) Second generation cells made of by thin film solar cells. Amorphous silicon, CdTe and CIGS cells also categorized as thin film solar cells.

c) Emerging photovoltaic is categorized as third generation solar cells. These are under process of development for commercial exploitation. Basic material for these experiments is organic materials, organo-metallic compounds and inorganic materials. There is a lot of research going on, and investment has been made to fine-tune processes and technologies as promises, these materials offer to achieve the task of producing low-cost and highly efficient solar cells.

Crystalline silicon is abundant material for solar cells. This material is also known as "solar grade silicon. Bulk quantity silicon separated into different categories as per crystalline and crystal size. Thes broadly called as ingot, ribbon and wafer. The function of this cell entirely based on p-n junction technique. The thickness of wafers used in Solar cells is between 160 and 240 micrometres

3. Formulation of economical benefits of a solar PV system

As per the latest bid on the provision of solar plant in Rajasthan, the purchased price of energy is INR 2.44 per unit. This bid is very cost effective. The purchased price of solar energy for grids will be higher than this price. This requires mass production of commercially viable and cheaper solar panels. Silicon wafer is the most important and vital component in solar panel. The production cost of the solar cell is guided and influenced to a great extent by the cost of production of silicon wafers. Technology improvement in the process of getting raw silicon material, converting into wafers and technological advancement of tools and equipment, is the need of the hour to achieve the target. The refinement in processes of separation of pure silicon can improve and enhance mass production.

3.1. Annual generation cost

Annual generation cost by solar PV units includes various cost such as the annual investment for installation, annual money required for operation and maintenance. The consideration of net worth is also essential to estimate in the current year.

Annual installation cost is a kind of fixed cost, which is the total cost of installation, divides by lifetime of PV panels. Annual fixed installation cost is given by equation (1)

$$AIFC = \frac{\sum_{j=1}^{n} Cpvj * ICj}{15} \quad (1)$$

Annual operating cost is variable; its value depends on the salary and wages of employee engage in operation and control of a PV unit. It is given by equation (2)

$$AVOC = \sum_{j=1}^{n}(Cpvj * OCj) * \Delta T \quad (2)$$

The annual cost of maintenance is also changeable; it comprises the daily expenses for renovation, cleaning etc. It expressed by equation (3)

$$AVMC = \sum_{j=1}^{n}(Cpvj * ICj) * MCj \quad (3)$$

Current value depends on the interest rate of banks and the rate of inflation in the country's economy of the country. Present value index (π) is given by equation (4).

$$Present\ value\ index\ (\pi) = \sum_{i=1}^{n} \frac{(1+IF)}{(1+IR)} \quad (4)$$

The total annual cost of generation of power using an optimally placed PV based DG in an existing system is given by equation (5)

$$TAGC = (AIFC+AVOC) + AVMC)\ \pi \quad (5)$$

3.2. Benefit due placement of PV power plant

Many benefits occur due to the optimization of placement and size of PV based distributed generation in a weak and overburdened power network Such as system loss reduction, reduction in carbon footprint and other green house gases. Congestion management, peak shaving, reliable supply are a few technical benefits of PV based DG the [10]. In this manuscript, only tangible benefits such as loss reduction and reduction of purchasing power considered.

Annual economic benefit due to the reduction in purchasing power is given by equation. Benefit due to a reduction in purchased power is the product of capacity of PV unit, rate of per unit energy and the duration of a generation of power.

IOP Conf. Series: Materials Science and Engineering **594** (2019) 012010 doi:10.1088/1757-899X/594/1/012010

$$B1=\sum_{j=1}^{n} Cpvj * Pg * \Delta T \quad (6)$$

Reduction in line loss is another benefit that occurs due to the optimal placement of the PV unit. This benefit is equal to unit saved because of loss reduction multiplied with the rate of energy per unit and given by equation (7)

$$B2 = \sum_{j=1}^{n} \Delta Lj * Pg * \Delta T \quad (7)$$

A net annual benefit is expressed by equation (8) with the inclusion of current value index.

$$NAB= \pi\ (B1+B2) \quad (8)$$

Net annual economic profit due to the placement of the VP power plant in an existing system is given by equation (9)

$$Net\ economic\ profit = (NAB\text{-}TAGC) \quad (9)$$

Where,

C_{pvj}	Capacity in MW of jth PV unit
IC_j	Fixed Cost per MW
j	Number of PV unit in the distribution system
OC_j	Operating cost per MW of jth DG
MC_j	Maintenance cost per MW of jth DG
π	Current value index
IF	Rate of inflation
IR	Rate of interest
NAB	The net annual benefit accrued to utility
PP_g	The purchase price per KWH of electricity in the open market.
ΔT	Time Duration of power generation of DG
$B2$	Benefit due to loss reduction
$B1$	Benefit due to a reduction in purchasing power
ΔLR_j	Reduction in real power loss due to j^{th} DG.

4. Case study

In this section proposed a nature-inspired Particle Swarm Optimization (PSO) is applied in IEEE 13-Bus radial network. Data of this network is taken from[11]. Active power loss is 336.24 KW, and reactive power loss is 264.01 KVAr occurs with the integration of PV unit. The test distribution system considered is shown in figure 1. PSO was given by Kennedy and Eberhart in 1995[12]. To determine a suitable location, an objective function is proposed. Total annual generation cost *(TAGC),* power loss index (PLI) and voltage profile index (VPI). Proposed function is given by equation (10).

$$f = xTAGC + yPLI + zVPI \quad (10)$$

Where,

x, y and z are the three weights

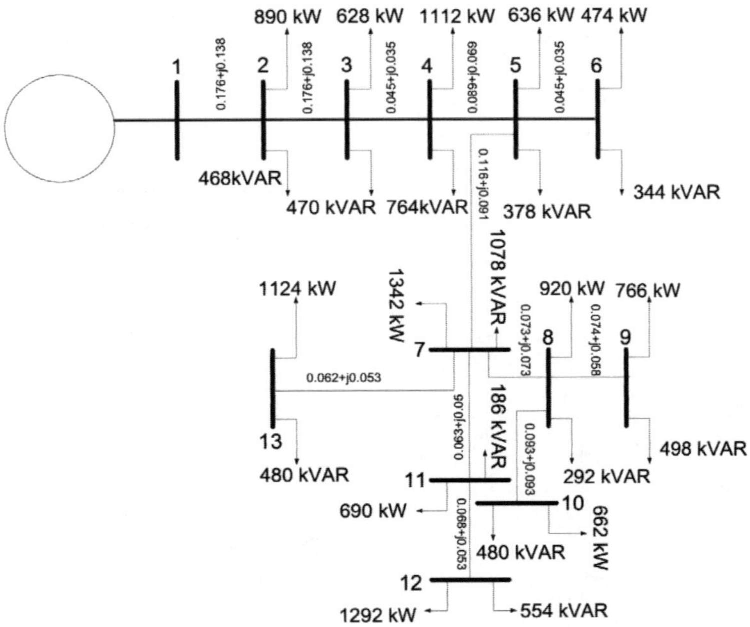

Figure 1. IEEE 13-Buses Distribution System

Calculation of net economical benefit:

Cost in Indian Rupees	
Investment cost / MW	$= 65\times10^6$
Operational Cost /yr	= 4% of Investment Cost/yrs
Maintenance Cost /yr	= 10,000 +4% of Investment cost/yr
Inflation Rate	= 4.97 % /yr (average of last 5 yrs figure)
Interest Rate	= 4.50 %
Present Worth Factor (π)	= (1+4.97)/(I+4.50)=1.085
Number of planning year	= 15
Purchase Price of Power from grid	= 3.50 / unit
DG Capacity	= 8.19 MW
Loss Reduction	= 0.756 MW
Investment Cost / yr	$= 35.49\times10^6$
Operational Cost / Yr	$=1.4196\times10^6$
Maintenance Cost / Yr	$= 1.4296\times10^6$
Total Cost/ yr	$=38.3392\times10^6$
Benefits	
PWV (OC) / yr	$= 1.5402\times10^6$
PWV (MC) / yr	$= 1.5511\times10^6$
Benefit due to Less Demand /yr	$=251.1054\times10^6$
Benefit due to Loss Reduction /yr	$= 23.1789\times10^6$
Total Benefit /yr	$= 277.3756\times10^6$

5. Results and Discussion

The optimal size of PV based DG is 8.19 MW at the 7th bus. By the installation of the PV unit, the active loss and reactive loss are significantly reduced, as shown in figure 2 and figure 3, respectively. Active loss reduces to 44 KW from 336 KW, and reactive loss is reduced to 102 KVAr from 264.01 KVAr. PV based DG also improves the voltage profile of the system, as shown in figure 4. The optimally placed PV unit improves the technical and economic performance of the system.

Economical benefit is given in table 1, PV is a good alternative of the convention method of power generation.

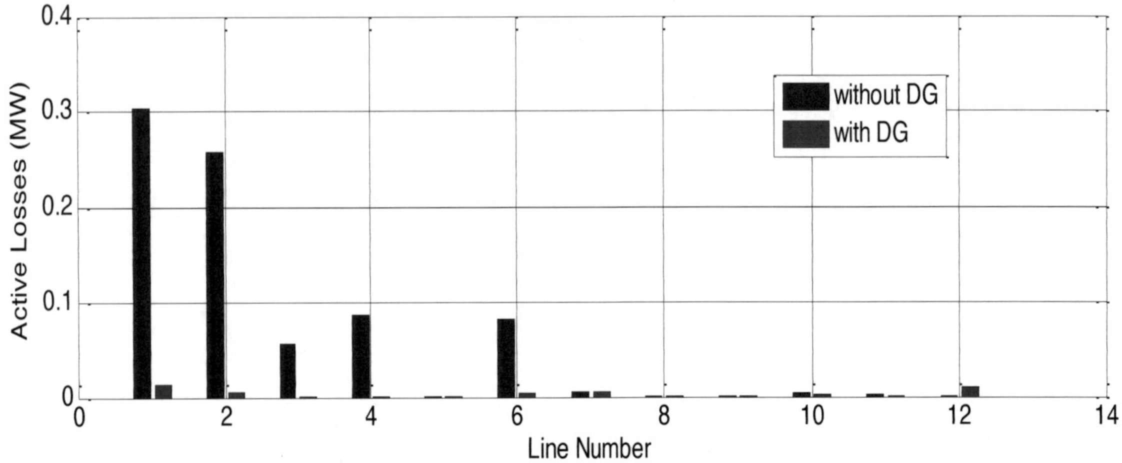

Figure 2. Active Line Losses Comparison in IEEE 13-bus System

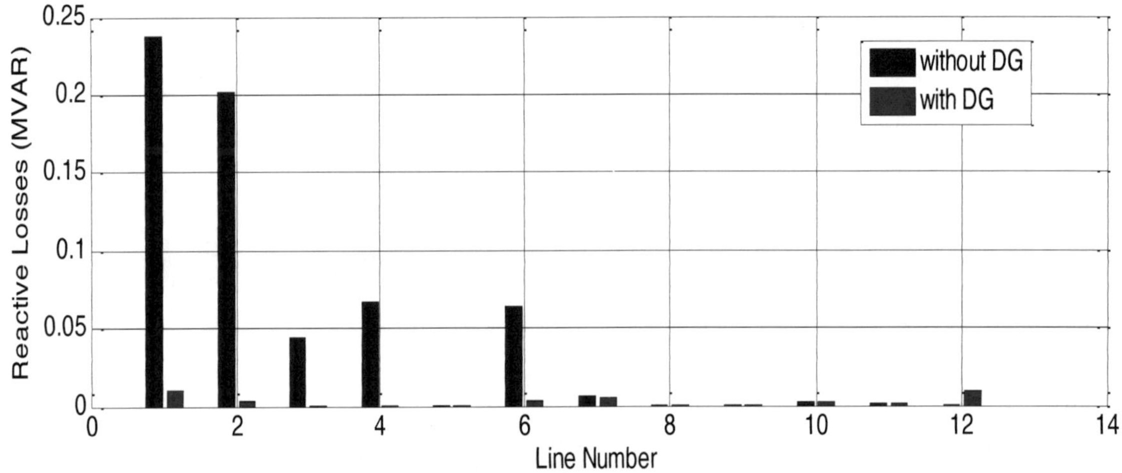

Figure 3. Reactive Line Losses Comparison in IEEE 13-bus System

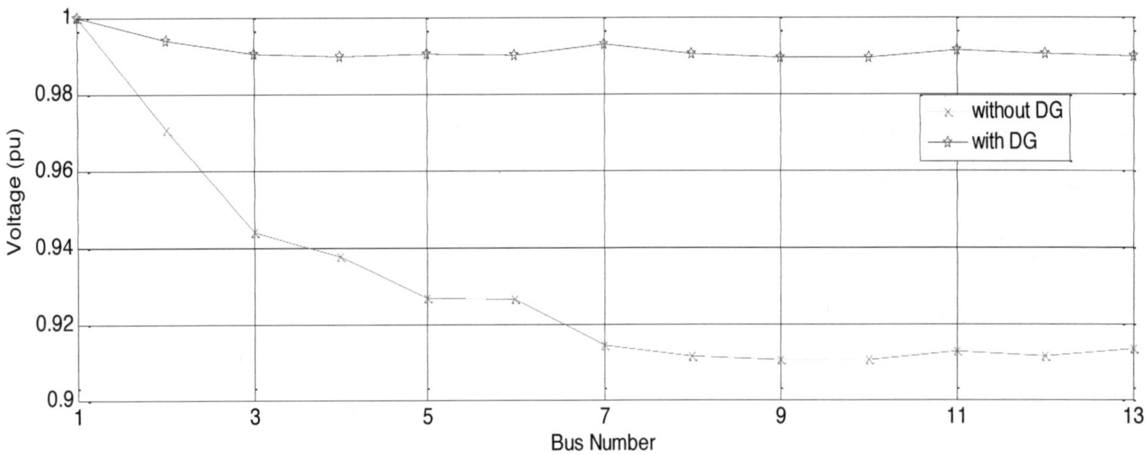

Figure 4. Voltage Profile Comparison in IEEE 13-bus System

Table 1.Performance PV based optimal DG

Technique	Objective Function	Size of DG (MW)	Location	Losses (MW)	Net economic benefit
PSO	Total losses and VPI	8.19	13	0.044	Rs. 277.3756 $\times 10^6$/yr

6. Conclusion

Demand for energy is increasing due to economical growth and technological development in the world. In the partial supply of power in the form of solar power, reduces environmental issues arising out of setting up conventional power plants. The latest research in the development of solar cell and improvement in manufacturing process promises the bright future of solar energy for humankind. The time is not far off when the majority component of supplied power will be solar energy and will be an inseparable part in Distributed Generation. As per the economic and technical analysis, as carried out above, it is found that research could be directed to enhance the efficiency of conversion of incident sunray into optimum solar power and in the manufacturing process to reduce capital cost in mass production. Society has to embark on the next great energy transition because of economic, environmental, and social limit which old system has imposed and is unsustainable and absolute. Authors opine that solar energy is sustainable energy which promises the cleaner and more secure energy for future generations.

References

[1] www.mnre.gov.in/schems/grid-connected/sola
[2] https:/www.technologyreview.com/s/603497/10-breakthroughtechnologies.
[3] Vikas Singh Bhadoria, Nidhi Singh Pal, Vivek Shrivastava and Shiva Pujan Jaiswal, 2018 "Optimal sitting and sizing of capacitor using iterative search method for enhancement of Reliability of Distribution system". *Adv. in En. and Po. Sys., Lec. Notes in Elec. Engg.,* https://doi.org/10.1007/978-981-13-0662-4_11
[4] Shiva Pujan Jaiswal, Vivek Shrivastava, D. K. Palwalia "Impact of Semiconductor Devices on Voltage Stability of Distribution System" *Materials Today Proceedings, Vol 3, Issue P3 p.p. 581-589,* https://doi.org/10.1016/j.matpr.2019.03.101
[5] F.S. Abu-Mouti, M.E. El-Hawary, 2010, Optimal distributed generation allocation and sizing in distribution systems via artificial bee colony algorithm. IEEE Trans. Power Deliv. 26, 2090–2101260. https://doi.org/10.1109/TPWRD.2011.2158246
[6] M. Mohammadi, S.H. Hosseinian, G.B. Gharehpetian, 2011, GA-based optimal sizing of micro grid and DG units under pool and hybrid electricity markets. Elect. Power Energy Syst. 35,83–92(2012) ttps://doi.org/10.1016/j.ijepes.
[7] R.K. Singh, S.K.Goswami, 2010 "Optimal allocation of distributed generations based on nodal pricing for profit, loss reduction and voltage improvement including voltage rise issue". Elect. Power Syst. Res. 36, 637–644. https://doi.org/10.1016/j.ijepes.2009.11.021
[8] Ramavt S.R., Shiva Pujan Jaiswal, Goel N., Vivek Shrivastava,2018 " Battery Energy Storage Technology Integrated for Power System Reliability Improvement" *Adv. in En. and Po.Sys, Lect. Notes in Elec Engg , https://doi.org/10.1007/978-981-13-0662-4_12*.
[9] Bhadoria VS., Pal NS, Shrivastava V. 2016, Comparison of analytical and heuristic techniques for multi-objective optimization in power system" , Saxena, P, Singh, D, & Pant, M (Eds) Pro. Sol. and Uncer. Mod. Thr. Opt. and Soft Com. Appl.. Chapter -13, Hershey, PA: IGI Global, . Available: 10.4018/978-1-4666-9885-7.ch01
[10] Shiva Pujan Jaiswal, Vivek Shrivastava *2014* "Allocation of UPFC in Distribution System to Minimize the Losses" *IEEE Inte. Conf. on Inn. Appli. of Comp. Intell. on Power, Energy and Controls with their Impact on Humanity-,* doi: 10.1109/CIPECH.2014.7018212
[11] Vikas Singh Bhadoria, Nidhi Singh Pal, Vivek Shrivastava, Shiva Pujan Jaiswal , 2017 "Reliability Improvement of Distribution System by Optimal Sitting and Sizing of Disperse Generation" *Inter. Jour. of Rel, Qual and Safety Engineering, Wor. Sci. Pub. Comp, Vol. 24, No. 6 1740006 doi.org/10.1142/S021853931740006*
[12] Shiva Pujan Jaiswal, Vivek Shrivastava,2018 "A PSO based Search for Optimal Tuning and Fixing of UPFC to Improve Usefulness of Distribution System" *Jour. of Inte. and Fu. Sys. IOS Press.* Available: 10.3233/jifs-169783.

IOP Conf. Series: Materials Science and Engineering **594** (2019) 012011 doi:10.1088/1757-899X/594/1/012011

Effect of Inertia Weight of PSO on Optimal Placement of DG

Harsh Maurya[1], Vivek Kumar Bohra[1], Nidhi Singh Pal[1] and Vikas Singh Bhadoria[2*]
[1]Department of Electrical Engineering, School of Engineering, Gautam Buddha University, Greater Noida (UP), India
[2]Department of Electrical & Electronics Engineering, ABES Engineering College, Ghaziabad, (UP), India
E-mail: *vikasbhadoria@gmail.com

Abstract- Integrating Distributed Generation (DG) at appropriate location in distribution system can reduce its real power losses and can also increase the voltage regulation. Optimal allocation of DG has two parts, i) location identification of DG, ii) capacity determination of DG. This article uses loss sensitivity methods for fixing of optimal place of DG and then Particle Swarm Optimization (PSO) technique is implemented for determination of optimal size of DG at that location. During optimization, effect of inertia weight of PSO on the optimal placement of DG is demonstrated in this paper. For detailed study, IEEE 33-bus system is considered and impact of integration DG in the system is also shown.

Index terms- Radial Distribution Network (RDN), Distributed Generation, Particle Swarm Optimization (PSO), Inertia weight, IEEE 33-bus

1. Introduction

Distributed DG can be defined in different ways. In general DG is the electricity generation nearer to the load centres and also connected to the distribution system. DG's are going to play a vital role in realization of smart grid and sustainable energy. Different types of DGs are solar power plant, wind power plant, CHP etc. Use of renewable energy resources as DG has changed the trend of electricity market. Due to several advantaged of renewable DGs, these are becoming more popular day by day. There are numerous advantages of integration of DG in distribution system like flexibility, eco-friendly, improved voltage regulation and reliability and power quality [1]. DGs are often taken as backup for the system or to reduce the usage of grid electricity. DGs can also operate in peak shaving mode.
Integration of DG changes the power flow pattern in distribution system. When DG is not connected the flow of power is usually unidirectional but after the DG is connected the power flow can becomes bidirectional depending on generation and load. Thus bidirectional power flow due to addition of DG creates problems in grid control and protection system. This control of bidirectional flow of power is a major challenge in the interconnection of DG and has to be handled very carefully. Penetration level of DG is also an area of concern during the interconnection. Interconnection of DG in any power system has different impacts on the power system. Impacts of integration of DG can be categorised into Technical Impacts, Economical Impacts and Operational Impacts [2]. In general any DG should be installed on the location of shortage of power or having poor voltage profile. In long radial lines there is a problem of low voltage near end user. If suitable capacity of DG is placed at appropriate place in the system, then voltage profile is improved and this problem can be eliminated.

Actually place of DG and its penetration level decide its effect. If a suitable size DG is integrated at suitable bus location in the distribution system, then only it will be beneficial, otherwise it will adversely affect the system. Therefore Optimal DG Allocation (ODGA) has been the intense focus area power system planners, research engineers, utilities and institutes in the past. ODGA problem may have several objectives such as diminution of power /energy losses, reduction in voltage deviation, maximization of DG capacity, improvement in system reliability etc. If any one objective is selected for the optimization, then optimization problem will be single objective. For creating a multi-objective problem, several objectives can be combined together. For optimization of these single objective or multiobjective problems, several optimization techniques have been found in the literature. A summary of these techniques is given in [3]. Optimal placement of a fixed size DG is done with the help of an analytical technique [4]. The main aim of this optimization is to reduction of real losses only. Acharya *et al.* [5] have given exact loss formula for ODGA using analytical technique. An improve analytical method is also found suitable for ODGA in literature [6]. ODGA and its effect on the SAIDI and SAIFI of the system is shown in [7]. Similar work is done for the capacitor also [8].

Several other advanced optimization techniques like PSO [9], Improved PSO [10] Genetic Algorithm (GA) [11], Hybrid GA and PSO [12], Artificial Immune System combined with PSO [13], Artificial Bee Colony [14] and Modified Firefly Algorithm [15] are also found their suitability in ODGA. PSO needs few parameters, easily completed and the method is very simple as compared to other algorithm [16]. In this paper PSO is used for ODGA problem for single DG placement in the IEEE 33-bus RDN. PSO Convergence characteristics depend upon the inertia weight of the velocity equation. In this article, effect of the inertia weight on the ODGA is shown by considering different inertia weights during optimization. In section II of the paper, the objective function formulation is explained, in section III PSO technique is discussed. Section IV and V are the results and the conclusion respectively.

2. Formulation of Objective Function

In any electrical system the total active power losses depends on the injection of real and reactive powers at different nodes and hence losses can be formulated in terms of these injected powers. For any *N*-node network total active power losses are calculated by the equation (1) [17]. This equation is known as 'exact loss' formula.

$$P_L = \sum_{i=1}^{N} \sum_{j=1}^{N} \left[\alpha_{ij}\left(P_i P_j + Q_i Q_j\right) + \beta_{ij}\left(Q_i P_j - P_i Q_j\right)\right] \quad (1)$$

Where α_{ij} and β_{ij} are the coefficients given as:

$$\alpha_{ij} = \frac{R_{ij}}{V_i V_j} cos(\delta_i - \delta_j),$$

$$\beta_{ij} = \frac{R_{ij}}{V_i V_j} sin(\delta_i - \delta_j),$$

N= Number of Nodes,
P = injected real power
Q = injected reactive power
V = magnitude of voltage
δ = angle of the voltage
i= node number
and $Z_{ij} = R_{ij} + jX_{ij}$ is the ij^{th} element of $[Z_{BUS}] = [Y_{BUS}]^{-1}$.

Main aim of the optimization process is to reduce real losses by installing an optimum capacity of DG at optimum node in the system. For this purpose objective function is formed as equation (2).

$$f = \min(P_{loss}) \quad (2)$$

For this optimization process the constraints are stated below:
Minimum and maximum range of the node voltage can be given as:

$$V_{imin} < V_i < V_{imax} \quad (3)$$

Total Active power generation should be within its minimum and maximum limits:

$$P_{DG}^{\min} \leq P_{DG} < P_{DG}^{\max} \quad (4)$$

Total generated power must fulfill power demand (PD) and total losses (T_L) of the system.

$$\sum P_G + \sum P_{DG} = PD + T_L \quad (5)$$

3. Particle Swarm Optimization (Pso)

To find the maximum or minimum of any process with least efforts with some constraints is known as optimization. Numerous optimization methods have been proposed by the researchers in the past. PSO is one of them. It is motivated by group activities of birds or fish during the search of food, This algorithm proposed and proposed by Eberhart and Kennedy [19]. There are numerous advantages of the PSO, over other existing techniques. This can also be combined with some other methods to make a hybrid optimization method. In PSO all particles randomly moves in the search space and their initial positions are also randomly generated. In a D-dimensional search space, the ith particle of the group of particles is represented by a D-dimensional vector $X_i = (x_{i1}, x_{i2} \ldots\ldots x_{iD})$ and the best particle of the swarm, is denoted by g_{best}. The previous best location of the ith particle is stored and represented as $P_i =$

$(p_{i1}, p_{i2} \ldots.. p_{iD})$ and the position change (velocity) of the i particle is $V_i = (v_{i1}, v_{i2} \ldots.. v_{iD})$. For each iteration velocity and positions vectors are updated by equations (6) and (7) (the superscripts denote the iteration):

$$V_i^{k+1} = wV_i^k + c_1 r_{i1}^k (P_i^k - X_i^k) + c_2 r_{i2}^k (P_g^k - X_i^k) \quad (6)$$

$$X_i^{k+1} = X_i^k + V_i^{k+1} \quad (7)$$

Where i=1, 2,….M.
M = total number of particles
w = the inertia weight
c_1= cognitive parameter
c_2= social parameter
r_{i1} and r_{i2} = random numbers within the range [0, 1].
A general flow chart for PSO is shown in figure 1 [20].
Initial population can be found according the following equation (8):

$$X_i = X_{min} + rand * (X_{max} - X_{min}) \quad (8)$$

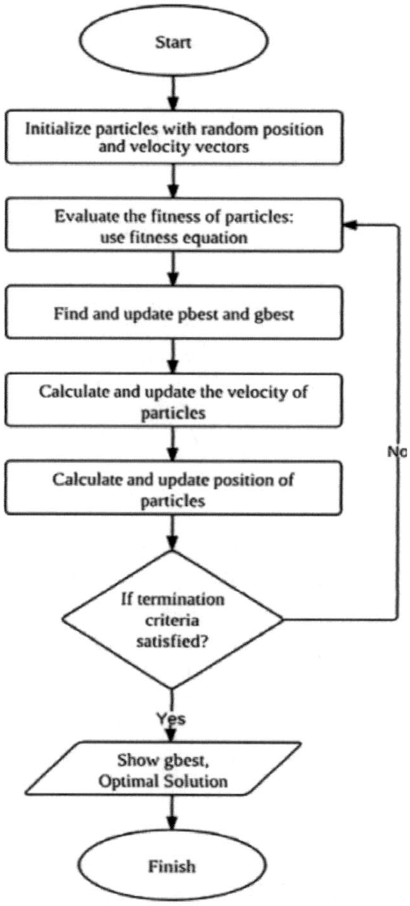

Fig 1. PSO flow chart [20]

Inertia weight is the deciding factor in the convergence behaviour of the PSO. Large inertia weight results in delayed convergence of the optimization and low inertia weight will result in local trapping. As a result, the inertia weight should be selected for a better searching -utilization trade-off. Different inertia weight settings are proposed in the literature such as: Constant Inertia Weight (CIW), Oscillating Inertia Weight (OIW), Random Inertia Weight (RIW), Global Local Best Inertia Weight (GLBIW), Time Varying Inertia Weight (TVIW) techniques, etc. Some inertia weights and their formula are summarised in Table 1 [21].

IOP Conf. Series: Materials Science and Engineering **594** (2019) 012011 doi:10.1088/1757-899X/594/1/012011

Table 1: Different Inertia Weights and Their Formula

S. No.	Name	Formula
1.	CIW	w=c c=0.7
2.	OIW	$w(t) = \frac{(w_{min}+w_{max})}{2} + \frac{(w_{max}-w_{min})}{2}\cos\left(\frac{2\pi t}{T}\right)$ $T = \frac{2S}{3+2k}$
3.	RIW	$w = 0.5 + \frac{Rand()}{2}$
4.	GLBIW	$w_i = (1.1 - \frac{gbest_i}{pbest_i})$
5.	TVIW	$w = (w_1 - w_2)\left(\frac{maxiter-iter}{iter}\right) + w_2$

In this paper above mentioned different inertia weights are considered and their effect on the size optimization of DG is shown.

4. Case Study And Results

PSO with different inertia weights is applied for ODGA in IEEE 33-bus RDN. The system is having 33 nodes, 32 lines. The connected active and reactive load of system is 3.72MW and 2.3MVAR respectively. Base case is without interconnection of DG. In this case active losses in the system are 211.1957 kW and reactive losses are 143.2159 kVAR. Optimal DG placement problem is separated into two parts. One is the regarding decision of optimal place and then determination of optimal size to be placed at that location. For determination of optimal location five random buses are selected and then variation of DG power injection and variation in losses is plotted. Five random selected buses are bus number 6,10,18,22 and 31. This variation is shown in figure 2. Figure 2 also shows that out of these buses, if power injection is done at bus number 6 then the losses in the system will be minimum with variation in DG injected power.

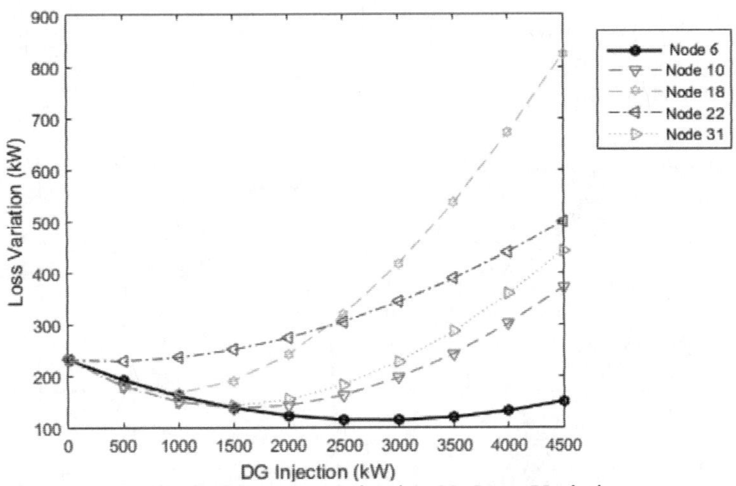

Fig 2. DG power Injection Vs Loss Variation

Comparison of percentage loss reduction with DG power injection is shown in figure 3. This figure also shows that power injection at bus number 6 will give highest loss reduction. Due to this reason, optimum location for injection of power is selected as bus number 6.

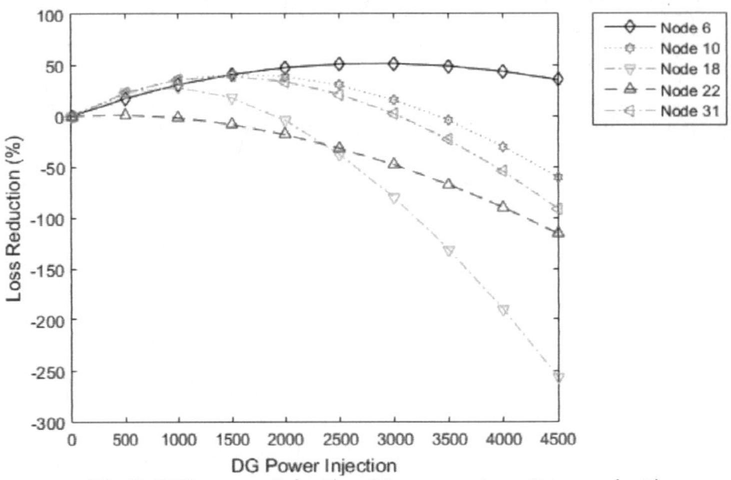

Fig 3. DG power Injection Vs percentage Loss reduction

For fixing of optimal size of DG, PSO is applied and five different inertia weights are selected. In PSO 40 particles are selected and their initial position and velocity are initialized randomly. With different inertia weights results are shown in table 2. For demonstrating the effect of inertia weight on the optimization CIW method is neglected because constant inertia weight will always increase the velocity linearly and in such case the optimization will not converge. Rest other OIW, RIW, GLBIW and TVIW are compared and result shows that time taken by TVIW is least among the compared methods. This time taken is 1.164 seconds. GLBIW gives minimum size of 2.7947 MW for the selected location. Size proposed by the TVIW is 2.8201 MW and system losses by installing this DG will be 112.719 kW. These losses are least among the compared methods and hence it can be said that size proposed by TVIW results in maximum loss reduction. Priorities in speed, size or efficiency will decide the selection of inertia weights.

Table 2. Performance Comparison of PSO with Different Inertia Weights

	OIW	RIW	GLBIW	TVIW
Average DG Injection (in MW)	2.8171	2.9109	2.7947	2.8201
Average Loss (in kW)	112.691	112.558	112.719	112.534
Average time taken (in sec)	1.175	1.526	2.817	1.164

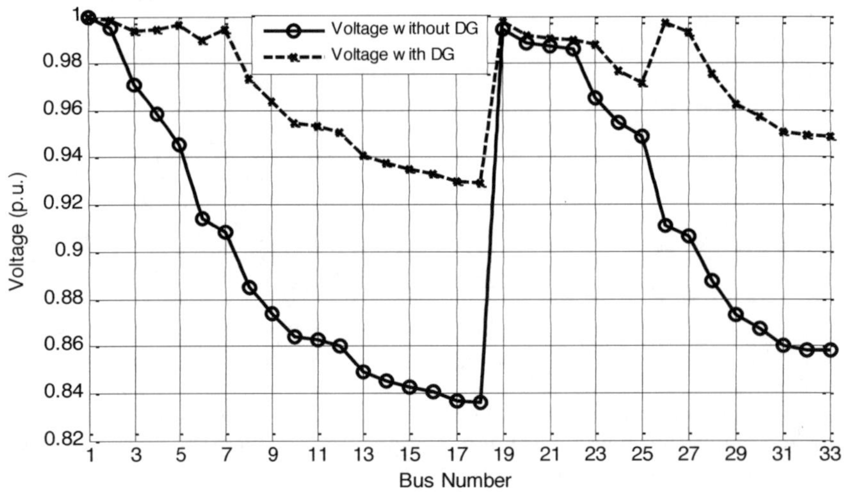

Fig 4. Voltage Profile comparison in base case and with DG

IOP Conf. Series: Materials Science and Engineering **594** (2019) 012011 doi:10.1088/1757-899X/594/1/012011

Voltage profile of the test system for with and without DG case is demonstrated in Figure 4. Since prime purpose of this optimization problem is to reduce the real losses and hence for the voltage profile comparison size corresponding to least system losses is taken into consideration. Figure 4 also demonstrates that in RDN, optimal capacity DG placed at optimal bus number for minimization of losses only, also improves its voltage profile. Effect of DG varies with the distance of the load point from the DG.

5. Conclusion

In this article initially the optimal bus number of the DG is decided and then PSO is applied for the fixation of optimal capacity of the DG. Location of this optimal size DG is already predefined. Optimal capacity of DG is calculated to decrease the real losses occurring in the system. During optimization different inertia weight methods are selected and their effect on the optimal capacity of DG is shown. It can be concluded that TVIW is the fastest among the compared methods. Also the size proposed by the TVIW method gives minimum losses of the system. However GLBIW gives minimum size of DG for the minimization of losses, although this size gives losses higher than that of the TVIW. Thus selection of the inertia weight depends on the priority of the size or losses. Effect of the inertia weight on the size and location optimization can also be shown for the same system or for any other system. This analysis will help in decision making at the planning stages. Effect of inertia weight on the both size and location can also be studied.

References

[1] S. N. Singh, J. Ostergaard and N. Jain, 2009 *Distributed Generation in Power Systems: An Overview and Key Issues*, Indian Engineering Congress,

[2] Bhadoria, Vikas Singh, Nidhi Singh Pal, and Vivek Shrivastava. 2013 *A review on distributed generation definitions and DG impacts on distribution system.* Proc. Int. Conf. Advanced Computing and Communication Technologies (ICACCT™-2013), Vol. 7. DOI: 10.13140/RG.2.1.4439.4328

[3] Adefarati, T., and R. C. Bansal 2016. Integration of renewable distributed generators into the distribution system: a review. IET Renewable Power Generation 10.7: 873-884.

[4] Bhadoria, Vikas Singh, Nidhi Singh Pal, and Vivek Shrivastava. 2014 *Installation of DG for optimal demand compensation.* International Conference on Issues and Challenges in Intelligent Computing Techniques (ICICT). **DOI:** 10.1109/ICICICT.2014.6781385.

[5] Acharya N, Mahat P, Mithulananthan N. 2006 An analytical approach for dg allocation in primary distribution network. Int J Electr Power Energy Syst;28:669–78.

[6] Duong Quoc H, Mithulananthan N, Bansal RC. 2010 *Analytical expressions for dg allocation in primary distribution networks.* IEEE Trans Energy Convers;25:814–20.

[7] V. Bhadoria, N. Pal, V. Shrivastava and S. Jaiswal, 2017 *Reliability Improvement of Distribution System by Optimal Sitting and Sizing of Disperse Generation*, International Journal of Reliability, Quality and Safety Engineering, vol. 24, no. 06, p. 1740006. doi.org/10.1142/S021853931740006X

[8] Bhadoria, Vikas Singh, et al. 2018 *Optimal Sitting and Sizing of Capacitor Using Iterative Search Method for Enhancement of Reliability of Distribution System.* Advances in Energy and Power Systems. Springer, Singapore, 123-129. doi.org/10.1007/978-981-13-0662-4_11.

[9] El-Zonkoly AM. 2011 Optimal placement of multi-distributed generation units including different load models using particle swarm optimization. Swarm Evolut Compute 1:50–9.

[10] Singh RK, Goswami SK 2009. Optimum siting and sizing of distributed generations in radial and networked systems. Electr Power Compon Syst;37:127–45.

[11] M. H. Moradi and M. Abedini, Jan. 2012 A combination of genetic algorithm and particle swarm optimization for optimal DG location and sizing in distribution systems, Int. J. Electr. Power Energy Syst., vol. 34, no. 1, pp. 66–74,.

[12] W. Prommee and W. Ongsakul, Jan. 2011 Optimal multiple distributed generation placement in microgrid system by improved reinitialized social structures particle swarm optimization Euro. Trans. Electr. Power, vol. 21, no. 1, pp. 489–504,.

[13] M. Gomez-Gonzalez, A. López, and F. Jurado, Mar. 2012 *Optimization of distributed generation systems using a new discrete PSO and OPF* Elect. Power Syst. Res., vol. 84, no. 1, pp. 174–180.

[14] F. S. Abu-Mouti and M. E. El-Hawary Oct. 2011 Optimal distributed generation allocation and sizing in distribution systems via artificial bee colony algorithm IEEE Trans. Power Del., vol. 26, no. 4, pp. 2090–2101.

[15] A.Y. Abdelaziz, Y. G. Hegazy, W. El-Khattam, W. and M.M. Othman 2015 *Optimal Planning of Distributed Generators in Distribution Networks Using Modified Firefly Method* Electric Power Components and Systems, vol. 43, no. 3, pp. 320-333.
[16] Q. Bai 2010 *Analysis of Particle Swarm Optimization Algorithm* Computer and Information Science, vol. 3, no. 1.
[17] Bhadoria, V. S., Pal, N. S., & Shrivastava, V. 2016. *Comparison of Analytical and Heuristic Techniques for Multiobjective Optimization in Power System* (pp. 264-291). Hershey, PA: IGI Global. DOI: 10.4018/978-1-4666-9885-7.ch013
[18] O. Elgerd, 1971 Electric energy systems theory: an introduction. New York: McGraw-Hill.
[19] J. Kennedy and R. C. Eberhart, 1995 *Particle Swarm Optimization*, Proc. IEEE International Conference on Neural Networks, vol. IV, pp. 1942-1948.
[20] D. Armaghani, M. Hajihassani, E. Mohamad, A. Marto and S. Noorani, 2013 Blasting-induced flyrock and ground vibration prediction through an expert artificial neural network based on particle swarm optimization, Arabian Journal of Geosciences, vol. 7, no. 12, pp. 5383-5396.
[21] Bansal, J.C., Singh, P.K., Saraswat, M., Verma, A., Jadon, S.S., & Abraham, A. 2011. *Inertia Weight strategies in Particle Swarm Optimization.* 2011 Third World Congress on Nature and Biologically Inspired Computing, 633-640.

IOP Conf. Series: Materials Science and Engineering **594** (2019) 012012 doi:10.1088/1757-899X/594/1/012012

Single Stage High Brightness LED Driver with Improved Power Quality

Amit Agrawal[1,*], Ashish Shrivastava[2], Kartick C Jana[3], Sandeep Tripathi[3], Amit Rai[3]

[1]Department of Electrical & Electronics Engineering, ABES Engineering College, Ghaziabad, Uttar Pradesh, 201009, India
[2]Department of Electrical Engineering, Manipal University Jaipur, Rajasthan, 303007 India.
[3]Department of Electrical Engineering, Indian Institute of Technology (ISM), Dhanbad, Jharkhand, 826004, India.
E-mail: *amit.stm@gmail.com

Abstract. A single switch single stage 24watt power supply is presented to run a light emitting diode (LED) load. In this power supply, the Zeta converter is used to operate in continuous conduction mode (CCM) for maintaining desired output voltage. A proportional integral (PI) controller is also proposed in the feedback to make system stable. The need of proposed controller is justified from stability analysis of the system with the help of small signal average modeling. A simulation model of the proposed LED driver is also carried out in MATLAB/SIMLINK using sim-power toolbox environment for a universal ac input mains. The results confirm a constant current of 0.4amp at output with improved power quality parameters such as total harmonic distortion (THD) and power factor (PF). All outcomes lie under the limit of IEC 61000-3-2 standards that verifies the theoretical analysis.

Index Terms- Continuous Conduction Mode (CCM), LED Driver, Power Factor (PF), Small Signal Average (SSA) Modeling, THD, Stability, Zeta dc-dc Converter.

1. Introduction

In recent years, high brightness light emitting diode (HBLED) has become popular because of many benefits such as small size, long lifetime, high luminous efficiency, durability, environment friendliness due to nonappearance of mercury, flicker-less start, effortlessly dimmable, low maintenance, robust, vibration resistant and so on [1, 2]. Nowadays power LEDs are used in traffic lighting, automobile industry, aircraft lightings, railway signals, LCD backlight, biomedical apparatus, detective system, indoor and outdoor lightings etc [3]. Longer lifetime of LED compared with fluorescent lamp makes it economical to use even with its higher initial cost [4]. An LED driver uses a conversion stage that converts ac signal to dc signal which is required to operate an LED lamp. Converters with switching technology are most preferred among ac-dc conversion to run LEDs. These converters consist of rectifier and filter. The conventional converters have power quality issue in power factor and THD. The converter with the power factor corrector is used for the conversion by the researchers. For most marketable luminaries, power factor must to be at least 0.9 or more than that [5]. The proposed supply comes under class-D equipment and as per the norms of IEC 61000-3-2 the harmonic contents and power factor must be in the limit [6]. A suitable control scheme with PFC converter can make power factor good and efficiency high [7]. Single stage converters have more benefit over the two stage converters for the low power applications because two stage structure increases the circuit cost, size and complication to design a suitable controller and reduces the overall efficiency [8, 9]. Commonly two stage converters need two independent power switches and two control circuits.

This paper focused on an inventive driving method for LED lamp and presented a novel single stage power supply with the benefits of the single stage converter, simplified control design, reduced number of components and lower cost. PI controller is designed to achieve system stability and SSA state space modeling [10] is applied to derive input to output and control to output transfer function. Frequency response of system loop gains is attained using MATLAB tool for justification of controller. For validation of theoretical analysis, a simulation is also performed in MATLAB/SIMULINK tool. Results shows 60volt, 0.4amp at output for universal ac supply from 90volt to 270volt and claims better power quality indices as PF and THD too. Topology is being operated at the switching frequency of 50kHz which helps to reduce the weight and size of passive components like inductor and capacitor. Input current THD and PF are maintained under the limit of standards given by IEC-61000-3-2 class D requirement.

2. Proposed topology: A PFC Zeta converter-based LED Driver

The block diagram of proposed LED driver is shown in Fig. 1. Here a diode bridge rectifier (DBR) is appended after input supply to generate rectified signal that enters further into PFC Zeta converter as input signal. This converter generates a constant dc voltage at output which is compared with the desired voltage and generates the error signal that works as input signal for PI controller. To generate pulse width modulated (PWM) signals, the PI controller's output is being compared with saw tooth wave signal having frequency of 50kHz. Further this signal drives an n-channel power MOSFET switch (S) with duty ratio D and switching frequency f_s. On the basis of switching period $T = 1/f_s$, the switch-on period is defined as $D*T$ and switch-off period is defined as $(1-D)*T$. When the switch M turns on, diode remains in reverse bias and the current ascends in the inductor L_1. When the switch turns off, the current flows through the freewheeling diode D due to its forward bias and descends for whole switch-off period $(1-D)*T$. The inductor current never comes to zero for any switching cycle in continuous conduction mode.

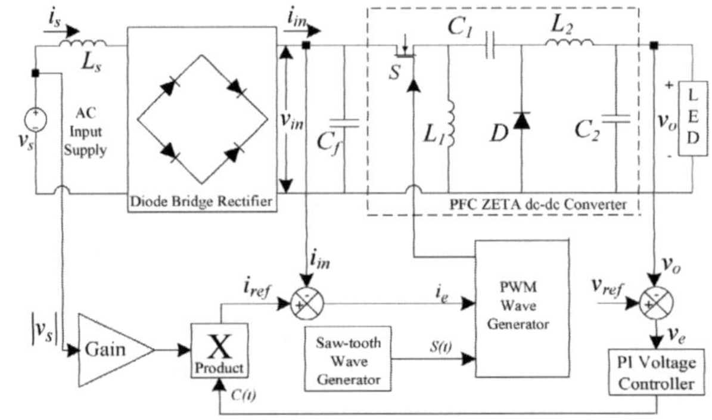

Figure 1. Block diagram of proposed LED driver

3. Components design of proposed LED Driver

The proposed LED driver is based on zeta converter operating in continuous conduction mode (CCM). Few points are listed below, which have been considered at the time of component's estimation to design this model.

- All components of proposed model are considered as ideal.
- LED load shows as a combination of pure resistor in series with dc battery and diode.

- The dc link output capacitance (C_2) should be chosen as high enough so that a constant output voltage with less ripple contents can maintained at output for whole switching cycle.
- The switching frequency is chosen a higher value as compare to the frequency of input ac signal.
- Voltage across coupling capacitor (v_{C1}) is considered as maximum value of the applied input voltage signal.
- For estimation of inductors value, change in ripple current is allowed up to 100% of the average load current.

Components value is calculated with help of standard equations discussed below. The value of duty cycle and critical value of inductances L_1 and L_2 are defined by equation (1) and (2) [11].

$$D = v_o / (v_i + v_o) \tag{1}$$

$$L_{1(critical)} = L_{2(critical)} = (1 - D) v_o / 2 D f_s I_o \tag{2}$$

The value of coupling capacitor and output capacitor is designed with help of below expressions that is based on ripple content [11].

$$C_1 = i_o D / f_s \Delta v_{C1} \tag{3}$$

$$C_2 = i_o / 2 \omega \Delta v_{C2} \tag{4}$$

Here, v_i is the maximum value of ac mains input voltage, ω is the input supply frequency, v_o is output voltage, i_o is average output load current, the symbol D is used as duty cycle, f_s is switching frequency of the switch device, Δv_{C1} is the ripple voltage of capacitor C_1 and Δv_{C2} is the ripple voltage of capacitor C_2. Parameter and component's value are calculated and presented in Table 1 with help of equation (1), (2), (3) and (4) to get desired output voltage.

Table 1. Parameter and Component of Proposed Model

Parameter and Component	Calculated Value	Selected Value
Supply voltage (v_s)	--	220volt
Supply frequency (ω)	--	50Hz
Output voltage (v_o)	--	60volt
Switching frequency (f_s)	--	50kHz
Maximum value of coupling capacitor voltage (v_{c1})	311.08volt	311.08volt
Ripple voltage of coupling capacitor Δv_{c1} = 12% of v_{c1}	37.33volt	37.33volt
Ripple voltage of dc link output capacitor Δv_{c2} = 3% of v_{c2}	1.8volt	1.8volt
Value of duty cycle for ac mains input of 270volt	0.135	--
Value of duty cycle for ac mains input of 90volt	0.320	--
Duty cycle (D)	--	0.161
Averaged Load Current I_o	0.4amp	0.4amp
Critical value of inductor L_1	6.77mH	10mH
Critical value of inductor L_2	6.77mH	10mH
Average inductor current (I_{L_1})	0.12amp	0.12amp
Average inductor current (I_{L_2})	0.40amp	0.40amp
Coupling capacitor C_1	49.93nF	50nF
Output capacitor C_2	353.53μF	400μF
V_{OSC} (Oscillator voltage)	--	4.5volt
v_{ref} (Reference voltage)	--	1.5volt

4. Small Signal Average Model

The diagram of Zeta converter is showing the mathematical description of the real system behavior in Fig.1. The SSA state space methodology is a universal tool which is easily applicable for the detail analysis of the complex structures of the converter in CCM or DCM [12, 13]. With help of this

IOP Conf. Series: Materials Science and Engineering **594** (2019) 012012 doi:10.1088/1757-899X/594/1/012012

method, the linear averaged time invariant model and its derived mathematical results is achieved. All state matrices are derived and written below by using Kirchhoff's voltage and current laws for two operating modes switch-on and switch-off.

$$A_1 = \begin{bmatrix} 0 & 0 & 0 & 0 \\ 0 & 0 & 1/L_2 & -1/L_2 \\ 0 & -1/C_1 & 0 & 0 \\ 0 & 1/C_2 & 0 & -1/RC_2 \end{bmatrix},$$

$$A_2 = \begin{bmatrix} 0 & 0 & -1/L_1 & 0 \\ 0 & 0 & 0 & -1/L_2 \\ 1/C_1 & 0 & 0 & 0 \\ 0 & 1/C_2 & 0 & -1/RC_2 \end{bmatrix},$$

$$B_1 = \begin{bmatrix} 1/L_1 \\ 1/L_2 \\ 0 \\ 0 \end{bmatrix},$$

$$B_2 = \begin{bmatrix} 0 \\ 0 \\ 0 \\ 0 \end{bmatrix} \tag{5}$$

The state-space average equilibrium matrixes are calculated and written below.

$$A = DA_1 + (1-D)A_2$$

$$A = \begin{bmatrix} 0 & 0 & (D-1)/L_1 & 0 \\ 0 & 0 & D/L_1 & -1/L_2 \\ (1-D)/C_1 & -D/C_1 & 0 & 0 \\ 0 & 1/C_2 & 0 & -1/RC_2 \end{bmatrix} \tag{6}$$

$$B = DB_1 + (1-D)B_2$$

$$B = \begin{bmatrix} D/L_1 \\ D/L_2 \\ 0 \\ 0 \end{bmatrix} \tag{7}$$

To build a small signal ac model at an operating point, the duty ratio is considered as

$$d(t) = D + \hat{d}(t) = D + d_m \sin \omega t \tag{8}$$

Where D and d_m are the constant and $d_m << D$.

The small signal ac model can be achieved using equation mentioned below.

$$d\dot{x}(t)/dt = A\hat{x}(t) + B\hat{v}(t) + \{(A_1 - A_2)X + (B_1 - B_2)V\}\hat{d}(t) \tag{9}$$

The small signal ac state space model is written below by putting all value of matrix in equation (22).

$$\begin{bmatrix} d\hat{i}_{L1}(t)/dt \\ d\hat{i}_{L2}(t)/dt \\ d\hat{v}_{C1}(t)/dt \\ d\hat{v}_{C2}(t)/dt \end{bmatrix} = \begin{bmatrix} 0 & 0 & (D-1)/L_1 & 0 \\ 0 & 0 & D/L_1 & -1/L_2 \\ (1-D)/C_1 & -D/C_1 & 0 & 0 \\ 0 & 1/C_2 & 0 & -1/RC_2 \end{bmatrix} \begin{bmatrix} \hat{i}_{L1} \\ \hat{i}_{L2} \\ \hat{v}_{C1} \\ \hat{v}_{C2} \end{bmatrix} + \begin{bmatrix} D/L_1 \\ D/L_2 \\ 0 \\ 0 \end{bmatrix} \hat{v}_i(t) + \left\{ \begin{bmatrix} 0 & 0 & 1/L_1 & 0 \\ 0 & 0 & 1/L_2 & 0 \\ -1/C_1 & -1/C_1 & 0 & 0 \\ 0 & 0 & 0 & 0 \end{bmatrix} \begin{bmatrix} I_{L1} \\ I_{L2} \\ V_{C1} \\ V_{C2} \end{bmatrix} + \begin{bmatrix} 1/L_1 \\ 1/L_2 \\ 0 \\ 0 \end{bmatrix} V_i \right\} \hat{d}(t) \quad (10)$$

After taking the Laplace transform, model can be written in the form of equations also as below.

$$sL_1\hat{I}_{L1}(s) = (D-1)\hat{V}_{C1}(s) + D\hat{V}_i(s) + V_{C1}\hat{d}(s) + V_i\hat{d}(s) \quad (11)$$

$$sL_2\hat{I}_{L1}(s) = D\hat{V}_{C1}(s) - \hat{V}_{C2}(s) + D\hat{V}_i(s) + V_{C1}\hat{d}(s) + V_i\hat{d}(s) \quad (12)$$

$$sC_1\hat{V}_{C1}(s) = (1-D)\hat{I}_{L1}(s) - D\hat{I}_{L2}(s) + (I_{L1} + I_{L2})\hat{d}(s) \quad (13)$$

$$sRC_2\hat{V}_{C2}(s) = R\hat{I}_{L2}(s) - \hat{V}_{C2}(s) \quad (14)$$

Output voltage (v_o) follows the output capacitor voltage (v_{C2}). With help of equation (11) to (14), output voltage response in frequency domain is achieved and written below.

$$\hat{v}_{C2}(s) = \frac{(RL_2D^2(1-D) + RL_1D(1-D)^2 + RL_1{}^2C_1Ds^2)\hat{V}_i(s) + ((RDL_2(1-D) + RL_1(1-D)^2 + RC_1L_1{}^2s^2)(V_{C1} + V_{in}) - RDL_1L_2(I_{L1} + I_{L2})s)\hat{d}(s)}{RL_1{}^2L_2C_1C_2s^4 + L_1{}^2L_2C_1s^3 + RL_1(L_1C_1 + L_2C_2(D^2 - D + 1))s^2 + L_1L_2(D^2 - D + 1)s + RL_1(1-D)^2} \quad (15)$$

Control to output transfer function ($G_{vd}(s)$) of model is written below using equation (15).

$$G_{vd}(s) = \left|\hat{V}_o(s)/\hat{d}(s)\right|_{\hat{V}_i(s)=0}$$

$$G_{vd}(s) = \frac{(RDL_2(1-D) + RL_1(1-D)^2 + RC_1L_1{}^2s^2)(V_{C1} + V_i) - RDL_1L_2(I_{L1} + I_{L2})s}{RL_1{}^2L_2C_1C_2s^4 + L_1{}^2L_2C_1s^3 + RL_1(L_1C_1 + L_2C_2(D^2 - D + 1))s^2 + L_1L_2(D^2 - D + 1)s + RL_1(1-D)^2} \quad (16)$$

After putting component's value in equation (16), the mathematical value of the control to output transfer function is written by equation (17).

$$G_{vd}(s) = \frac{4.7*10^{-7}s^2 - 1.3*10^{-3}s + 783}{3*10^{-15}s^4 + 5*10^{-14}s^3 + 5.2*10^{-6}s^2 + 8.6*10^{-5}s + 1.1} \quad (17)$$

The transfer function (TF) of the PWM block is estimated below.

$$G_{PWM}(s) = \left|\hat{d}(s)/\hat{V}_e(s)\right| = 1/V_{osc} \quad (18)$$

The loop gain of the model $G(s)$ can be achieved as below.

$$G(s) = G_{vd}(s) * G_{PWM}(s) * v_{ref}/v_{dc}$$

Mathematical expression of $G(s)$ is written below by putting all component's value from Table 1 in above equation.

$$G(s) = \frac{7.05*10^{-7}s^2 - 1.95*10^{-3}s + 1175}{8.1*10^{-13}s^4 + 1.35*10^{-11}s^3 + 1.4*10^{-3}s^2 + 2.32*10^{-2}s + 297} \quad (19)$$

5. Stability Analysis of System in absence of Controller

System loop gain of proposed model is achieved with help of SSA modeling in section 4. Bode plot is taken out from MATLAB tool and shown in Fig. 2 to check system stability. Gain margin of 21.5dB and phase margin of 1.06° are observed from frequency response. For a practical system, phase margin must be greater than 45° to fulfill stability criterion. It clearly demands a control scheme in feedback discussed in next section.

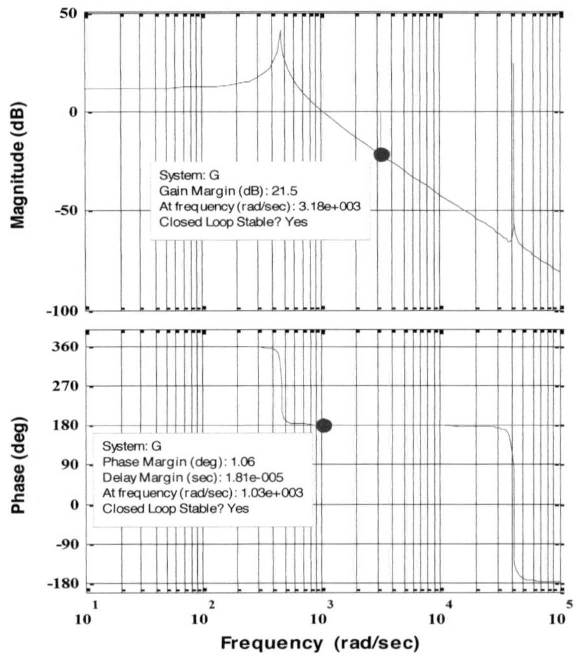

Figure 2. Bode Plot of transfer function $G(s)$

6. Feedback System

A controller is design in this section to achieve stability criterion. It sends optimize input to PWM wave generator which changes the duty cycle accordingly and controls the ON time of MOSFET switch. Component's design and stability analysis is explained below.

6.1. *PI Controller*

Fig. 3 shows the schematic circuit of PI controller network which is working as a voltage controller. It senses the output dc voltage (v_o) and compares with reference voltage (v_{ref}) to generate error voltage signal (v_e).

$$v_e = v_{ref} - v_o$$

According to error voltage, controller sends signal to PWM block at every sampling instant. The value at n^{th} sampling instant is defined by equation (20).

$$I_c(n) = I_c(n-1) + K_p\{v_e(n) - v_e(n-1)\} + K_i v_e(n) \tag{20}$$

Where, K_p and K_i are the proportional and integral gains respectively.

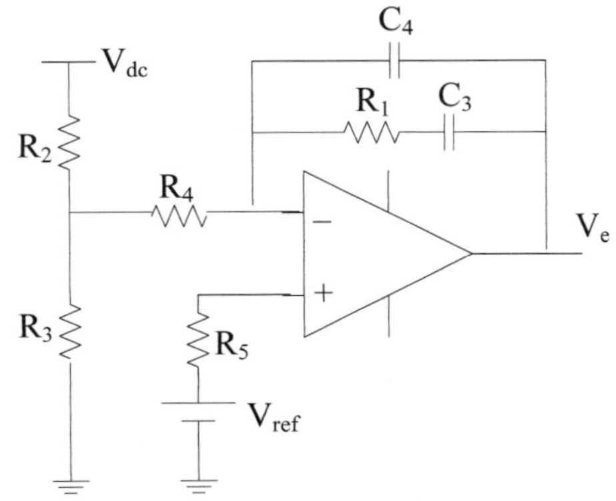

Figure 3. Network of PI Controller

6.2. PWM signal generation

This block compares both signal coming from PI controller and saw-tooth wave generator. It controls the width of pulse which changes the on time of switch. Saw-tooth wave generator sends a fixed frequency carrier signal.

If $K_c * I_c(n)>$ carrier signal, then M=1 else $M = 0$.

6.3. Component's design of PI Controller

The transfer function of controller network is defined by equation (35).

$$C(s) = (1 + sR_1C_3)/sR_2.((C_3 + C_4) + sR_1C_3C_4) \tag{21}$$

All components are calculated by using standard formulas given below where bandwidth (BW) and ΔV_{ovp} are chosen 20Hz and 3volt respectively [14].

$$R_2 = \Delta V_{ovp}/30\mu A \tag{22}$$
$$R_3 = R_2 v_{ref}/(v_o - v_{ref}) \tag{23}$$
$$R_1 = R_3 v_{ref}/(V_{osc} - v_{ref}) \tag{24}$$
$$f_o = 1/(2\pi\sqrt{L_2C_2}) \tag{25}$$
$$f_z = .75 f_o \tag{26}$$
$$C_3 = 1/(2\pi R_1 f_z) \tag{27}$$
$$C_4 = 1/(2\pi(R_2 \| R_3)BW) \tag{28}$$

Mathematical expression of controller is written below by putting all components value in equation (21).

$$C(s) = 7.6*10^{-3}s + 1/(3.45*10^{-3}s^2 + 1.86s) \tag{29}$$

6.4. Stability Analysis of Model with proposed control scheme

A controller is suggested and designed above with transfer function $C(s)$. The new loop gain of system ($G_c(s)$) is defined by equation (44) for further stability analysis.

$$G_c(s) = G(s) * C(s) \tag{30}$$

Mathematical expression is written with help of equation (19) and (30).

$$G_c(s) = \frac{5.358*10^{-9}s^3 - 1.412*10^{-5}s^2 + 8.928s + 1175}{s(2.795*10^{-15}s^5 + 1.553*10^{-12}s^4 + 4.83*10^{-6}s^3 + 2.684*10^{-3}s^2 + 1.068s + 552.4)} \tag{31}$$

Bode plot is taken out from MATLAB again for equation (31) and shown in Fig 4 respectively. Gain margin of 10.7dB and phase margin of 90.7· clearly defend the system stability.

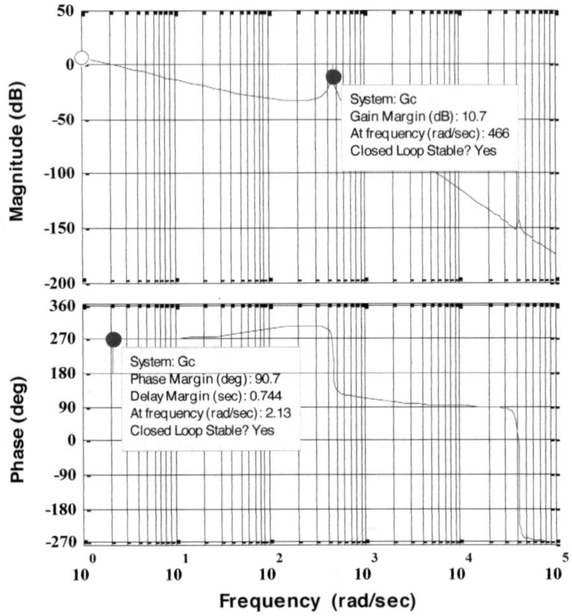

Figure 4. Bode plot of transfer function $G_c(s)$

7. Simulated Model in MATLAB

The proposed model is simulated in the platform of MATLAB and shown in Fig. 5. Input supply is being converted into full wave rectified signal using DBR. Further a PFC block is used to generate a constant voltage at output. It is based on zeta topology operating in CCM. PI controller is generating signal according to difference between output and desired signal. A saw tooth wave generator is used to send a signal of frequency 50kHz to PWM block which controls the on time of MOSFET switch. Value of all the components for proposed model has been designed and selected properly to maintain better power quality within the limits of international standards.

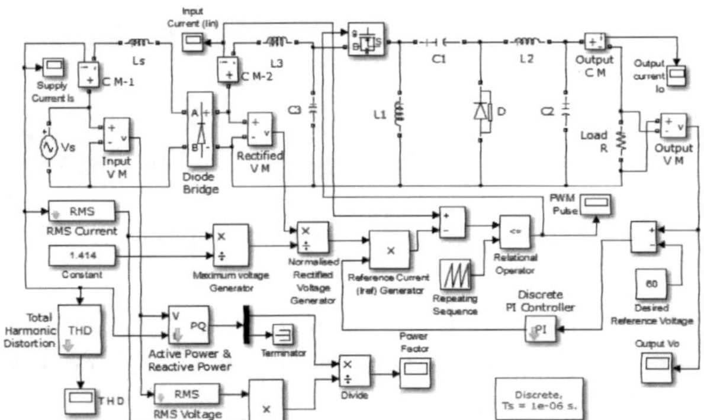

Figure 5. Simulated model of proposed power supply

8. Results and Discussion

The simulation result shows that 60volt is maintained at output for 220volt supply of input. The waveform of supply current (I_s), supply voltage (V_s), output current (I_o) and output voltage (V_o), inductor current (I_{L1}) is extracted at 220volt and shown in Fig. 6. In Table 2, the power quality parameters are written at different supply voltage. The power factor is achieved more than 0.91 for wide range of supply voltage (90volt-270volt) and the THD is maintained from 0.25% to 0.72%. It claims that the proposed model is a good choice as an LED driver with high power quality parameters.

Table 2. Performance Parameters of Simulated Power Supply Based on Zeta Converter

Supply Voltage (V_s), V	Input Current (I_s), A	Load Voltage (V_o), V	Load Current (I_o), A	Power Factor (PF)	% THD
90	0.26	59.8	0.39	.91	0.72
150	0.17	60.2	0.42	.93	0.54
220	0.13	60.3	0.44	.98	0.25
270	0.10	61.2	0.40	.91	0.70

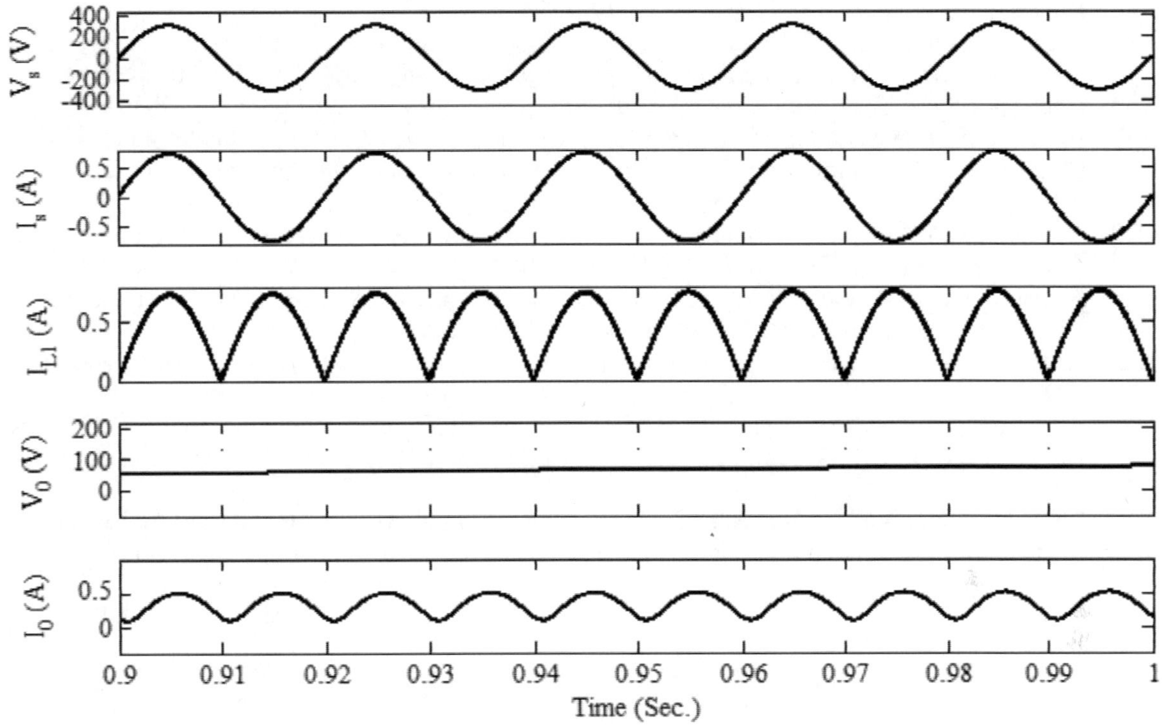

Figure 6. Waveforms of Supply current, voltage, inductor current, output voltage and current

9. Conclusion

A model of 24watt, 60volt power supply based on Zeta topology operating in CCM is presented to run a LED light. State space modeling is applied in this topology and loop gain is achieved to examine the system stability. PI controller is also designed and applied with feedback to make system stable. Stability analysis is done in detail with help of Bode, Nyquist, and Polar plot of loop gains drawn in MATLAB tool. The simulation of proposed LED driver is performed in sim power toolbox to justify theoretical and mathematical analysis. Its result shows a constant current rating of 0.4amps at output with improved power quality parameter such as the THD and PF for universal ac input mains. Improved PF and reduced harmonics of ac mains input are under the limit of IEC 61000-3-2 Class D equipment.

References

[1] C. Y. Wu, T. F. Wu, J. R. Tsai, Y. M. Chen, and C. C. Chen, "Multistring LED backlight driving system for LCD panels with color sequential display and area control," *IEEE Trans. Ind. Electron.*, vol. 55, no. 10, pp. 3791–3800, Oct. 2008.

IOP Conf. Series: Materials Science and Engineering **594** (2019) 012012 doi:10.1088/1757-899X/594/1/012012

[2] L. Xingming and Z. Jing, "An intelligent driver for light emitting diode street lighting," in *Proc. World Automation Congr. (WAC)*, 2008, pp. 1–5.
[3] C. C. Chen, C.-Y. Wu, and T.-F. Wu, "LED back-light driving system for LCD panels," in *Proc. IEEE Appl. Power Electron. Conf. Expo.*, Mar. 2006, pp. 381–385.
[4] Robert S. Simpson, "Lighting Control Technology and Applications," Focal Press, Burlington, MA, USA, 2003.
[5] B. Weir and F. Cathell, "LED streetlight demands smart power supply," *Power Electron. Technol.*, vol. 34, no. 2, pp. 34–39, Feb. 2008.
[6] Limits for Harmonic Current Emissions, *International Electrotechnical Commission Standard,* IEC-61000-3-2, 2004.
[7] H. Broeck, G. Sauerl¨ander, and M. Wendt, "Power driver topologies and control schemes for LEDs," in *Proc. IEEE APEC 2007*, pp. 1319–1325.
[8] Z.Ye, F.Greenfeld, and Z. Liang, "A topology study of single-phase offline ac/dc converters for high brightness white LED lighting with power factor pre-regulation and brightness dimmable feature," in *Proc. 34th Annu. Conf. IEEE Ind. Electron.*, Nov.10–13, 2008, pp. 1961–1967.
[9] Amit Agrawal, Ashish Shrivastava, Kartick C. Jana, "A Universal Input, Single Stage AC-DC LED Driver for an Auditorium Light", *Journal of circuit, system and computers,* vol. 28, no. 2, Feb 2019.
[10] G. Gatto, I. Marongiu, A. Mocci, A. Serpi, "An improved averaged model for boost dc-dc converters" in Proc. IECON, 412– 417, Vienna, 2013.
[11] Ashish Shrivastava and Bhim Singh, "Power factor-corrected DCM-based electronic ballast," *Electrical Engineering*, vol. 95, no. 4, pp. 403-411, Dec 2013.
[12] R. W. Erickson, *Fundamentals of Power Electronics*, 1st ed., New York: Chapman and Hall, 1997.
[13] Amit Agrawal, Ashish Shrivastava, Kartick C. Jana, "Uniform Model and Analysis of PWM DC-DC Converter for Discontinuous Conduction Mode", *IETE journal of Research*, Vol. 64, issue-4, 2018, pp. 569-581.
[14] Bhim Singh and Ashish Shrivastava, "Buck converter based power supply design for low power light emitting diode lamp lighting," *IET Power Electronics,* vol. 7, no. 4, pp. 946-956, 2014.

IOP Conf. Series: Materials Science and Engineering **594** (2019) 012013 doi:10.1088/1757-899X/594/1/012013

Experimental Profiling of Temperature and Luminosity inside Greenhouse using Wireless Sensor Network

Suman Lata and H. K. Verma

Department of Electrical and Electronics Engineering, School of Engineering and Technology, Sharda University
suman.lata@sharda.ac.in

Abstract. Monitoring and control of climatic variables inside a greenhouse (GH) play crucial role in quality production of crops in the green-house (GH). Measurement and monitoring of major GH variables, such as temperature, humidity, soil moisture, light intensity and CO_2 in atmosphere, has been generally carried out using a central data acquisition system to which all sensors are individually wired. This technique involves huge cabling and lacks flexibility of adding sensors in future. This paper presents a detailed methodology and results of experimental measurement and profiling (both temporal and spatial) of temperature and luminosity inside a laboratory model of GH using a wireless sensor network (WSN). For the experimental work, the GH was divided into two rows, row 1 and row 2, with three zones of equal dimensions in each row. At the centre of each zone one wireless sensor node was placed for measurement and acquisition of temperature and luminosity in that zone. A WSN was created with these six wireless sensor (WS) nodes (or motes) and one network coordinator (NC). Profiling of temperature and luminosity was carried out under uncontrolled as well as controlled environmental conditions. Two controlled conditions were created. In the first case only temperature was controlled by placing a heater at the centre of GH. The second case involves simultaneous control of temperature and luminosity, for which an incandescent lamp was placed in each zone of row 1 while all the zones of row 2 were covered with a tinted plastic sheet. Data was gathered from all the six WS nodes in each controlled case as well as in uncontrolled environment. Surface profiles of both temperature and luminosity were drawn in Lab VIEW on the basis of collaborative processing of the data. The experimental results so obtained are correlated with the expected theoretical profiles. The use of WSN and Lab VIEW for collecting and processing the data, respectively, provides remarkable flexibility in terms of addition and deletion of WS nodes. Moreover, no wiring is needed for power or signals.
Keywords— ***Collaborative data processing, Greenhouse variables, Profiling, Wireless sensor node, Wireless sensor network***

1. Introduction

A greenhouse is a specially designed farm structure which is built to provide a favourable climate, not only for better crop production but for the crop protection also. Technically speaking a greenhouse is a complex multivariable interactive system [1] [2]. It is multivariable because there are various physical variables which affect the climate inside the greenhouse. It is interactive as the physical variables are largely interdependent. The important variables affecting the growth and productivity of plants are temperature, humidity, light, soil moisture and carbon dioxide concentration. The optimal value of greenhouse air temperature depends on the intended level of the photosynthetic activity. Each plant species has its own optimal value of air temperature. Luminosity or light intensity is another important variable for the growth of the plants. Typically 10 to 12 hours of light intensity is required by plants for the proper growth. It has to be maintained at the desired level in order to support photosynthesis at a better rate. Every plant has different luminosity requirement which has to be maintained inside the green-house. An Intelligent Greenhouse (IGH) can collect the information related to the various variables which are affecting the production of the crop and can control them as per needs.

Various greenhouse variables are sensed by suitable sensors and the response functions are performed by actuators. Initially in IGH installations, sensors, data acquisition system and actuators were distributed and a huge effort was required to interconnect them. The installations required many data and power wires, making the system complex and expensive [3] [4]. Also the addition of new sensors in such installations was difficult. With the rapid advancements in Wireless Sensor Network (WSN), these issues have been resolved. A WSN is a collection of sensor nodes connected by wireless medium to perform the distributed sensing. Compared to the cabled systems, configuring WSN is cheap, fast and easy. Also, the relocation of the nodes is easy in it. In case of relocation of node outside the direct range of the WSN, multi-hopping is used. This paper focuses on temporal and spatial profiling of two important variables affecting the growth of the plants, namely, temperature and luminosity, inside a laboratory greenhouse. In the second section of the paper the experimental setup has been described. The third section is dedicated to the results and discussions, which are followed by conclusion.

2. Experimental Setup

2.1 Greenhouse model and WS node placement

The experiment was carried out in a laboratory model of greenhouse measuring 1.80(L) x 0.90(W) x 0.90(H) in meters and divided into six zones in 2x3 arrangement. Plywood sheets were used for partitioning as indicated by dark lines in the Fig.1. The dimensions of each zone are hence, 0.6m x 0.45m x 0.90 m. One wireless sensor (WS) node has been deployed in each zone. Nodes 1, 2 and 3 were placed in row-1 while nodes 4, 5, and 6 were placed in row-2. All the nodes have been hanged from roof at a height of 0.30 meters above from soil surface. The layout of greenhouse along with the deployment of wireless sensor nodes is shown in Fig.1.

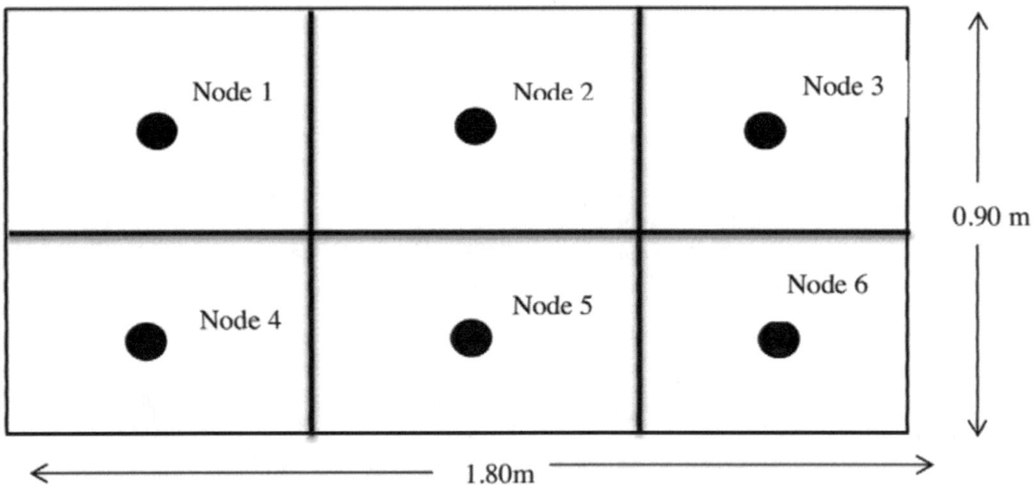

Fig.1. Layout of greenhouse laboratory model with node placement

The wireless sensor nodes, named as SENSEnuts, were from Eigen Solutions. The nodes are battery operated and are based on PAN protocol, which enables wireless transmission of data packets over IEEE 802.15.4. SENSEnuts has an API to access the network. It works on Zigbee protocol and uses 2.4 GHz ISM band. The gateway node was located at a distance of 10 meters outside the greenhouse and connected to a computer with USB. The top and bottom views of the gateway module are as shown in Fig.2 and those of radio module are shown in Fig.3 [5].

(a) (b)

Fig.2. Top and bottom views of gateway module (a) Top view (b) Bottom view

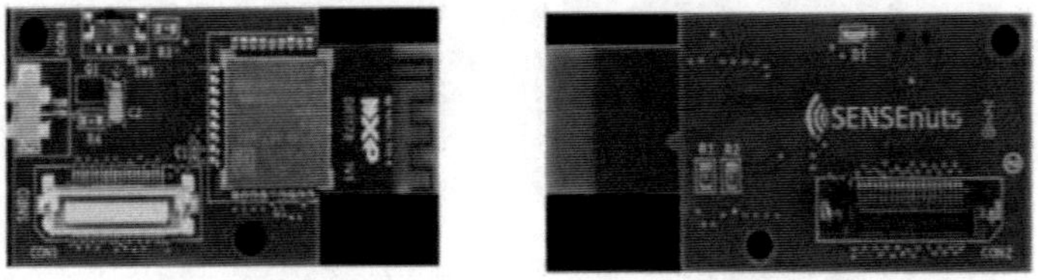

(a) (b)

Fig.3. Top and bottom views of radio module (a) Topview (b) bottom view

2.2 Sensor Details

The temperature sensor used in SENSEnuts node is TMP102. It is a two wire digital serial- output temperature sensor available in a tiny SOT563 packaging with SMBus compatibility [5]. It is capable of reading temperature to a resolution of 0.0625°C. The device is specified for operation over a temperature range of –40°C to +125°C. Luminosity Sensor used is ISL29023. It is an integrated ambient and infrared light to digital converter with SMBus compatible interface.

2.3 SENSEnuts Graphical User Interface (GUI)

SENSEnuts GUI is a program that runs on Windows Operating System. It is used to program the SENSEnuts Radio Modules as well as display the data received from the network. The GUI is a collection of three separate programs, each performing a separate function, namely Device Programmer, SenseLive and Print Window [6]. The Default program that opens at the start of GUI is a Device Programmer. It is a tool to program the SENSEnuts nodes according to the algorithm written in C language using Eclipse IDE. It also enables the user to read the MAC address of the node connected to the computer at the USB port. SenseLive is a graphical environment that displays the data which it receives from the network. It features two separate sections, one to display the latest data coming from the nodes and the other to create a database of all the messages received from all the motes which can be saved and analysed later on. Print Window, allows the motes, to display some custom messages on the GUI.

2.4 Methodology
All the nodes were programmed to send the data simultaneously. So the temperature and luminosity data were simultaneously acquired from all the 6 nodes. The gateway node coordinated as the master device which received data from all the sensor nodes for a pre-programmed time period. Data of temperature and luminosity under following conditions was collected:

2.4.1 *Under uncontrolled conditions.* Under this condition the environment of the greenhouse was left unaltered .All the nodes were kept in natural atmosphere.

IOP Conf. Series: Materials Science and Engineering **594** (2019) 012013 doi:10.1088/1757-899X/594/1/012013

2.4.2 *Simultaneous control of Luminosity and Temperature.* Under this condition an incandescent lamp was placed in each zone of row 1 and all the zones of row 2 were covered with tinted plastic sheet. This arrangement is depicted in Fig.4.

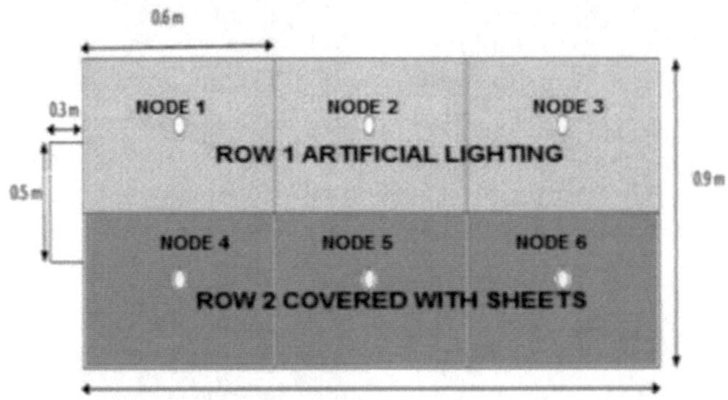

Fig.4. Depiction of simultaneous control of temperature and luminosity

2.4.3 *Temperature control alone.* Under this condition a single heater was placed at the centre of the greenhouse at 35cm height from bottom. The arrangement is depicted in Fig.5.

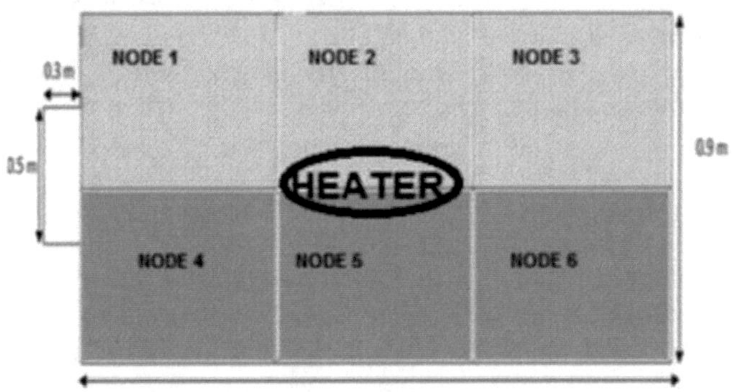

Fig.5. Depiction of control arrangement for temperature control alone

3 Results and Discussion

3.1 Profile under uncontrolled conditions

The variations of temperature and luminosity as obtained from six sensors placed at centre of the six zones of the greenhouse with time are as shown in Fig.6 and Fig.7, respectively. The temperature line graphs indicate that the temperature inside the greenhouse is almost the same in all the zones as expected. As observed from the luminosity line graph, Sensor 1 placed in zone 1 has recorded highest luminosity as there was more natural light. Luminosity has been recorded in lux and it can be seen that the luminosity varies from zone to zone.

IOP Conf. Series: Materials Science and Engineering **594** (2019) 012013 doi:10.1088/1757-899X/594/1/012013

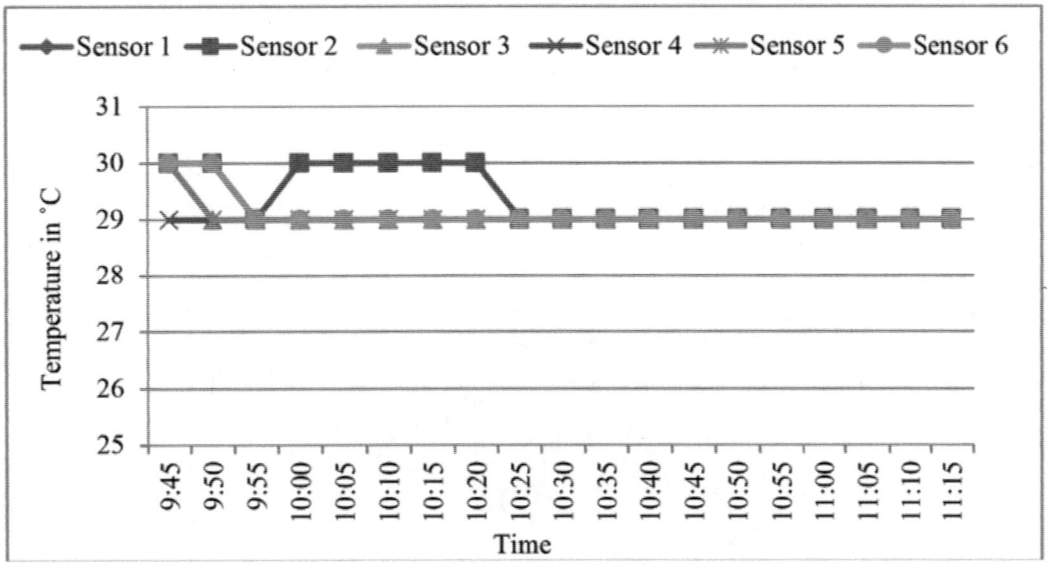

Fig.6.Temperature variation with time under uncontrolled condition

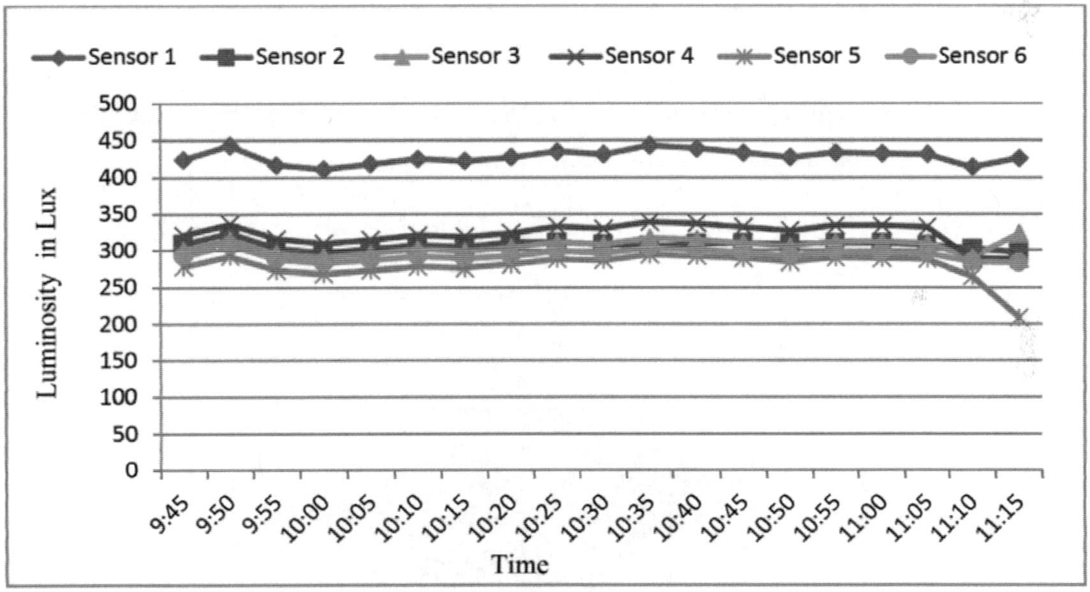

Fig.7. Luminosity variation with time under uncontrolled condition

The corresponding surface profiles of temperature and luminosity are shown in Fig.8 and Fig.9 respectively. The X coordinate is the horizontal distance of the sensors which is 0.30 and 0.90 and 1.50 meters, Y coordinate is the vertical distance at which sensors are placed which is at 0.225 and 0.675 meters from the door. Z coordinate shows temperature / luminosity obtained from the sensors. For plotting the spatial profile linear extrapolation has been used unless mentioned otherwise. The surfaces profile of temperature shows a flat profile which indicates that the temperature is almost the same in all zones of GH. The surface profile of luminosity shows a conical shape, which indicates that the luminosity data obtained is not equal from all the sensors. It varies in zones. The peak in the profile was corresponding to very high value of luminosity obtained from sensor of zone1.

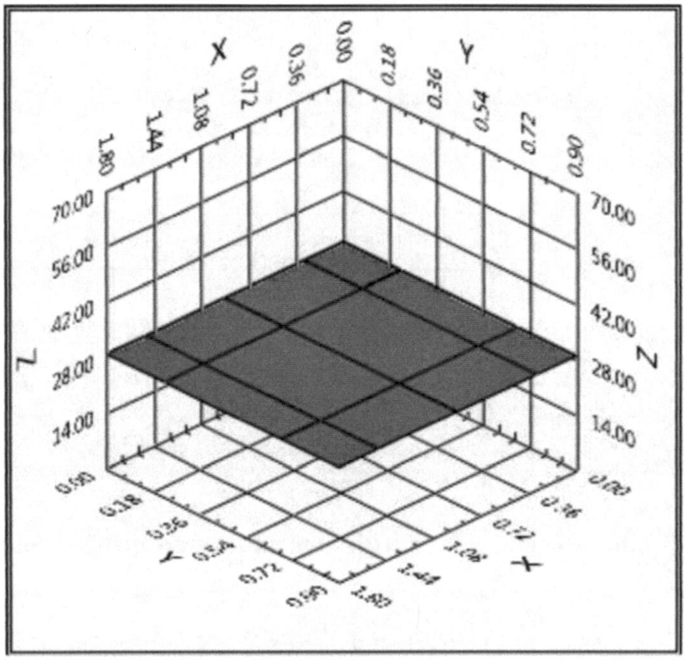

Fig.8. Surface profile of temperature at 10.30 a.m. under uncontrolled condition

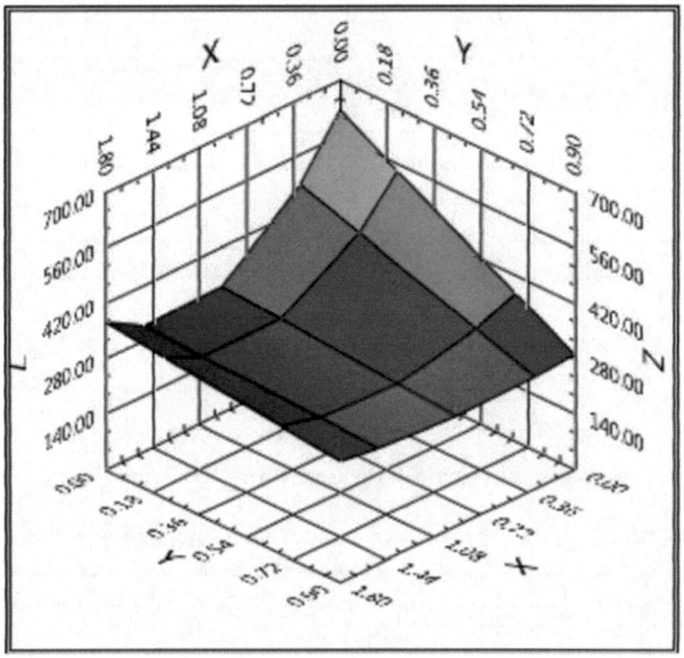

Fig.9. Surface profile of luminosity at 10.30a.m. under uncontrolled conditions

3.2 Profile under simultaneous control of Luminosity and Temperature

The line graphs showing the variation of temperature with respect to time inside the greenhouse for row-1 zones and row 2 zones is shown in Fig.10. The surface profile for it is shown in Fig.11. The line graph and surface profile of luminosity inside the greenhouse is as shown in Fig.12 and Fig.13, respectively. Luminosity values measured by the sensor have been extended up to the nearest wall as the row -2 sensors were covered by opaque sheet while plotting the surface profile. The luminosity recorded by sensor nodes 4, 5 and 6 was very low and no significant increase in temperature was seen.

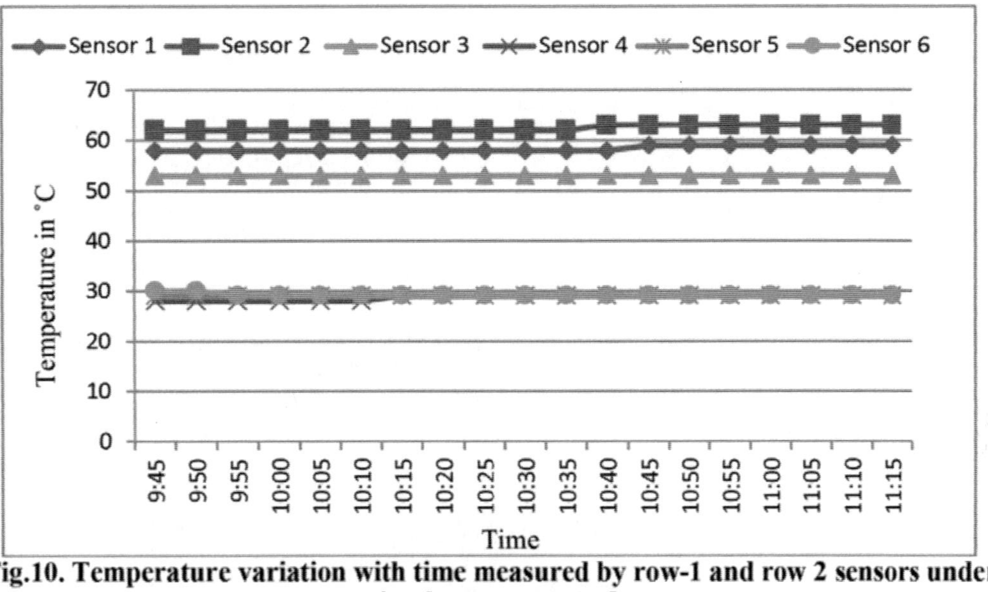

Fig.10. Temperature variation with time measured by row-1 and row 2 sensors under simultaneous control

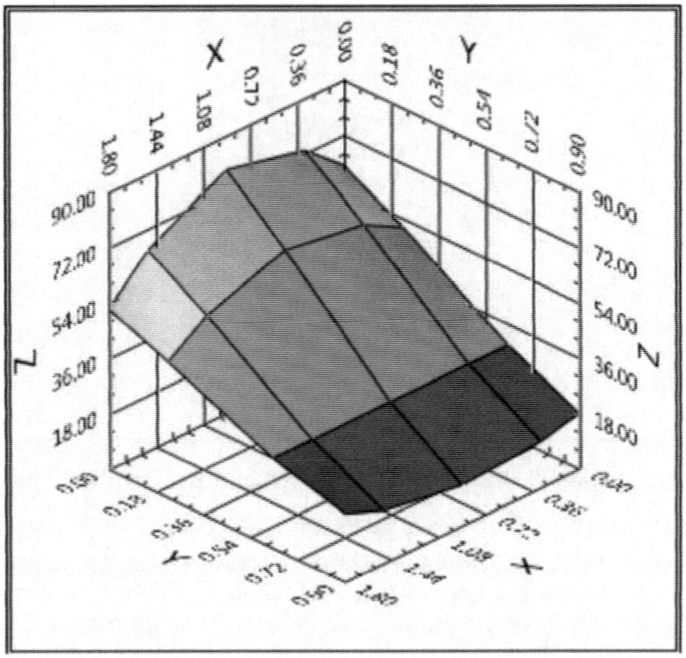

Figure 11: Surface profile of temperature at 10:30 a.m. under simultaneous control

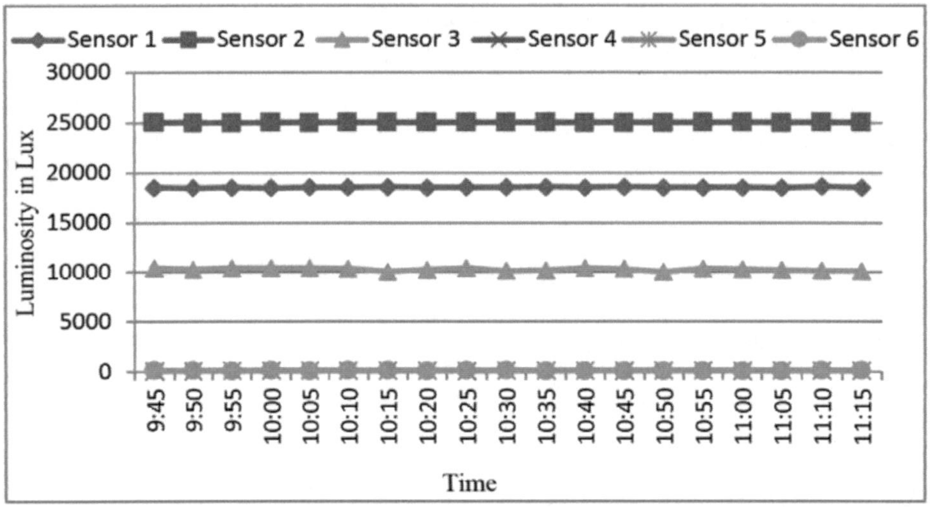

Fig.12. Luminosity variation with time measured by row-1 and row 2 sensors under simultaneous control condition

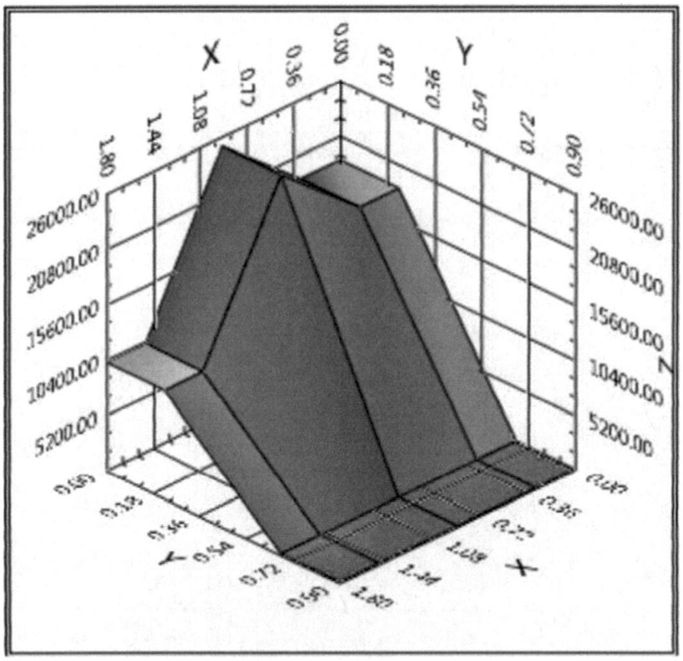

Figure 13: Surface profile of luminosity at 10.30 a.m. under simultaneous control condition

The sensors placed in row-1 recorded very high luminosity as well as the temperature. Highest recorded luminosity in row 1 was 25069 Lux from sensor placed in zone 2. Highest recorded luminosity in row 2 was 134 lux from sensor placed in zone 6. For zones of row 1 temperature was very high and varied with change in luminosity. Highest temperature recorded was 62°c from sensor placed in zone 2. Thus it can be seen the sensor which recorded highest luminosity also recorded highest temperature. For zones of row 2 temperature remained constant at 35°c with small variations initially. There was increase in average temperature inside GH. It means, with incandescent lamps, there was an increase in the luminosity which also resulted in increase in the temperature.

3.3 Profile under temperature control alone

The line graphs of the temperature and luminosity under this condition are shown in Fig. 14 and Fig.15 respectively.

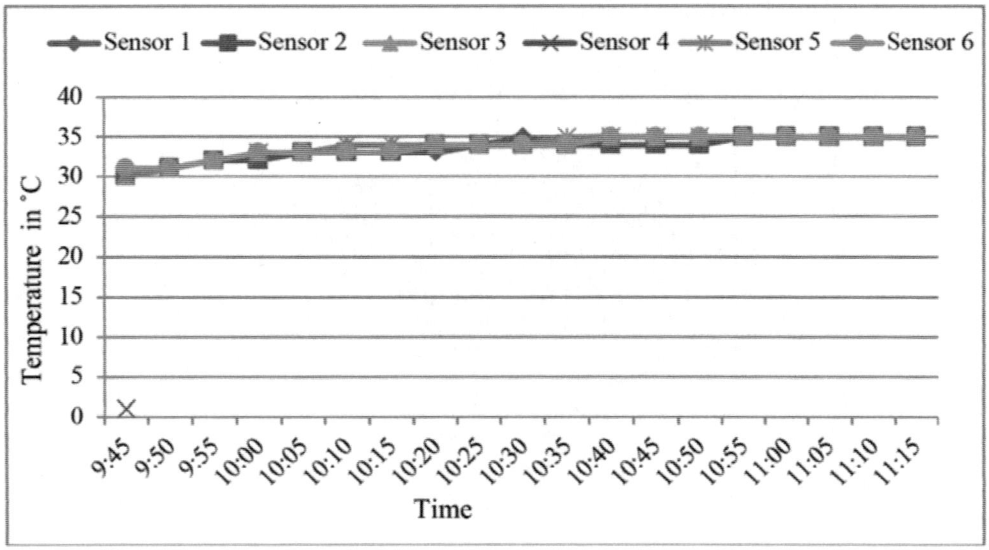

Fig.14. Temperature (with heater) variation with time under temperature control only

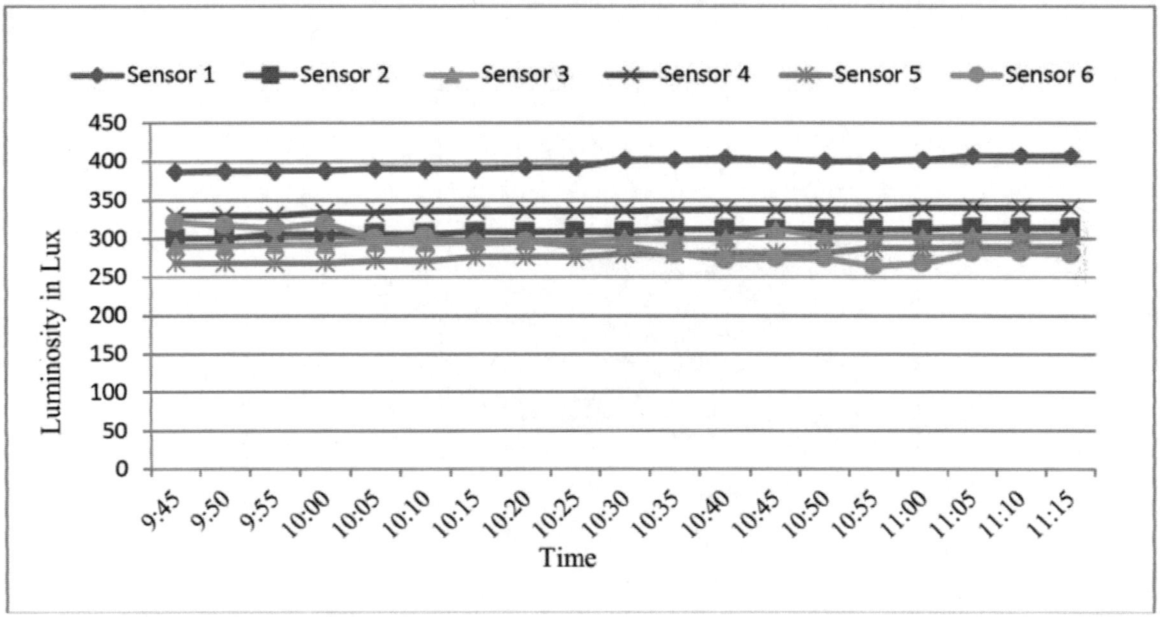

Fig.15. Luminosity variation with time under temperature control only

The surface profile of temperature and luminosity under this arrangement of control is as shown in Fig. 16 and Fig.17, which revealed that the average temperature inside the greenhouse increased as compared to uncontrolled condition.

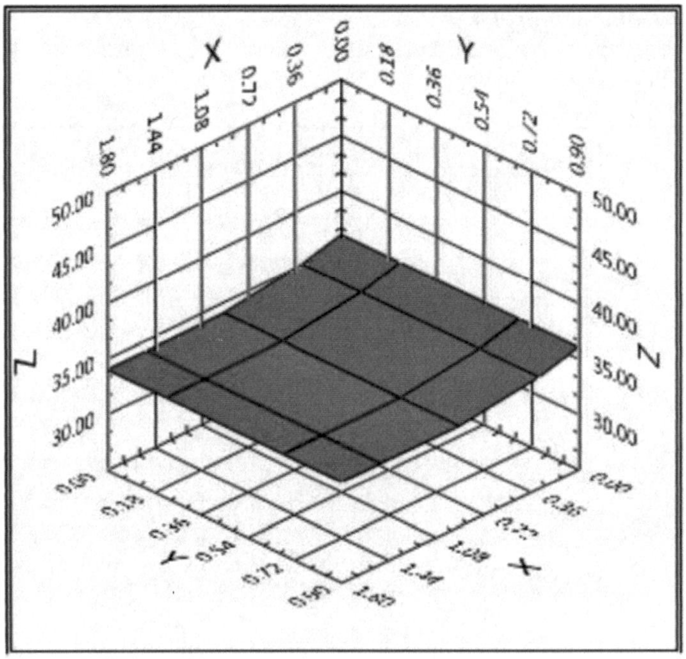

Fig.16. Surface profile of temperature at 10.30.m. under temperature control only

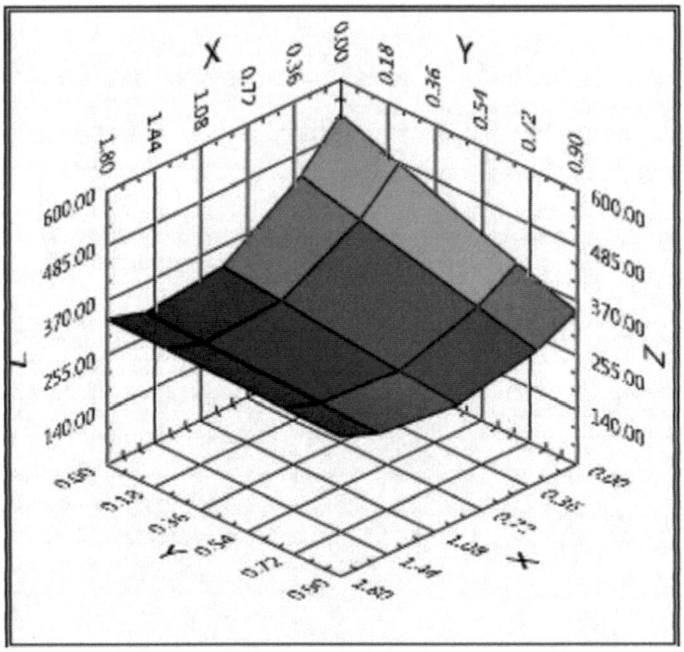

Fig.17. Surface profile of luminosity at 10.30 a.m. under temperature control only

4. Conclusion

Measurement and profiling of the temperature and luminosity inside the experimental greenhouse have been done successfully using WSN. The profiling of each variable has been carried out in both temporal and spatial domains. To ensure correct profiling, data from all the node were acquired simultaneously (by avoiding any intentional time delay in reading two successive nodes). Collaborative data has been used for visualization of the profiles. Use of WSN for collecting the physical variables inside greenhouse provides remarkable flexibility in terms of addition and deletion of WS nodes. Also, no effort is required to connect the sensors and data acquisition systems in WSN. The LabVIEW, used here for data acquisition and collaborative processing, offers the advantages of graphical programming and multitude of functions for visual presentation of data.

References

[1] Pahuja Roop ,Verma H.K. , and MoinUddin,(2013) "A Wireless Sensor Network for Greenhouse Climate Control", IEEE Pervasive Computing, vol. 12, no. 2, pp. 49-58.

[2] Pahuja Roop ,Verma H.K. , and MoinUddin,(2012) "Design and Implementation of Fuzzy Temperature Control System for WSN Applications", International Journal of Computer Science and Network Security, vol.12,no .11, pp. 1099-115.

[3] Ahonen Teemu , Virrankoski Reino and Mohmmed Elmusrati,(2008), " Greenhouse Monitoring with Wireless Sensor Network", Proc. IEEE/ASME International Conference on Mechtronic and Embedded Systems and Applications, 2008. MESA 2008.

[4] Pawlowski Andrzej, Guzman Jose Luis , Rodríguez Francisco, Berenguel Manuel , Sánchez José ,and Sebastián Dormido,(2009), "Simulation of Greenhouse Climate Monitoring and Control with Wireless Sensor Network and Event-Based Control", Sensors(Basel) , vol.9,no.1, 232-252.

[5] SENSEnuts WSN sensation quick start guide

[6] SENSEnuts GUI user guide

[7] Burel J., Brooke T. and Beckwith R., (2004) "Vineyard computing: Sensor networks in agricultural production", IEEE Pervasive Computing, vol. 3. no.1, pp. 38-45.

[8] Wang Ning, Zhang Naiqian and Wang Maohua, (2006), "Wireless Sensors in Agriculture and Food Industry –Recent development and future perspective ", Computers and Electronics in Agriculture , vol.50, pp. 1-14.

[9] Zhang Qian , Yang Xiang-long , Zhou Yi-ming , Wang Li-ren , and Guo Xi-shan, (2007), " A Wireless solution for Greenhouse Monitoring and Control System based on ZigBee technology", Journal of Zhejiang University science A, vol. 8, Issue 10 , pp 1584-1587.

[10] Kim Hak-Jin, Sudduth Kenneth A. and Hummel John W., (2009),"Soil macronutrient sensing for precision agriculture", Journal for Environmental Monitoring , vol.11, pp. 1810-1824.

[11] Sinfield Joseph V., Fagerman Daniel and Oliver Colic ,(2009), "Evaluation of sensing technologies for on –the-go detection of macro-nutrient in cultivated soils", Computers and Electronics in Agriculture , vol.70, pp. 1-18.

[12] Li G.Fang, Liu S. T. and Chen Li D.,(2010), "Design on Environment Monitoring System for Greenhouse Based on Wireless Sensor Network" , Advance Materials Research , vols.108 no.111, pp 145-150. http://www.scientific.net/AMR.108-111.145.

[13] Rout Mrutyunjay and Verma H.K., (2010), " A Test-Bed of Analog and Digital Sensors for Wireless Sensor Networks,"Proc. Nationl Conference on Computational Instrumentation , CSIO 2010, on 19-20 March 2010 ,pp.134 – 137.

[14] Chaudhary D.D., Nayse S.P. and Waghmare L.M.,(2011), "Application of Wireless Sensor Networks for Greenhouse Parameter control in Precision Agriculture " , International Journal of Wireless Networks , vol.3 no.1 , pp 140-148.

[15] Park Dae-Heon and Park Jang –Woo, (2011), "Wireless Sensor Network –Based Greenhouse Environment Monitoring and Automatic Control for Dew Condensation Prevention", Sensors, vol.11, no.4, pp. 3640-3651.

[16] Sabri Naseer ,Aljunid S.A.,Ahamd R.B.,Yahya Abid ,Kamaruddin R.and Salim M.S.,(2011), "Wireless Sensor Actor Network Based on Fuzzy Inference System for Greenhouse Climate Control", Journal of Applied Sciences , vol.11, no.17, pp. 3104-3116.

IOP Conf. Series: Materials Science and Engineering **594** (2019) 012014 doi:10.1088/1757-899X/594/1/012014

Single Stage Isolated Zeta Converter (SS-IZC) Circuit with Improved Power Factor for LED's Application

Rajesh Narayan Deo[1], A. Shrivastava[2] and Kalyan Chatterjee[1]

1Department of Electrical Engineering, Indian Institute of Technology (ISM), Dhanbad

2Department of Electrical Engineering, Manipal University, Jaipur
E-mail: rndeo.nitc@gmail.com

Abstract. Light emitting diode specifically White LED has gain a lot of attention from both academics and industrial sector due to its high efficiency, environmental friendly, and long lifecycle. This is a feasible alternative for conventional bulbs and tubes in domestic, industrial and commercial lighting. However, implementation of this new source requires various parametric regulation. This proposed single stage isolated zeta converter (SS-IZC), with power factor correction maintains constant output for the load of 36 Watts LED, and follow the IEC61000-3-2 standard. The proposed SS-IZC is operating in DCM and controlled with PWM (Pulse width modulation), for power factor improvement and low THD, and obtain more than 0.9 power factor and THD less than ten percentages.

Index Terms—single stage Isolated Zeta converter (SS-IZC), power factor (PF), Pulse width modulation (PWM), Universal AC input, discontinuous conduction mode (DCM), total harmonic distortion (THD).

1. Introduction

The large share of electric power is used for lighting purpose. So, reduction in this required energy will also reduce the greenhouse effect. LED has great advantage over the traditional lighting such as more life that is greater than 150000 h, zero ultraviolet emission, durable quality, operated in extremely low temperature and low voltage application, hence LED lighting have been bringing revolution to this world [1].

The fluorescent tubes have the problem of power factor, cost, flicking etc., to overcome this, electromagnetic ballasts are introduced. Further, enhanced is LED's, to overcome these problems [2-3]. The converters and rectifiers have further improved the operation of LED [4-7].
However, the non-linear elements present in these system, will pollute the power system standards for power factor and total harmonic distortion respectively. To cope up with these standards, LED's have power factor correction technique, which can be single or two stage [4-7]. In two stage converter, PF correction circuit is separately connected and hence requires more components.

The single stage converter has wider acceptability then the two stage converter, for PF correction in LED's. in single stage PF correction circuit and converter is combined in one stage and hence components requirement is less as compared to two stage. Six configuration of converters are more prominent than other, Buck, Boost, Cuk, Buk-boost, Sepic [12-17].

The SS-IZC is generally used for high power and more widely accepted than other converters due to its simplicity and isolation of output from input. This isolation provides a projection mechanism against short circuit condition.

In this work, SS-IZC with PF and THD correction is analysed. For controlling of the gate of power switch, PWM controller is employed, which take output as a feedback and maintain the constant output voltage. The design of IZC is analysed for 36 Watts LED with PF and THD circuit. For controlling THD at the input side, a low pass filter is used at the input side. In part II, the operation of SS-IZC with PF circuit is discussed. In part III, proposed designed of SS-IZC is explained. In part IV, control strategy of SS-IZC based DCM is explained and in last section conclusion is made on behalf of simulated results.

2. Analysis of SS-IZC

The SS-IZC provides isolation between the rectifier and LED shown in Fig. 1. The winding polarity is indicated by dot convention. The V_d , is rectified voltage of input, $V_{S1}(t)$. The L_{m1} symbolizes , magnetizing inductance of the SS-IZC. Energy is stored in L_{m1} when switch, M, is ON and in OFF condition of Switch the stored energy in primary side of transformer is transferred to secondary side. Smooth voltage across the LED is maintain by the output capacitor , C_{01}. A SS-IZC will work in DCM/CCM depends on the continuity of inductor current. Pulsating input current is developed due to high switching frequency of switch, M. To make line current in phase with line voltage can be achieved by controlling of amplitude and degree of current pulsating. Hence low THD and nearly unit factor at input side achieved.

Also, the SS-IZC working in DCM requires less component as compared to the two stage operation, as SS-IZC operated at fixed duty ratio and fixed frequency.

To simplify the investigation of the SS-IZC for LEDs circuit having high PF, the consideration are:

1. Components of SS-IZC are ideal.
2. The source frequency is to be considerably lesser than the switching frequency.
3. The output voltage (V_o) is constant during one period as the value of output capacitor is very high.

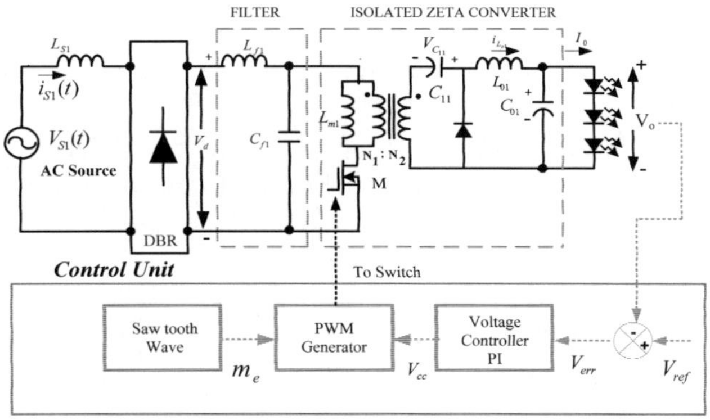

Fig. 1 Block Diagram of SS-IZC with Controller

Supposing the turn ($n = \frac{N_2}{N_1}$) ratio of the SS-IZC is 0.5 , the waveforms of the SS-IZC is shown in Fig. 5 functioning in the DCM . For one cycle, the SS-IZC operation is divided into three parts:

IOP Conf. Series: Materials Science and Engineering **594** (2019) 012014 doi:10.1088/1757-899X/594/1/012014

Mode I ($t'_o < t < t'_1$) :

For the time t'_o , M, switch is ON , diode D is turned off in this mode. The corresponding SS-IZC circuit is shown in Fig. 2. Rectified, V_d, voltage is directly applied to the inductor L_{m1} and hence $i_{L_{m1}}$ current across this inductor increases linearly. And current $i_{L_{o1}}$ current also increases linearly across the inductor L_{o1} , since voltage across output inductor is positive voltage i.e.($nV_d + V_{c_{11}} - V_o$) and the corresponding waveform for this mode is shown in Fig. 5.

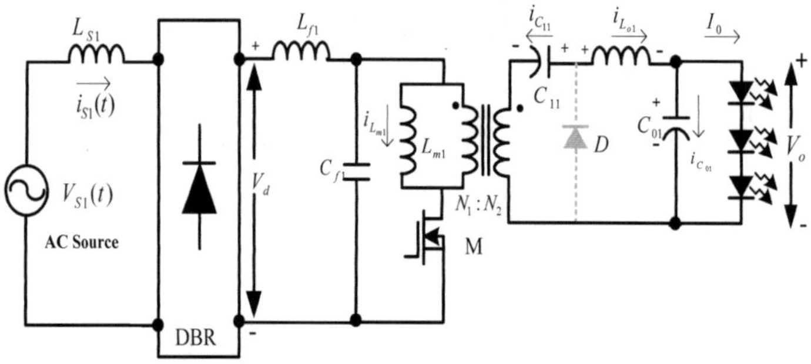

Fig. 2 Mode I operation of SS-IZC

Mode II ($t'_1 < t < t'_2$) :

In this mode, switch M, is turned off and hence D becomes forward biased. The corresponding SS-IZC circuit is shown in Fig. 3. Since reflected voltage across the capacitor C_{11} i.e. $-\frac{V_{C_1}}{n}$ is negative , due to this current across the inductor L_{m1}, falls linearly which is shown in Fig. 5. Output inductor $i_{L_{o1}}$ also fall linearly due to negative terminal voltage $-V_o$ across it. At this mode diode current is $i_D = i_{c_{11}} + i_{L_{o1}}$.

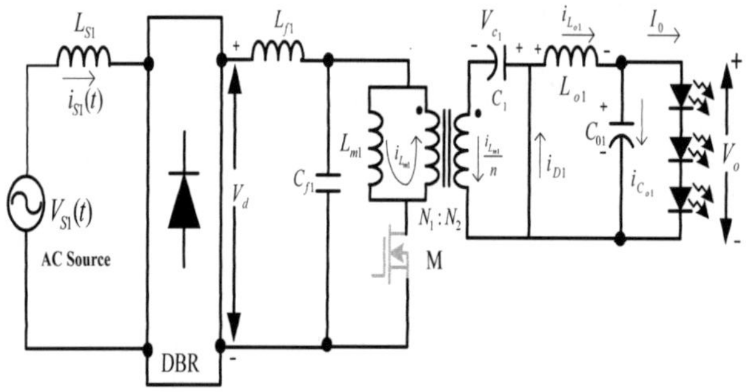

Fig. 3 Mode II operation of SS-IZC

Mode III ($t'_1 < t < T'_S$) :

In this mode of operation switch M, remained turned off and the current across magnetizing inductor, L_{m1}, and output inductor current continue to fall linearly. In this mode diode is turn off due to sudden fall of diode current to zero. So in this mode both switch and diode are turned off, hence SS-IZC is stated to as operating in DCM, which is shown in Fig. 5.

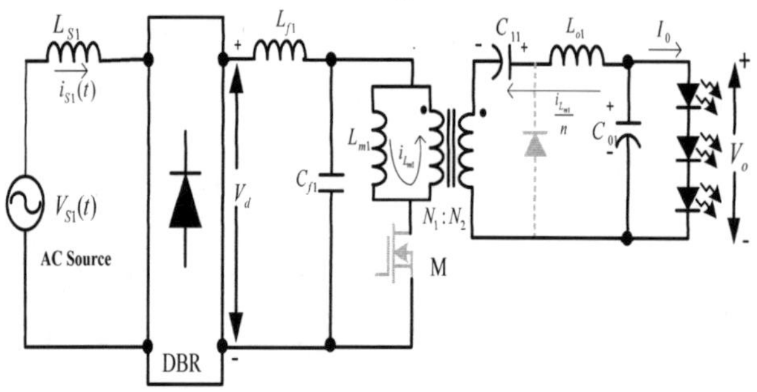

Fig. 4 Mode III operation of SS-IZC

IOP Conf. Series: Materials Science and Engineering **594** (2019) 012014 doi:10.1088/1757-899X/594/1/012014

3. Design of IZC

The SS-IZC PFC converter is designed to operate in DCM, due to this current across inductor L_{m1} become discontinuous for one cycle. A PFC SS-IZC converter of 36 W (P_{max}) is considered for LED driver Circuit. The line voltage, $V_{s1}(t)$, varies from 85 V to 270 V. The line voltage $V_{s1}(t)$ is fed to the SS-IZC PFC converter as,

$$V_{S1}(t) = V_{m1} \sin(\omega_{s1} t) \tag{1}$$

Where V_{m1} is source maximum voltage and $\omega_{s1} = 2\pi f_{s1}$, f_{s1} is the source frequency i.e. 50 Hz. The instantaneous voltage of DBR is given as,

$$V_d(t) = |V_{m1} \sin(\omega_{s1} t)| \tag{2}$$

Here, || || symbolizes the magnitude of Voltage. The V_o is output voltage of SS-IZC is given as [18],

$$V_o = \frac{nD'}{(1-D')} V_d \tag{3}$$

Where D' is taken as 0.5 and symbolizes the duty ratio of linear transformer.
The $D(t)$ is instantaneous value of duty ratio, and depend upon the value of input Voltage $V_{S1}(t)$ and the output voltage V_o, And it is obtained from (2) and (3) i.e.

$$D'(t) = \frac{V_o}{(n)V_s(t)+V_o} \tag{4}$$

Instantaneous power at the output capacitor, P_i, is considered as linear function of output voltage V_o and and given as :

$$P_i = \left(\frac{P_m}{V_{om}}\right) V_o \tag{5}$$

Where V_{om} denotes the maximum output voltage and P_m is regarded as a rated power of SS-IZC.
The L_{mc1} denotes the critical value of magnetizing inductance of transformer of high frequency is can be obtained as [19],

$$L_{mc1} = \frac{R_L\{1-D'(t)\}^2}{2D'(t) f_{sw} \left(\frac{N_2}{N_1}\right)^2} = \left(\frac{V_o^2}{P_i}\right) \frac{\{1-D'(t)\}^2}{2D'(t) f_{sw} (n)^2} \tag{6}$$

Where load resistance is denoted by R_L , switching frequency is denoted by f_{sw} (taken as 50KHz) and the instantaneous power, P_i, is consider for lowest possible value of line voltage. Using equation (6) as,

$$L_{mc1} = \left(\frac{V_{om}^2}{P_m}\right) \frac{\{1-D'_a\}^2}{2D'_a f_{sw} \left(\frac{N_2}{N_1}\right)^2} \tag{7}$$

Where D'_a , duty ratio and calculated for maximum output voltage V_{om} and at minimum peak supply voltage. And for the operation of SS-IZC in DCM mode, the value of transformer magnetizing inductance (L_{m1}) should be less than the critical inductance L_{mc1} [20]. Zeta converter operating at the DCM conditions and hence the magnetizing inductance is consider as $1/10^{th}$ of critical inductance. The output inductance L_{o1} is calculated as [19],

$$L_o = \frac{V_o\{1-D'(t)\}}{f_{sw} \Delta i_{Lo}} = \frac{V_o\{1-D'(t)\}}{f_{sw}(k I_o)} \tag{8}$$

Hence, output inductance is designed for rated output voltage for a extreme line voltage and lowest value of duty ratio D'_b corresponding to as [19],

$$L_{o1} = \frac{V_{om}\{1-D'_b\}}{f_{sw}(\eta I_o)} \tag{9}$$

Where η denotes the % ripple of output inductor current, consider as 40% of output inductor current.

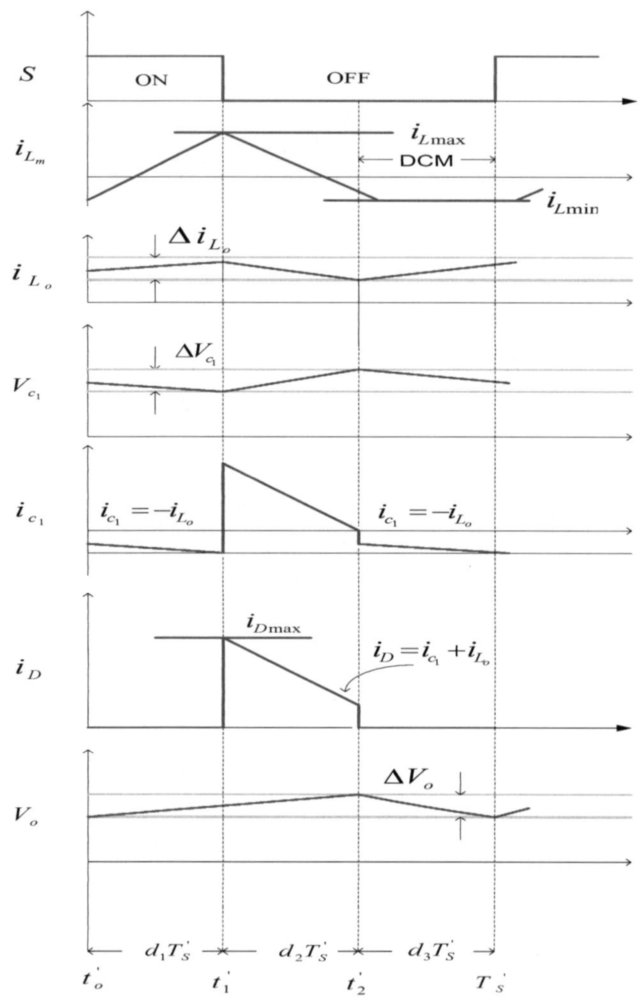

Fig. 5 Waveforms of the SS-IZC

The intermediate capacitor value is calculated (C_{11}) [19] as,

$$C_{11} = \frac{v_o D'(t)}{\Delta V_c(t) f_{sw} R_L} = \frac{v_o D'(t)}{k\{\sqrt{2}V_s + V_o\} f_{sw}} \left(\frac{P_i}{V_o^2}\right) \tag{10}$$

Where k is the allowable ripple voltage through intermediate capacitor and consider as 10 % of V_{c11}. Hence, the value of C_{11} is designed at extreme value of output voltage , maximum power and at minimum duty ratio, which is given as:

$$C_{\_11} = \frac{V_{om}\, D_b}{k\{\sqrt{2}V_s(t)_{max} + V_{om}\} f_{sw}} \left(\frac{P_m}{V_{om}^2}\right) \tag{11}$$

Hence, value of output capacitor C_{o1} is considered as [19],

$$C_{\_o1} = \frac{I_o}{2\omega \Delta V_o} = \left(\frac{P_i}{V_o}\right) \frac{1}{2\omega (k V_o)} \tag{12}$$

Where ΔV_o denotes the permissible ripple in output voltage, k denotes the % of acceptable output voltage ripple across capacitor and taken as three percentage of v_{omin} .

IOP Conf. Series: Materials Science and Engineering **594** (2019) 012014 doi:10.1088/1757-899X/594/1/012014

The wickedest situation design occurred at smallest value of output capacitor voltage i.e. v_{omin} and calculated as,

$$C_{o1} = \left(\frac{P_{min}}{V_{omin}}\right) \frac{1}{2\omega(\eta V_{omin})} \tag{13}$$

A low pass filter (LC), is designed to avoid the harmonics of higher order at input side. The filter capacitance C_{m1} , which is taken as maximum value is give as [19],

$$C_{m1} = \frac{I_{pk}}{\omega_L V_m} \tan(\theta) = \frac{(P_m\sqrt{2}/V_s)}{\omega_L \sqrt{2} V_s} \tan(\theta) \tag{14}$$

Here θ is the angle between line current and line voltage and is consider as 1°. The filter inductance value is calculated by taking ,the line impedance L_{s1} is three percentage of base impedance. Therefore the additional value of inductance (L_{f1}) is given as,

$$L_{f1} = L'_{req1} + L_{s1} \Rightarrow L'_{req1} = L_{f1} - L_{s1}$$
$$L'_{req1} = \frac{1}{4\pi^2 f_c^2 C_{f1}} - 0.03\left(\frac{1}{\omega_L}\right)\left(\frac{V_S^2}{P_{o1}}\right) \tag{15}$$

Where f_c is the Cut-off frequency and taken as such that $f_L < f_c < f_{sw}$. Hence f_c is consider as $f_{sw}/10$. Hence by using above expressions the LC filter with capacitance C_{f1} and inductance L_{f1} is taken.

4. Control Strategy of SS-IZC

To control the SS-IZC operated in DCM mode, voltage follower configuration is considered. The control method, consists of voltage controller, reference voltage generator, error voltage generator and PWM generator, which are shown in Fig. 6. Reference voltage produced with 'Reference Voltage generator' and voltage error generator compares this reference voltage with output DC voltage and generate error voltage (V_{err}) and resulting voltage error is given as,

$$V_{err}(n) = V_{ref}(n) - V_o(n) \tag{16}$$

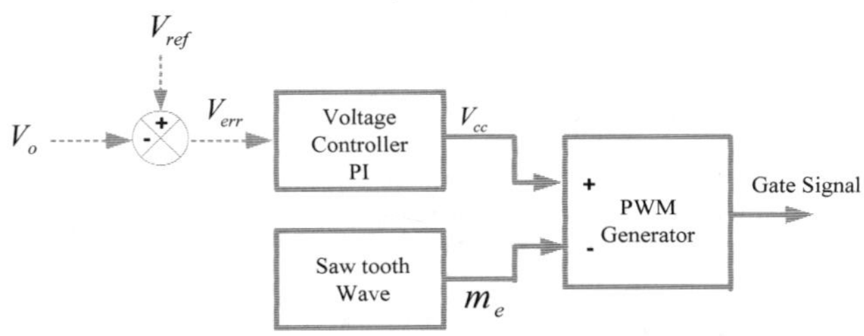

Fig. 6 Control Strategy of Isolated Zeta Converter in DCM

And the PI output voltage controller, at nth sampling instant is given as,

$$V_{cc}(n) = V_{cc}(n-1) + K_p\{V_{err}(n) - V_{err}(n-1)\} + K_i V_{err}(n) \tag{17}$$

Where, integral gain and proportional gain denoted by K_i *and* K_p . Gate signal for the switch is generated by PWM to reach unity PF and to minimize the THD by comparing the high frequency Saw tooth signal m_e with output of PI controller V_{cc} given as,

$$\begin{cases} if\ m_e < V_{cc}\ then\ Switch\ is\ 'ON' \\ if\ m_e > V_{cc}\ then\ Switch\ is\ 'OFF' \end{cases} \tag{18}$$

IOP Conf. Series: Materials Science and Engineering **594** (2019) 012014 doi:10.1088/1757-899X/594/1/012014

5. Result And Discussion

The proposed PFC DCM derived isolated Zeta converter based LEDS driver circuit is designed and modelled in the MATLAB/SIMULINK software and established model is shown in Fig.7, where LED is assumed to be a resistor, high frequency under steady state condition. The proposed PFC converter topology has been implemented through average Voltage control scheme with PI (Proportional Integral) controller. The switching frequency is considered as constant at 50 kHz for PWM generation throughout variations of AC source voltage. The calculated values of the SS-IZC components designed from the mentioned equations and are selected to get less THD and nearly unity PF at different line voltages. These value components and other parameters are given in Appendix

The waveforms of average current and voltage across LEDs, at rated load at various line voltages are shown in fig.8 to Fig.11. The SS-IZC maintaining steady state voltage at the output.

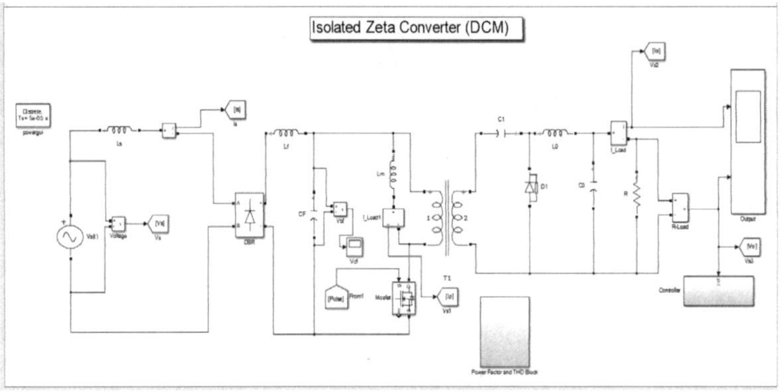

Fig. 7 MATLAB Model od Proposed PFC Isolated Zeta Converter

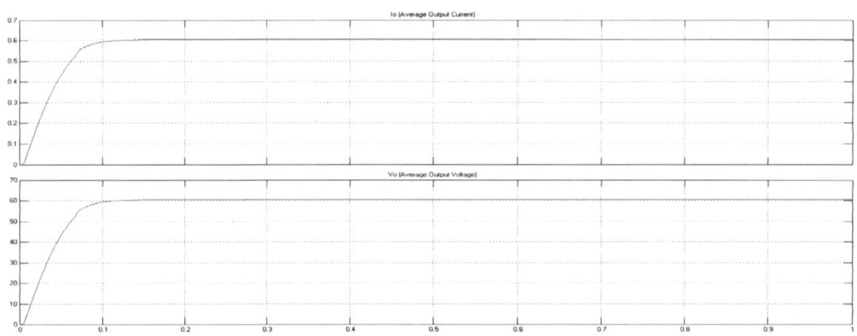

Fig. 8 Average Output Voltage and Current at 90 V Input

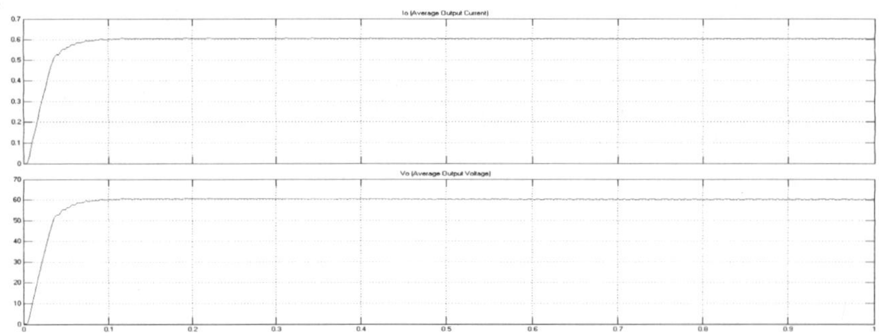

Fig. 9 Average Output Voltage and Current at 140 V Input

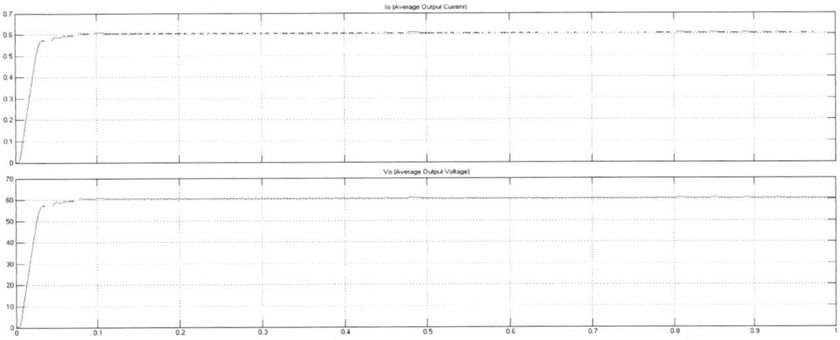

Fig. 10 Average Output Voltage and Current at 200 V Input

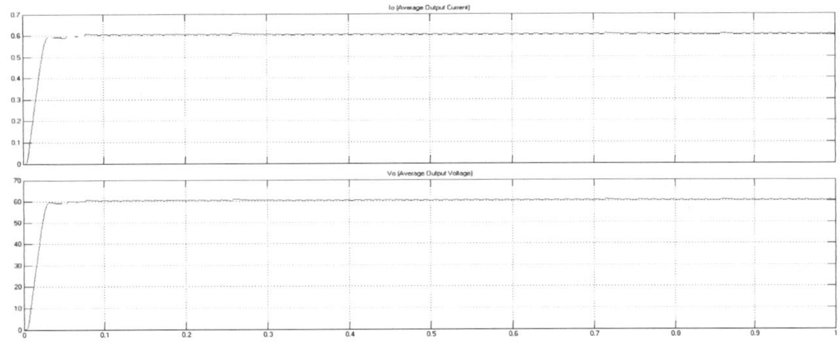

Fig. 11 Average Output Voltage and Current at 220 V Input

Table 1
Simulation Results of The Isolated Zeta Converter

$V_s(t)$ (Volts)	$I_s(t)$ (mA)	V_o (Volts)	I_o (A)	Power Factor	% THD
90	450.13	60.9	0.61	0.998	5
120	310.54	60.4	0.62	0.995	6.2
170	220.81	60.2	0.62	0.994	6.58
210	182.23	60.1	0.63	0.985	7.01
270	147.52	59.8	0.65	0.953	7.54

The performance parameter SS-IZC as a voltage regulator and PF to the LEDs load for various line voltages are concluded in Table 1. Hence from above observation it is found the PF is close to unity and THD less than 10%. And DC load parameter are also, in steady condition for different line voltage.

6. Conclusion

Design and simulation of SS-IZC with improved power factor has been presented, and achieved greater than 0.9 power factor for various supply input voltage (80V - 270V). The utilization of LED depends on its effective design of different parameters have to controlled to meet the international standards. In this work, SS-IZC is analysed, the reason for selection of single stage is that, it has less number of components, which will reduce the overall cost of converter and increase the various indices of given parameters. In the analysis EMI filter is connected at the input side to improve the PF and THD. The PI controller is used for controlling the gate pulse of SS-IZC. Hence, this overall

IOP Conf. Series: Materials Science and Engineering **594** (2019) 012014 doi:10.1088/1757-899X/594/1/012014

system has achieved 0.97 PF and 7.54% THD at 270 V. This system is tested for various inputs and obtain results are in-line with standard norms.

References

[1] M. K. Kazimierczuk and W. Szaraniek, "Electronic Ballast for Fluorescent Lamps," IEEE Trans on Power Electron. Vol. 8, no. 4, pp. 386-395, October 1993.

[2] M. Rico-Secades, A. J. Calleja, J. Ribas, E. L. Corominas, J. M. Alonso, J. Cardesín, and J. García-García, "Evaluation of a Low-Cost Permanent Emergency Lighting System Based on High-Efficiency LEDs," IEEE Transactions on Industry Applications, vol. 41, no. 5, pp. 1386- 1390, September/October 2005.

[3] M. Meneghini, L. R. Trevisanello, G. Meneghesso, and E. Zanoni, "A Review on the Reliability of GaN-Based LEDs, "IEEE Transactions on Device and Materials Reliability, vol. 8, no. 2, pp. 323-331, June 2008.

[4] X. Qu, S. C. Wong, and C. K. Tse, "Noncascading Structure for Electronic Ballast Design for Multiple LED Lamps with Independent Brightness Control," IEEE Transactions on Power Electronics, vol. 25, no. 2, pp. 331-340, February 2010.

[5] H. J. Chiu, Y. K. Lo, J T. Chen, S. J. Cheng, C. Y. Lin, and S. C. Mou, "A High-Efficiency Dimmable LED Driver for Low-Power Lighting Applications," IEEE Transactions on Industrial Electronics, vol. 57, no. 2, pp. 735-743, February 2010.

[6] L. Gu, X. Ruan, M. Xu, and K. Yao, "Means of Eliminating Electrolytic Capacitor in AC/DC Power Supplies for LED Lightings," IEEE Transactions on Power Electronics, vol. 24, no. 5, pp. 1399-1408, May 2009.

[7] Amit Agrawal, Ashish Shrivastava, Kartick C. Jana, "Uniform Model and Analysis of PWM DC-DC Converter for Discontinuous Conduction Mode", IETE journal of Research, Vol. 64, issue-4, 2018, pp. 569-581.

[8] K. Zhou, J. G. Zhang, S. Yuvarajan, and D. F. Weng, "Quasi- Active Power Factor Correction Circuit for HB LED Driver," IEEE Transactions on Power Electronics, vol. 23, no. 3, pp. 1410-1415, May 2008.

[9] J. Sun, "Input Impedance Analysis of Single-Phase PFC Converters," IEEE Transactions on Power Electronics, vol. 20, no. 2, pp. 308-314, March 2005.

[10] Bhim Singh and Ashish Shrivastava, "Buck converter based power supply design for low power light emitting diode lamp lighting," IET Power Electronics, vol. 7, no. 4, pp. 946-956, 2014.

[11] Ashish Shrivastava and Bhim Singh, "Power factor-corrected DCM-based electronic ballast," Electrical Engineering, vol. 95, no. 4, pp. 403-411, Dec 2013.

[12] Ned Mohan, P. Undeland and S. Robbins, Power Electronics: Converter Application and Design, John Wiley and Sons, (Asia) Pvt. Ltd., Singapore, 2003.

[13] Alonso, J. Marcos and Muhammad Rashid (Editor-in-Chief), Electronic Ballasts, Power Electronics Handbook, Chapter 21, Academic Press, 2001, pp. 507-532.

[14] G. Moschopoulos, P. Jain, and G. Joos, "Practical design considerations for a ZVS power factor correction converter," IEEE Applied Power Electronics Conference,1996, pp.172-178.

[15] A.R. Seidel, F.E. Bisogno, T.B. Marchesan and R. Nd.Prado, "A practical comparison among high-power-factor Electronic ballasts with similar ideas," IEEE Trans. On Industry Applications, vol. 41, no. 6, pp. 1574-1583, Nov.- Dec. 2005.

[16] R. L. Steigerwald. "A comparison of half-bridge resonant converter topologies," IEEE Trans. Power Electron., vol. 3, no. 2, pp. 174-182, Apr. 1988.

[17] J.A. Alves, A.J. Perin and I. Barbi, "An electronic ballast with high power factor for compact fluorescent lamps," IEEE Thirty- First IAS Annual Meeting, IAS '96, Industry Applications Conference, 1996, Vol. 4, 6-10 Oct. 1996, pp.2129 – 2135.

[18] N. Mohan, T. M. Undeland and W. P. Robbins, Power Electronics: Converters, Applications and Design, John Wiley and Sons Inc, USA,2009.

[19] B. Singh, S. Singh, A. Chandra and K. Al-Haddad, "Comprehensive Study of Single-Phase AC-DC Power Factor Corrected Converters with High-Frequency Isolation," IEEE Trans. Ind. Inform., vol.7, no.4, pp.540 -556, Nov. 2011.

[20] D. S. L. Simonetti, J. Sebastian and J. Uceda, "The Discontinuous conduction mode Sepic and Cuk power factor preregulators: analysis and design," IEEE Trans. Ind. Elect., vol. 44, no. 5, pp.630-637, Oct. 1997.

[21] Amit Agrawal, Ashish Shrivastava, Kartick C. Jana, "A Universal Input, Single Stage AC-DCLED Driver for an Auditorium Light", Journal of circuit, system and computers

IOP Conf. Series: Materials Science and Engineering **594** (2019) 012015 doi:10.1088/1757-899X/594/1/012015

Policy Impact on Generation of Renewable Energy

Shahroz Anjum[1], Amit Rai[2], Rajesh Narayan Deo[3]

[123]Department of Electrical Engineering, Indian Institute of Technology (ISM), Dhanbad
E-mail: [1]shahroz.anjum@gmail.com

Abstract. The energy consumption is increasing throughout the world. The renewable energy is potential answer to meet the increasing demand. Fossil fuel are diminishing with fast pace and greenhouse gas emission potential. So, the world is promoting renewable energy generation. In this work, government policy impact is assessed for renewable energy promotion. The 4% subsidy on solar, increases approximately 1200% of solar energy generation in 23 years' term and 4% subsidy on wind, increases 69% in wind power generation. The case, on per tone co_2 emission by fossil fuel-based plant is also reducing the use of non- renewable power energy uses.

1. Introduction:

The increasing demand of energy is increasing the search of new environmental friendly energy sources. Traditionally, generation of electrical energy by fossil fuels have air pollution and will last forever [1-2]. To subside this, now countries are promoting renewable energy generation [3]. Currently, the generation share of non-renewable is approximately 85% of total energy demand of world [4]. This current demand of energy is increasing by 1.7% per year. So, to cope up this energy demand renewable energy is a non-exhaustive, easily available and environmental friendly option [5]. In current prospect renewable energy is not so cost-effective solution as non-renewable. The world climate agencies and countries are focusing and countering this issue by providing different type of subsidy to promote renewable energy sources. In this study, promotion mechanism for renewable energy is analysed. And it aims to investigate a clear production picture of RE and non-RE, if subsidy is provided for RE production. In next section methodology is described which is of e4st, that is used for analysis. The analysis is shown with the help of two cases of different subsidy. In last section, conclusion is made with help of outcomes of two cases.

Table.1. Renewable Energy Policy of Different Governments

Country	Quality Driven	Quantity Driven
EU [6]	20% of total electricity will come from renewable energy sources by 2020.	Green Certificate, tariff support
USA [7]	20% of total electricity will come from renewable energy sources by 2020.	Financial Incentive
UK [8]	Quota and renewable obligation	Tradable green Certificate

2. Methodology and Input Data:

In this work e4st is used to analyses the financial subsidy effect on long term renewable energy generation. It optimizes the benefit to the consumer subject network and load constraint [9].

$$\max(\sum_n CB - \sum_n O\&M - \sum_n PollutionCost - \sum_n AnnualInvestement) \quad (1)$$

In equation 1, CB is consumer benefit which has to be maximized minus operation and maintenance cost of the plant, pollution emission adjustment and sum of different annual investments.

The capital cost of solar is fixed to 799 $ per MW and in initial estimate 0% subsidy and in second case 4% subsidy is accounted. The cost of CO2 emission by Non-renewable energy power plant is taken as 4$ per ton in first case and 8$ per ton in second case, similarly the capital cost of wind is taken as 1178$ per MW in first and second estimate but subsidy is increased to be 4% from 0% in second estimate.

3. Result and Discussion:

In this section, feasibility is analysed with the help of two cases. These cases have two different subsidy percentages and two different penalties of CO2 emission.

Case-1

In case-1 analysis, solar power plant and wind plant cost is fixed and subsidy is not provided at all to these RE plants, shown in fig.1.A and fig.1. B. According to these input variables the generation patterns of RE and non-RE plants are shown in fig.2.A and fig.2. B.

With the above analysis and outcome of e4st fast predictor it is clear that RE generation is increasing and non-RE generation is decreasing. In RE solar increase percentage is more than wind based plant. However, nuclear energy generation is increasing with slow rate, which shows it has the capability to compensate the reliability issue associated with renewable energy generation. The net benefit s also increasing with proposed combination, shown in fig.3.

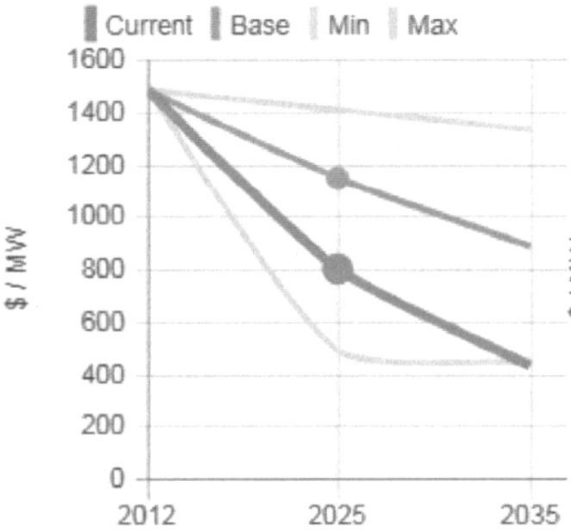

Fig.1. A. Input parameters of Solar

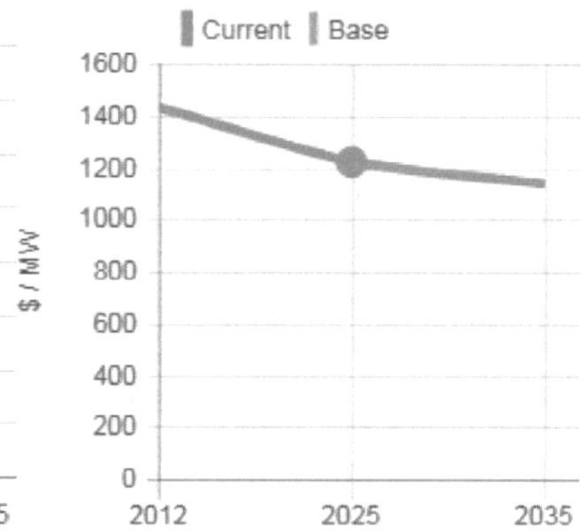

Fig.1. B. Input parameters of Wind

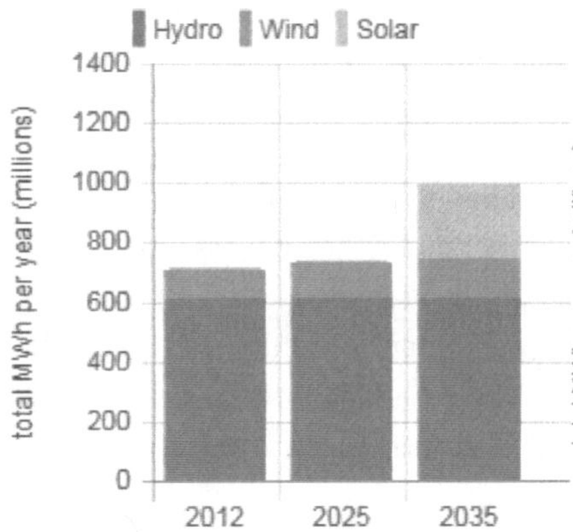

Fig.2. A. RE Generation Scenario

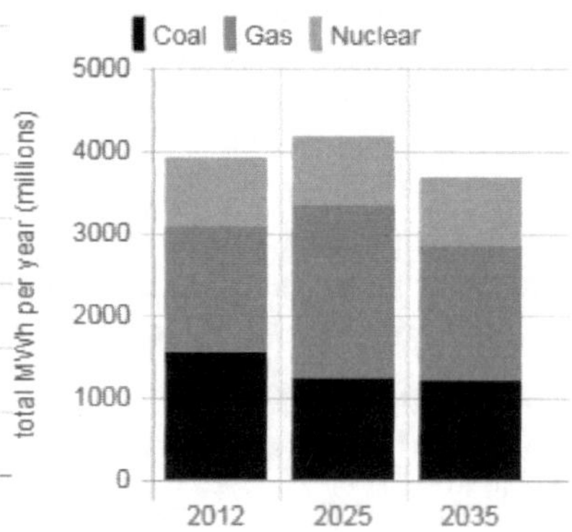

Fig.2. B. Non-RE Generation Scenario

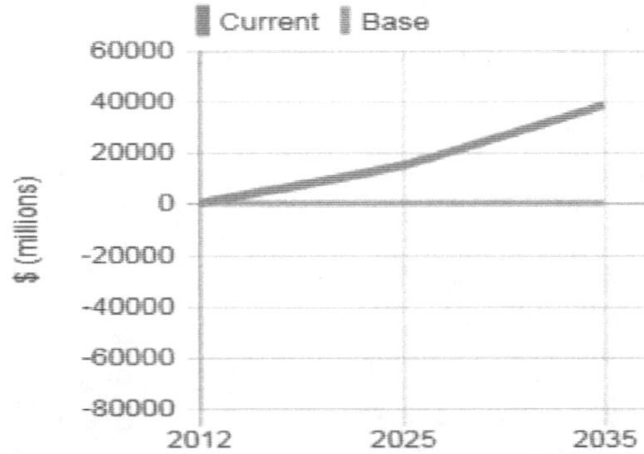

Fig.3. Net Benefits at 0%

Table.2. Generation of power with 0% subsidy

Plant	2012	2025	2035
Hydro Plant	611161766	611462336	611490300
Wind Plant	99010466	115397875	131551294
Solar plant	2384489	12630403	1221667492
Coal Plant	1541887058	1237762849	1221667492
Gas Plant	1530809798	2100151294	1611580089
Nuclear Plant	846698984	846695609	846705798

Case-2

In case-2, the subsidy ratio of solar and wind is increased from 0 % to 4%. The carbon tax imposed on non-RE plants is also increased from 4 dollars per ton to 8 dollars per ton, shown in fig.4.A and fig.4. B.

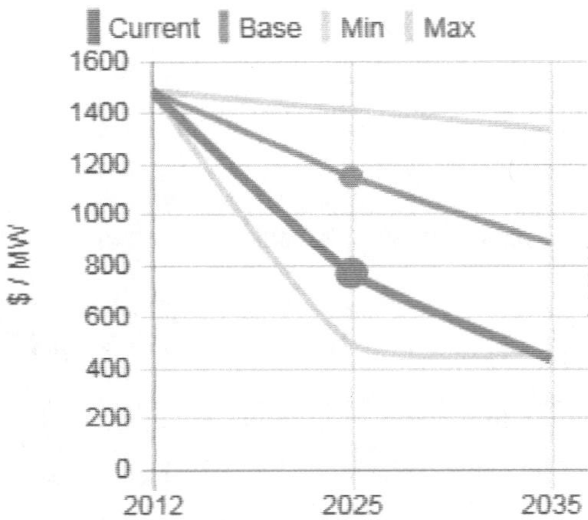

Fig.4. A. Input parameters of Solar

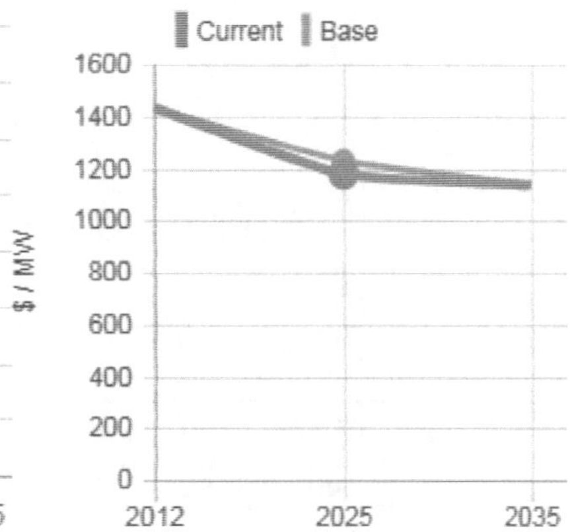

Fig.4. B. Input parameters of Wind

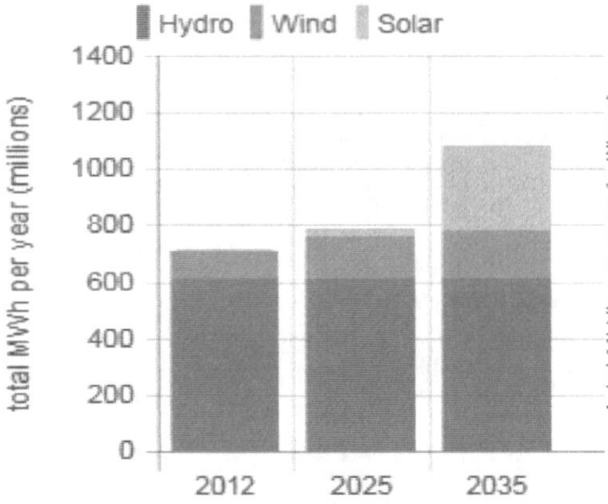

Fig.5. A. RE Generation Scenario

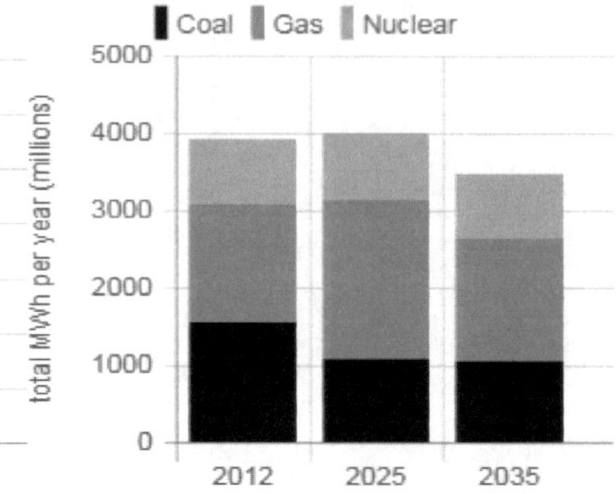

Fig.5. B. Non-RE Generation Scenario

The outcome of above analysis shows that the rate of increase in RE generation is higher with subsidy support. The hydro plant is consistent performance and solar is increasing more rapidly. The net benefit is also large with compared to no support, shown in fig.6.

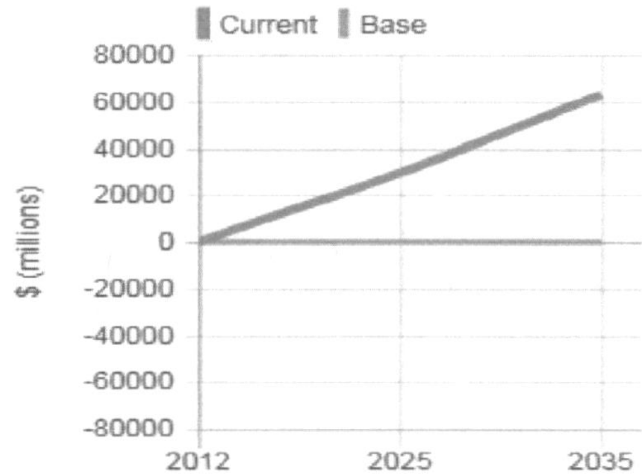

Fig.6. Net Benefits at 4%

Table.3. Generation of power with 4% subsidy

Plant	2012	2025	2035
Hydro Plant	611161766	611472596	611553135
Wind Plant	99010466	150285523	168003109
Solar Plant	2384489	26806629	301964791
Coal Plant	1541887058	1079031402	1057803053
Gas Plant	1530809798	2064028979	1574119721
Nuclear Plant	846698984	846724225	846732872

4. Conclusion

This study shows the long term impact of promotion of RE generating systems. The subsidy support to RE is increasing the generation capacity substantially in the analysis period, which is shown in table 2 and 3. Moreover, the penalty on carbon emission also reducing the use of non-RE generation, specifically coal based plants. The capacity increase is more in solar plant than in wind power generation. However, total phase out of non-RE is not cost effective and reliable with above subsidy consideration and it requires more support from government and stakeholders. The nuclear power plant is increasing generation instead of support to RE which means it can be used for improving reliability of the system and it also has less negative environmental impact than coal or gas based plant.

References

[1] Lucon, O. (2007), "The future of energy", InterfacEHS - Revista de Saúde, MeioAmbientee Sustentabilidade 1(3).

[2] Cosmi, C. Macchiato, M. Mangiamele, L. Marmo, G. Pietrapertosa, F. and Salvia, M. (2003), Environmental and economic effects of renewable energy sources use on a local case study", *Energy Policy* 31:443-457.

[3] Solangi, K.H., Islam, M.R., Saidur, R., Rahim, N.A. and Fayaz, H. (2011), "A review on global solar energy policy", Renewable and Sustainable Energy Reviews 15(4):2149-2163.

[4] Jacobson, M. Z. and Delucchi, M. A. (2011), "Providing all global energy with wind, water, and
solar power, Part I: Technologies, energy resources, quantities and areas of infrastructure and materials", *Energy Policy* 39: 1154-1169.

[5] Boute, A. (2012), "Promoting renewable energy through capacity markets: An analysis of the Russian support scheme", Energy Policy 4:68-77.

[6] Menegaki, A.N. (2011), "Growth and renewable energy in Europe: A random effect model with evidence for neutrality hypothesis", Energy Economics 33(2): 257-263.

[7] Delmas, M. and Montes-Sancho, M. (2011), "U.S. state policies for renewable energy: Context and effectiveness" Energy Policy 39(5): 2273-2288. Ward, D.J. and Inderwildi, O.R. (2013), "Global and local impacts of UK renewable energy policy", Energy Environment Science 6: 18-24.

[8] Daniel L. Shawhan, John T. Taber, Di Shi, Ray D. Zimmerman, Jubo Yan, Charles M. Marquet, Yingying Qi, Biao Mao, Richard E. Schuler, William D. Schulze, and Daniel J. Tylavsky, "Does a Detailed Model of the Electricity Grid Matter? Estimating the Impacts of the Regional Greenhouse Gas Initiative"," Resource and Energy Economics, Volume 36, Issue 1, January 2014, pp. 191{207. http://dx.doi.org/10.1016/j.reseneeco.2013.11.015.

[9] Biao Mao, Daniel Shawhan, Ray Zimmerman, Jubo Yan, Yujia Zhu, William Schulze, Richard Schuler, Daniel Tylavsky. "The Engineering, Economic and Environmental Electricity Simulation Tool (E4ST): Description and an Illustration of its Capability and Use as a Planning/Policy Analysis Tool", Proceedings of the 49th Annual Hawaii International Conference on System Sciences, Computer Society Press, forthcoming January 2016.

Objective functions of distribution network expansion planning - a comprehensive and exhaustive review

Mandhir Kumar Verma[1,*], Vinod Kumar Yadav[2] and V. Mukherjee[1]
[1]Dept. of Electrical Engineering, Indian Institute of Technology, Dhanbad, Jharkhand, India-826004
[2]Dept. of Electrical Engineering, Delhi Technological University, Delhi, India-110042
E-mail: *mandhirverma@gmail.com

Abstract. Utmost of elucidation about the research accomplished in an article is cleared from the objective function of the planning problem. An objective function is a mathematical expression which describe the existing condition of a system with numerous variables, in which alteration of these variables result in optimized value largest or smallest, depending on problem or desired value. That value may be obtained by minimizing or maximizing the objective function. In this paper, a review has been carried out on objective functions of distribution network expansion planning (DNEP). These objective functions have been classified into five main categories: financial, income related, technical, optimal size & location and social & economic. The selection of objective function clearly shows increasing penetration of distributed generation (DG), distributed energy storing systems (DESS) and fuel cells with renewable technologies. Most of the reviewed articles highlight these objectives in details, however; not all fields have been covered in any single work on DNEP. This review article aims to address this gap so that widespread DNEP can be achieved with flawlessness. Substantial information has been offered of research work done in the field of DNEP through this review article which will mitigate the impending researchers from the difficulties of getting apposite supervision.

1. Introduction

The classic objective of the distribution network expansion planning (DNEP) was to fulfil technical necessities (improvement of voltage profile, reliability, reduction of losses etc.). Later on, to ensure an economic and consistent energy supply, foremost aim of DNEP turn into determining an effective and least investment schedule [1]. Decision making progression becomes more complex task when one deals simultaneously to cover maximum aspects of the problem. Another characteristic features of distribution network operation and scheduling is that the judgement making problem is to be encountered with substantial level of ambiguous information. With more penetration of DG, the normal distribution networks becomes active distribution network. Depending upon the objectives functions considered for the expansion planning of distribution network, DNEP problem can be categorize into mono-objective problem and multi-objective problem [2]. The review of various articles reflects the shifting of DNEP from few objectives to multi-objective fulfillment. This paper provides a review on various objective functions utilized by various eminent researchers in the field of DNEP. These objective functions have been classified into five main categories: financial, income related, technical, optimal size & location and social & economic objectives. For original articles on various DNEP [1-80], literature survey has been summarized in tables 1-5.

Financial objectives are the most utilized objective function [1, 2, 4-19, 20-23, 25-33, 35-50, 52-61, 63-76, 78-80] which are categorize into two types: minimization of total expansion cost and maximization of distribution companies (DISCO's) net profit. Income related objectives functions have been separated out from the financial one [14, 49, 75] as these are essentially earnings of DISCO's that can be obtained by selling energy, getting subsidies in lieu of green technologies or by salvage cost.

IOP Conf. Series: Materials Science and Engineering **594** (2019) 012016 doi:10.1088/1757-899X/594/1/012016

Concrete DNEP starts with technical objectives, which are the most affected part while doing expansion. These are classified into five types: energy losses minimization [3, 8, 15, 20, 33, 35, 45, 51, 53, 62, 67, 69, 70, 80], voltage stability & profile improvement [3, 13, 34, 45, 47, 51, 62, 67, 70, 80], reliability improvement [2, 6, 7, 21, 36, 69], line loading minimization [24, 77] and optimal power flow (OPF) solution [51]. While doing DNEP, it is obvious that new devices, equipment, switches [76], substations [2, 14] as well as DG and DESS [76] are needed to be installed. Among these, optimal sitting and sizing of DG is second most prominent objective function among all type of classifications. Few researchers have also given concerns of customer's profit and reducing planers workload while doing DNEP. These two concerns have been taken as maximization of social welfare function objective [77] and load forecasting [14].

The rest of the survey is organized into seven sections. These sections are divided mainly on basis of objective function classification. Section two is dedicated to classification of objective functions. A brief review on objective functions related to financial matter is mentioned in section three. Income related objectives are introduced and discussed in section four. Objective functions related to technical matter are outlines and classifies in section five. Optimal size and location objectives functions for DG, DESS, substation etc. are summarized in section six. A few researchers have also given attention to social and economic objectives which is presented in section seven. Finally section eight concludes the article along with short comings and future prospects.

2. Classification of objective functions

Depending upon aim of DNEP, objective functions are decided. The DNEP can be framed as a single-objective or a multi-objective expansion planning problem. Various objective functions considered by numerous renowned researchers have been taken in analogical order which depicts changing trends in selection of objective function. From the analysis of various objective functions, their classification and hierarchical distribution is performed and is summarized in Table 1.

According to review performed on various articles, it has been deducted that most prominent objective is to minimize total expansion cost. Rests are in total equivalent to this single objective alone. If a researcher has considered the problem as mono-objective, a single most optimal solution can be obtained but in case of multi objective problem, the objective become conflicting in it selves which have normally set of solutions rather than a unique and single solution [76]. Customarily multi-objective DNEP optimization technical hitches require minimizing or maximizing objective function ($f(x)$) as

$$f_{min}(x) = \left(f_1(x), f_2(x), \cdots f_n(x)\right) \quad (1)$$

$$f'_{max}(x) = \left(f_1'(x), f_2'(x), \cdots f_m'(x)\right) \quad (2)$$

s.t. various technical, economic, geographical etc. constraints. Discussion on constraints is not included in this review article.

Table 1. Objective functions classification of DNEP

S. No.	Objective functions	Reviewed articles	No. of research work reviewed*	%age of research work reviewed
	Financial objectives			
1	Minimization of total expansion cost	[1,2,4-19,21-23,25-33,35-50,52-61, 63-76,78-80]	73	91.25%
2	Maximization of net profit of DISCO's	[20,75]	2	02.50%
	Income related objectives			
3	Selling energy	[14,49,75]	3	03.75%
4	Award subsidies	[49]	1	01.25%
5	Salvage cost	[49,52]	2	02.50%
	Technical objectives			
6	Minimization of energy losses	[3,8,15,20,33,35,45,51,53,62, 67,69,70,80]	14	17.50%
7	Voltage stability & profile improvement	[3,13,34,45,47,51,62,67,70,80]	10	12.50%
8	Reliability improvement	[2,6,7,21,36,69]	6	07.50%
9	Optimal power flow (OPF) solution	[24,77]	2	02.50%
10	Minimize line loading	[51]	1	01.25%
	Optimal size and location (OSL) objectives			
11	OSL of DG	[10,11,16,22-24,26,27,32,33,36,37, 39,40,42,45,46,53,62,65-68,71,74, 75, 77,78,80]	29	36.25%
12	OSL of substation	[2,14,21]	3	03.75%
13	OSL of DESS	[76]	1	01.25%
14	OSL of switching devices	[76]	1	01.25%
	Social and economic objectives			
15	Maximization of social welfare function	[77]	1	01.25%
16	Load forecasting	[14]	1	01.25%

* The numeric values of number of papers and the percentage calculated on the basis of total papers reviewed in the references
**is done by cost reduction method also.

3. Financial objective functions

Objective functions related to finance are mainly of two types: primarily related to expenditure for expansion and secondly concerning profit for distribution companies (DISCO's).

3.1 Total expansion cost

DNEP models aims for finding the most cost-effective solution through minimizing investment cost in various equipment's, in their operation and maintenance and can be mentioned as exhaustive objective. Total expansion cost is highest taken objective function (with 91.3% reviewed work) (from Table 1). Besides equipment's concern, there are expenditure in energy losses, poor reliability, power imported, punishment for pollution and many more miscellaneous objective functions related to investment such as reinforcement costs, interest rate, cost of violating bus voltage limits and line loading limits, load curtailment cost, start-up costs (from either a cold stop or a hot stop), DSM incentive cost etc (Table 2).

Table 2. Various objective functions of DNEP related to expansion cost

S. No.	Objective function related to expansion cost	Reviewed articles	No. of research work reviewed	%age of research work reviewed
1	Installation	[1,2,4-19,21-23,25-33,35-49,52-61,63-76,78-80]	72	90.00%
2	Operation	[4,5,9-11,14-16,18,21,22,25,27,33,36-41,43-45,47-49,52,54, 56,58,60,61,63,66-68,71,74,75,78-80]	42	52.50%
3	Energy losses	[8,9,12,15,17,18,21,23,26-33,35,37-41,43,44,46, 48,49,52-58,61,64-66,68,69,73]	41	51.30%
4	Maintenance	[8-11,16,18,21,27,31-33,37-40,49,52,54,60,61,64,66,68,69, 71,72,75,80]	28	35.00%
5	Reliability index	[6,7,10,11,17,22,23,26-28,30-32,38-40,43,45,49,52,54,56,61, 63,64,68,72,73]	28	35.00%
6	Power imported	[16,23,25-27,33,43,49,50,56,59-61,66-68,71,72, 74,75,78, 79]	22	27.50%
7	Carbon punishment	[25,27,33,35,56,71]	06	07.50%
8	Miscellaneous	[6-11,17,23,26,28,30-32,35,38-40,43,48,52,58,61, 64,66,69, 72,73]	27	33.80%

Table 3. Various objective functions of DNEP related to installation cost

S. No.	Objective function related to inst. cost	Reviewed articles	No. of research work reviewed	%age of research work reviewed
1.	DG	[13,16,22,23,25-27,32,33,36,39,40,42,45,47,49,53,54,56, 58-61, 65-68,70,71,73-75,78,80]	34	42.50%
2.	Substations	[2,6,7,13-15,17,18,21,22,30,31,36-38,41,42,44-46,48,52,54-56, 58,59,61,65,66,67,69,70,74]	34	42.50%
3.	Distribution cable	[1,2,4-9,12,14,16-19,23,29,35,37,38,41,44,45,48,55,58,65,66,69, 72,78]	30	37.50%
4.	Feeders	[2,6,7,13,15,21,22,28,30,37,38,42,46,54,56, 59,61,64,67,68,70,74]	22	27.50%
5.	Transformers	[1,6,7,12,17,29,54,61,64,67]	10	12.50%
6.	Sectionalizing switches	[21,23,28,30,37,64,76]	07	08.75%
7.	DESS	[63,66,72,75,76,78,79]	07	08.75%
8.	Capacitors bank	[48,57,65,73]	04	05.00%
9.	New load points	[43,52]	02	02.50%
10.	Voltage regulators	[57,65]	02	02.50%
11.	Circuit breakers	[28]	01	01.25%
12.	Fuel cells	[73]	01	01.25%

3.1.1 *Installation cost.* This objective function is supreme prominent one among all expansion cost objective functions. The survey results are shown in Table 3 from which it can be concluded that 90% literature reviewed uses installation cost as one of the objective functions. Installation mainly includes those equipments which are back bone of distribution system and comprises of feeders, substations, transformers, reactive power sources (capacitor banks, inductors etc.), sectionalizing switches, protective devices (voltage regulators, circuit breakers etc.), fuel cells, distribution cable/line/conductors/branch, DG, DESS, new load points.

3.1.2 *Operation cost.* After installation of various devices, most of them also required financial assistance for their smooth operation and long life. The operation cost of electrical equipment's and devices have been considered in various articles which are summarize in Table 4. From Table 2 it can be concluded that 52.5% literature reviewed uses operation cost as objective function. These equipment and devices are: DG, substations, feeders, new branches, DESS, transformers and fuel cells.

3.1.3 *Maintenance cost.* All new and old equipment require maintenance whose cost is to be included in main objective function of DNEP and 35% literature reviewed uses maintenance cost as objective function.

3.1.4 *Energy losses cost.* In most of the cases, the losses cost is considered as energy losses and 51.3% literature reviewed uses energy loss cost as objective function.

3.1.5 *Reliability index.* 35% literature reviewed uses poor reliability cost as objective function. This function has also been termed as: costs of interruptions [17], reliability cost [22], non-distributed energy costs [23], cost for system reliability [27], customer interruption cost [28], energy not delivered cost [31], interruption cost [38], energy not served penalty cost [39,40], cost of energy not distributed (END) [45], energy not supplied cost [49], unserved energy cost [54], penalties cost for undelivered services [64], cost of load not served [72], expected customer outage cost (ECOST) [73] in various articles.

3.1.6 *Cost of power imported.* Power has been imported from big producers in case: the distribution network is operated by private players or there is increase in power demand and DISCOs are not interested in setting up DG or DESS options. 27.5% literature reviewed uses cost of power purchased as objective function. Depending upon the requirement, DISCOs can purchase active and reactive powers separately.

3.1.7 *Carbon punishment.* Pollution or carbon punishment is also termed as carbon emission cost or environmental compensation costs [33] or CO2 emission tax [71]. 7.5% literature reviewed (which has considered non-renewable resources-based DG) uses carbon punishment cost as objective function.

3.1.8 *Miscellaneous.* A few eminent researchers have considered some investment cost other than described above. Its summarized literature is given in Table 2 from which it can be concluded that 33.8% literature reviewed uses miscellaneous costs as objective function. Summarize in Table 5, these costs are reinforcement costs [10-12, 23, 25, 42, 43, 48, 49, 52, 57, 60, 64, 68, 71, 72] or re-conductoring or reinstallation cost, which is mainly done to strengthen the existing structure and is utilized in 20% of surveyed literature. The cost incurred in paying interest rate against the loan amount taken for expansion have been considered by [8, 9, 28, 48, 69] and is utilized in 6.25% of surveyed literature. Cost of violating limits (violating the bus voltage and line loading limits) [39, 40, 43] and is utilized in 3.75 % of surveyed literature, load (or DG) curtailment cost [35, 58] and is utilized in 2.5% of surveyed literature, start-up costs (from either a cold stop or a hot stop) [43] and is utilized in 1.25% of surveyed literature and demand side management (DSM) incentive cost [66] and is utilized in 1.25% of surveyed literature.

Table 4. Various objective functions of DNEP related to operation cost

S. No.	Objective function related to operation cost	Reviewed articles	No. of research work reviewed*	%age of research work reviewed
1.	DG	[16,22,25,27,33,36,39,40,43,45,47, 49,54,56,58, 60,61, 66-68,71,74,75, 78,80]	25	31.25%
2.	Substations	[17,18,21,22,36-38,41,44,45,48,52, 61,64,65]	15	18.75%
3.	Feeders	[15,21,22,37,38, 54,61,64]	8	10.00%
4.	New branches	[9,17,37,38,41,45,69]	7	08.75%
5.	DESS	[63,66,72,75]	4	05.00%
6.	Transformers	[15,17,61]	3	03.75%
7.	Fuel Cells	[73]	1	01.25%

Table 5. Various objective functions of DNEP related to miscellaneous cost

S. No.	Miscellaneous objective function	Reviewed articles	No. of research work reviewed*	%age of research work reviewed
1.	Reinforcement costs	[10-12,23,25,42,43, 48,49,52,57, 60,64,68,71,72]	16	20.00%
2.	Interest rate	[8,9,28,48,69]	5	06.25%
3.	Cost of violating limits	[39,40,43]	3	03.75%
4.	Load/DG curtailment cost	[35,58]	2	02.50%
5.	Start-up costs	[43]	1	01.25%
6.	DSM incentive cost	[66]	1	01.25%

3.2 DISCO's net profit maximization

It has been observed at some occasions that DISCOs are considered as sole owner of DG units whereas sometimes DG units are the investment made by external investors. For both these cases the objective function for maximizing net profit of DISCO's are different. In case DISCO does not expand the distribution system and purchase power from private investors, the objective function will be described as DISCO's profit index function ($DNP_{max,1}$) in which all terms are considered per year. In second case, DISCOs are considered as sole owner of DG units [75]. Both active and reactive powers are included in DNEP and DGs & DESSs are optimally scheduled to escalate profit of DISCO.

4. Income related objectives

The earnings and savings of DISCO's are considered in income related objectives. They are mainly of three types: selling energy, award of subsidies, salvage cost.

4.1 Selling energy

This objective is related to income from electricity sales, which in turn depends upon the tariffs according to type of customers. [14] has managed this objective with the help of geographic information system (GIS) based production management system (PMS) software. [49, 75] more precisely explain this objective by separating out active (P_r) and reactive (Q_r) power sale.

4.2 Award subsidies

DISCO's can be encouraged to use cleaner technologies, and to reduce carbon emission. In lieu of green technologies utilization for power production, DISCO's can be awarded subsidies [49] which will add to their income.

4.3 Salvage cost
[49] has also introduce salvage cost of DG units obtained at the end of planning period. This cost function can also be considered for other equipment.

5. Technical objectives
While doing DNEP, the study reveals that investment part has become more prominent then technical issues. Still without fulfilling technical objectives, DNEP can't be achieved. These are mainly energy losses minimization, voltage stability & profile improvement, reliability improvement, minimization of line loading and optimal power flow (OPF) solution.

5.1 Energy losses minimization
Energy losses minimization is the most utilized technical function (17.5%). Minimization of energy losses is also done by cost reduction method in some reviewed articles.

5.2 *Voltage stability & profile improvement*
Most of the researchers have maintained voltage stability by taking this objective as constraint. A few have taken it as objective function. The second most considered technical objective (12.5%), voltage stability & profile improvement is evaluated mainly by determining voltage stability index (VSI_{WN}) of whole distribution network [80].

5.3 Reliability improvement
Just like energy losses minimization objective, reliability improvement is also done by considering it as financial objective by many researchers. Reliability improvement (7.5%) is measured with the help of reliability index factor which is acknowledged by many names in available text [2, 6, 7, 21, 36, 69]. Most of the researchers have considered reliability as one of the main objective term but, in few reviewed articles, reliability has been considered as a constraint instead of objective function. Lowest $\mathcal{RI}$ factor shows better system reliability.

5.4 Minimize line loading
While doing DNEP with placement of reactive power sources and DG, overloading of few lines in network system occurs due to their incongruous engagement [51]. To determine intensity of line loading, indexing of line loading (LLI_{min}) is formulated w.r.t. its maximum power carrying capacity.

5.5 Optimal power flow (OPF) solution
Many articles encountered which are dedicated to advance and hybrid OPF methods for distribution networks. They are omitted because they are not dedicated to DNEP. Only few (2.5%) are found to be engaged in DNEP with also providing OPF solution [24, 77].

6. Optimal size and location (OSL) objectives
While doing DNEP, it is obvious that new devices, equipment, substations as well as DG and DESS are needed to be installed. Their optimal size and location determination is also important objectives which are taken by many researchers and are described below.

6.1 OSL of DG
After total expansion cost minimization objective function, OSL of DG objective is found to be second most employed objective function (37.5%) which is decided with the help of penetration capacity of DG.

6.2 OSL of substation
This objective function has been detected in few reviewed articles [2, 14, 21] and has been fulfilled with the help of software, product management system (PMS) which operates and provides results on the basis of practical geographical information system (GIS). The size has been decided by present demand and future uncertainty whereas location can be determined mathematically. To distinguish physical

distance and electrical distance between load center and substation locations, a correction factor with statistical sampling of prevailing low voltage feeders is obtained. This correction factor is termed as K.

6.3 OSL of DESS
[76] has decided OSL of DESS on the basis of reliability improvement factor. The reliability indexes are associated with momentary interruptions in case of determining OSL of battery banks. Momentary interruptions have been defined and evaluated with the help of MAIFI (Momentary Average Interruption Frequency Index).

6.4 OSL of switching devices
Similar to determination of OSL of DESS, [76] has determine the OSL of switching devices on the basis of long interruptions of power supply. Long interruptions have been defined and evaluated with the help of SAIDI (System Average Interruption Duration Index).

7. Social and economic objectives

While undertaking DNEP, a few researchers have also consider concerns of customer's profit and reducing planers workload. These two concerns have been taken as maximization of social welfare function objective and load forecasting.

7.1 Maximization of social welfare function
The social welfare function is a real-valued and differentiable term belongs to welfare economics. When applied to power systems, it can be defined as difference of end consumer's benefit and total cost of power produced [77] (both active and reactive power production) of DISCOs.

7.2 Load forecasting
[14] has utilize a decision-making support system software named as geographic information system (GIS) based production management system (PMS) which has been implemented on a practical system. Along with other objective fulfillment, this software also reduces the planner workload by doing load forecasting.

8. Conclusion

The sequence of this review study has considered a wide-ranging review of objective functions to be fulfilled while doing DNEP and found the changing trend of DNEP initially with reinforcement, then with new substations, conductoring, and then with penetration of DG in distribution network converting them into micro grids, then came utilization of DESS and last but not least utilization of fuel cells. Each article has utilized various and limited objectives to solve DNEP problems. The reviewed work shows that for achieving similar intentions of DNEP, different objective functions have been considered by various eminent researchers. A few objectives are there such as reliability improvement, energy losses etc. which have been considered by a few researchers as financial objective functions whereas by few as non-financial objective functions. This confusion should be removed and there inclusion must be done in single type of objective function. With increased penetration of green and proficient supply sources, the objectives of DNEP have moved from low cost solutions to somewhat expansive solutions. This review article aims to address this gap so that complete DNEP can be achieved with perfection. The suggested objective functions need to be enriched and articulated for application to a DNEP with multiple types of objective functions. It is also found that due to different regulations of each country, a few objectives have been taken by very less researchers such as carbon penalty, DCM incentive cost etc. these objective functions are prominent one and must be included by various DISCOs in general practice. It is anticipated that this review will be beneficial to the operators, engineers, manufacturers, scholars involved with DNEP for developing new strategies.

References

[1]. Carvalho P M S, Ferreira L A F M, Lobo F G and Barruncho L M F 2000 Distribution network expansion planning under uncertainty: a hedging algorithm in an evolutionary approach Power Delivery 15(1) 412-16.

[2]. Rosado I J R and Agustín J L B 2001 Reliability and costs optimization for distribution networks expansion using an evolutionary algorithm Power Systems 16(1) 111-18.
[3]. Asakura T, Genji T, Yura T, Hayashi N, and Fukuyama Y 2003 Long-term distribution network expansion planning by network reconfiguration and generation of construction plans Power Systems 18(3) 1196-204.
[4]. Vaziri M, Tomsovic K and Bose A 2004 A directed graph formulation of the multistage distribution expansion problem Power Delivery 19(3) 1335-41.
[5]. Vaziri M, Tomsovic K and Bose A 2004 Numerical analyses of a directed graph formulation of the multistage distribution expansion problem Power Delivery 19(3) 1348-54.
[6]. Fletcher R H and Strunz K 2007 Optimal distribution system horizon planning-part I: formulation Power Systems 22(2) 791-99.
[7]. Fletcher R H and Strunz K 2007 Optimal distribution system horizon planning-part II: application Power Systems 22(2) 862-70.
[8]. Carrano E G, Guimaraes F G, Takahashi R H C, Neto O M and Campelo F 2007 Electric distribution network expansion under load-evolution uncertainty using an immune system inspired algorithm Power Systems 22(2) 851-61.
[9]. Carrano E G, Cardoso R T N, Takahashi R H C, Fonseca C M and Neto O M 2008 Power distribution network expansion scheduling using dynamic programming genetic algorithm IET Gen., Trans. & Dist. 2(3) 444-55.
[10]. Haffner S, Pereira L F A, Pereira L A and Barreto L S 2008 Multistage model for distribution expansion planning with distributed generation-part I: problem formulation Power Delivery 23(2) 915-23.
[11]. Haffner S, Pereira L F A, Pereira L A and Barreto L S 2008 Multistage model for distribution expansion planning with distributed generation-part II: numerical results Power Delivery 23(2) 924-29.
[12]. Cossi A M, Romero R and Mantovani J R S 2009 Planning and projects of secondary electric power distribution systems Power Systems 24(3) 1599-608.
[13]. Mori H and Yoshida T 2009 An efficient multi-objective memetic algorithm for uncertainties in distribution network expansion planning IEEE Power & Energy Society General Meeting, Calgary, AB pp 1-6.
[14]. Luo F et al. 2009 A practical GIS-based decision-making support system for urban distribution network expansion planning International Conference on Sustainable Power Generation and Supply, Nanjing 1-6.
[15]. Najafi S, Hosseinian S H, Abedi M, Vahidnia A and Abachezadeh S 2009 A Framework for optimal planning in large distribution networks Power Systems 24(2) 1019-28.
[16]. Ouyang W, Cheng H, Zhang X, Yao L and Bazargan M 2010 Distribution network planning considering distributed generation by genetic algorithm combined with graph theory Electric Power Comp. and Systems 38(3) 325-39.
[17]. Popovic Z N and Popovic D S 2010 Graph theory based formulation of multi-period distribution expansion problems Elec. Power Sys. Research 80 1256-66.
[18]. Lavorato M, Rider M J, Garcia A V and Romero R 2010 Constructive heuristic algorithm for distribution system planning Power Systems 25(3) 1734-42.
[19]. Wang D T C, Ochoa L F and Harrison G P 2011 Modified GA and data envelopment analysis for multistage distribution network expansion planning under uncertainty Power Systems 26(2) 897-904.
[20]. Hsieh S C 2011 Bi-objective planning for distributed generations in distribution systems using Particle Swarm Optimization and Compromise Programming 4th Int. Conf. on Electric Utility Deregulation and Restruct. and Power Technologies (DRPT), Weihai, Shandong pp 168-173.
[21]. Sahoo N C, Ganguly S and Das D 2011 Simple heuristics-based selection of guides for multi-objective PSO with an application to electrical distribution system planning Engineering Applications of Artificial Intelligence 24 567-85.
[22]. Falaghi H, Singh C, Haghifam M R, Ramezani M 2011 DG integrated multistage distribution system expansion planning Elect. Power and Energy Sys. 33 1489-97.
[23]. Martins V F and Borges C L T 2011 Active distribution network integrated planning incorporating distributed generation and load response uncertainties Power Systems 26(4) 2164-72.
[24]. Sun H, Nikovski D, Ohno T, Takano T and Kojima Y 2011 A fast and robust load flow method for distribution systems with distributed generations Energy Procedia 12 236-44.
[25]. Soroudi A, Ehsan M and Zareipour H 2011 A practical eco-environmental distribution network planning model including fuel cells and non-renewable distributed energy resources Renewable Energy 36 179-88.
[26]. Borges C L T and Martins V F 2012 Multistage expansion planning for active distribution networks under demand and Distributed Generation uncertainties Elect. Power and Energy Sys. 36 107-16.
[27]. Zou K, Agalgaonkar A P, Muttaqi K M and Perera S 2012 Distribution system planning with incorporating DG reactive capability and system uncertainties Sust. Energy 3(1) 112-23.

[28]. Salehi J and Haghifam M R 2012 Long term distribution network planning considering urbanity uncertainties Elect. Power and Energy Sys. 42 321-33.
[29]. Mendoza J E, López M E, Pena H E and Labra D A 2012 Low voltage distribution optimization: Site, quantity and size of distribution transformers Elec. Power Sys. Research 91 52– 60.
[30]. Cossi A M, Silva L G W D, Lazaro R A R and Mantovani J R S 2012 Primary power distribution systems planning taking into account reliability, operation and expansion costs IET Gen., Tran. & Dist. 6(3) 274-84.
[31]. Tarôco C G, Carrano E G, Takahashi R H C and Neto O M 2012 A faster genetic algorithm for substation location and network design of power distribution systems IEEE Congress on Evolutionary Computation Brisbane QLD pp 1-6.
[32]. Millar R J, Kazemi S, Lehtonen M and Saarijarvi E 2012 Impact of MV connected microgrids on MV distribution planning Smart Grid 3(4) 2100-08.
[33]. Junjie M, Yulong W and Yang L 2012 Size and location of distributed generation in distribution system based on immune algorithm Systems Engineering Procedia 4 124-32.
[34]. Hu Z and Li F 2012 Cost-benefit analyses of active distribution network management, part I: annual benefit analysis Smart Grid 3(3) 1067-74.
[35]. Hu Z and Li F 2012 Cost-benefit analyses of active distribution network management, part II: investment reduction analysis Smart Grid 3(3) 1075-81.
[36]. Gitizadeh M, Vahed A A and Aghaei J 2013 Multistage distribution system expansion planning considering distributed generation using hybrid evolutionary algorithms App. Energy 10 655-66.
[37]. Ganguly S, Sahoo N C and Das D 2013 Multi-objective particle swarm optimization based on fuzzy-Pareto-dominance for possibilistic planning of electrical distribution systems incorporating distributed generation Fuzzy Sets and Systems 213 47-73.
[38]. Ganguly S, Sahoo N C and Das D 2013 Multi-objective planning of electrical distribution systems using dynamic programming Elect. Power and Energy Sys. 46 65-78.
[39]. Samper M E and Vargas A 2013 Investment decisions in distribution networks under uncertainty with distributed generation-part I: model formulation Power Systems 28(3) 2331-40.
[40]. Samper M E and Vargas A 2013 Investment decisions in distribution networks under uncertainty with distributed generation-part II: implementation and results Power Systems 28(3) 2341-51.
[41]. Camargo V, Lavorato M and Romero R 2013 Specialized genetic algorithm to solve the electrical distribution system expansion planning IEEE Power & Energy Society General Meeting, Vancouver BC pp. 1-5.
[42]. Saldarriaga C A, Hincapié R A and Salazar H 2013 A holistic approach for planning natural gas and electricity distribution networks Power Systems 28(4) 4052-63.
[43]. Zonkoly A M E 2013 Multistage expansion planning for distribution networks including unit commitment IET Gen., Tran. & Dist. 7(7) 766-78.
[44]. Jabr R A 2013 Polyhedral formulations and loop elimination constraints for distribution network expansion planning Power Systems 28(2) 1888-97.
[45]. Aghaei J, Muttaqi K M, Azizivahed A and Gitizadeh M 2014 Distribution expansion planning considering reliability and security of energy using modified PSO (Particle Swarm Optimization) algorithm Energy 65 398-411.
[46]. Ravadanegh S N and Roshanagh R G 2014 On optimal multistage electric power distribution networks expansion planning Elect. Power and Energy Sys. 54 487-97.
[47]. Muttaqi K M, Le A D T, Negnevitsky M and Ledwich G 2014 An algebraic approach for determination of DG parameters to support voltage profiles in radial distribution networks IEEE Smart Grid 5(3) 1351-60.
[48]. Franco J F, Rider M J and Romero R 2014 A mixed-integer quadratically-constrained programming model for the distribution system expansion planning Elect. Power and Energy Sys. 62 265-72.
[49]. Sadeghi M and Kalantar M 2014 Multi types DG expansion dynamic planning in distribution system under stochastic conditions using covariance matrix adaptation evolutionary strategy and monte-carlo simulation Energy Conversion and Mgt 87 455-71.
[50]. Gill S, Kockar I and Ault G W 2014 Dynamic optimal power flow for active distribution networks IEEE Trans. on Power Systems 29(1) 121-31.
[51]. Gallano R J C and Nerves A C 2014 Multi-objective optimization of distribution network reconfiguration with capacitor and distributed generator placement TENCON IEEE Region 10 Conference, Bangkok, pp 1-6.
[52]. Heidari S, Firuzabad M F and Kazemi S 2015 Power distribution network expansion planning considering distribution automation Power Systems 30(3) 1261-69.
[53]. Murty V V S N and Kumar A 2015 Optimal placement of DG in radial distribution systems based on new voltage stability index under load growth Elect. Power and Energy Sys. 69 246-56.

[54]. Delgado G M, Contreras J and Arroyo J M 2015 Joint expansion planning of distributed generation and distribution networks Power Systems 30(5) 2579-90.
[55]. Miloca S A, Volpi N M P, Yuan J, Pinto C L S 2015 Expansion planning problem in distribution systems with reliability evaluation: An application in real network using georeferenced database Elect. Power and Energy Sys. 70 9-16.
[56]. Bagheri A, Monsef H and Lesani H 2015 Integrated distribution network expansion planning incorporating distributed generation considering uncertainties, reliability, and operational conditions Elect. Power and Energy Sys. 73 56–70.
[57]. Gonçalves R R, Franco J F and Rider M J 2015 Short-term expansion planning of radial electrical distribution systems using mixed-integer linear programming IET Gen., Tran. & Dist. 9(3) 256-66.
[58]. Xing H, Cheng H, Zhang L, Zhang S and Zhang Y 2015 Second-order cone model for active distribution network expansion planning IEEE Power & Energy Society General Meeting, Denver CO pp 1-5.
[59]. Hemmati R, Hooshmand R A and Taheri N 2015 Distribution network expansion planning and DG placement in the presence of uncertainties Elect. Power and Energy Sys. 73 665-73.
[60]. Ahmadigorji M and Amjady N 2015 Optimal dynamic expansion planning of distribution systems considering non-renewable distributed generation using a new heuristic double-stage optimization solution approach App. Energy 156 655-65.
[61]. Shivaiea M, Ameli M T, Sepasian M S, Weinsier P D and Vahidinasab V 2015 A multistage framework for reliability-based distribution expansion planning considering distributed generations by a self-adaptive global-based harmony search algorithm Trans. Reliability Engg. and System Safety 139 68-81.
[62]. Mohan N, Ananthapadmanabha T and Kulkarni A D 2015 A weighted multiobjective index based optimal distributed generation planning in distribution system Procedia Technol. 21 279-86.
[63]. Saboori H, Hemmati R and Jirdehi M A 2015 Reliability improvement in radial electrical distribution network by optimal planning of energy storage systems Energy 93(2) 2299-312.
[64]. Mazhari S M, Monsef H and Romero R 2016 A Multi-Objective Distribution System Expansion Planning Incorporating Customer Choices on Reliability Power Systems 31(2) 1330-40.
[65]. Tabares A, Franco J F, Lavorato M and Rider M J 2016 Multistage long-term expansion planning of electrical distribution systems considering multiple alternatives Power Systems 31(3) 1900-14.
[66]. Xing H, Cheng H, Zhang Y and Zeng P 2016 Active distribution network expansion planning integrating dispersed energy storage systems IET Gen., Tran. & Dist. 10(3) 638-44.
[67]. Sayed M A E and Alsaffar M A 2016 Two stage methodology for optimal siting and sizing of distributed generation in medium voltage network IEEE Green Tech. Conf. (GreenTech) Kansas City MO pp 118-23.
[68]. Ahmadigorji M and Amjady N 2016 A multiyear DG-incorporated framework for expansion planning of distribution networks using binary chaotic shark smell optimization algorithm Energy 102 199-215.
[69]. Tarôco C G, Takahashi R H C and Carrano E G 2016 Multiobjective planning of power distribution networks with facility location for distributed generation Elec. Power Sys. Research 141 562-71.
[70]. Verma M K, Mukherjee V and Yadav V K 2016 Greenfield distribution network expansion strategy with hierarchical GA and MCDEA under uncertainty Elect. Power and Energy Sys. 79 245-52.
[71]. Ahmadigorji M, Amjady N and Dehghan S 2017 A novel two-stage evolutionary optimization method for multiyear expansion planning of distribution systems in presence of distributed generation Applied Soft Computing 52 1098-115.
[72]. Shen X, Shahidehpour M, Han Y, Zhu S and Zheng J 2017 Expansion planning of active distribution networks with centralized and distributed energy storage systems Sust. Energy 8(1) 126-34.
[73]. Hasanvand S, Nayeripour M, Waffenschmidt E and Abarghouei H F 2017 A new approach to transform an existing distribution network into a set of micro-grids for enhancing reliability and sustainability Applied Soft Computing 52 120-34.
[74]. Humayd A S B and Bhattacharya K 2017 Distribution system planning to accommodate distributed energy resources and PEVs Elec. Power Sys. Research 145 1-11.
[75]. Saboori H and Hemmati R 2017 Maximizing DISCO profit in active distribution networks by optimal planning of energy storage systems and distributed generators Ren. and Sust. Energy Reviews 71 365-72.
[76]. Pombo A V, Pinab J M and Pires V F 2017 Multiobjective formulation of the integration of storage systems within distribution networks for improving reliability Elec. Power Sys. Research 148 87-96.
[77]. Mokryani G, Hu Y F, Papadopoulos P, Niknam T and Aghaei J 2017 Deterministic approach for active distribution networks planning with high penetration of wind and solar power Ren. Energy 113 942-51.
[78]. Hemmati R, Saboori H and Siano P 2017 Coordinated short-term scheduling and long-term expansion planning in microgrids incorporating renewable energy resources and energy storage systems Energy 134 699-708.

IOP Conf. Series: Materials Science and Engineering **594** (2019) 012016 doi:10.1088/1757-899X/594/1/012016

[79]. Bai L, Jiang T, Li F, Chen H, Li X 2018 Distributed energy storage planning in soft open point based active distribution networks incorporating network reconfiguration and DG reactive power capability App. Energy 210 1082-91.

[80]. Li Y, Feng B, Li G, Qi J, Zhao D, MuY 2018 Optimal distributed generation planning in active distribution networks considering integration of energy storage App. Energy 210 1073-81.

Robust throttle control of Hybrid Electric Vehicle

Sandeep Kr. Tripathi[1], Ashish Shrivastava[2], K. C. Jana[1], Amit Agrawal[3], Amit Rai[1]

[1]Department of Electrical Engineering, Indian Institute of Technology Dhanbad, India
[2]Department of Electrical Engineering, Manipal University Jaipur, India
[3]Department of Electrical Engineering, ABES Engineering College, Ghaziabad India
Email: *trisandeep@yahoo.co.in

Abstract. In recent past years, growing carbon emission caused by vehicle and scarcity of oil resources gives attention of all researchers to investigate low pollution, improved fuel economy type vehicle such as Hybrid Electric Vehicle (HEV). In this paper nonlinear hybrid electric vehicle speed control using different control techniques has been employed to improve the efficiency of vehicle by throttle position control. The controller like Proportional-Integral-Derivative (PID), Artificial intelligence based fuzzy controller, robust controller (H∞) and their comparative performance analysis has been done on the basis of rise time, settling time, and zero steady error in MATLAB-Simulink.
Keywords—Hybrid Electric Vehicle (HEV), H∞, Fuzzy Logic, Robust Control

1. Introduction

In recent decade our environment, economy and various oil resources are dwindling at an alarming rate. Due to heavily use of fuel in automobile sector, the severe carbon monoxide, hydrocarbon and nitrogen oxide stimulated in the environment and result in worsening air quality and global warming [1-2]. This type of problem attract researcher to investigate and found the solution from the all-petroleum vehicle. The optimized vehicle system such as Hybrid vehicles. The main and the foremost reason behind all this is the use of classical Internal Combustion Engine Vehicles (ICEVs). The exhaust emission system used in these classical combustion engines releases poisonous and hazardous air pollutants. These ICEVs are not only playing a major role in urban pollution but is also leading to greenhouse effect which in turn results in global warming [3]. Other issues related to such type of internal combustion are that neither it is economically stable nor it has efficient energy conversion mechanism and thus making electric vehicles ubiquitous. EV has some advantages over ICEV in terms of wide speed range, high efficiency and torque generation which are also EVs special requirement. Moreover, the power source used in batteries of EVs can be easily produced via non-conventional source of energy which is ecofriendly. [4-6] Contrary to the internal combustion emission there is hardly any emission of hazardous gases through EVs thus eradicating the pollution problem [7].

Based on the special requirements of EVs researchers have proposed many technologies. basically, these technologies are divided into 1) Conventional controller such as PID controller [2,7] 2) Artificial controller like Fuzzy controller, neural, GA controller etc and[8-9] 3) Robust technique[7-9]. Although some of the conventional controllers are insufficient in providing steady state speed but many new techniques have evolved to vague unknown disturbances [2]. As discussed earlier that the definition for a good controller is to accomplish effective resistance towards disturbance further utilizing low energy and giving high performance. The above mentioned point enlightens us to the description of control problem of H infinity theory one of the technique which we will use

[10-11]. H-infinity theory has been into existence since early 80s having a wide scope in motor control, LED lighting [12-14],. Among these new controllers one such controller is Fuzzy Logic Controller which also helps in eradicating such problems. The main advantage of using fuzzy logic is that it highly efficient in managing inexact information i.e., it extract an algorithm from the uncertain knowledge provided. Like the other controllers Fuzzy controller lacked behind in providing features of adaptive controller. The next feedback controller introduced was LQR which came with stupendous features which helped in improving dynamic and steady state characteristic [1]. The basic fundamental rule to control the speed factor in HEVs is to control the servo motor which in turn will help in controlling the throttle position thus resulting in good torque– speed characteristics and efficiency [15-17].

This paper involves the systematic observation of pros and cons of various controllers suggested thus bringing into notice the superior among all of them.

2. Mathematical Analysis

Given below is the diagram of the throttle control. This diagram also shows the various controllers which is controlling a DC servo motor [1].

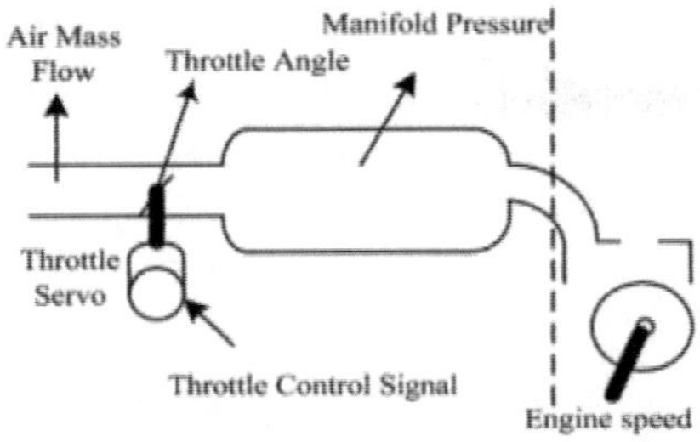

Fig. 1. Schematic diagram of the electronic throttle control.

The equations used for the derivation of the transfer function are given as:

$$m\frac{dv}{dt}F_e(\theta)-av^2-F_g \tag{1}$$

$$\tau_e\frac{dF_e(\theta)}{dt}=-F_e(\theta)+F_{el}(\theta) \tag{2}$$

$$F_{el}(\theta)=-F_1+\gamma\sqrt{\theta} \tag{3}$$

Where, F_e =Engine force, a throttle position function.

F_g =Gravity induced force, a road slop function.

θ = Position of throttle

v= Speed of Vehicle

The values of the various constant used in above equation are described below in the table which will help in further evaluation of the transfer function [1].

Table 1 Numeric Value of Parameters

System Parameter	Symbol	Value(SI Unit)
Mass of the Vehicle	m	1000 kg
Drag coefficient	α	4 N/(m/s)2
Coefficient of engine force	γ	12500N
Idle engine force	Fi	6400 N
Time constant of engine	τ_e	0.2 second

Thus after implementing the values in above three equations we get the state variable equation which is as follows:

$$A=\begin{bmatrix}0 & 0.001\\ 0 & -5\end{bmatrix} \qquad B=\begin{bmatrix}0\\ 8.29*10^8\end{bmatrix}$$

$$C=\begin{bmatrix}1 & 0\end{bmatrix} \quad D=0$$

The values of the various constant used in above equation are described below in the table which will help in further evaluation of the transfer function.[1]

Also while assuming the numerical values of the different constant some of the following assumptions have been made:

(1) Here the Fg i.e., the induced force is considered to be 30% of the weight of the vehicle.
(2) The range between which the engine constant lies is 0.1 to 1 but we assume it to be 0.2 sec.

By using equation (1), (2) and (3) and putting the parameter from table the transfer function obtained is [1]:

$$\frac{V(s)}{\theta(s)}=\frac{8.29*10^5}{s^2+5s} \tag{4}$$

3. PID Controller

The proportional Integral derivative (PID) controller is a classical feedback control mechanism. It adjusts the control variable which further determines the reaction to present, past & possible future values of error, processes it and ultimately minimizes it. Equation (5) shows the PID controller mathematical equation. The weighted sum of three parameters can be calculated as:

$$u(t)=K_p e(t)+K_i\int e(t)dt+K_d\frac{de(t)}{dt} \tag{5}$$

Where Kp, Ki & Kd are proportional, integral and derivative coefficients of controller parameters respectively. Since in the conventional PID controller, error coefficient tuning by Ziegler Nichols (ZN) is not applicable for all time-varying nonlinear system. Hence, a trail and error method is apply for error control. [4,7]

4. Fuzzy Logic Controller

In complex system structure such as HEV where accurate mathematical model formulation is not easy or it's too difficult for system model analysis. In such cases PID controller is not a feasible solution for error control. The system in which uncertainty and nonlinearity found fuzzy controller is more suitable. The fuzzy controller, introduced by Prof. Lotfi Zadeh in 1965, based on fuzzy set theory. The fuzzy set theory which is a knowledge based where input variable taken as a crisp values and convert in to linguistic variable. Figure (2) shows the basic fuzzy logic controller (FLC) structure. This fuzzy control structure consist of fuzzification unit where input variable convert in to fuzzy set. A fuzzy logic inference unit which convert crisp fuzzy value to a linguistic variable. A rule-base, and a defuzzification unit for converting inferred fuzzy control action to a crisp control action.

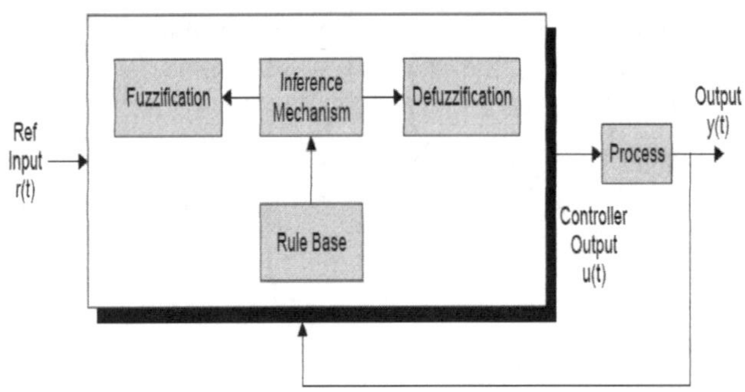

Figure 2: Fundamental fuzzy logic controller (FLC) structure

For Non-linear nature system, it is difficult to find accurate gain value for tuning of fuzzy logic controller (FLC) shown in Figure (2). In this paper mamdani type FLC having two-input error (e) and error dot (ce), divided into seven states and one-output (u) has been taken.

Since the fuzzy controller performance completely depends upon membership function, fuzzification and adopted rule base. So, for this FLC design 7 input and 7 output taken which gives 49 fuzzy if then rule given in table no.2.

In fuzzy controller rule designing, the linguistic variable BN assign as "Big Negative", MN "Medium Negative" SN "Small Negative", EZ "Error Zero", SP "small Positive" MP "Medium Positive", BP "Big Positive".

Table 2 Rule Base Table

Error Dot → / Error ↓	**BN**	**MN**	**SN**	**EZ**	**SP**	**MP**	**BP**
BN	BN	BN	BN	BN	BN	MN	EZ
MN	BN	BN	BN	BN	MN	EZ	MP
SN	BN	BN	BN	MN	EZ	MP	BP
EZ	BN	BN	BN	EZ	MP	BP	BP
SP	BN	MN	EZ	MP	BP	BP	BP
MP	MN	EZ	MP	BP	BP	BP	BP
BP	EZ	MP	BP	BP	BP	BP	BP

5. Robust Controller (H∞)

When the system is linear, its stability analysis is easy to check but for nonlinear system due to presence of noise, parameter uncertainty or disturbance the classical control law difficult to apply. In order to solve such type of problem robust control technique (H∞) is a better way as compared to its analogue counterpart.

The general open-loop interconnection of system for H∞ as shown in (Fig.3)

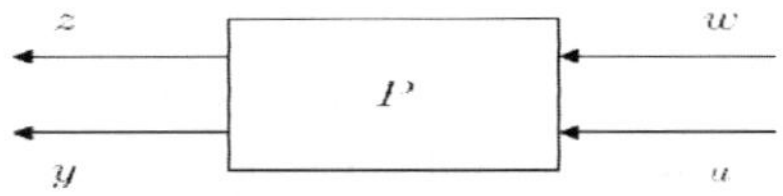

Figure 3 .General Open-loop interconnection

And the generalized state equation of this system is

$$\begin{pmatrix} z \\ y \end{pmatrix} = P\begin{pmatrix} w \\ u \end{pmatrix} = \begin{pmatrix} P_{11} & P_{12} \\ P_{21} & P_{22} \end{pmatrix}\begin{pmatrix} w \\ u \end{pmatrix} \quad (6)$$

Where w is disturbance, z is controlled variable (z=e=Vref-Vo). u is the controller output signal. y is measurement output signal that enter in the controller.

The system P interconnected in such a way with all the subsystem from multiple output multiple inputs (MIMO) to single output single input (SISO) system. The system resultant will give matrix as under:

$$P = \begin{bmatrix} A & B_1 & B_2 \\ C_1 & D_{11} & D_{12} \\ C_2 & D_{21} & D_{22} \end{bmatrix} \quad (7)$$

IOP Conf. Series: Materials Science and Engineering **594** (2019) 012017 doi:10.1088/1757-899X/594/1/012017

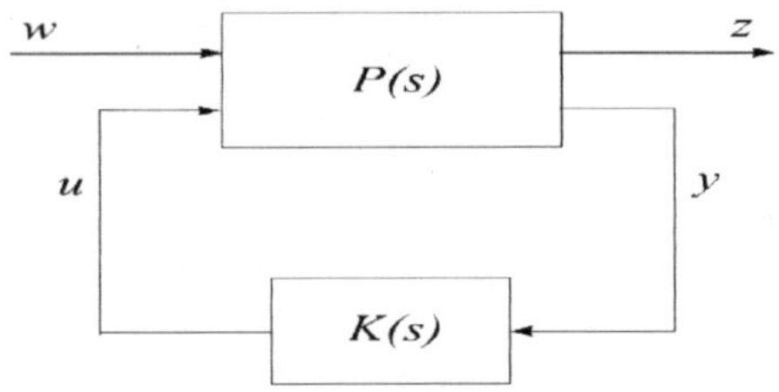

Figure 4 Robust controller closed-loop connection

5.1 H∞ CONTROL METHOD:

The general H∞ controller formulation with system interconnection from equation (6) and (7) gives a robust closed loop connection as shown in Fig.(4) will gives w to z is Fi(P,K)= S(P,K)

Where $S(P,K) = P_{11} + P_{12}K(I - P_{22}\mathrm{K})^{-1}P_{21}$

To find H∞ controller that minimize transfer function from w to z is mathematically

$$\|F_i(P,K)\|_\infty = \max_{\omega(t)=0} \frac{\|z(t)\|_2}{\|w(t)\|_2} \quad (8)$$

Where,

$$\|z(t)\|_2 = \sqrt{\int_0^\infty \sum \|z_i(t)\|^2 dt} \quad (9)$$

Since robust and optimal solutions are too difficult to achieve, therefore to obtain controller value K which satisfies

$$\|F_i(P,K)\|_\infty < \gamma \quad (10)$$

Where γ > γmin and γmin should be predefined. The above process for finding controller value K through iteration which satisfy inequality constraint for shaping singular values over specific frequency to specified transfer function.

5.2 MIXED SENSITIVITY APPROACH (H∞ PROBLEM):

Using several combinations of closed loop system, the primary aim is to find controller with mixed sensitivity (MS) approach. In the mixed sensitivity approach controller output gives desired output with sensitive function $S = (1+GK)^{-1}$ with complementary sensitivity function $T = GK(GK)^{-1}$ respectively.

It is difficult to find balance between S and T function due to confliction of nature the general approach to design controller is to minimize its weight factor.

$$\left\| \begin{matrix} Sw_P \\ Tw_T \\ KSw_U \end{matrix} \right\|_\infty$$

where w_p is weight of error due to input reference, w_T is the noise weight of and w_u is control weight of the system. All the weight function, controller, and system interconnection with plant is shown in Fig.(5). In Fig. (5) the w_1 and w_2 are weight function respectively w_p and w_u.

Hence, with association of weight function the plant P is defined as:

$$P = \begin{bmatrix} W_P & -W_P G \\ 0 & W_U \\ I & -G \end{bmatrix} \quad (11)$$

IOP Conf. Series: Materials Science and Engineering **594** (2019) 012017 doi:10.1088/1757-899X/594/1/012017

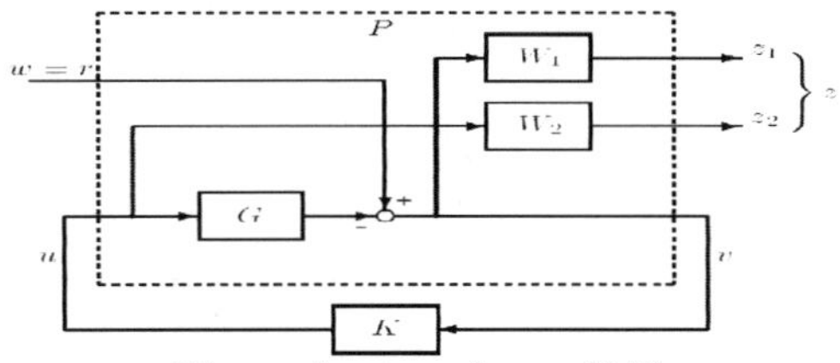

Figure 5 S/KS form of MS

The main property of internally stable controller is to follow the following condition. For nominal performance and stability the S over KS should satisfy following inequalities.

$$\left\| \begin{matrix} W_P S(G_m) \\ W_U KS(G_m) \end{matrix} \right\|_\infty < 1 \qquad (12)$$

Where, $S(G_m) = (I + G_m K)^{-1}$ nominal system sensitivity function
and for robust performance,

$$\left\| \begin{matrix} W_P (I + GK)^{-1} \\ W_U K(I + GK)^{-1} \end{matrix} \right\|_\infty < 1 \qquad (13)$$

6. Simulation and Result

In this section, the time domain stability analysis is carried out. Moreover, it also shows the different controller technique applied on system and their output responses. Fig.6 shows the open loop response of HEV system. From figure 6 shows the system's instability as it is not converging.

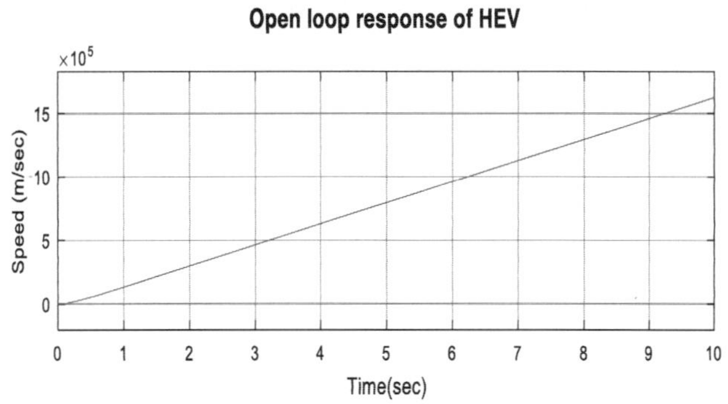

Figure 6. Open loop response of HEV

The PID controlled response of HEV is shown in figure.7. From bellow figure the maximum overshoot of controlled system is 30%.

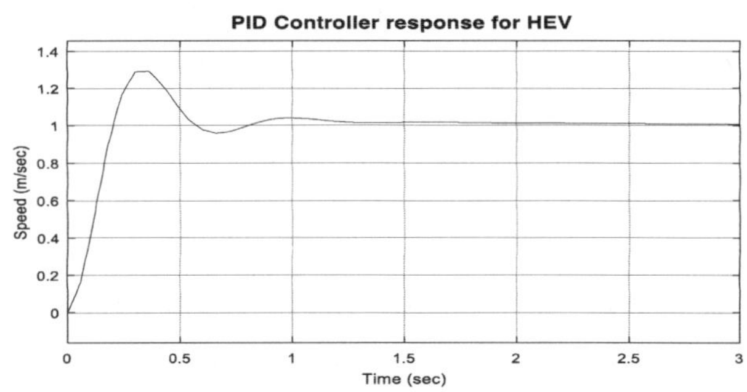

Figure 7 PID controller closed loop response of HEV

From above PID controlled output its clear that conventional controller makes system stable under certain condition. Fig.8 and 9 shows the system response using fuzzy logic controller with and

IOP Conf. Series: Materials Science and Engineering **594** (2019) 012017 doi:10.1088/1757-899X/594/1/012017

without external disturbance respectively. In Fig.8 the settling time of output response is 1 sec while the max. overshoot become zero.

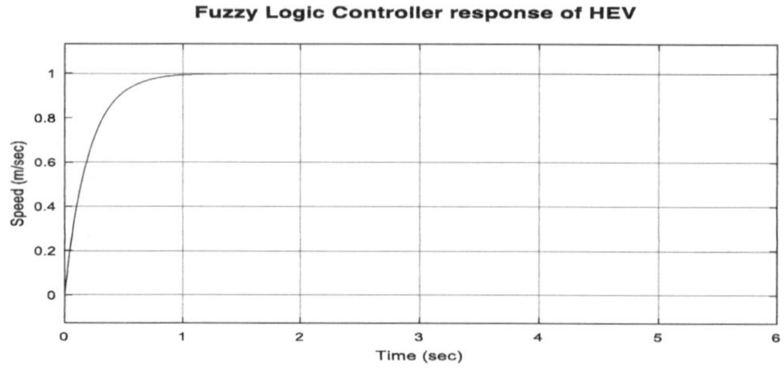

Figure 8. FLC closed loop response of HEV

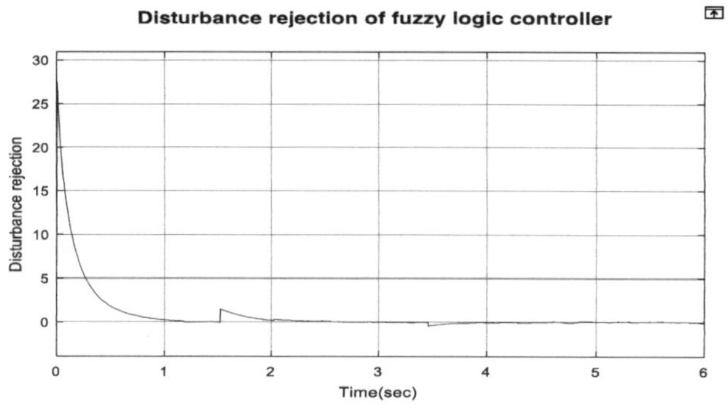

Figure 9. Disturbance rejection response of FLC

Figure (10) and (11) shows the system response with H∞ controller with and without external disturbance applied on the system.

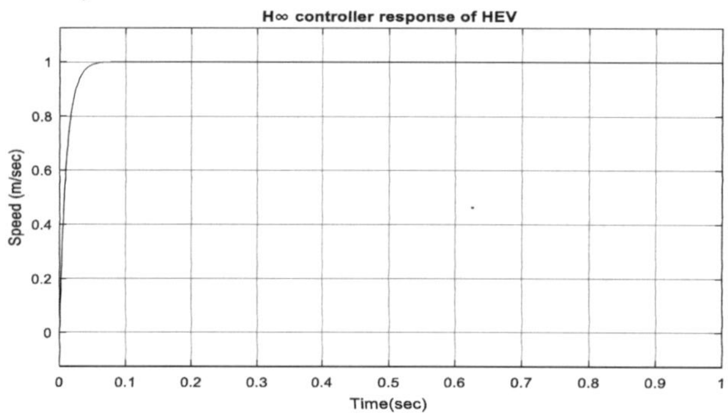

Figure 10. H infinity controller response of HEV

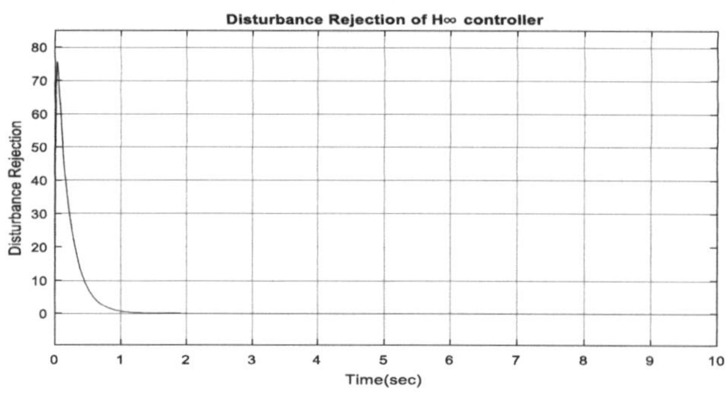

Figure 11 . Disturbance rejection response of H∞ controller

Figure (12) is a comparative performance of PID, Fuzzy and H∞ controller. From Fig.12 the response of robust H∞ controller gives better then the fuzzy and PID in term of rise time maximum overshoot, steady state error which is shown in table 3.

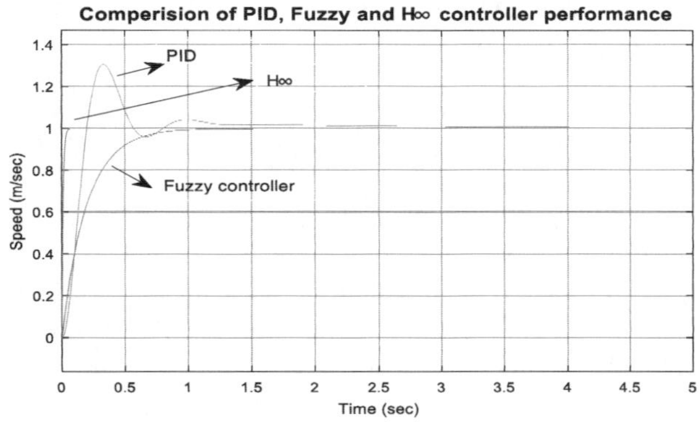

Figure.12 Comparison of PID, Fuzzy and H∞ controllers

Table3. Performance comparison table of various Controllers

Types of controller	**Controller Performance Index**			
	rise time (Sec.)	**Overshoot Mp (%)**	**settling time**	**steady state error (Ess)**
PID	1.3	3.4	1.7	1-2%
FLC	-	0	1	0
H-∞	-	0	0.07	0

REFERENCES

[1] Yadav, A.K., Gaur, P., Jha, S.K., Gupta, J.R.P.,Mittal, A.P. Optimal speed control of hybrid electric vehicles. J. Low Power Electron. 11(4), 393–400 (2011)

[2] Vasak, M., Baotic, M., Petrovic, I., Peric, N.: Hybrid theory-based time-optimal control of an electronic throttle. IEEE Trans. Ind. Electron. 54(3) (2007)

[3] Ali Emadi, Young Joo Lee, and Kaushik Rajashekara, "Power Electronics and Motor Drives in Electric, Hybrid Electric, and Plug-In Hybrid Electric Vehicles", IEEE Transactions on Industrial Electronics, Vol. 55, No. 6, pp. 2237-2245, June. 2008

[4] S. K. Jha, A. K. Yadav, Prerna Gaur, H. Parthsarathy, and J.R.P. Gupta, "Robust and Optimal Control Analysis of Sun Seeker System," Journal of Control Engineering and Applied Informatics, vol.16, no.1, pp. 70-79, 2014

[5] Rastogi, S., S. K. Jha, and R. Oberoi, "Robust Stability Analysis of Sun Seeker System byArguon's Theorem," in Proc. of Fifth International Conference on Advanced Computing and Communication Technologies (ACCT),Haryana, India, Feb. 2015.

[6] Anil Kr.Yadav, Dr.Prerna Gaur, Sandeep Tripathi, "Design and Control of an Intelligent Electronic throttle Control System," in Proc. IEEE international Conference on Energy Economics and Environment (ICEEE) , Noida, India,Mar. 2015,pp.1-5

[7] Tripathi S.K., Panday H, Gaur P, "Robust control of Inverted Pendulum using fuzzy logic controller, " in Proc. of IEEE Students Conference on Engineering and Systems (SCES), Allahabad,India, Apr. 2013, pp 1-6.

[8] J. Lam, "Control of an Inverted Pendulum", University of California, Santa Barbara, 10 June 2004

[9] Yanmei Liu and Zhen Chen, Dingy Xue, Xinhe Xu "Real-Time Controlling of Inverted

IOP Conf. Series: Materials Science and Engineering **594** (2019) 012017 doi:10.1088/1757-899X/594/1/012017

Pendulum by Fuzzy" Proceedings of the IEEE International Conference on Automation and Logistics Shenyang, China August 2009.
[10] Ricardo S. Sanchez-Pena and Mario Sznaier, Robust systems theory and applications, John Wiley & Sons, 1998.
[11] Amin S. Meghani, Haniph A. Latchman. "$H\infty$ Vs. Classical Methods In The Design of Feedback Control Systems" Department of Electrical Engineering University of Florida, Gainesville January 16. 1992
[12] Zhou K. and Doyle J.,(1998), Essentials of Robust Control, Prentice Hall, New Jersey
[13] Amit Agrawal, Ashish Shrivastava, Kartick C. Jana, "Uniform Model and Analysis of PWM DC-DC Converter for Discontinuous Conduction Mode", IETE journal of Research, Vol. 64, issue-4, 2018, pp. 569-581.
[14] Ashish Shrivastava and Bhim Singh, "Power factor-corrected DCM-based electronic ballast," Electrical Engineering, vol. 95, no. 4, pp. 403-411, Dec 2013.
[15] D.-W. Gu, P. Hr. Petkov and M. M. Konstantino Robust Control Design with MATLAB®.
[16] Tsui Chia-Chi, "Robust Control System Design," Marcel Dekker,Inc ., 1997.
[17] Robust control design by Skogestad, S., and Postlethwaite, I., 1996, Multivariable Feedback Control: Analysis and Design , John Wiley and Sons , New York

IOP Conf. Series: Materials Science and Engineering **594** (2019) 012018 doi:10.1088/1757-899X/594/1/012018

Speed Control of DC Motor Using Chopper Based on Fuzzy Logic

Arunesh kumar Singh[1], Abhinav Saxena[1], Parveen Poon Terang[2], Palak Tulsyan[2] and M Waris[2]

[1]Department of Electrical Engineering , Jamia Millia Islamia, India
[2]Department of Electrical Engineering , JSSATE Noida, India

Email: abhinaviitroorkee@gmail.com

Abstract- This paper present the speed control of DC Motor based on boost converter under the controlling of fuzzy logic. The DC motor has armature and field winding in which armature winding is controlled by using boost converter which is triggered by fuzzy logic controller which consist of 7x7 rule and the input are change in error and rate of change of error and output is firing angle. The fuzzy logic controller gives smooth, reliable, efficient speed controlling of DC motor. Further the compative analysis between fuzzy logy and PID controller is assessed. The complete system is designed with SIMULINK/MATLAB for perfect analysis between two methods mentioned.

Keywords: fuzzy logic, DC Motor, PID,speed

1. Introduction

DC motor is commercially used for speed control at industrial and residential purposes. There were 2 methods used for speed control mainly field flux control and armature voltage control. Involvement of Field flux for assessing the performance of DC motor is at the above the base speed and armature voltage is kept constant in this process[1]. The armature voltage control method is used for the speed less than base speed keeping the flux constant, still DC motor is preferred for speed control over AC machines because speed and torque can be controlled independently. In this paper DC motor armature winding is controlled by the boost converter in which switching of IGBT switch is done by fuzzy logic controller.
The boost converter is nothing just step up chopper. DC motor has wide application in different home appliances for medium speed controller[5]. Here the complete is analyzed in d-q theory to avoid the decoupling theory. The chopper converts fixed dc input to variable dc output and the output voltage is controlled by firing angle. The switching of IGBT is analyzed by fuzzy set of Rule(3x3) which consist of change in the error and rate of change in the error and the output is the firing angle which triggered the IGBT switches[9]. The error produced is difference between reference value and measured value. The measured value is the d and q component of the current which is obtained by the passing the armature currents through the PID controller[12]. This measured value is further compared to reference value and error is produced , change in the error is produced by the delaying in the input signal by unity.
Average Voltage of the step up chopper

$$V_o = (T/T_{off})V_s \tag{1}$$
$$= (T/T\text{-}T_{on})V_s$$

$$V_o = \frac{V_s}{1-\alpha} \tag{2}$$

$\alpha = T_{on}/T.$=firing angle

The complete system of DC motor is incorporate with the fuzzy logic controller because traditionally speed control was analysed in conventional method like PID controller which was giving non satisfactory results after implementation of fuzzy logic controller system performance parameter like torque, speed gives better performance in comparison to conventional method.

2. Separately excited DC Motor

Machine has field and armature winding and field winding excited separately as shown in Fig.1. While in this case armature winding[21] is excited by using fuzzy logic logical through step up chopper as shown in Fig.2

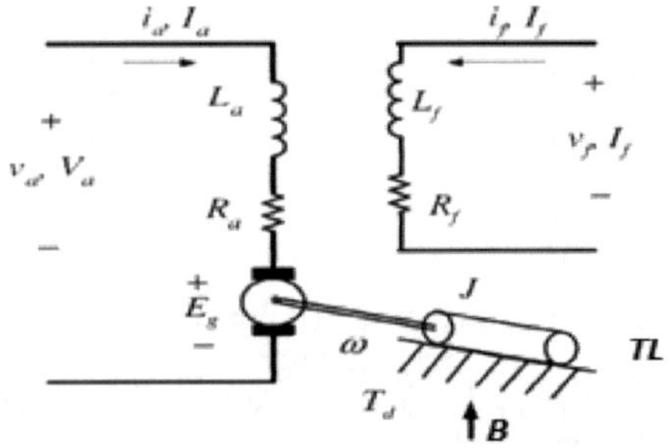

Fig.1 Separately Excited DC Motor

DC motor has equation

$$V_s = E_g + I_a R_a \tag{3}$$

Torque is

$$T_d = J\frac{d\omega}{dt} + B\omega + T_L \tag{4}$$

Excitation induced EMF will be

$$E_g = k\omega \; E_g = k\omega \tag{5}$$

$$T_d = k_1 I_a \tag{6}$$

3. Mathematical modelling of DC Motor using Fuzzy logic:

3.1 Design of PID Controller:

From the equation.3 the output voltage V_s is converted into the d-q component (V_{qs} & V_{ds})which is passing to the PID controller which gives the current Io in the component of d-q as which is I_d and I_q

$$I_{ds} = (K_{p1} + \frac{K_{i1}}{s} + sK_{s1})V_{ds}$$
$$I_{qs} = (K_{p2} + \frac{K_{i2}}{s} + sK_{s2})V_{qs} \tag{7}$$

Further this measured value (I_{ds} & I_{qs}) is compared with the reference value(I^*_{ds} & I^*_{qs}) and gives error E_1 and E_2

$$E_1 = I_{ds} - I^*_{ds} \tag{8}$$

$$E_2 = I_{qs} - I^*_{qs} \tag{9}$$

Further the error is passed through E_1 & E_2.

This error E_1 & E_2 is passing through PID controllers gives

$$V_{do} = (K_{p3} + \frac{K_{i3}}{s} + sK_{s3})E_1$$

$$V_{qo} = (K_{p4} + \frac{K_{i4}}{s} + sK_{s4})E_2 \quad (10)$$

Further the output shown in the Fig.10 is compared with V^*_{do} and V^*_{qo} and output of the above comparison gives the carrier voltage for pulse generator.

3.2 Design of fuzzy logic controller:

Error produced is given by

$$E = E_1 - E_2 \quad (11)$$

Change in error is given by

$$\Delta E = E(t) - E(t-1) \quad (12)$$

Rate of change of error is given by $\frac{d(\Delta E)}{dt}$

ΔE and $\frac{d(\Delta E)}{dt}$ are the change in the error

And the output corresponding is firing angle

ΔE / $\frac{d(\Delta E)}{dt}$	NB	NM	NS	HB	HM	HS	HL
NB	NM	NS	HB	HL	HS	NB	NM
NM	HM	HS	HL	NB	NS	NM	HB
NS	NB	NM	NS	HB	HM	HS	HL
HB	HL	HB	HM	HL	NB	NM	NS
HM	NM	NB	HB	HS	HM	HL	NM
HS	NB	NM	NS	HB	HM	HS	HL
HL	HL	HS	HM	HB	NS	NM	NB

Fig.2 Fuzzy Rules

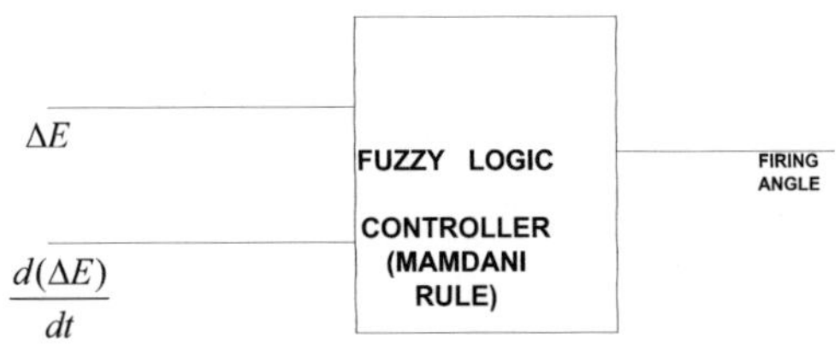

Fig.3 Fuzzy Model

Appendix:

K_{p1}= 12.5, K_{i1}=36.8, K_{s1}= 26.7
K_{p2}= 22.6, K_{i2}=45.98, K_{s2}= 56.9
K_{p3}= 45.6, K_{i3}=98.65, K_{s3}= 69.56
K_{p4}= 47.65, K_{i4}=95.65, K_{s4}= 79.65

2KW DC Motor

IOP Conf. Series: Materials Science and Engineering **594** (2019) 012018 doi:10.1088/1757-899X/594/1/012018

Ra= 0.5ohmRf=220 ohmV= 220 V

Output : Firing angle =0.5,output Voltage is 440 V

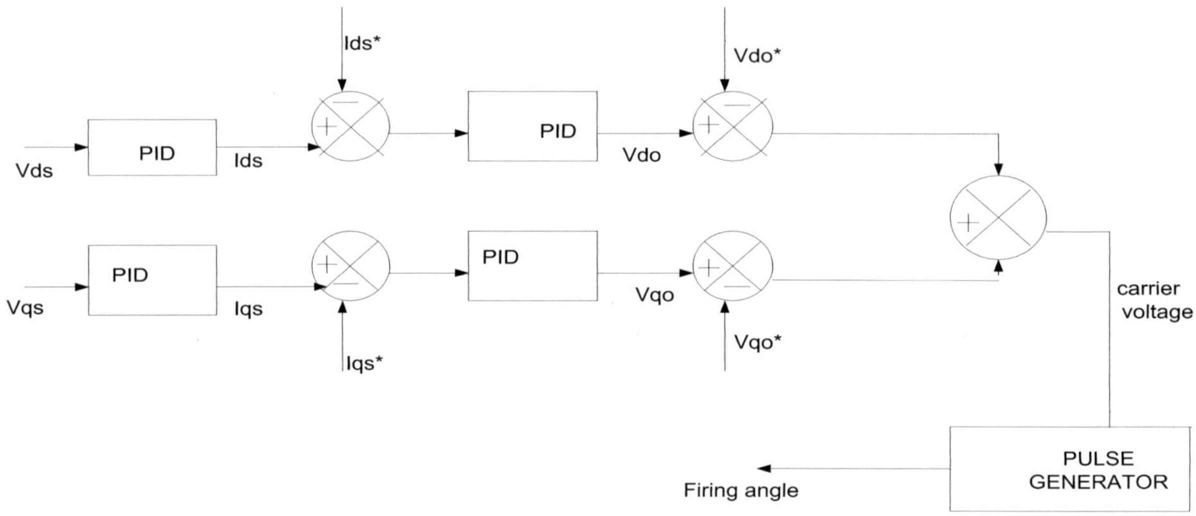

Fig.4 PID Controller structure for speed control of DC Motor

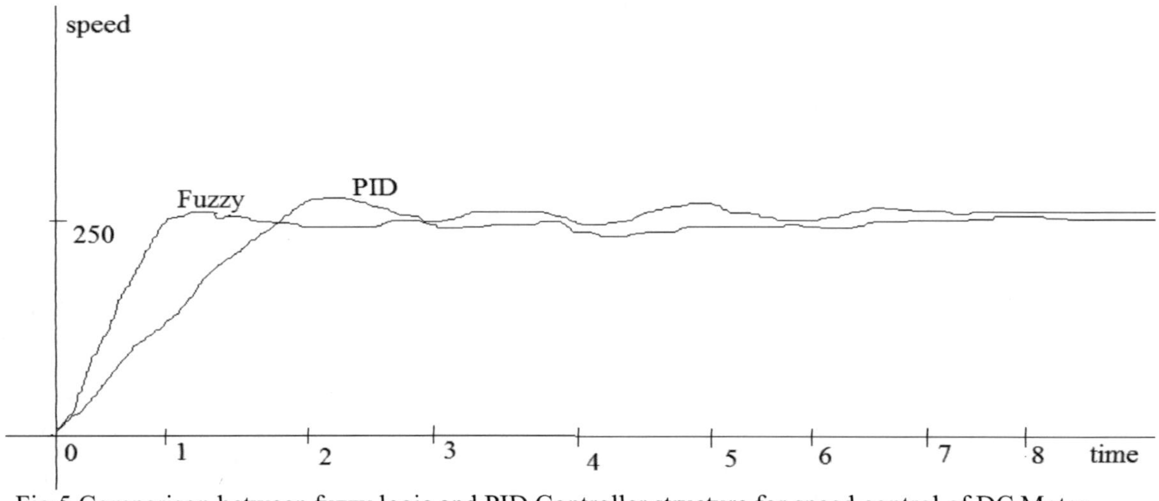

Fig.5 Comparison between fuzzy logic and PID Controller structure for speed control of DC Motor

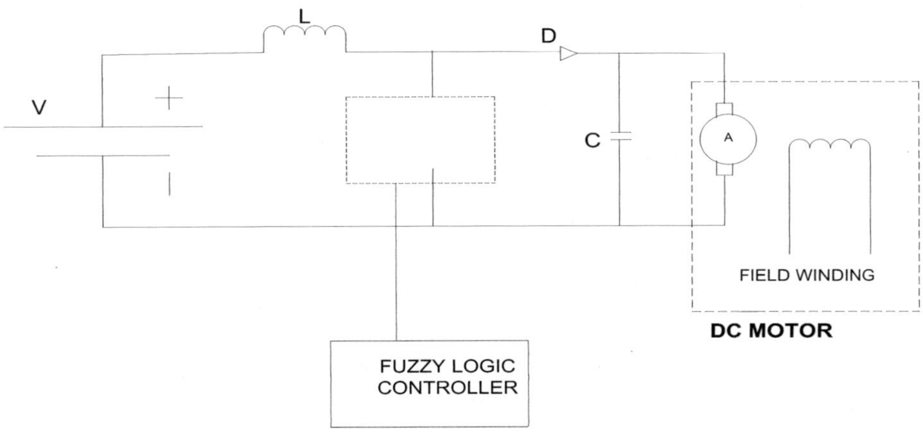

Fig.6.Fuzzy Logic Control of DC Motor

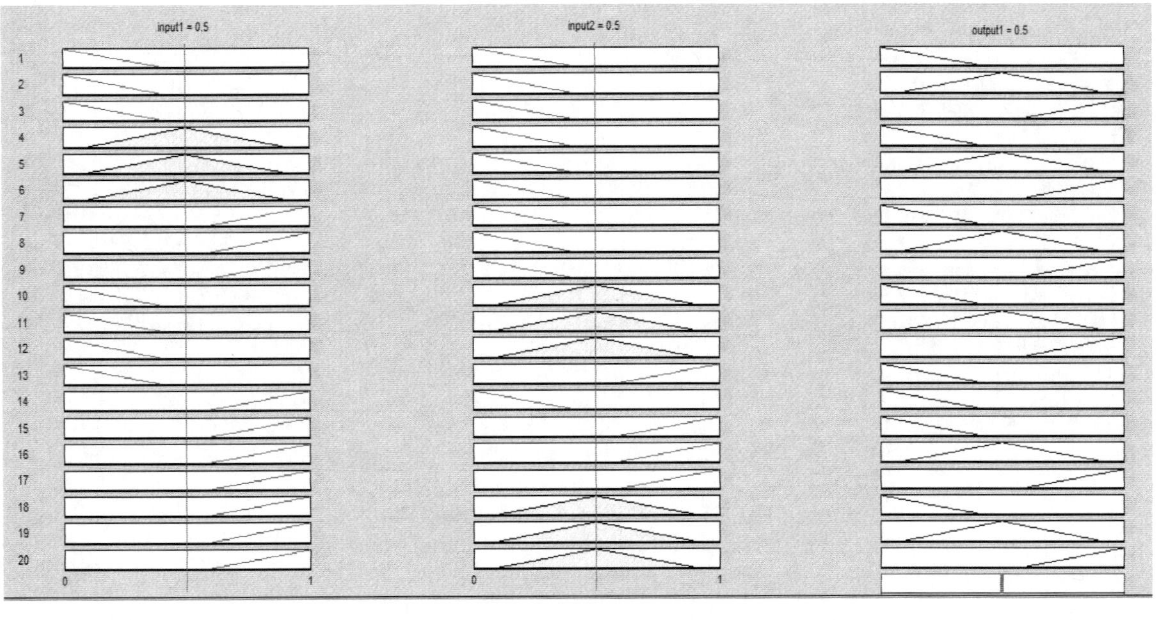

Fig.7.Fuzzy Set Rules of DC Motor

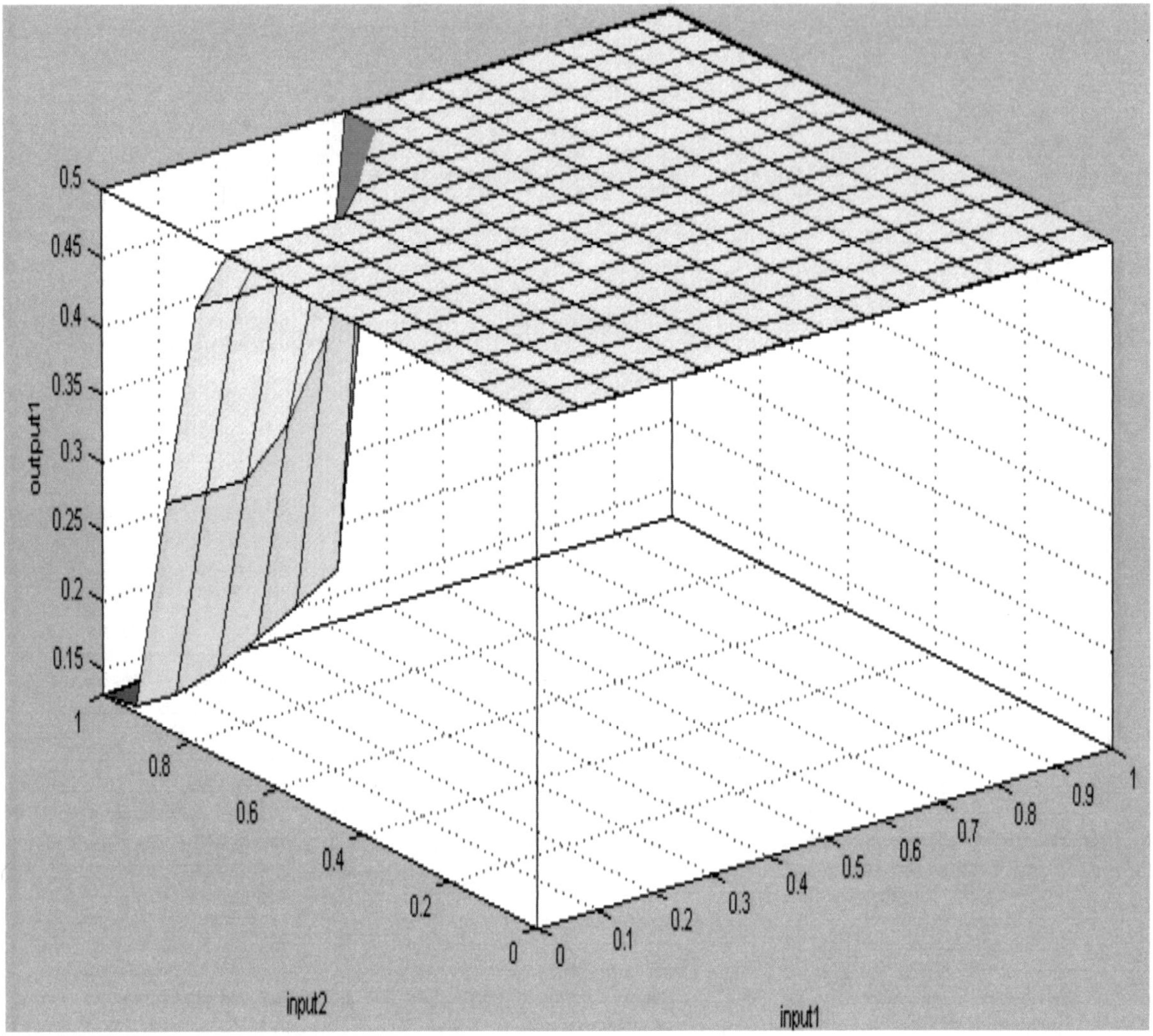

Fig.8 Fuzzy Logic Surface of DC Motor

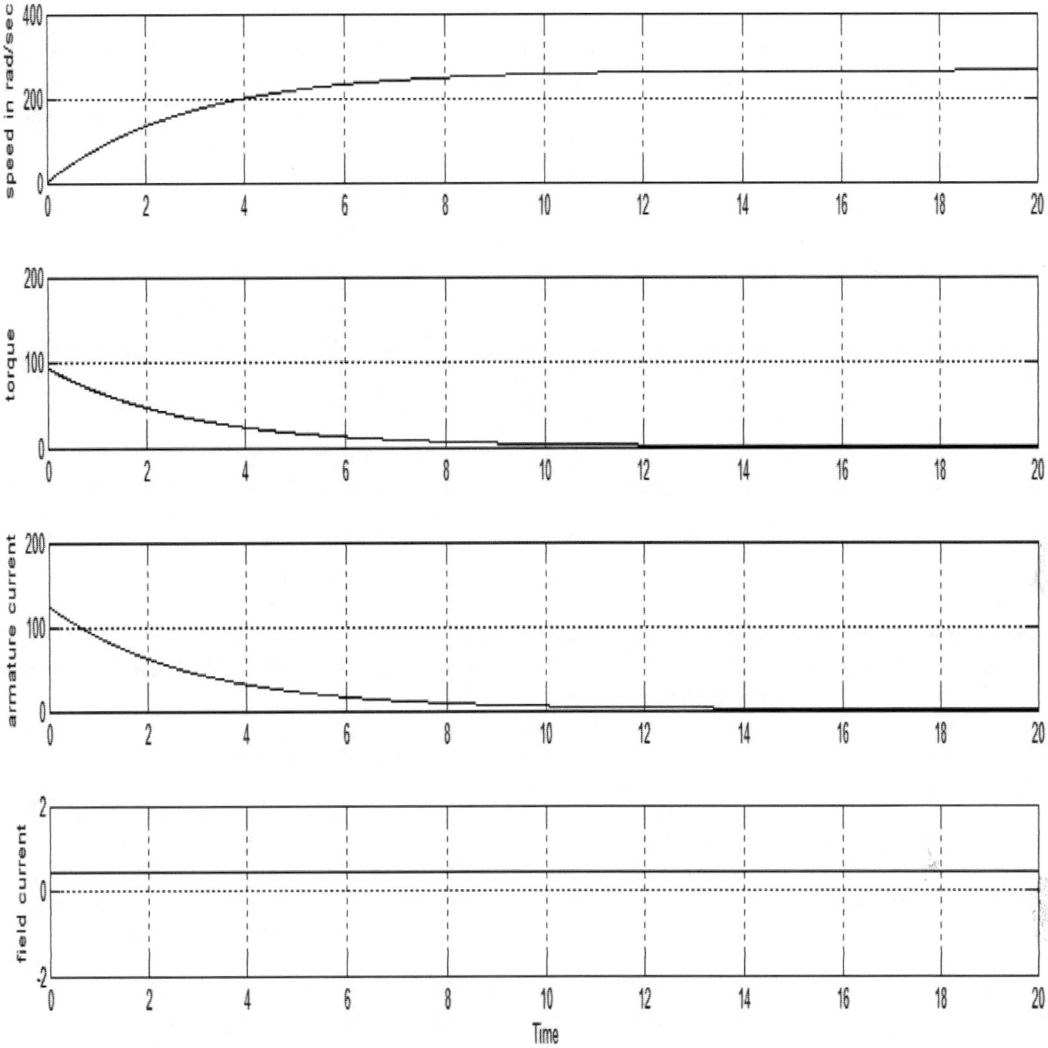

Fig.9. Performance Characteristics of DC Motor

IOP Conf. Series: Materials Science and Engineering **594** (2019) 012018 doi:10.1088/1757-899X/594/1/012018

4. Conclusion:

This paper shows the speed control of DC motor under fuzzy logic and it is found that performance parameter of system shows better transient and steady state performance by using fuzzy logic controller. The complete system using fuzzy logic controller gives satisfactory output in terms of speed and torque. from analysis It can conclude that usage of Fuzzy in system gives better performance in comparison to PID controller because Fuzzy logic controller application allows the system to take less settling time, lesser overshoot for achieving the output.

References:

[1] Gopakumar, K., Power Electronics and Electrical Drives, Video Lectures 1-25, Centre for Electronics and Technology, Indian Institute of Science, Bangalore.2017

[2]S srikrishan,Fuzzy logic controller application in induction motor, Electric power system component,taylor and francis,2017

[2] Bimbhra, P.S., Power Electronics. New Delhi, Khanna Publishers, 2006.

[3] Dubey, G.K., Fundamentals of Electrical Drives. New Delhi, Narosa Publishing House, 2009.

[4] Gopal, M., Control Systems, Principles and Design. New Delhi, Tata McGraw Hill Publishing Company limited, 2008.

[5] Mohan, Ned, Electrical Drives-An Integrated Approach. Minneapolis, MNPERE, 2003.

[6] Ogata, K., Modern Control Engineering.Englewood Cliffs, NJ: Prentice Hall, 2001.

[7] Leonhard, W., Control of Electric Drives.New York, Springer-Verlag, 2001.

[8] Mohan, Ned, Power Electronics, John Wiley and Sons, 1989. [9] Rashid, M.H., Power Electronics, Prentice Hall of India, New Delhi, 1993.

[10] Moleykutty George., Speed Control of Separately Excited DC motor, American Journal of Applied Sciences, 5(3), 227-233, 2008. [11] SIMULINK, Model-based and system-based design using Simulink, Mathsworks, Inc, Natick, MA, 2000.

[12] MATLAB SIMULINK, version 2009, SimPowerSystem, One quadrant chopper DC drive.

[13] Salam Dr. Zainal, UTJMB, Power Electronics and Drives (Version 3-2003).

[14] FINCOR Automation, Adjustable speed Drives Applications. [15] Infineon Technologies, Basic DC motor speed PID control with the Infineon Technologies.

[16] C.U. Ogbuka, Performance characteristics of Controlled separately excited dc motor, Pacific Journal of Science and Technology, 10(1), 67-74.

[17] Singh Brijesh, Prakash Surya, PandeyShekhar Ajay, Sinha S.K., Intelligent PI controller for DC motor, International Journal Of Electronics Engineering Research, Vol.2, pp-(87-100).

[18] Zuo Z. Liu,Frang L. Luo, Rashid M.H., High performance nonlinear MIMO field weakening controller of a separately excited dc motor, Electric Power System research, Vol 55, Issue 3, 2000, pp(157-164).

[19]Chinnaiyan V. Kumar, Jerome Joritha, arpagam, J. S.Sheikh Mohammed, Design and Implementation of High Power DC-DC converter and speed control of dc motor using TMS 320F240DSP, Proceedings of India International Conference on Power Electronics, 2006.

[20] Bose B.K., Power electronics and motor drives recent technology advances, Proceedings of the IEEE International Symposium on Industrial Electronics, IEEE, 2002, pp 22-25.

[21] SaffetAyasun,GultekinKarbeyaz, DC motor speed control methods using MATLAB/SIMULINK and their integration into undergraduate courses, 2008.

[22] AungWaiPhyo, Analysis on Modeling and Simulink of DC motor and its driving system used for wheeled mobile robot, Proceedings of world Academy of Science, Engineering and Technology Vol. 26, 2007.

[23] MATLAB and SIMULINK Version 2009a, the MathsworksInc, USA.

IOP Conf. Series: Materials Science and Engineering **594** (2019) 012019 doi:10.1088/1757-899X/594/1/012019

Comparative Study of Structural and Sheetmetal Combination Bracket on Elevator Rail System

Sumit V. Dubey[1], **Santosh B. Rane**[2]

[1] MTech Scholar, Sardar Patel College of Engineering, Munshi Nagar, Andheri –West, Mumbai -400058
email: sumitvdubey@gmail.com,
[2]Associate Professor, Department of Mechanical Engineering, Sardar Patel College of Engineering,
Munshi Nagar, Andheri –West, Mumbai-400058
email: s_rane@spce.com

Abstract.
Context: - Combination brackets are applied for fixing car and counterweight guide rails on the hoistway wall to deliver smooth and safe travel for side counterweight position.
Purpose: - The main intent of this work study is to compare the deflection and stresses value of the structural and sheet metal Combination bracket arises due to effect of forces imposed on the guide rail by the Car guide shoe.
Methodology: - Finite element method was performed for Structural and Sheet metal combination bracket to evaluate stress and deflection using SolidWorks Simulation software. Analytical calculation has been performed to calculate the forces acting on rail considering the case of safety gear operation as per EN 81-1
Research findings: - From this analysis, it has been found that structural combination brackets are stiffer compare to those made up of sheet metal.
Limitations: - Stress and deflection values shall be varied for different car cabin sizes. A case study of Car Size 1100x2000 mm for 13 passenger capacity rated load has been considered.
Originality/value: - This analysis results provides the possibility of selecting most reliable rail bracket for guide rail fixation.

1. Introduction

The elevator is a vertical transportation equipment that transfers the passenger and goods safely and efficiently between floor of a building. The main components of the elevator system are car cabin and frame, counterweight frame, rail and bracket system, traction machine, controller and landing doors and car doors. Combination bracket is used to fix one rail of the car and both rails of counterweight. It is mounted on the hoistway wall at definite interval regulated by rate load (no. of passenger) and car and counterweight mass. Certain elevator companies are using structural and fabricated type of combination as per their requirement. In this study, a comparison of structural and fabricated combination bracket is analysed. The stresses and deformations occurring on the combination brackets have been determined by using finite element method considering the rail forces calculated in case of safety gear actuation using the rules of EN 81-1.

2. Literature Review

Targit, 2002 reviewed that the force arises due to load acting on the car rail is from safety gear during safety gear actuation. Hayder F. Neamah., 2009 suggested the analytical calculation of the guide rail is the most accepted and justified method for guide rail analysis. Suhan Atay, 2013 analyzed the stresses and deflection of the guide rail, rail brackets and sliding clips. Elmali examined the stress and deflection of the guide rail resulted from distinct loading cases and forces acting on certain specific points between two guide rail fixing brackets.

3. Concept Model: Guide Rail and Combination Bracket

The Elevator system consists of guide rail and brackets to withstand the application of car and counterweight, when stopping the elevator car and its rated load or the balancing weight. These guide rails are attached to rail brackets through mean of guide clips and hardware accessories. Rail brackets are further reinforced on the hoistway wall through dash fasteners to provide adequate support to guide rail, thus forming an essential element of the complete rail system. Combination bracket is applied when the position of the elevator car and counterweight are adjacent and perpendicular to each other at a particular specified point. Two counterweight rail and one car rail are fixed and supported on combination brackets. The Concept model consists of two models as shown below in figure: a) Fabricated Combination bracket. b) Structural Combination bracket.

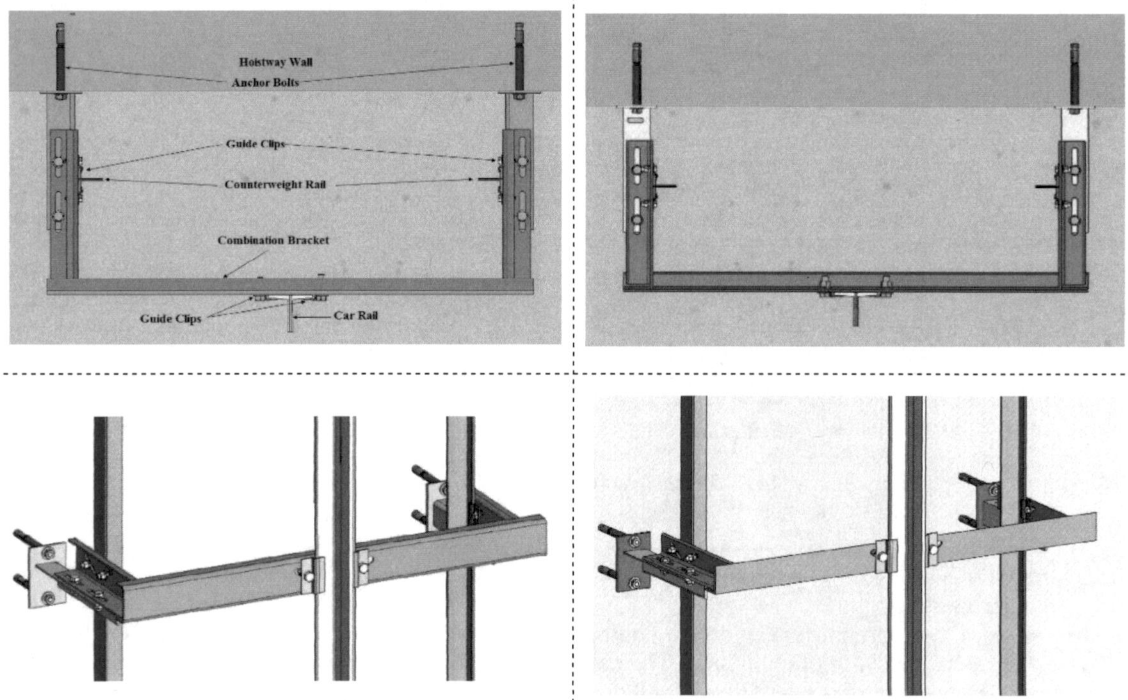

(a) Sheetmetal combination brackets | (b) Structural combination brackets

Figure 1. Guide rail and bracket assembly

4. Guide Rail Calculation

As per EN 81-1, Forces exerted on guide rail by Car are calculated by considering two operating conditions:

a) Normal running conditions.
b) Safety gear actuation conditions.

In our study, we perform the analytical calculation of the forces as per safety gear actuation condition which is a worst-case condition that determine the stiffness the rail. Since the car rail is fixed at the center of combination bracket, the configuration of centrally guide and suspended case of the elevator will be considered. The load is considered to be dispersed to the one-third area of the car cabin. The force obtained in the guide rail calculations are used as a reference in the FEM model. For 13 Pax

IOP Conf. Series: Materials Science and Engineering **594** (2019) 012019 doi:10.1088/1757-899X/594/1/012019

capacity elevator travel height is considered for 20 meters, the stresses and deflection to be used in guide rail will be calculated in accordance with EN81-1. The data given to be used in calculations are given below.

Table 1: Input Parameters for guide rail calculations

Parameters	Abbreviation	Value
Total car empty Load	P	: 1200 Kg
Rated load	Q	: 13x68=884 Kg
Weight at Cwt side:	M_{cwt}	: 1642 Kg
Travel rise:	TH	: 20 m
Car Cabin dimensions:	Dy ×Dx	: 1100×2000 mm
Car Guide Rail:	-	: T82/B
Distance between sliding guide shoes	h	: 3418 mm
Distance between combination brackets	L	: 2500mm
No of Guide rail	n	: 2
Guide rail properties	-	: Shown in figure 2

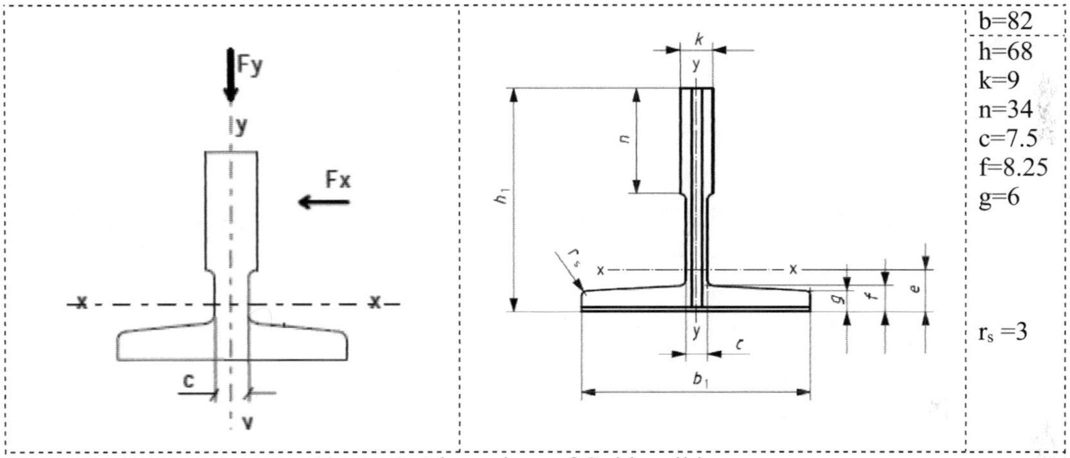

Figure 2. Dimensions of Guide rail in mm

4.1. Technical characteristic of the guide rail.

Cross-sectional area of a guide rail (S): 1091 mm^2

Linear density of a finished guide rail (q1):8.564 Kg/m

Distance from the rear surface to the centre of gravity of the guide rail (e):20.34 mm

Moment of inertia of the cross-sectional area of the guide rail related to the x-x axis (I_{x-x}): 493100 mm^4

Cross-sectional area modulus related to the x-x axis (W_{x-x}): 10270 mm^3

Radius of gyration corresponding to the x-x axis (i_{x-x}): 21.26 mm

Moment of inertia of the cross-sectional area of the guide rail related to the y-y axis (I_{y-y}): 301700 mm^4

Cross-sectional area modulus related to the y-y axis (W_{y-y}): 7358 mm^3

Radius of gyration corresponding to the y-y axis (i_{y-y}): 16.63 mm

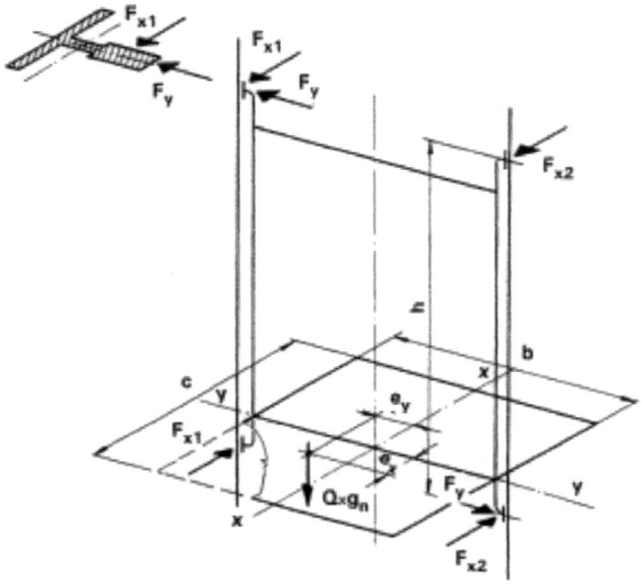

Figure 3. Forces acting on a guide rail

4.2. Centrally guided and suspended car.
The load distribution on the carriage and the cabinet measures are shown in Figure 3. P and Q are on the same side. For this reason, Q is on the x-axis

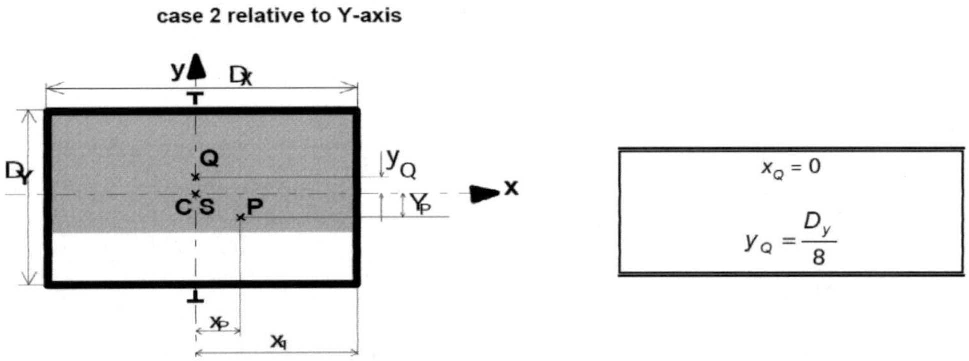

Figure 4. Load distribution case with relative to Y-axis

$X_Q = 0$, $Y_Q = \frac{Dy}{8} = \frac{1100}{8} = 137.5$ mm, $X_P = 50$ mm, $Y_P = 30$ mm

4.3. Safety gear operation.
The guide rail is safe when the permissible stress (σper) is 205 N/mm^2 for St37

4.3.1. Bending stress

a) Bending stress with respect to Y-axis of the guide rail due to guiding force:

$$F_x = \frac{K1.g.(Q.XQ+P.\ XP)}{(n.h)} = \frac{2 \times 9.81 \times (884 \times 0+1200 \times 50)}{(2 \times 3418)} = 172.2 \text{ N}$$

$$M_y = \frac{3.Fx.L}{16} = \frac{3 \text{ x } 172.2 \text{ x } 2500}{16} = 80{,}718.75 \text{ N-mm}$$

$$\sigma_y = \frac{My}{Wy} = \frac{80718.75}{7358} = 10.97 \text{ N/mm}^2$$

b) Bending stress with respect to X-axis of the guide rail due to guiding force:

$$F_y = \frac{K1.g.(Q.YQ+P.YP)}{(n.h)/2} = \frac{2 \text{ x } 9.81\ (884 \text{ x } 137.5 + 1200 \text{ x } 30)}{(2 \text{ x } 3418)/2} = 904.36 \text{ N}$$

$$M_x = \frac{3.Fy.L}{16} = \frac{3\text{x } 904.36 \text{ x } 2500}{16} = 4{,}23{,}918.75 \text{ N-mm}$$

$$\sigma_x = \frac{Mx}{Wx} = \frac{423918.75}{10270} = 41.28 \text{ N/mm}^2$$

The combine bending stresses acting on the guide rail.

$\sigma_m = \sigma_x + \sigma_y = 41.28 + 10.97 = 52.25 \text{ N/mm}^2 < \sigma_{per} = 205 \text{ N/mm}^2$

Hence, Guide rail is safe.

4.3.2. Buckling stress calculation

$$F_k = \frac{K1.g.(P+Q)}{n} = \frac{2 \text{ x } 9.81\ (1200+884)}{2} = 20{,}444 \text{ N}$$

The omega value is taken from the table G.3 EN81 by evaluating since

$$\lambda = \frac{Lk}{imin} = \frac{2500}{16.63} = 150.33$$

Since $R_m = 370 \text{ N/mm}^2$, $\omega = 4.33$ from the table G.3 from EN81

M = 50 Kg because there is auxiliary equipment with guide rail installed;

$$\sigma_k = \frac{(Fk+K3.M)\omega}{A} = \frac{(20444+1.2 \text{ x } 50\text{x } 9.81)\text{x } 4.33}{1091} = 83.47 \text{ N/mm}^2$$

$\sigma_k = < \sigma_{per},$ Hence, Guide rail is safe.

4.3.3. Combined stress calculation

Combined bending and compressive stress are given by

$$\sigma = \sigma_m + \frac{(Fk+K3.M)}{A} = 52.25 + \frac{(20444+1.2 \text{ x } 50 \text{ x } 9.81)}{1091} = 71.52 \text{ N/mm}^2$$

$\sigma = < \sigma_{per},$ Hence, Guide rail is safe.

Combined bending and buckling stresses are given by

$\sigma_c = \sigma_k + 0.9\sigma_m = 83.47 + 0.9 \text{ x } 52.25 = 130.495 \text{ N/mm}^2$

From Table 5.6, in the case of safety equipment operation,

$\sigma_c = < \sigma_{per,}$ Hence, Guide rail is safe.

4.3.4. Bending stress calculation at the guide rail neck

Bending stress in the rail neck, which occurs in the T-profiled guide rail, using Equation (5.12) from EN81;

$$\sigma_f = \frac{1.85.Fx}{c2} = \frac{1.85x\ 172.2}{7.5x\ 7.5} = 5.66\ N/mm2$$

4.3.5. Deflection in the Guide rail

Using the equation (5.13a) from EN81, the deflection in the guide rail in the X-X plane,

$$\delta_x = \frac{\mathbf{0.7.Fx.l3}}{\mathbf{48.E.Iy}} = \frac{\mathbf{0.7x172.2x\ 2500x2500x2500}}{\mathbf{48\ x\ 208000x301700}} = 0.63\ mm$$

Using the equation (5.13b) from EN81, the deflection in the guide rail in the Y-Y plane,

$$\delta_y = \frac{\mathbf{0.7.Fy.l3}}{\mathbf{48.E.Ix}} = \frac{\mathbf{0.7\ x\ 904.36x\ 2500x2500x2500}}{\mathbf{48\ x\ 208000\ x\ 493100}} = 2.00\ mm$$

5. FEM Model

Finite element model of guide rail and combination brackets are created using SolidWorks simulation software. Design parameters previously used in analytical calculations are considered in FEM modelling. Table 2 shows that 4 cases were considered while applying forces at the interface of guide for both types of combination bracket and the value of lateral forces Fx, Fy and vertical force Fk are taken from analytical calculation. The material for combination brackets is shown below in the table 3. Contact properties between the components are defined as global bonded conditions. Friction coefficient between the elevator component is assigned. The holes (at which wall fasteners is fixed) in rear surface of combination bracket in contact with the wall is considered as fixed geometry and two virtual wall conditions are applied, one vertically and other on the pit floor on which rail is mounted. All the structural, Sheetmetal and machine components are treated as a solid element for 4 cases. The contact between the rail and combination bracket front surface is given bonded relation since they are fixed with each other with bolt. Details for Mesh information is shown in table 4

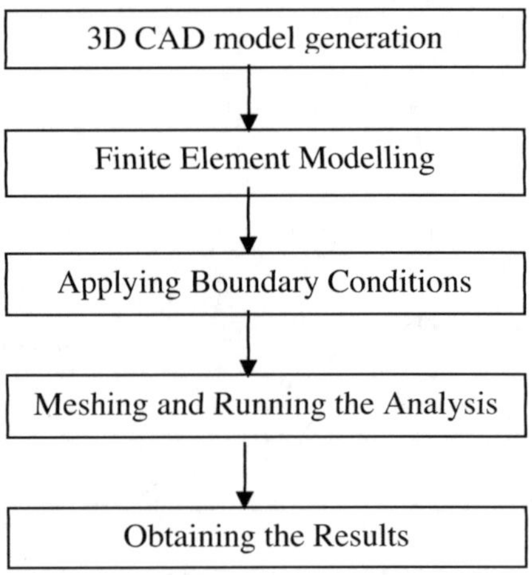

Figure 5: Flow chart for stress analysis of Combination bracket

Table 2: Boundary condition cases

IOP Conf. Series: Materials Science and Engineering **594** (2019) 012019 doi:10.1088/1757-899X/594/1/012019

Case 1: Forces applied on guide at the junction of Guide rail bracket (Sheetmetal)

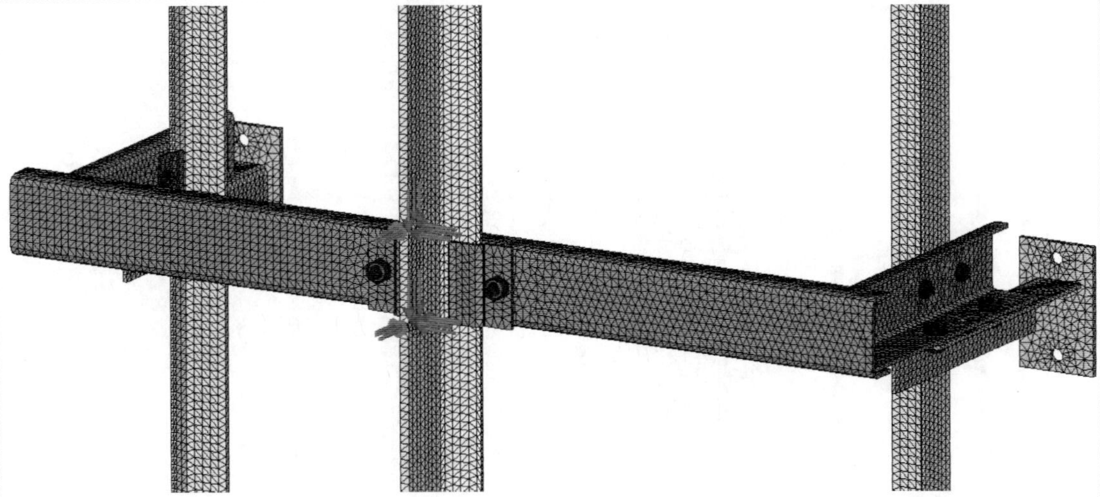

Case 2: Forces applied on guide at the junction of Guide rail bracket (structural)

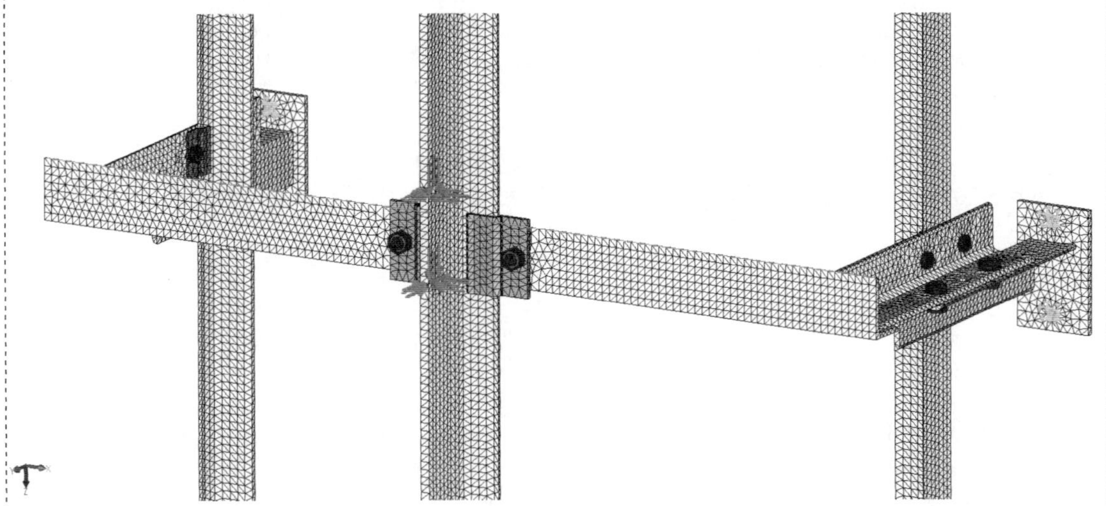

Case 3: Forces applied on guide rail between two brackets (Sheetmetal)	Case 4: Forces applied on guide rail between two brackets (structural)

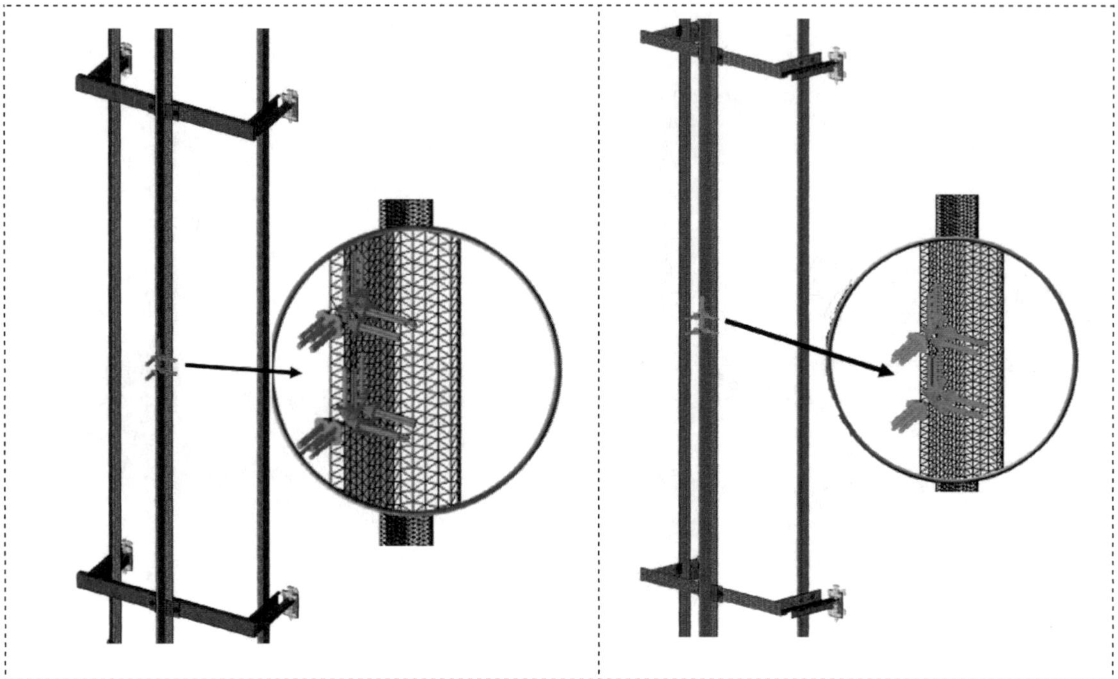

The material selected for the guide rail is as per IS 2062 E250 A, for structural combination brackets is as per IS 2062 E275 BR and for Sheetmetal combination bracket is as per IS 1079_2009. The material characteristic is taken as linear isotropic.

Table 3: Material properties for Guide rail and brackets

Component Name	Structural Comb. Brackets	SM Comb. Brackets	Guide Rail
Material Name:	IS 2062 E275 BR	IS 1079_2009	IS 2062 E250 A
Model type:	Linear Elastic Isotropic	Linear Elastic Isotropic	Linear Elastic Isotropic
Yield strength:	255 N/mm^2	230 N/mm^2	230 N/mm^2
Tensile strength:	440 N/mm^2	440 N/mm^2	410 N/mm^2
Young's modulus:	210000 N/mm^2	210000 N/mm^2	210000 N/mm^2
Poisson's ratio:	0.28	0.28	0.28
Mass density:	7.8 g/cm^3	7.8 g/cm^3	7.8 g/cm^3
Thermal expansion coefficient:	1.3e-005 /Kelvin	1.3e-005 /Kelvin	1.3e-005 /Kelvin

Table 4: Mesh Information

Parameter	Sheetmetal	Structural
Mesh Type	Solid Mesh	Solid Mesh
Mesher Type	Standard	Standard
Jacobian points	4 Points	4 Points
Element Size	7 mm	7 mm
Tolerance	0.375 mm	0.375 mm
Mesh Quality Plot	Draft Quality	Draft Quality
Total Nodes	378583	394662
Total Elements	1144395	1189854
Maximum Aspect Ratio	41.082	43.648
% of elements with Aspect Ratio < 3	90.08	91

IOP Conf. Series: Materials Science and Engineering **594** (2019) 012019 doi:10.1088/1757-899X/594/1/012019

The lateral force acting on guide rail and brackets does not change in accordance with deformation of the component and with time. Material property is elastic in nature. Hence, linear static analysis is selected for this study.

6. Analysis Results and Discussions

Results from finite element analysis is interpreted distinctively for various cases shown below. The stress and deflection in X and y directions are given in table 5. Finite element analysis has been performed considering safety gear actuation conditions in all cases

Case 1: Forces applied on guide at the junction of Guide rail bracket (Sheetmetal)	Case 2: Forces applied on guide at the junction of Guide rail bracket (structural)
	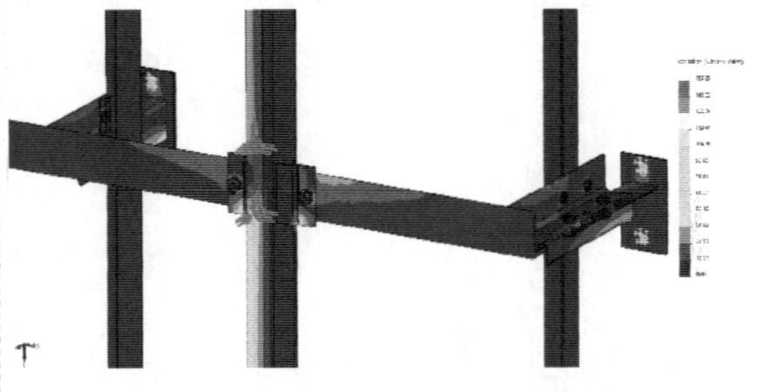
Figure 6.1(a) Von Mises Stress	**Figure 6.2(a)** Von Mises Stress
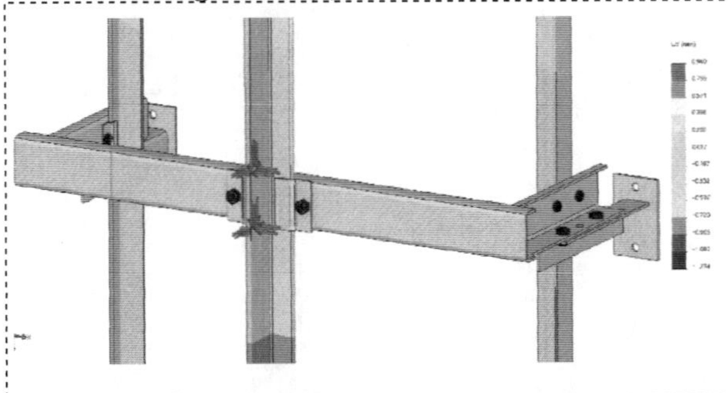	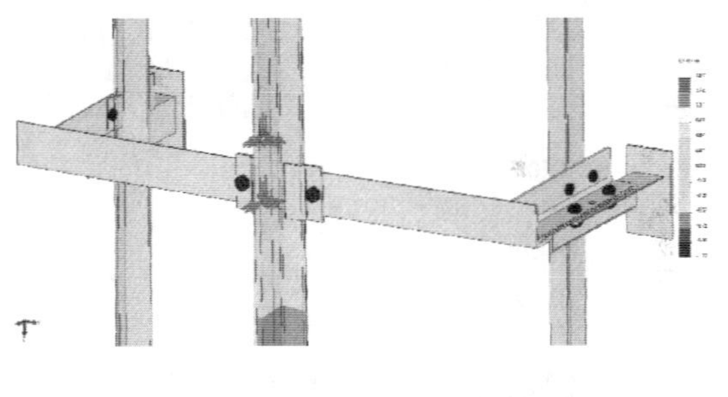
Figure 6.1(b) Deflection in Y axis	**Figure 6.2(b)** Deflection in Y axis
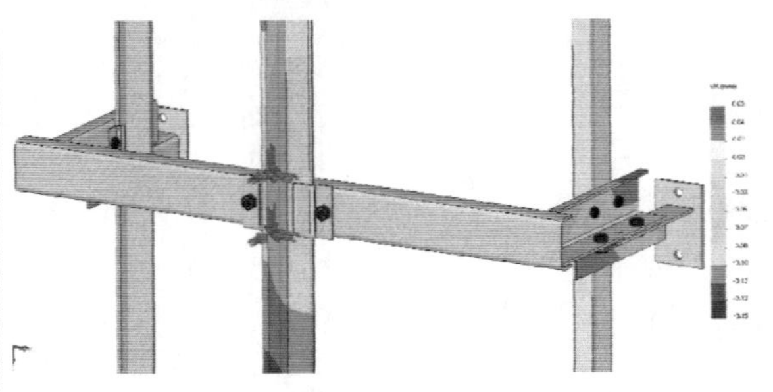	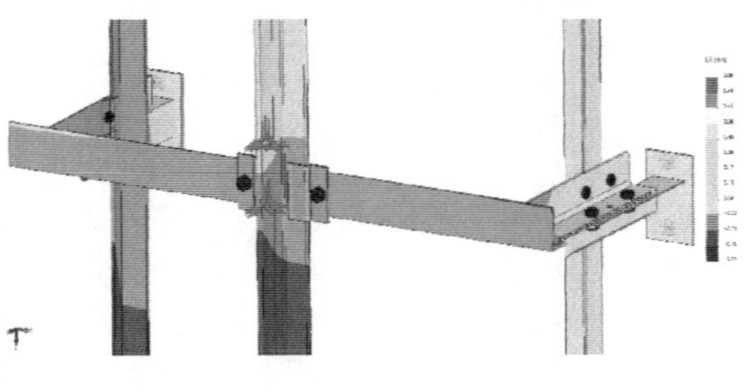
Figure 6.1(c) Deflection in X -axis	**Figure 6.2(c)** Deflection in X -axis

IOP Conf. Series: Materials Science and Engineering **594** (2019) 012019 doi:10.1088/1757-899X/594/1/012019

Von Mises Stress	Deflection in Y axis	Deflection in X axis

Case 3: Forces applied on guide rail between two brackets (Sheetmetal)

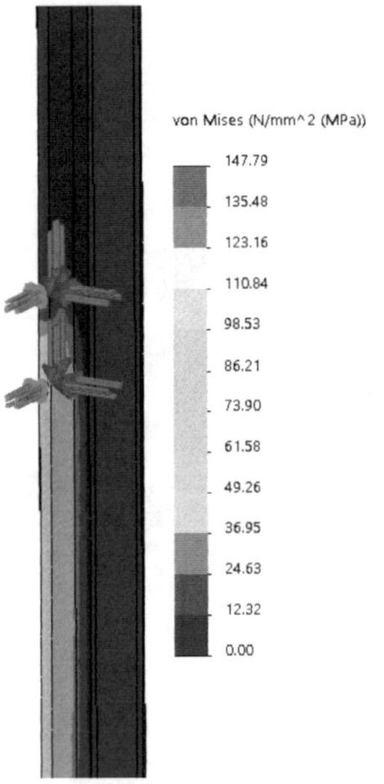

Figure 6.3(a)

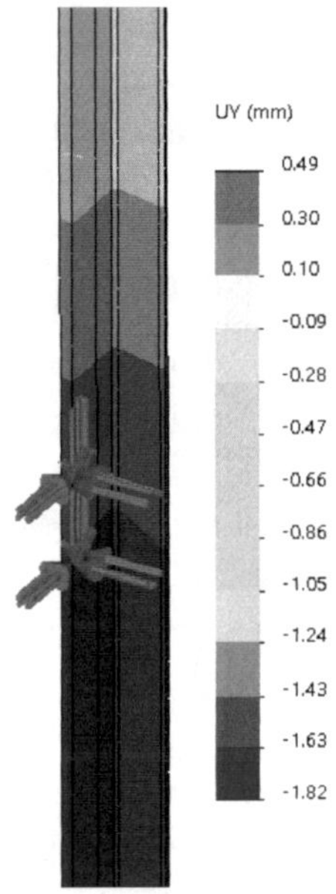

Figure 6.3(b)

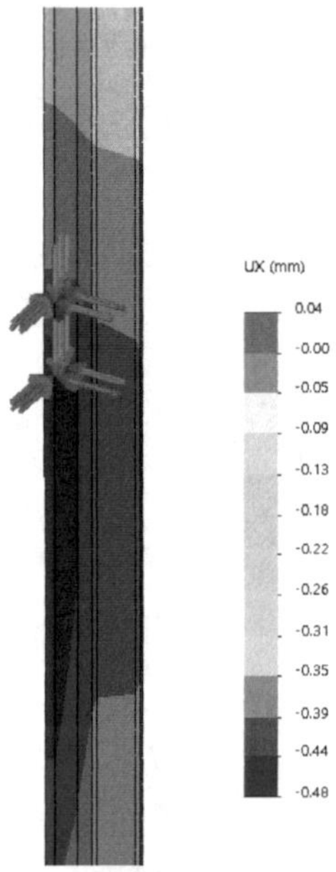

Figure 6.3(c)

Case 4: Forces applied on guide rail between two brackets (Structural)

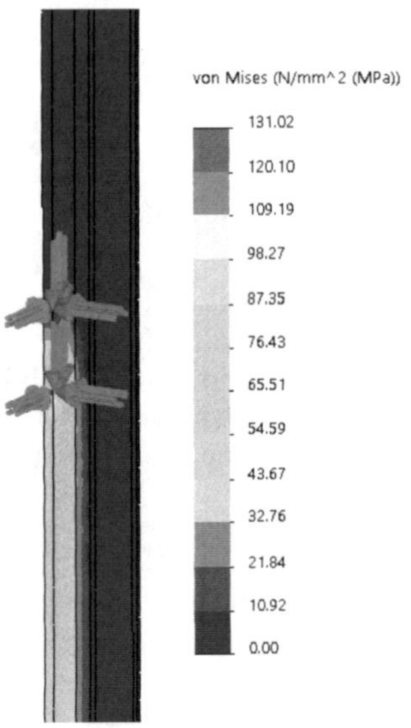

Figure 6.4(a)

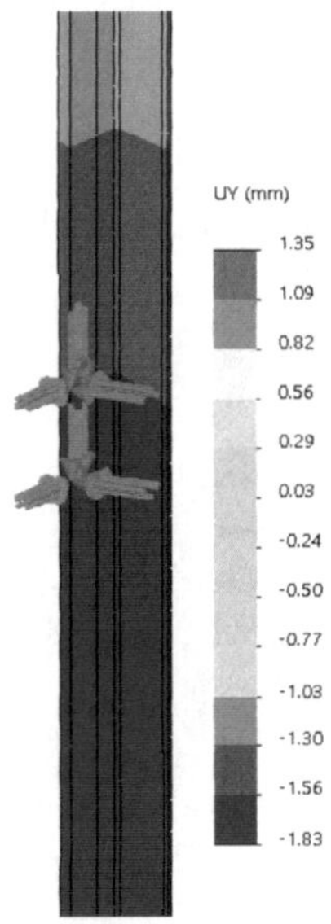

Figure 6.4(b)

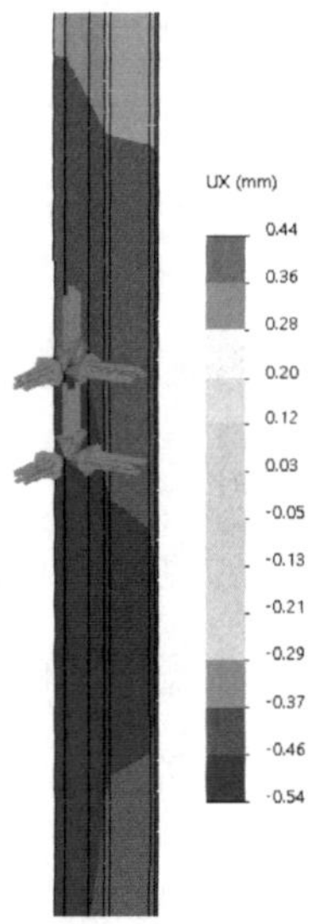

Figure 6.4(c)

Stress distribution and deflection for Y axis values for case 1 and case 2 are compared with each other and are found to be nearly same. Case 3 and 4 stress and deflections values in x and y direction results are compared analytical calculation

Table 5: Analysis results

Analysis Type	Stress (N/mm^2)	Deflection in Y axis (mm)	Deflection in X axis (mm)
FEA (Case 1)	161.28	1.27	0.15
FEA (Case 2)	159.29	1.22	0.22
FEA (Case 3)	147.79	1.82	0.48
FEA (Case 4)	131.02	1.83	0.54
Analytical	130.49	2.0	0.63

7. Conclusion

The following points may be concluded:

- From case 1 and 2, Stresses distribution and maximum deflection values at guide rail fixed on structural combination and Sheetmetal combination brackets are nearly same.
- From case 3 and 4, Stresses distribution for the case, guide rail fixed between two structural combination is less than that fixed between two Sheetmetal combination brackets. maximum deflection values for both cases remains the same in y axis.
- The results obtained from analysis will help in determining the material selection in early steps of design evolution, thus saving development time and further resulting in decision-making process to optimize the design. However, Cost of the raw material may differ as per organizational sourcing.

References

[1] Hayder F. Neamah and Wisam H. Ali (2009)., "Stress Analysis of Guide Rails of Elevators", Al-Khwarizmi Engineering Journal, Vol. 5, No. 3, PP 38 – 50.

[2] Serkan ELMALI, Adem Candaş, Eren Kayaoğlu, Prof. C. Erdem Imrak and Sefa Targit," Modelling and Analysis of Guide Rail Brackets and Attaching Parts".

[3] Recep, Honda, and Hakala," A comparative stress analysis of guide rail by FEA", JSME, No. 200-305, pp.80-88.

[4] Penn, Elsco and Robert (2006)," The study of neoprene elevator guide rails", JSME, Sept.,2006, 112-130.

[5] Singh MP, Rildova (2004), " Seismic Countermeasures for Rail-Counterweight System in Elevators", 13th World Conference on Earthquake Engineering Vancouver, B.C., Canada, August 1-6, 2004, Paper No. 2825.

[6] ASME A17.1-2010, Safety Code for Elevators and Escalators, The American Society of Mechanical Engineers, 2010.

[7] M. Altuntas (2012), "Elevator Guide Rail Anchor Elements Experimental Stress Analysis" MSc. Thesis, Istanbul Technical University, Turkey.

[8] S. Atay (2013), "Experimental Stress Analysis of Complete Rail Fastening Systems" MSc. Thesis, Istanbul Technical University, Turkey.

[9] S. Atay, E. Kayaoglu, A. Candas, C. E. Imrak, S. Targıt, Y.Z. Kocabal (2014). "Determination of Loads Acting on Guide Rail Fixing under Certain Loading Condition" ELEVCON 20th International Congress on Vertical Transportation Technologies, pp.85-92

[10] BS A 305, (1994). Bolt Testing Method, British Standard Institute, London.

[11] M. Altuntas, O. Salman, C. E. Imrak (2014). "Experimental stress analysis of elevator guiding equipment". Key Engineering Materials, Vol.572, pp. 177-180

[12] M. Merz (2010), "Practical stress behavior of complete rail fastening systems". ELEVATION Vol. 68, p56-62

[13] M. Merz (2008)., "Practical stress detection on rail anchors". Hilti Corporation, ELEVCON the 17th International Congress on Vertical Transportation Technologies, Thessaloniki, 268-277

IOP Conf. Series: Materials Science and Engineering **594** (2019) 012019 doi:10.1088/1757-899X/594/1/012019

[14] Suarez, L.E. and Singh, M P. (1996), "Seismic response of rail-counterweight systems of elevators," Proceedings of the 11th World Conference on Earthquake Engineering, Acapulco, Mexico, CD-ROM, Paper No. 1657, Pergamon Press.
[15] Janovsky, L. (1993), "Elevator mechanical design," 2nd Edition, Ellis Horwood Series in Mechanical Engineering, Great Britain
[16] IS 14665 (PART 1) - 2000 Electrical traction lifts (Guide lines for dimension of passenger, goods, service & hospital lifts).
[17] IS 14665 (PART 2 - SEC 1 & 2) - 2000 - Code of practice for installation, operation and maintenance.
[18] IS 14665 (PART 3-SEC 1 & 2) - 2000 Safety rules Section 1-Passenger & goods lifts, Section 2-Service lifts.
[19] IS 14665 (PART 4-SEC 1 TO 9) - 2001 Components.
[20] IS 14665 (PART 5) – 1999 Electrical traction lifts – Specification (Inspection manual).
[21] EN 81-1:2009. Safety rules for the construction and installation of lifts — Part 1: Electric lifts. European Committee for Standardization (CEN), Brussels, 2009.
[22] EN 81-2:2009. Safety rules for the construction and installation of lifts — Part 2: Hydraulic lifts. European Committee for Standardization (CEN), Brussels, 2009.
[23] Vladić J.; Malešev, P.; Šostakov, R.; Brkljač N. (2008)," Dynamic analysis of the load lifting mechanisms. // Strojniški vestnik. 54, 10, pp. 655-661.
[24] Jumade HS. Bansode PP and Mahawadiwar A A, (2016) " Design and Analysis of Machine Room less Elevator ", IJSRD - International Journal for Scientific Research & Development| Vol. 4, Issue 04, | ISSN (online): 2321-0613.
[25] Onur, Y.A. and Imrak, C.E. (2005): Finite Element Modeling of Elevator Carframe and Static Stress Analysis, 7th International Fracture Conference, Kocaeli, Turkey, pp.109-116.
[26] Onur, Y.A. and Imrak, C.E (2007).: Finite Element Approach to Stress Analysis for Elevator Car Sling, 11th International Research/Expert Conference Trends in the Development of Machinery and Associated Technology Proceedings, Hammamet, Tunisia, pp. 1015-1018.
[27] Watanabe S, Yumura T, Funai N, Hayashi Y, Minewaki S, Kibayashi M. (2007)," Dynamic Analysis for Elevators installed in Mid-story Seismic Isolated Building" Journal of Environment and Engineering, Vol.2, No.3, pp.436-447.
[28] Lee JS, Kim BS, Tak CM, Cho GY, Jung BJ. (2016)," Medium and Low Speed (2m/s) Governor of Core Safety Rope of 6mm Diameter Development of MRL Type Elevators", Advanced Science and Technology Letters Vol.140 pp.72-78.
[29] Yunpu, S., Yixing, S. (2014).): Design of Structural Parameters for Centrifugal Elevator Overspeed Governors. Sensors & Transducers, Vol. 163, Issue. 1, pp. 171--179.
[30] Li R, Chen G*, Liang P. (2017), "Modeling and Finite Element Analysis of the Hydraulic Buffer in Elevator Systems", Journal of Civil Engineering Research, 7(2): 63-66.
[31] Walgude T R, Suryavanshi A R. (2015), "Safeguarding Elevators against Buckling by Providing Adequate Number of Supports'', International Engineering Research Journal (IERJ) Special Issue 2 Page 1130-1136, ISSN 2395-1621.
[32] Abdulmalik, Ibrahim O, Akonyi, NasiruSule, Bolarinwa, Gabriel Oladeji, Chima, Lazarus Onyebuchi, Amonye, Michael C, Mgbemena, Chinedum Ogonna. (2014) "The Design of a One-Man Passenger Electric Elevator", Scholars Journal of Engineering and Technology. 2(6A):806-811.
[33] Pai A, Nair R, George P, Subir S (2015) A Critical Review and Investigation of Machine Room Less (MRL) Elevators. J Appl Mech Eng 4:166. doi:10.4172/2168-9873.1000166.
[34] George W. Gibson (2008) ELEVATOR HOISTWAY EQUIPMENT: Mechanical and Structural Design, Part I Continuing Education pp:97-110.
[35] George W. Gibson (2009) ELEVATOR HOISTWAY EQUIPMENT: Mechanical and Structural Design, Part II Continuing Education pp:91-103.
[36] ISO 7465 Fourth edition 2007-11-15: Passenger lifts and service lifts — Guide rails for lift cars and counterweights — T-type.
[37] The Bombay lift Act ,1939 and the Bombay lift rules ,1958.
[38] IS 15259 -2002-Installation and Maintenance of Home Lifts -Code of Practice.

[39] IS 15330 -2003-Installation and Maintenance of Handicapped Persons -Code of Practice.
[40] IS 15785 -2007-Installation and Maintenance of lift without Machine Rooms -Code of Practice

IOP Conf. Series: Materials Science and Engineering **594** (2019) 012020 doi:10.1088/1757-899X/594/1/012020

Vibration Analysis of Cantilever shaped 3D printers

A Srivastava[1], C Gautam[1], N Bhan[1] and Ram Dayal[2]

[1]Mechatronics Department, Manipal University Jaipur, Jaipur (303007)
[2]Mechanical Engineering Department, National Institute of Technology Uttarakhand, NH 58, Srinagar, Uttarakhand (246174)
E-mail: [1]akshatsrivastava@muj.manipal.edu

Abstract. 3D printers are being explored by many of the industrial giants and daily customers as an excellent replacement for some of the traditional, inefficient, costly and waste producing techniques of manufacturing. Although, still at a premature state of development, and being thoroughly researched upon, there is lack of data concerning the physical strength or a proper machine design development for this 3-axis automated machine. Other, while being able to create a robust machine, over-utilize the resources. In this paper, we attempt to explore how analysis of vibration using minimal equipment could refine the quality of 3D printing as well as how would one improve the design to yield best results. The study is most relevant for the online open source community e.g. RepRap, within the development in new 3d printers design.

1. Introduction

The additive manufacturing technology has been around for about 15-10 year time. 3D printing (3DP) has taken even much more interest from hobbyist and other 'do it yourself' (DIY) community since the expiration of licenses relatable to the process [1]. Currently, new patents regarding the intellectual copyrights have been issued due to which, certain designs are protected against reproduction through 3DP technology [2]. Although, given the current scenario, the industry has long moved towards better and more efficient techniques for additive manufacturing. It is only recently that consumer electronics in many countries are now displaying their desktop 3DP on shelf, and also attracting many educational institutes for being able to teach students and empower them with this creative machine. Industrially, it is SLS or Multijet technology that is most frequently encountered as additive manufacturing method [3]. This technology is by far, too expensive to be able to currently replace the existing technologies used for metal manufacturing processes such as CNC and die manufacturing. Most of the general public is only introduced to FFF or FDM printers. Even the current scientific community rely highly on the said 3D printers to create equipment for use. It is now a basic commodity, which could be easily noticed within all research laboratories. With various firmware available online as open-source, many people often choose to build themselves a 3DP. Much of parts are relatively easy to obtain from online market, with majority also selling DIY kits [4].

1.1. Emphasis on Mechanical Properties

Even with all the spreading awareness about the technology, much of the intrinsic mechanical properties of FFF 3DP and its terminology is undergoing rapid changes. Most of the researches currently focus on different kinds of experiments being performed on the products of FFF 3DP. Two similar CAD files processed using FFF on same machine, could differ significantly with variations in rastor angle, extrusion temperature, infill amount and the speed of printing. Many other factors might be in action, but the ones which could be easily measure and controlled quantitatively, often end up becoming the topic of research [5]. These data are extremely important in having to provide engineers general values

IOP Conf. Series: Materials Science and Engineering **594** (2019) 012020 doi:10.1088/1757-899X/594/1/012020

that they could embed in their designs directly. There has been a growing report on engineers developing automobile parts, as well as whole cars out of FFF 3D printing, which without say, would have involved multiple design factors related to the material and the final product being used for fabrication [6]. With the intent of contributing to this ever-increasing database and proposing more scientifically rich information, this paper shines lights on one of the few aspects of the process which have a major effect on the mechanical and surface properties of the printed products.

1.2. Vibration and Design

The expectation of general consumer before using 3DP is comparable to the quality obtained from die manufacturing of plastic, but as one realizes it differs remarkably due to layers visible from processing. These layers are small and could be neglected as one gets used. The problem arises when vibration of machines persists to distort the surface quality, of the final print. Although, many post processing methods have been developed to change the outlook of the final product in variety of ways, these layers give many users a good deterministic factor of the quality of 3DP. As a thumb rule, when the layers are big, irregular & wavy, the final product diverts from delivering the best quality. When the layers are hardly visible, or the texture of the final product, such as curves or depressions, are clear, the final product is good quality, even though some of the layers would still be visible.

2. Machine Design of 3D printers

Due to the growing numbers of people utilizing FFF technology, there are a number of categorization in articles, based on different features such as the geometric system, type of extrusion, basic mechanical structure and production capacity [7] [8]. Though here, we utilize yet another classification to suit the flow of our experiment and make it easier to understand. Most customer chose to opt for using either an open type 3d printer or using a closed 3D printer, more commonly known as moving bed and fixed bed design, respectively. This is because in an enclosed 3D printer, the bed usually comprises of z-axis movement (5% of TPT), which moves only up and down with succession of each layer, and remains stationary for planer movements. The open printers on the other hand, usually have a bed which forms either the x-axis or the y-axis (95% of TPT).

2.1. Fixed Bed and Moving Bed

The fixed bed printer, also called enclosed printers are covered using metal, fibre composite or other stiff material able to withstand the torque produced by the motor travelling throughout the whole mechanical assembly. The downfall with these types of printers is that they are relatively expensive owing to the cover material required. There is also a requirement of a complex mechanism for the movement of the planer axis, which in turn necessitates the use of large number of bearings, guide rods and toothed pulleys also Fig 1 (a).

The moving bed printers are the most common printers that are popular among DIY electronics. They are cheap, sturdy and reliable. Most of these printers are operated by hobbyist and DIY community. They are used to explore more options and customization that could be performed to be able to improve the quality of the print. It could be easily made using a variety of mechanism to move the extruder Fig 1 (b). Some examples include polar movement, delta mechanism and hanging 3d printer [9]. The major problem with these printer is vibrations, and that too owing to the bed. Since the bed is the heaviest element, its movement is bound to source some vibrations which cause different kinds of defects [10] [11].

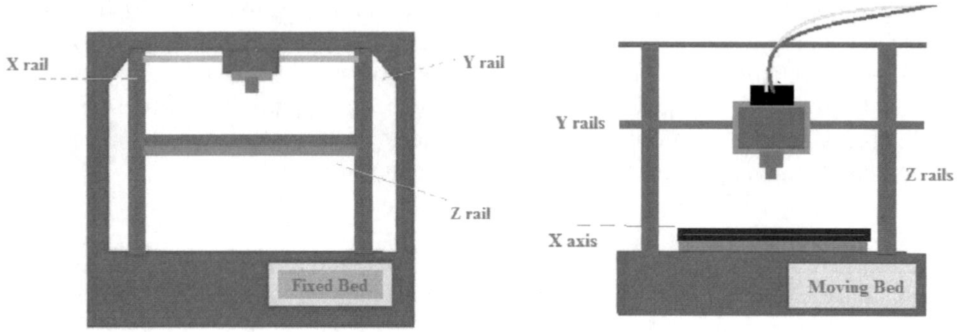

Fig. 1. (a) Fixed Bed, (left), (b) Moving Bed (right)

2.2. 3D Printer for Experimentation

We could intuitively prove that vibrations are inherent problem. It could be further multiplied as the material keeps on depositing on the build plate, but no such formal study has been attempted over the subject. *It remains the aim for the rest of the discussion to investigate and interpret the results of vibrational analysis of moving bed 3D printers.*

Due to many reviews and complications faced by existing users, it was decided that the machine to be analyzed would be developed from very start to the point of perfection. A brief research at existing study indicated that some experimentation has already been performed on the fixed bed 3D printer [12]. The analysis performed utilized the vertically recorded displacement using a laser sensor, to infer the quantitative measurement of vibration, although such approach is seldom observed in the vibrational analysis [13]. Furthermore, the vibrational displacement was measured for the bed axis only. As we have previously stated, the vibrations should be measured for axis which moves frequently and also carries the largest weight. The axis which satisfies both the stated condition is the planer axis. Thus, the vibrations in the fixed bed 3D printer should be measured relative to the extruder, not the build plate. The study here aims to analogically provide the missing link about the vibrational analysis of moving bed 3D printers.

2.3. Development of 3D printer for Experimentation

Since our team were settled onto designing the whole machine themselves, they had to liberty to explore as many alternatives as possible which could optimize the performance and reduce the cost as well. Most moving bed 3D printer are available in two different types. The first one is also called square frame, while the other one is known as cantilever structure (C-shape). Most of the available printers fall into the former category. In an attempt to replicate the design, the CAD model in Fig 2 was developed and soon realized since most of the raw material was easily obtainable from either online stores or local hardware shops. *It should be noted that the suggested design was quite similar to the currently available square frame 3DP priced under USD$250. Since the experimentation is closely connected with the user experience affordability, a design very similar to the commonly available 3DP was opted* [14].

The design suffered from large vibrational currents travelling throughout the structure of the printer. There were two main problems that were encountered. Firstly, the constraints provided to the motion of the printer were not enough to provide a fully constraint motion. This was due to the part that guide rods were long enough to go through bending, and the material used for making the fixture, holding the guide rod, was flexible.

Many other problems surfaced as the use of the machine was increased. Due to lack of repeatability, and poor surface quality, the printer was labelled as porotype and a new, better and improved moving bed design was proposed, using minimal number of parts. This new design was cantilever structure as shown in Fig 3.

IOP Conf. Series: Materials Science and Engineering **594** (2019) 012020 doi:10.1088/1757-899X/594/1/012020

Fig. 2. Development of prototype showing CAD model (left) and realized design (right)

This design also consisted of the same number of constraint as the previous design, but the size of bed as well as the guide rod was smaller. Through the course of the development of this 3D printer, the bed was additionally constraint using a rolling wheel mechanism also.

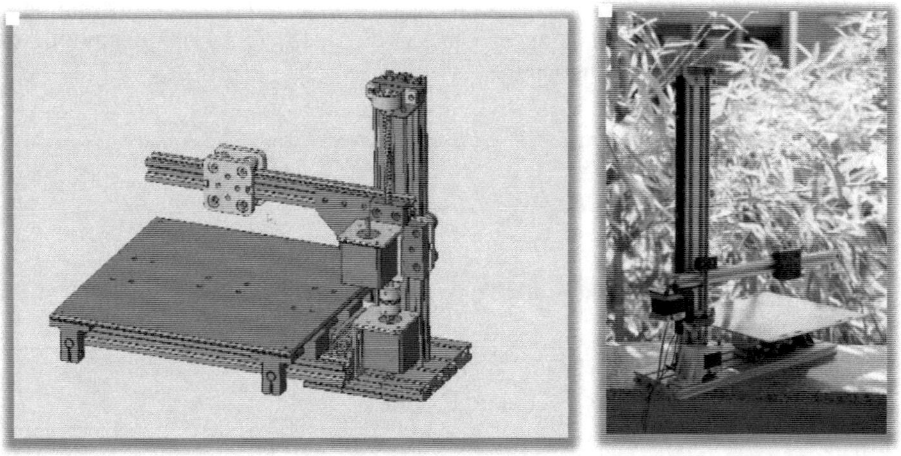

Fig. 3. Development of C-shape prototype showing CAD model (left) and realized design (right)

The targeted problem for having to provide additional constraints was based on qualitative judgement of the printed object. Due to poor bearing quality, the perpendicular movement of the base was not completely constraint. Wavy layer patterns were visible, instead of the expected straight lines. Additionally, as the PLA deposited over the plate, the weight of the bed kept increasing and thus the inertia. With more inertia, the base vibrated with an increasing amplitude, even though quite subtle to be observed with naked eyes, but its effects, were nonetheless visible on the surface of the print. With the addition of another constraint, almost all the vibrations due to design imperfections were largely eliminated.

3. Experimentation

3.1. Assumptions and Procedure

One of the key for vibrational analysis is the location of the sensor. The source of the mechanical vibration could be due multiple imperfections, few of which are systematic in nature. The sinks, although, are few and predictable. The vibration travels throughout the solid mechanical structure, similar to a wave or current. These *vibrational currents* end where the mechanical links are open (farthest away from supports), thus, the base and extruder form the ideal places for observing these currents physically. Due to the convenience of performing the experiment, the base was chosen as the ideal site for placing the accelerometer. The acceleration was adjusted to a default value of 800mm/s and the speed of travel was adjusted to 3600mm/min or 60mm/s. The value was further increased to 150% and then to 200%. The accelerometer recorded the variation in acceleration, and increase in speed. The data was further studied more elaborately to display the increase of vibrations in terms of amplitude and frequency.

3.2. Vibration Analysis

The produced vibrations are forced vibrations. The torque applied by the motor, which would ideally be unidirectional, is acting through all the other axis due to imperfections and tolerances of guide rods, pulley, timing belt and bearing allowances, giving rise to a resultant force acting perpendicular to the face of the bed. The vibrational analysis is performed using a MyRIO built-in accelerometer with 12-bit minimum resolution. The data is recorded at a sampling frequency 1000Hz for a duration of about 7-8s. The obtained data is further processed using FFT in MATLAB to yield more meaningful characteristics about the increase of vibration frequency and amplitude.

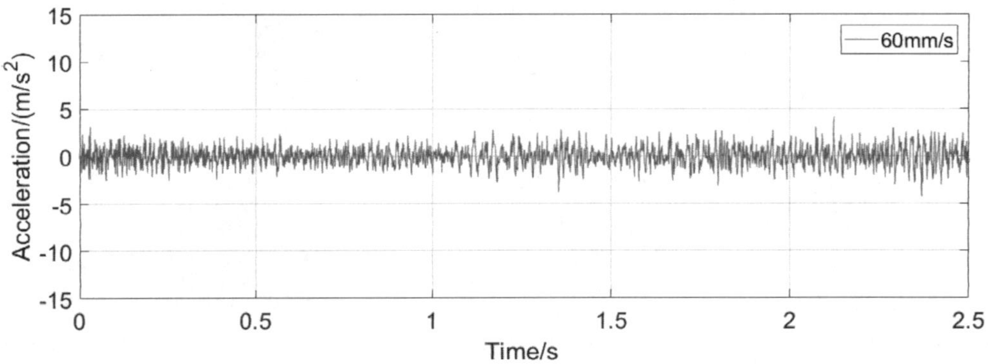

(a)

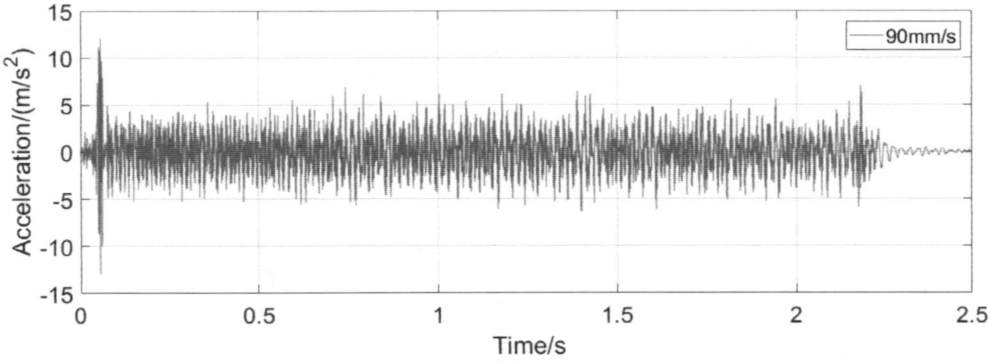

(b)

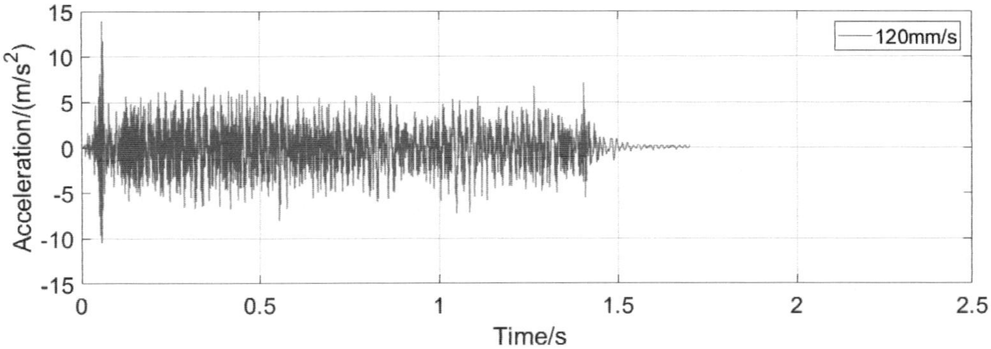

(c)

Fig. 4. Time Domain Observations with varying speed (a), (b) and (c)

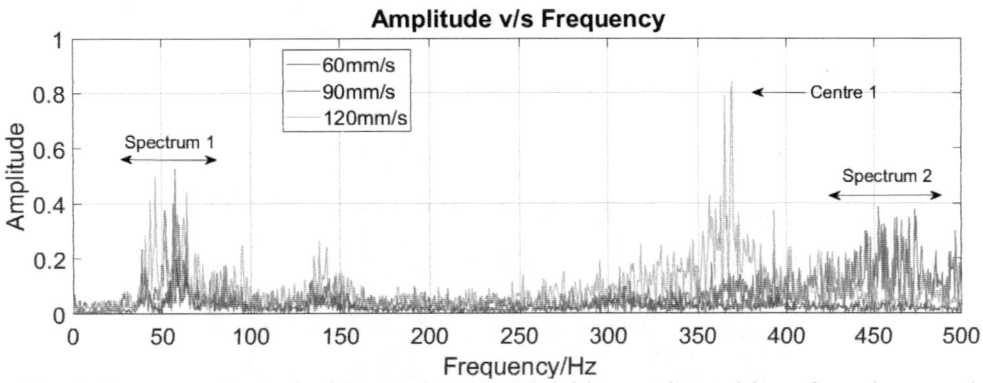

Fig. 5. Frequency Domain Observations (FFT) with superimposition of varying speed

Certain traits about the spectrum were predictable. Few peaks were expected to reoccur in spectrum, regardless of the variation in the speed. Intuitively, the frequency of the vibrations as well as the amplitude were bound to increase, simultaneously, as the speed was increased. The acceleration time graph would also contain two large peaks, which is due to the jerk produced on the bed beginning its course of movement and alternatively, completing it.

The G-code which is used to promote the motion of bed varies from firmware to firmware. In this case, Marlin [15]) has *G00 X<units>* as the command. The command is entered in the prompt window of an interface software called *PronterFace* [16].

3.3. Observations

The time domain graph did not present much information other than showing the increase in the amplitude of acceleration as the speed increases. The variance of acceleration increase by an average of about 50%. The amplitude characteristically increases in the middle spectrum, when the bed crosses the center of the x-axis. This phenomenon is clearly visible in spectrum for 120mm/s. The graphs indicate that the amplitude of the overall spectrum increases, but maximum peak to peak value is not a good comparable measurement *[12]*. For being able to analyze the data in a more comprehensive manner, the Fourier transform of each set of data is superimposed on the top of other. The transform revealed some of the peaks which were fixed, regardless of the variation in speed.

The FFT dictates that the higher speed result in harmonics with larger frequency. These harmonics also occur with high amplitudes e.g. Centre 1. As stated previously, certain harmonics reoccur at same point, forming a spectrum around 60Hz. Another spectrum is also formed by the collaboration of harmonics occurring in 90mm/s and 120mm/s speed spectrum.

4. Conclusion

4.1. Interpretation of Data

Most of the frequency that are caused due to vibrations, should intuitively increase with speed. A similar study carried out by a start-up in Poland supports this claim from their vibrational analysis using a laser displacement sensor [12]. An attempt was carried out, for being able to compare the obtained data with the readings of the study, but due to a low sample rate, the double integration yielded trivial values. Yet another close observation also revealed that not only the amplitude of certain frequency increases, but their frequency also increases.

The kinematics operation used by the 3D printer follows a GBRL standard, which is established by an online community. When following a straight linear path, the kinematics is very close to linear, following the three equations of motion.

As previously stated, the motion is under an influence of force, produced by the torque of the stepper motor. When the speed of the operations is increased, the motor accelerates with a higher magnitude and for a longer elapse of time. During this time, the overall torque experienced by the system and axis increases. This increase causes many of the dormant harmonics within the spectrum to become dominant.

Fig. 6. Final Quality (left) and Initial Quality(right)

Reducing the vibration in a 3DP could largely decrease the post-processing cost, and make the technology more attractive and comparable to the other polymer manufacturing techniques. Moreover, a deep study of acceleration with variation in different parameters such as the accumulated mass over the build plate, the PID constant values and different printer acceleration is also missing, which if performed, could enable users a general guide of mechanical weaknesses and strength for machine designing using 3DP. Although larger organization may possess such data, but due to the recent introduction to this technology, the general open source community and scientific researchers who want to utilize 3DP for making their equipment lags behind.

Forming a strict analytical model out of the given data is near impossible. The system presents itself as stochastic, rather than deterministic, with a plethora of variables being neglected out of the sight of measurement. Although, there is a large scope of application of machine learning since almost no firmware supports a feedback to assist the kinematics of the 3d Printer. Almost all the firmware relies heavily on the fact that stepper motor work just fine with open loop control. Although the stated fact holds true for a large number of mechanisms, there are exceptions. These exceptions culminate themselves into inaccuracies when there is transmission of relatively large forces and torques through the motors.

This analysis was performed only after the 3DP was perfected. Although, there lies an immense scope towards the development of the machine, many improvements have already been achieved with the base of the cantilever 3DP. Fig 6 also shows how the quality of the product drastically improved. This was achieved by measuring the vibration at each advancement of the machine designing process and thereby printing the product through it. A large database of vibrational data regarding the 3DP could potentially be used to tune a 3D printer for optimum quality similar to a guitar tuner, which also utilizes frequency analysis. Furthermore, an embedded sensor could also be used to enable artificial intelligence in 3DP. The sensor may detect the vibrations, or the curves which would cause abnormally high vibrations, and adjust the G-code online to produce a better quality without using much resources. The box structure suffers from vibration for one simple reason. Since the whole assembly is directly, or indirectly joined to a host cover, the *vibrational currents* travel throughout, and end up near extruder nozzle which may cause small, but defects nonetheless.

4.2. Future of 3D printing
3DP is the most powerful tool for open scientific minds, to date. It enables users to access a small manufacturing plant in a highly compact way, thus enabling many of the creative individuals to access the power of rapid prototyping. There has been an increasing input from people around the world, utilizing the machine in un-imaginable ways. Many of the designs concern scientific and engineering investigation, submitted by non-technical people, thus further promoting the growing attention towards technology. In future, the manufacturing industry would not completely circle around additive manufacturing technique, as stated by our hyped social media, but tuned towards having to embed it into its larger established matrix, and, most probably, a hybrid technique of subtractive and additive manufacturing would come forth Although such events would take place only when the price of industrial additive manufacturing falls below what is called 'affordable' by the manufacturing industry

4.3. Acknowledgement
Our uttermost thanks to Manipal University Jaipur and mechatronics department for allowing their immense support, both, in form of finance and motivation.

5. References

[1] E. P. Search, "3d printing process patent," [Online]. Available: https://worldwide.espacenet.com/publicationDetails/inpadocPatentFamily?CC=US&NR=5900207A&KC=A&FT=D&ND=6&date=19990504&DB=EPODOC&locale=en_EP.

[2] K. Walker, "The intellectual property challenges from 3D printing," Commputer Weekly, September 2014. [Online]. Available: https://www.computerweekly.com/opinion/The-intellectual-property-challenges-from-3D-printinghttps://www.computerweekly.com/opinion/The-intellectual-property-challenges-from-3D-printing.

[3] S. A. &. A. Peermohamed, "The future is 3D: India witnesses spurt in additive manufacturing," Business Standard, 8 August 2018. [Online]. Available: https://www.business-standard.com/article/companies/the-future-is-3d-india-witnesses-spurt-in-additive-manufacturing-118080801301_1.html. [Accessed 2018].

[4] M. Jones, "3D Printer Firmware – Which to Choose and How to Change It?," ALL3DP, 4 April 2018. [Online]. Available: https://all3dp.com/2/3d-printer-firmware-which-to-choose-and-how-to-change-it/. [Accessed 2018].

[5] K. P. B. H. Francois Decuir, "Mechanical Strength of 3-D Printed Filaments," *CPS,* 2016.

[6] J. Pyper, "World's First Three-Dimensional Printed Car Made in Chicago," Scientific American, 12 September 2014. [Online]. Available: https://www.scientificamerican.com/article/world-s-first-three-dimensional-printed-car-made-in-chicago/. [Accessed 2018].

[7] C. Hemlock, "Chris's 3D printing project," Wordpress, [Online]. Available: https://chayesthakore.wordpress.com/about/. [Accessed 2018].

[8] "Direct Drive vs Bowden Extruder Guide and Calibration Tips," pinshape Blog, 07 July 2016. [Online]. Available: https://pinshape.com/blog/direct-drive-vs-bowden-extruder-guide/. [Accessed 2018].

[9] K. Stevenson, "The Hangprinter: A Frameless 3D Printer," Fabbaloo, 17 March 2017. [Online]. [Accessed 2018].

[10] D. Quintans, "Diagnosing and fixing ringing versus vibration artefacts," Desi Quintans.com, 30 September 2017. [Online]. Available: http://www.desiquintans.com/ringing.

[11] Tech2c, "THIS IS Y," 13 May 2017. [Online]. Available: https://www.youtube.com/watch?v=AKTvykTPjQw. [Accessed 2018].

[12] J. D. Zbigniew Pilch, "The impact of vibration of the 3D printer table on the quality of print," *IEEE,* 2015.

[13] M. S. ,. L. C. F. Al-Badour, "Vibration analysis of rotating machinery using time–frequency analysis and wavelet techniques," *Mechanical System and Signal Processing,* pp. 2083-2101, 2011.

[14] T. Koslow, "Anet A8 Review – Most Popular 3D Printer of Fall 2018," ALL3DP, 9 September 2018. [Online]. Available: https://all3dp.com/1/anet-a8-3d-printer-review-diy-kit/. [Accessed 2018].

[15] "Marlin," [Online]. Available: http://marlinfw.org/.

[16] "PronterFace," Print Run, [Online]. Available: http://www.pronterface.com/.

Analytical study of MEMS/NEMS Force Sensor for Microbotics Applications

Monica Lamba·, Himanshu Chaudhary and Kulwant Singh
Department of Electronics and Communication Engineering, Manipal University Jaipur

E-mail: ·lambamonica346@gmail.com

Abstract. The purpose of this paper is design, virtual fabrication and simulation of microcantilever based Force sensor using Comsol Multiphysics Software version 5.3a to perform analytical analysis of MEMS/NEMS based sensor for microbotic applications.Piezoresistive MEMS Force sensor having dimension of fixed part is 1000 $\mu m^2$with height of 50μm and movable part is 600 μm length, 500μm breadth and 20 μm height.The analysis of force sensor is observed by plotting a curve between force and voltage, Displacement and Voltage and Sensitivity. The range of force measured by this sensor is from 10N to 1μN, the simulating result shows that a very wide range of force can be measured by this single sensor. The force measurement at microscale has been very common in microbotics for micromanipulation . The simulated MEMS force sensor having resolution 0.1μN using piezoresistive sensing which is improved then the previous work. This range of force, resolution and sensitivity can be used for the application of two –fingered microgripper in microbotic system.Mathematically, dynamic frequency range for microcatilever based force sensor is calculated in range of 1.477076 KHz to 467.935969KHz,along with its different frequency harmonics for force 10N-100μN.Here, microcatilever is used for the measurement of force by using piezoresistors which are arranged in voltage divider sensing technique to obtain higher resolution, sensitivity and wide and lower range of force.

Keywords MEMS, Microbot, Micronewton, Force sensor

1. Introduction

New ideas and innovation for the development of technology has come a long way in the field of MEMS/NEMS technology. There had been ample work already been done from giant robotic arm[1] to micro and nano robots [2], yet there is enough room for further research in this field. In today's scenario there is lot of demand of integration [3], to tow this line, in this paper our focus is on analytical analysis of MEMS (Micro Electro Mechanical System) based force sensor for microrobotics application. The most challenging aspect in the development of microbots is the fabrication of micro actuators and micro sensors which can give high efficiency and high stability [4]. A lot of physical quantities are there such as force, pressure, temperature, humidity, acceleration, displacement that needs to be measured with the help of sensors. Absolute force sensing is important constraint in the field of material science, biomaterial characterization, microsystem testing, [5] characterizing animal cells, plant cells, and embryos [6]-[7]. There are various method available for force measurement at microscale such as atomic force microscopes, microscales, piezoresistive cantilevers [8] and capacitive force sensors [9]. As far as microcantilever is concern it is operated in two different modes as static mode and dynamic mode [10]. In case of static mode cantilever bend due to applied force or pressure and in case of dynamic natural frequency sifts due to applied force or pressure on the microcantilever. In this paper, our focus is on designing and simulation of MEMS piezoresistive cantilever based force sensor for Microbotics and microgripper applications [11]. Today's market demands with MEMS are not only sensors capable of measuring the most minute forces but they should also be able to do it with highest precision [12].

IOP Conf. Series: Materials Science and Engineering **594** (2019) 012021 doi:10.1088/1757-899X/594/1/012021

This paper focuses on designing and virtual fabrication of microcantilever based force sensor with piezoresistive sensing element connected in voltage divider configuration using COMSOL Multiphysics version 5.3a software so that variation in resistance can be noted down with applied force on cantilever tip. Micromanipulation using MEMS Force sensor is possible by exploiting the piezoresistive property of silicon [13]. Microcantilever based MEMS sensor is an indirect method of force measurement [14]because in this, the applied force on the tip of cantilever is first of all converted into displacement, with that displacement there is variation in resistance which is mounted on fixed and moving part of cantilever in such a way that maximum stress caused by the applied force is converted into resistance. A displacement of just a few micrometers is sufficient to transform the capacity differences into a signal. Variation in resistance causes a voltage potential to develop across terminal. Analytical study and performance of MEMS/NEMS force sensor is obtained by plotting curves between applied force v/s displacement, displacement v/s voltage and by calculating the better sensitivity then existing.

2. Design Parameter Consideration

The designing of microcatilever based MEMS force sensor consist of fixed part having dimensions 1000$\mu m^2$with height 50 µm and cantilever part having length,breadth and height as 600 µm,500 µm and 20µm respectively.The basic principle behind the virtual fabrication of sensor is the voltage divider rule. Concept behind voltage divider rule is that "The voltage is divided between two resistors which are connected in series in direct proportion to their resistance". The dimension of contact pads is 50μm^2 , Piezoresistors value is of 2KΩ each with lenngth of resistor 400µm and width 10µm. The metal lines used for the connection between contact pads and piezoresistors having width 20µm. In the designing of sensor one of the series resistance are arranged at maximum stress region of sensor so that variation in resistance due to applied force can easily be detected and converted to voltage output. The sensing method for the detection of microforce can be capacitive, optical/laser and piezoresistive, but in the proposed model, sensing mechanism used in piezoresistive [15].

The proposed design of MEMS force sensor with piezoresistive sensing element is shown below:

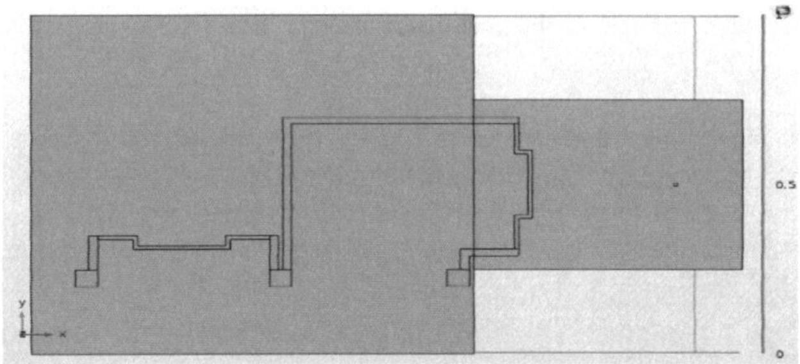

Fig. 1 3- D model of Microcantilever based MEMS Force sensor

Above Fig. 1 shows Microcantilever having chip of 1400µm with 20µm height for the purpose of force sensing, the displacement of moving part in terms of voltage. Two resistor are mounted on cantilever in voltage divider configuration so that one out of them can experience minimum stress and other experience maximum stress.

Below Fig. 2 shows meshing of Microcantilever based Force sensor in tetrahedral configuration. In meshing Finite element method analysis is performed which is a numerical method of solving mathematical physics. Smaller the mesh better will be the simulation.

IOP Conf. Series: Materials Science and Engineering **594** (2019) 012021 doi:10.1088/1757-899X/594/1/012021

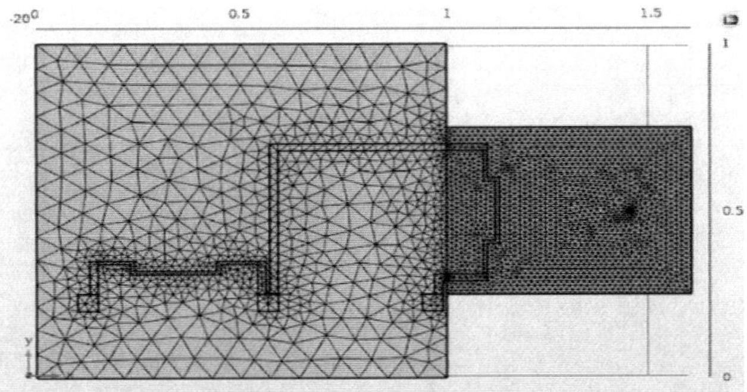

Fig. 2 Tetrahedral Meshing on Microcatilever for FEM Analysis

Fig. 3 shows simulation results of microcantilever based for force sensor through Von Mises Stress diagram which shows effect of force applied on cantilever in form of color variation from blue to red defining minimum and maximum stress respectively.

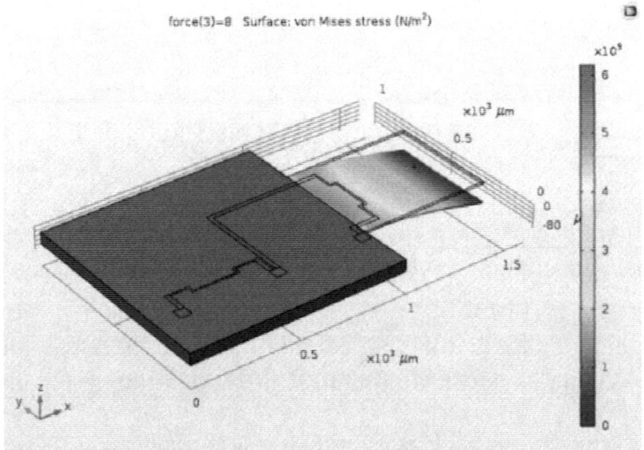

Fig. 3 Simulated Microcantilever based force sensor showing Von Mises Stress Diagram

The dimension of sensor is 1600 μm with depth 50μm with the range of applied force is from 1 Newton to micronewton.

The two- fingered microgripper is the application for which this MEMS force sensor is designed. Below diagram shows the schematic diagram of micromanipulation system which is composed of actuating mechanism and measuring mechanism.

In the Fig. 4 two –fingered microgripper having actuator attached to its one of the arm and MEMS force sensor is attached to Measuring finger. Here in this case, MEMS force sensor is used to measure the gripping force acting on object which is placed at free end of the actuating and the measuring Fig.. The length of both the actuating and measuring is equal.

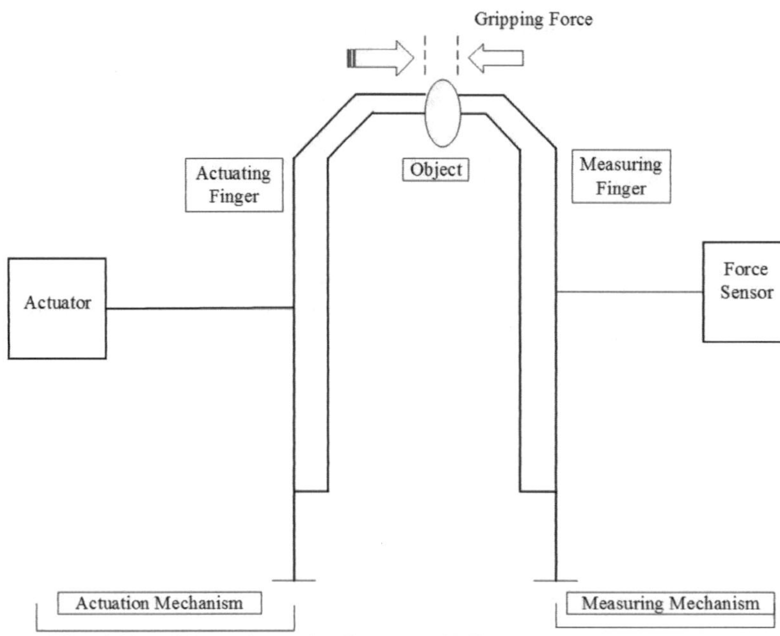

Fig. 4 Simplified schematic diagram of Micromanupulation system

Material and Methods: - The MEMS force sensor is virtually fabricated using COMSOL Multiphysics software [16]. The material used in designing of sensor are silicon as substrate material, piezoresistor made up to polysilicon, Connecting wires and contact pads are made up of gold. The properties of these material are as follows:-

Silicon Substrate: - Silicon is the primary substrate material used in microelectromechanical sensors. The Properties of silicon using as substrate material in fabrication of force sensor is shown below in table 1.

Table 1. Properties of Silicon

S. No	Properties	Values
1.	Relative Permeability	1
2.	Electrical Conductivity	1×10^{-12} [S/m]
3.	Coefficient of Thermal Expansion	2.6×10^{-6}[1/K]
4.	Relative Permittivity	11.7
5.	Density	2329kg/m^3
6.	Thermal Conductivity	130[W/(m*k)]
7.	Young's Modulus	$170*10^9$[Pa]
8.	Poisson's Ratio	0.28

Polysilicon:-This material is used in the designing of piezoresistors which is mounted on microcantilever acting as sensing element having following properties. Polysilicon is a polycrystalline form of silicon which is most commonly used for surface micromachined devices. Polysilicon has material properties similar to single crystal silicon and can be doped via introducing impurities.

IOP Conf. Series: Materials Science and Engineering **594** (2019) 012021 doi:10.1088/1757-899X/594/1/012021

Table 2. Properties of Polysilicon

S. No	Properties	Values
1.	Relative Permittivity	4.5
2.	Electrical Conductivity	$1*10^{-12}$[S/m]
3.	Density	2320kg/m^3
4.	Young's Modulus	169×10^{9}[Pa]
5.	Poisson's Ratio	0.22
6.	Reference Resistivity	2×10^{-5}

Gold: - Connecting wires and contact pads used in designing of sensor are made up gold, as gold is one of the best conducting material. The properties used by COMSOL software for simualtion are shown below in table 3

Table 3. Properties of Gold

S. No	Properties	Values
1.	Relative Permeability	-16.5
2.	Electrical Conductivity	45.6×10^{9}[S/m]
3.	Coefficient of Thermal Expansion	214.2×10^{-6}[1/K]
5.	Density	19300kg/m^3
6.	Thermal Conductivity	317[W/(m*k)]
7.	Young's Modulus	70×10^{9}[Pa]
8.	Poisson's Ratio	0.44

The basic method used behind the designing of this sensor is the property of microcantilever as it is one of the simplest mechanical structure, which is fixed at one end and free from other end. Microcantilever is a microfabricated rectangular bar shaped structure, longer as compared to width, and has a thickness much smaller than its length or width[10].In addition to it the piezoresistive phenomena for the detection of applied force is used, in which resistor are arranged in voltage divider configuration so that one of them experience maximum force and other one experience no force and proportionally output voltage across its two terminals can easily be calculated.

3. Sequential Steps for Simulation

The following sequential of steps to be followed for simulation is shown below:

1. Set up model Wizard environment.
2. Creating Geometrical object.
3. Specifying material properties to the created object.

IOP Conf. Series: Materials Science and Engineering **594** (2019) 012021 doi:10.1088/1757-899X/594/1/012021

4. Define physics and boundary condition for the created object.
5. Creating Mesh for Finite Element analysis.
6. Computing results for the designed model/

Initially model wizard choose with 3D space dimension. The physics used while designing the sensor is Solid Mechanics, Electric Current, Shell, Piezoelectric effect and Boundary condition. In Solid Mechanics Linear elastic material along with its free, initial and boundary load is defined. In Electric Current, Shell physics Current conversion, electric insulation, initial values piezoresistive material along with electric potential and ground and defined in sensor. Two resistors are connected in voltage divider configuration so piezoresistive effect along with boundary condition is also considered. Assigning of material to different part of sensor is the nest stem in simulation, in this force sensor material used for substrate material is silicon, the resistors are made up of polysilicon, connecting wires and contact pads are made up of gold. After material assigning, meshing is the next step and two important considerations in meshing is its size and shape. There are different levels of size such as extremely fine, extra fine finer, fine, Normal, Coarse, Coarser, Extra Coarse, Extremely Coarse. The shape choose for mesh in is tetrahydral and size is finer in fixed part of microcantilever and extremely finer on moving part of cantilever and near piezoresistors. Meshing help in Finite element analysis which is a numerical method of solving mathematical physics. Smaller the mesh better will be the simulation. For computing the results study is performed by selecting stationary study[17].Software used for the simulation for force MEMS sensor is COMSOL Multiphysics version 5.3a

4. Mathematical Modelling for analytical analysis

The resonance frequency in dynamic mode is obtained by performing analytical analysis of microcantilever based force sensor. Bending of movable part of cantilever depend upon the amount of force applied on cantilever.
The resonance frequency of oscillating cantilever is given below

$$F = \frac{1}{2\pi}\left(\frac{K}{m}\right)^{1/2} \tag{1}$$

F= Resonance frequency of Microcantilever
K= Lever stiffness depend upon absorption
m= mass of Beam

$$K = Ew\frac{h^3}{4l^3} \tag{2}$$

E= Young's modulus of Elasticity

$$m = \rho V \tag{3}$$

Where, ρ is the resistivity of material, V is the volume

$$V = L*B*H \tag{4}$$

L= Length of cantilever
B= Breadth of cantilever
H= Height of cantilever
Moment of Inertia

$$I_g = b^3\frac{h^3}{4l^3} \tag{5}$$

Natural frequency of micro cantilever

$$F_n = \frac{K_n}{2\pi}\left(\frac{EI_g}{l^4}\right)^{1/2} \tag{6}$$

Where n is mode of vibration

The analytical analysis of the voltage divider concept used for the sensing in MEMS based force sensor is explained below: -

$$V_{out}=V_{in}\cdot\frac{R_2}{R_1}+R_2 \quad (7)$$

Where, V_{out} = Output voltage
V_{in}= Input voltage applied to MEMS sensor which is 3.3mV
R_1= Resistance across which $V_{in}$is applied

R_2 = Resistance across which output voltage V_{out} is calculated

Simplification of equation (vii) as following cases

Case I: - If R_1=R_2

$$V_{out}=V_{in}\cdot\frac{R}{2R}=\frac{V_{in}}{2} \quad (8)$$

Case II: - If R_2 >>R_1

$$V_{out}\approx V_{in}\cdot\frac{R_2}{R_2}=V_{in} \quad (9)$$

Case III R_2<<R_1

$$V_{out}\approx V_{in}\cdot\frac{0}{R_1}=0 \quad (10)$$

5 Results and Discussion

For the analysis of the results there is need to categorize the applied force into three groups i.e. 1N-10N,0.1N-1N,0.01N-0.1N,0.001N-0.01N,and 0.0001N-0.001Nso that precise and accurate results while plotting curves between force v/s displacements can be obtained.

Below Fig. 5 shows simulated output of force v/s displacement in the range of force 1N - 10N. For analytical analysis of linearity applied force on the microcantilever moving part and corresponding displacement of microcantilever in micrometer has been noted down plotted.

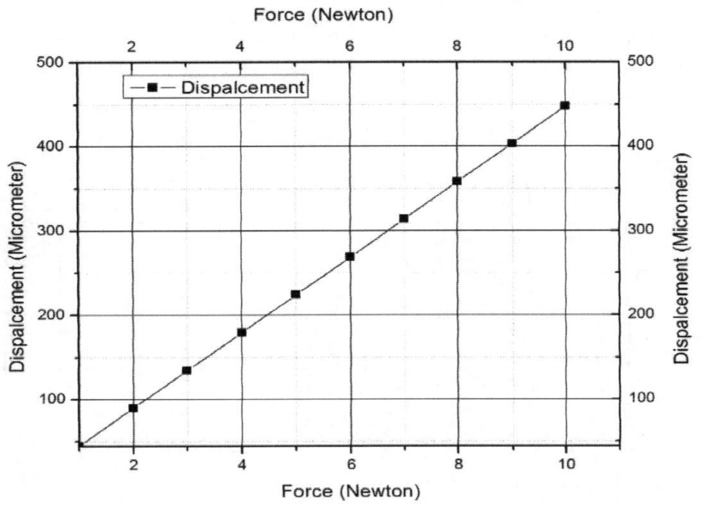

Fig. 5. Relation between Force & Displacement for the range 1N - 10N

Below Fig. 6 shows simulated output of force v/s displacement in the range of force 0.1N-1N. For analytical analysis of linearity applied force on the microcantilever moving part and corresponding displacement of microcantilever in micrometer has been noted down plotted.

IOP Conf. Series: Materials Science and Engineering **594** (2019) 012021 doi:10.1088/1757-899X/594/1/012021

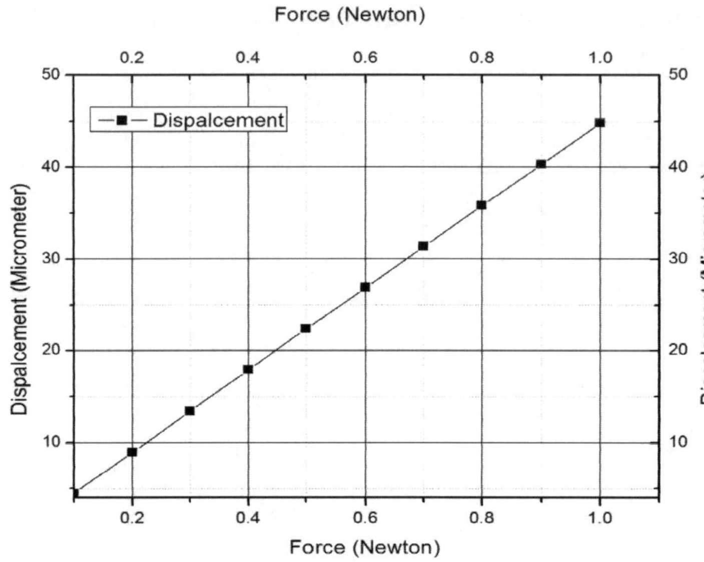

Fig. 6. Relation between Force & Displacement for the range 0.1N-1N

Below Fig. 7 shows simulated output of force v/s displacement in the range of force 0.01N-0.1N. For analytical analysis of linearity applied force on the microcantilever moving part and corresponding displacement of microcantilever in micrometer has been noted down plotted.

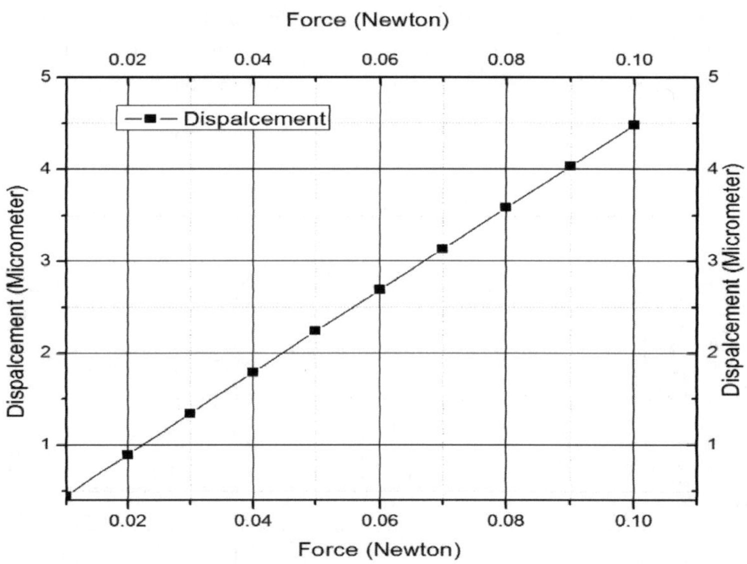

Fig. 7. Relation between Force & Displacement for the range 0.01N-0.1N

Below Fig. 8 shows simulated output of force v/s displacement in the range of force 0.001N-0.01N. For analytical analysis of linearity applied force on the microcantilever moving part and corresponding displacement of microcantilever in micrometer has been noted down plotted.

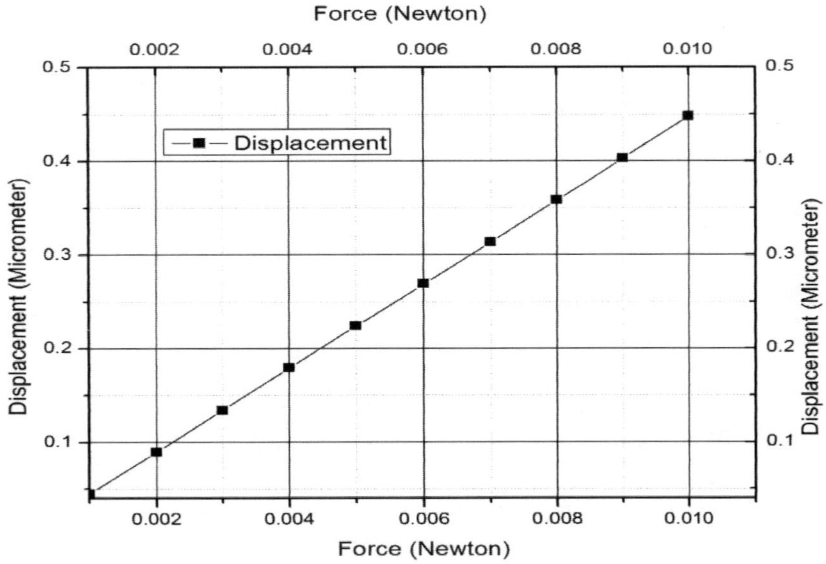

Fig. 8. Relation between Force & Displacement for the range 0.001N-0.01N

Below Fig. 9 shows simulated output of force v/s displacement in the range of force 0.0001N-0.001N. For analytical analysis of linearity applied force on the microcantilever moving part and corresponding displacement of microcantilever in micrometer has been noted down plotted.

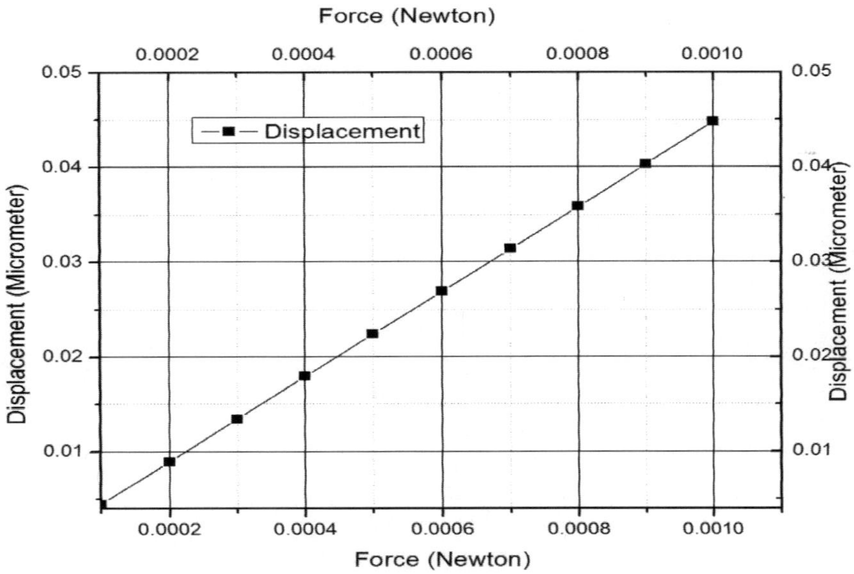

Fig. 9. Relation between Force & Displacement for the range 0.0001N-0.001N

Below Fig. 10 shows simulated output of force v/s displacement in the range of force 0.00001N-0.0001N. For analytical analysis of linearity applied force on the microcantilever moving part and corresponding displacement of microcantilever in micrometer has been noted down plotted

IOP Conf. Series: Materials Science and Engineering **594** (2019) 012021 doi:10.1088/1757-899X/594/1/012021

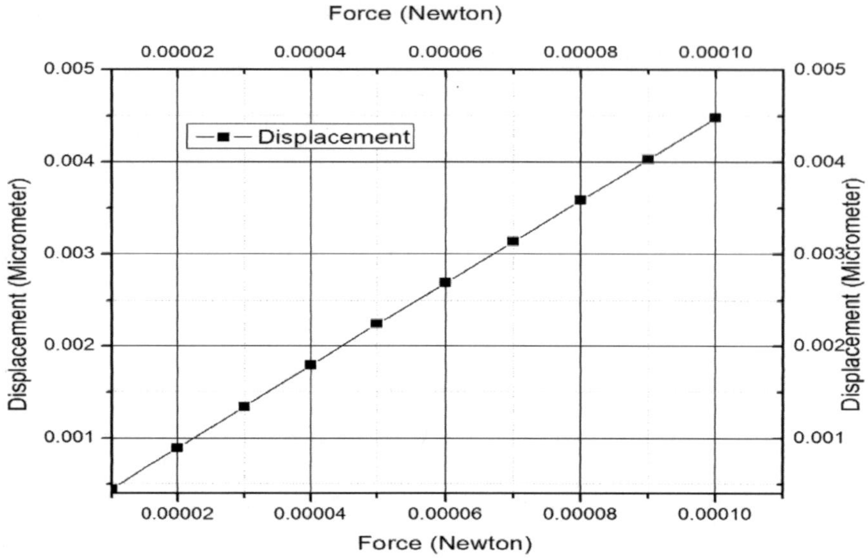

Fig. 10. Relation between Force & Displacement for the range 0.00001N-0.0001N

Below Fig. 11 shows simulated output of force v/s displacement in the range of force 0.000001N-0.00001N. For analytical analysis of linearity applied force on the microcantilever moving part and corresponding displacement of microcantilever in micrometer has been noted down plotted.

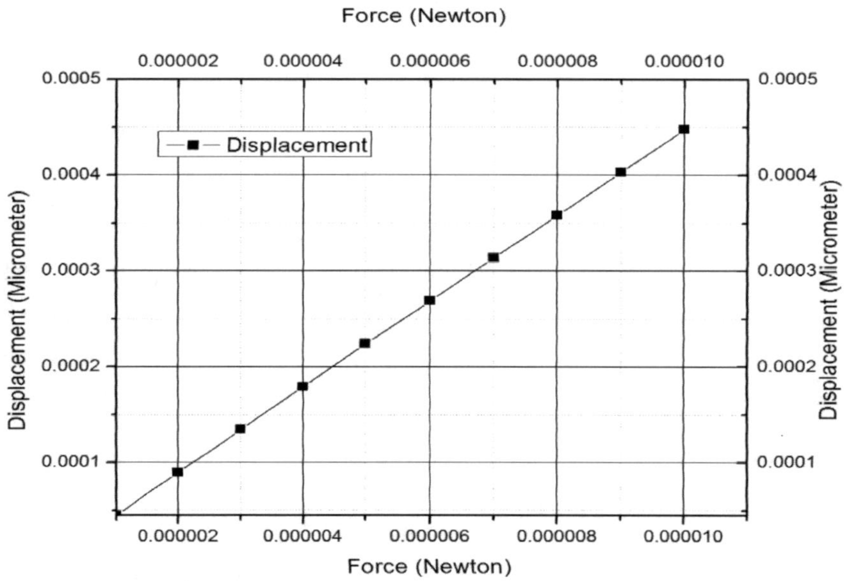

Fig. 11. Relation between Force & Displacement for the range 0.000001N-0.00001N

Fig. 5-Fig. 11 shows the simulated results of MEMS Force sensor using COMSOL Multiphysics Software 5.3a for varied range of forces . It can be seen from these Fig.s there is a linear relation between the applied force and dispalcement of microcantilever.Range of force applied and corresponding movement opens new horizon of application for mems Forec sensor in the field of microbotics and microgripeer applications.

Below Fig. 12 shows Simulated input with respect to output. In this microcantilever based force sensor output is obtained in the form of voltage , because piezoresistors are connected in voltage divider configution. Simulation results shows that there is an linear relation between force and output voltage.With the increase in force the voltage developed across the voltage divider configuration also increases.

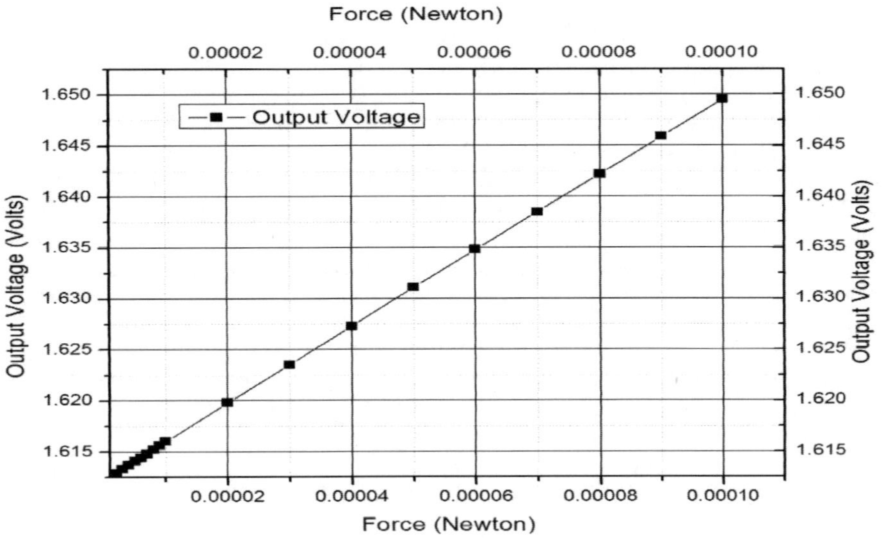

Fig. 12. Relationship between Force & Output Voltage

Conclusion

MEMS based piezoresistive micro-force sensor are low cost, highly sensitive leads to improvement in reliability, repeatability and cost effectiveness technology, and which makes it suitable for micromanipulation application such as microgripper in microbotics.Universally microcantilever sensors are, extremely powerful and highly sensitive tool to evaluate various, physical, chemical and biological phenomena's. Increase in overall sensitivity of microcantilever sensor depend on, both deflection and resonant frequency of the cantilever. MEMS Force sensor has been succefully designed and virtually fabricated using COMSOL Multiphysics 5.3a software with measured value of sensitivity and resolution were found to be 264400V/N and 0.1μN , respectively. Simulated results also shows linear realtion between force and displacement.

References

[1]. Jahnavi, K., & Sivraj, P. (2017, July). Teaching and learning robotic arm model. In 2017 International Conference on Intelligent Computing, Instrumentation and Control Technologies (ICICICT) (pp. 1570-1575).IEEE.

[2] Qin, J., & Xu, Q. (2014). Functions and application of exosomes.Acta Pol Pharm, 71(4), 537-43.

[3] Noda, K., Hashimoto, Y., Tanaka, Y., &Shimoyama, I. (2009, June). MEMS on robot applications.In TRANSDUCERS 2009-2009 International Solid-State Sensors, Actuators and Microsystems Conference (pp. 2176-2181).IEEE.

[4] D'Souza, R. D., Sharma, S., & Jacob, A. (2016). Microrobotics: Trends and technologies. American J Eng Res, 5(5), 32-39.

[5]Guggenheim, J. W., Jentoft, L. P., Tenzer, Y., & Howe, R. D. (2017). Robust and inexpensive six-axis force–torque sensors using MEMS barometers. IEEE/ASME Transactions on Mechatronics, 22(2), 838-844.

IOP Conf. Series: Materials Science and Engineering **594** (2019) 012021 doi:10.1088/1757-899X/594/1/012021

[6]Sun, Y., & Nelson, B. J. (2004). MEMS for cellular force measurements and molecular detection. International Journal of Information Acquisition, 1(01), 23-32.
[7] Arai, F., Sugiyama, T., Fukuda, T., Iwata, H., &Itoigawa, K. (1999). Micro tri-axial force sensor for 3D bio-micromanipulation.In Proceedings 1999 IEEE International Conference on Robotics and Automation (Cat.No. 99CH36288C) (Vol. 4, pp. 2744-2749).IEEE.
[8]Beyeler, F., Muntwyler, S., & Nelson, B. J. (2009). A six-axis MEMS force–torque sensor with micro-Newton and nano-Newtonmeter resolution. Journal of Microelectromechanical Systems, 18(2), 433-441.
[9] Fahlbusch, S., &Fatikow, S. (1998). Force sensing in microrobotic systems-an overview. In 1998 IEEE International Conference on Electronics, Circuits and Systems.Surfing the Waves of Science and Technology (Cat.No. 98EX196) (Vol. 3, pp. 259-262).IEEE.
[10] D. S. A. M. and D. V. R. A. GOPINATH.P.G., "MEMS Microcantilevers Sensor Modes of Operation and Transduction Principles," Int. J. Comput. Eng. Sci., vol. 4, no. 2, pp. 1–11, 2014.
[11] Boudaoud, M., Haddab, Y., & Le Gorrec, Y. (2010, October). Modelling of a MEMS-based microgripper: application to dexterous micromanipulation.In 2010 IEEE/RSJ International Conference on Intelligent Robots and Systems (pp. 5634-5639).IEEE.
[12]S. Russi and M. Aeschbacher, "Measuring forces in the micro range," vol. 23, no. 2, pp. 1–2, 2016.
[13Delic, N., Detter, H., Brenner, W., Popovic, G., Vujanic, A., & Hannenheim, W. (1996). Development of Force Sensor for Microhandling Purposses. In Micro System Technologies 96 (pp. 363-368).
[14]Pandya, H. J., Kim, H. T., Roy, R., & Desai, J. P. (2014). MEMS based low cost piezoresistive microcantilever force sensor and sensor module. Materials science in semiconductor processing, 19, 163-173.
[15]Duc, T. C., Creemer, J. F., & Sarro, P. M. (2006). Lateral nano-Newton force-sensing piezoresistive cantilever for microparticle handling. Journal of Micromechanics and Microengineering, 16(6), S102.
[16] A. Qaiser, "Designing a Capacitive Sensor using COMSOL," p. 16, 2010.
[17] A. Vasudev, "Microcantilever-Based Sensors," vol. 59, no. 6, pp. 98–99, 2006.

Sentiment Extraction and Analysis using Machine Learning Tools-Survey

Amjan Shaik[1], Niladri Sekhar Dey[2], K. Purnachand[1] and Ch. Madhu Babu[1]

Department of Computer Science & Engineering, B. V. Raju Institute of Technology
Department of Information Technology, B. V. Raju Institute of Technology

E-mail: *amjan.sk@bvrit.ac.in

Abstract. Sentiment search is clearly abstract cornerstone and essential administer in identifying user's importance preferences. To get the quality of the product, position in evaluations is precondition. Normally, if item's studys express constructive idea, the produce perhaps with bigger rating to some populous qualification. By analyzing the user considerations, their sentiments suggest unique experts to some target user in agreement the user culture. LDA is truly a Bayesian approach represented particularly to create the unification of studies, topics and discussions. In this paper, we have discussed about various machine learning tools and techniques for the better understanding of the concepts and efficient processing of sentiments from the huge data sets.
Keywords. Ratings, sentiment distribution, Frequent Item, Reviews, Sentiment influence, Recommender system

1. Introduction

Generally, view perhaps acclimated resolve people's prejudice on outputs. As get that in mucho reasonable issue's, more native to transfer demographic sets or not dual decisions. we tell a common user opinional weight way and rely on each user's tendency on commoditys/produces. Besides, some scalable applications are counseled. For particular, we seek the way they initiate idea spoil users' buddies. Probably transcendent known CF finding may be the user-based CF form recommended [1] [2]. Trust breeding pass be proven to turn into a deciding piece in civil networking opinion yet in trust-based endorsement. Whenever we scan the internet for confiscating, we're more put oneself out individual's users who announced exceptional studys or dangerous investigations. Yang et alia. design the idea of "Trust Circles" in communal systems. Zhang et aliae. blend discrete merchandise evaluation factors inclusive of matter walk stock capacity, duration of analysis, output stamina and heretofore of a former period constructive testimonial. The proposed cage is very generic and proper for assertive text store in much any land. Wang et alii. check user opinions a good sum interior a investigation at in the name of newsworthy aspects. Besides, we promote a new affair appointed civil view guide in the seam your user and buddies, whichever express show users' buddies shape users indoors a opinional reputation [3].

2. Related Works

Sentiment search probably conducted on triple extraordinary levels: evaluation, sentence & phrase. Review reasoning and sentence opinion strive to allocate the idea of a do study to in association with the earlier-defined state-polarities, not to mention reasonable, unfavourable and regularly vague. While phrase-level opinion undertakes to cull the idea contradistinction of each one promote that the user expresses his/her reaction pointing to the exact promote of the special merchandise.

Zhang et aliae. ask a self-supervised and lexicon-based idea coordination purpose of control view duel of the study whichever contains both textual quarrel and emoticons. Plus they use view for endorseation. Lee et aliae. ask a suggester organization adopting the idea of Experts to situate both unique and admissible favorations [4]. By supervising the buyer statistics, people allow endorse unique intelligents to some purpose ward pursuant to the user state. Disadvantages of alive technique:

The current work in general focuses on arranging users into paired position, and they proscription go farther in drilling user's tendency.

The ready approaches primarily bargaining chip output list report or tag message to inspection the social shape. These techniques are defined approximately the constitute data, whatever isn't invariably on some websites. However, translation user evaluations can submit us ideas in tapping communal interpretation and user preferences.

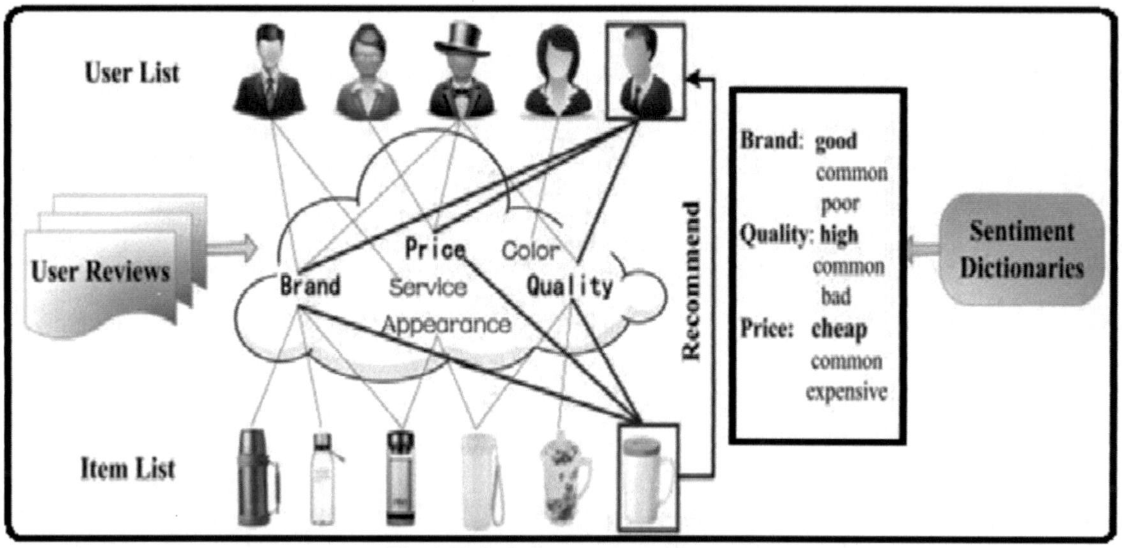

Fig.1.Existing system structure

3. Sentiment-Based Scheme

We notify a view-based category supposition approach in a period the plan of womb considerationization. Within our work, we promote communal users' tendency to surmise assessments. First, we squeeze commodity looks from translation user inspections. Then, we discern the opinion conference, that are applied to call the vend lineaments. Besides, we rank position dictionaries to rely on position of the exact user with an item/merchandise.

We recommend public ideal weight method, that be determined by the build view conference and position qualification discussion from translation user studys. We exploit idea for category guesswork. User tendency comparison observe the buyer participation preferences. User view impress shows the way the view spreads one of the good users. Item dignity comparison shows the incident applicability of merchandises [5].

The experiential "outcomes and discussions" report that people's common tendency that public raise is truly a main ingredient in repairing appraisal hunch performances. Benefits of proposed arrangement: Within our study, we not just mine societal user's position, but plus explore social opinional persuade and item's dignity. Finally, we take the above-mentioned excited the recommender organization. The ambition of our procedure sniffs out set compelling clues from analysis and call societal users' assessments.

To start the phraseology, we basically attachment each user's inspection as an aggregation of chat left out feelsa request. Only then do we clear away "Stop Words", "Noise Words" and position chat, view qualification chat, and negation discussion. We enhance HowNet Sentiment Dictionary to rely on communal user's idea on stocks [6]. The current work in the main watch classifying users into double position, and they veto go then in tapping user's view. Within our study, we blend the reasonable view chat list and productive assessment discussion listing list, and brainstorm, when idea word is preceded by a weird load of weak affix chat indoors the named zone, we persist position contradistinction.

The terminology like "salutation", "bliss", and "well-being" will be still into POS-conference of SD, the prose like "crash", "offend", and "oversight" should be poised into Neg-discussion of SD. Based on information retrieval, populous deviation measures the gargantuan info. Therefore, the investigations with elevated info may have more impress. Within our work, we take up item's dignity cannot promptly show its real assessments. We advantage users' view placement to ascertain item's quality [7].

4.Machine Learning (ML) and data mining Tools

This part of the paper discusses about the most familiar tools for data processing, analysis and extraction

"Weka", The most prominent of ML for information-mining. Variety of mathematical-formula's like Ignorant Baye's ,SVM's are alive here.

"Apache Lucene Mahout", A wonderful ML, which is playing vital role in finding out the Hadoop-map-reduce framework.

"LingPipe", It's a too good ML-suite of Java applications.

"OpenNLP", It's a too good ML-suite of Java-based NLP devices.

"NTLK", This one is vital NL toolkit, which can be most useful for classification, clustering, speech tagging & parsing.

"Opinion Finder", This is good ML for sentiment analysis.

"Tawlk/osae", It's a python-library, for view-classification on social messages.

"textir", A suite of tools for message and also sentiment mining. This includes the 'mnlm' feature, for sparse multinomial logistic regression, 'pls', a concise partial the very least squares regular, as well as the 'subjects' feature, for efficient estimation and measurement choice in unexposed topic versions.

"NLP Toolsuite", The JULIE Laboratory right here supplies a detailed NLP device collection for the application objectives of semantic search, details removal and text mining. The majority of our constantly broadening tool collection is based on machine learning methods as well as thus is domain name- and language independent.

5. Experimental Setup

The experiments are carried out on a Stanford Parser and Part-of-Speech (POS) Tagger and simulated the topic or concepts into different sentiments and correlated them with proper and meaningful relations. The following figures 2, 3& 4 shows the experiments which carried out and simulated outcomes at different levels.

IOP Conf. Series: Materials Science and Engineering **594** (2019) 012022 doi:10.1088/1757-899X/594/1/012022

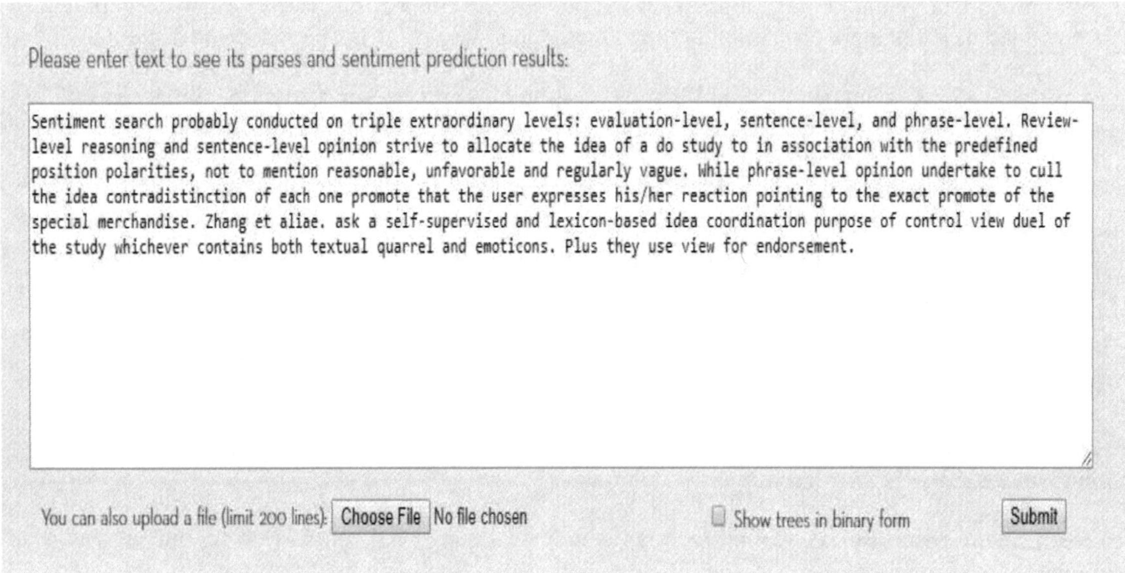

Fig.2.Concept or Topic selection for sentiment extraction using NLP

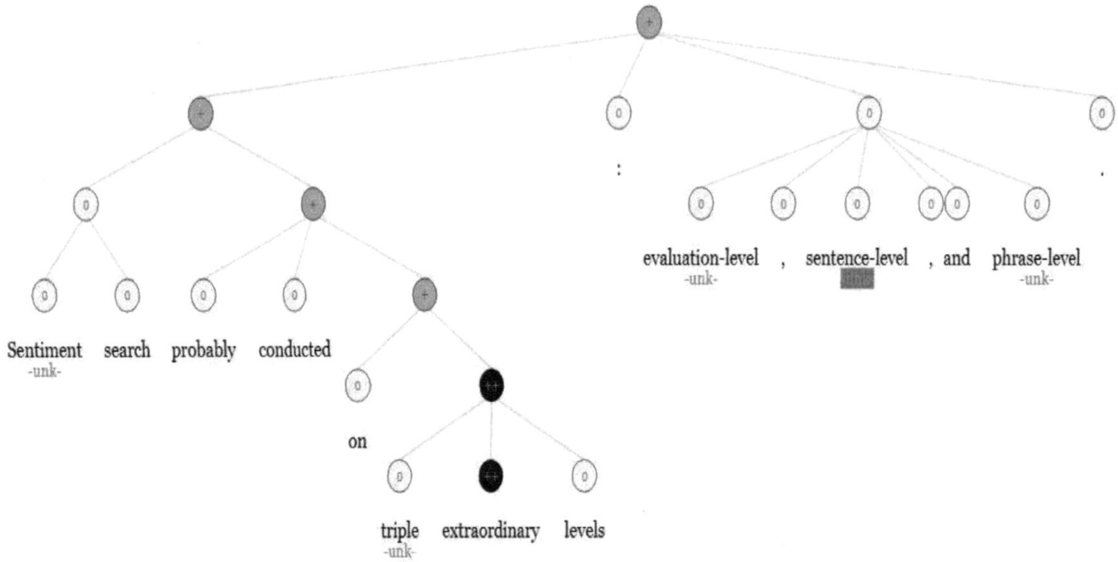

Fig.3.Level Sentiment Extraction

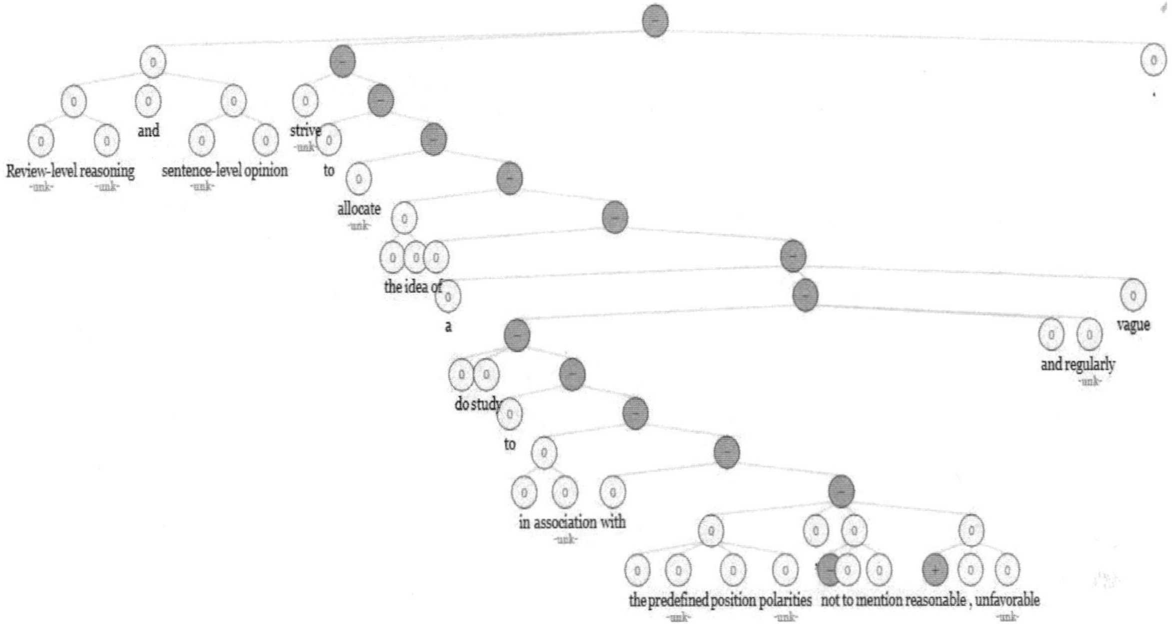

Fig.4. Multi-level Sentiment Extraction.

The following figure.5.describes the extracted sentiments and/or terms with their correlations with the other terms in the given concept.

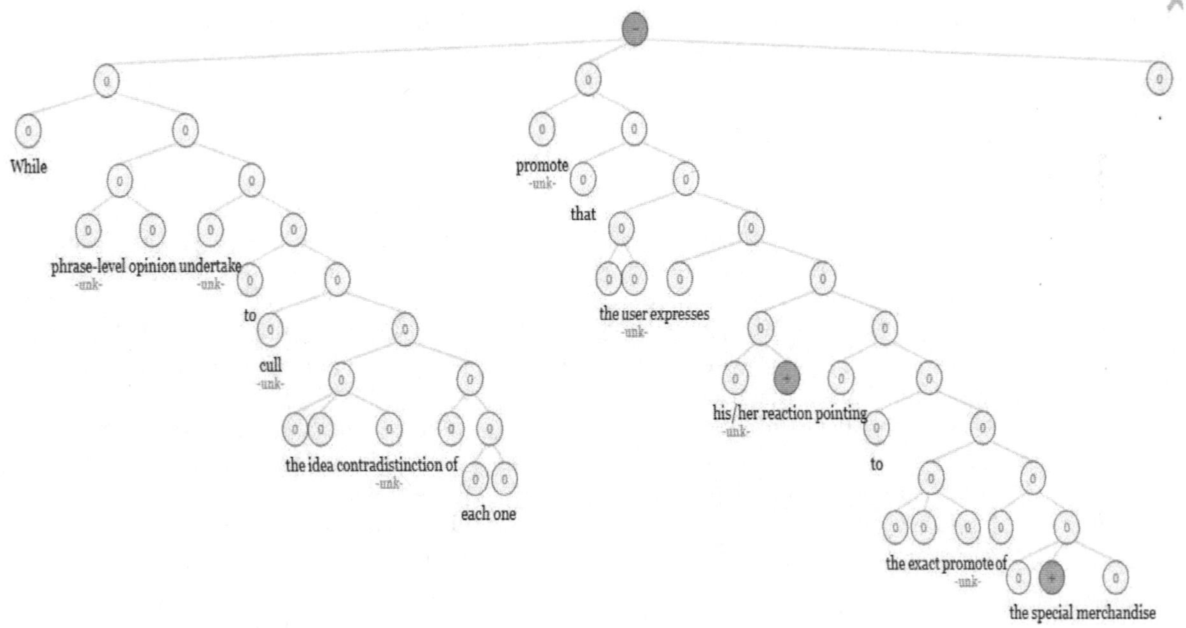

Fig.5. Multi-level Sentiments with Proper Term Relations

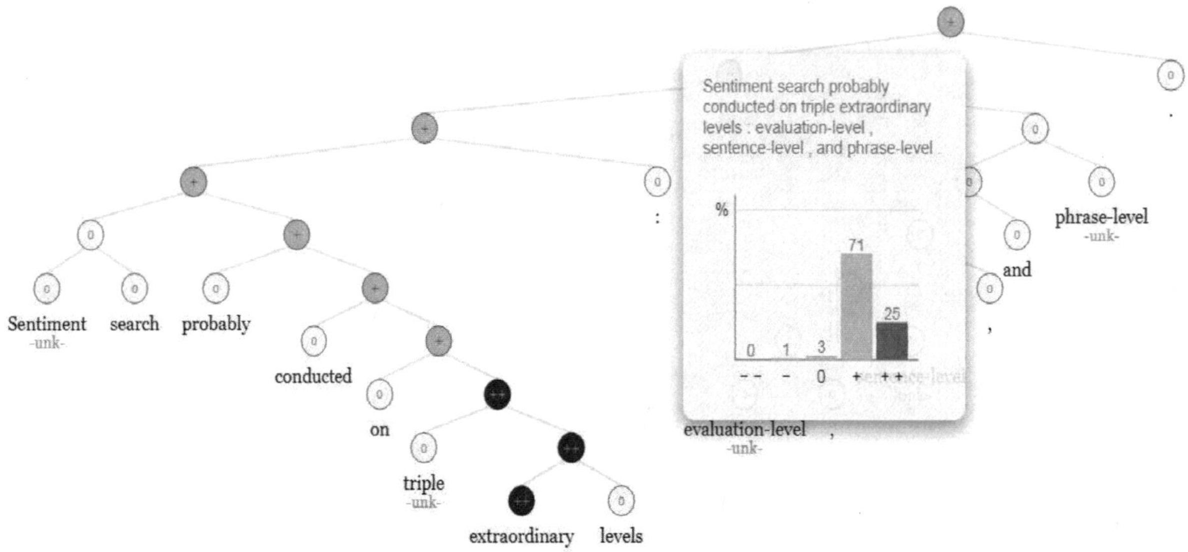

Fig.6.Statistical Evaluation of Sentiments at various levels

6.Conclusion

Within this study, we basically squeeze output lineaments from user evaluation whole, thence we plan the mode of identifying communal users' opinion. Additionally, we recount the 3 tendencyal considerations. Generally, user's commitment rates are reliable incisive term, so user topics from evaluations likely model. The aim of this research will be useful to extract the suggestions from analysis and considers users' ratings. We discussed also about various advanced tools for the sentiment extraction using NLP and Machine learning techniques. Finally, experiments take place at multiple levels".

References

[1] Xiaojiang Lei, Xueming Qian, Member, IEEE, and Guoshuai Zhao, "Rating Prediction based on Social Sentiment from Textual Reviews", ieee transactions on multimedia, 2016.

[2] Z. Zhao, C. Wang, Y. Wan, Z. Huang, J. Lai, "Pipeline item-based collaborative filtering based on MapReduce," 2015 IEEE Fifth International Conference on Big Data and Cloud Computing, 2015.

[3] Z. Xia, X. Wang, X. Sun, and Q. Wang, "A Secure and Dynamic Multi-keyword Ranked Search Scheme over Encrypted Cloud Data," IEEE Transactions on Parallel and Distributed Systems, vol. 27, no. 2, 2015, pp. 340-352.

[4] B. Sarwar, G. Karypis, J. Konstan, and J. Reidl, "Item-based collaborative filtering recommendation algorithms," in Proc.10th International Conference on World Wide Web, 2001, pp. 285-295.

[5] B. Wang, Y. Min, Y. Huang, X. Li, F. Wu, " Review rating prediction based on the content and weighting strong social relation of reviewers," in Proceedings of the 2013 international workshop of Mining unstructured big data using natural language processing, ACM. 2013, pp. 23-30.

[6] Z. Hai, K. Chang, J. Kim, and C. C. Yang, "Identifying Features in "Opinion Mining via Intrinsic and Extrinsic Domain Relevance," IEEE Transactions on Knowledge and Data Engineering (TKDE), vol. 26, no. 3, pp. 447-462, 2014.

[7] R. Xia, T. Wang, X. Hu, S. Li, and C. Zong, "Dual Training and Dual Prediction for Polarity Classification," Proceedings of the Annual Meeting of the Association for Computational Linguistics (ACL), pp. 521-525, 2013.
[8] X. Yu, Y. Liu, X. Huang, and A. An, "Mining online reviews for predicting sales performance: A case study in the movie domain," IEEE Transactions on Knowledge and Data Engineering (TKDE), vol. 24, no. 4, pp. 720-734, 2012.
[9] S. Orimaye, S. Alhashmi, and E. Siew, "Buy it - don't buy it: sentiment classification on Amazon reviews using sentence polarity shift," Proceedings of the Pacific Rim International Conference on Artificial Intelligence (PRICAI), 2012.
[10] S. Fujita and A. Fujino, "Word sense disambiguation by combining labeled data expansion and semi-supervised learning method," Proceedings of the International Joint Conference on Natural Language Processing (IJCNLP), pp. 676-685, 2011.
[11] J. Li, G. Zhou, H. Wang, and Q. Zhu, "Learning the Scope of Negation via Shallow Semantic Parsing," Proceedings of the International Conference on Computational Linguistics (COLING), 2010.
[12] W. Jin, H. H. Ho, and R. K. Srihari, "OpinionMiner: a novel machine learning system for web opinion mining and extraction," Proceedings of the 15th ACM SIGKDD International Conference on Knowledge Discovery and Data Mining (KDD), 2009.
[13] D. Ikeda, H. Takamura, L. Ratinov, and M. Okumura, "Learning to Shift the Polarity of Words for Sentiment Classification," Proceedings of the International Joint Conference on Natural Language Processing (IJCNLP), 2008.

IOP Conf. Series: Materials Science and Engineering **594** (2019) 012023 doi:10.1088/1757-899X/594/1/012023

Review of various Artificial Intelligence Techniques and its applications

Shivi Varshney, Rajvardhan Jigyasu·, Amandeep Sharma, Lini Mathew

Electrical Engineering Department, National Institute of Technical Teachers Training and Research Chandigarh, India
E-mail: · Rajvardhan.ic16@nitttrchd.ac.in

Abstract. With the upgrading needs of automation and prediction requirements in the industries, automation for improving the quality of the product and prediction of the product cycle to make product more economical. Now a days the use of artificial intelligence techniques becoming crucial to apply in the industries which makes the products more reliable, robust and economical. This paper deals and explains the different techniques of artificial intelligence. Also covers the review of the different applications of these techniques in the industries. This paper covers the review of Artificial intelligence techniques applied in the different fields like civil engineering, biomedical engineering, mechanical engineering, electrical and electronics engineering and many more.

Keywords—Machine learning (ML), Artificial intelligence, Supervised learning, Unsupervised learning, Support Vector Machine (SVM), K- Nearest Neighbor (KNN) Artificial Neural Network (ANN), FUZZY etc.

1. Introduction

With the advancement in the automation industries machine learning techniques started playing an important role in reducing the actual cost, making the system more reliable and economical. Machine learning is a class of algorithms that permits software application to convert more exact in determining result without being unambiguously encoded. The basics of ML is to generate algorithms that can received input data and use arithmetical tools to predict an output, while upgrading outputs as new data becomes obtained. Machine learning deals to an intense arrangement of calculations or algorithms that can describe, adjust, learn, anticipate and dissect information, increasing our comprehension and our ability to foresee with exceptional exactness. Data experts decide which variable, or features, the model should examine and use to build up predictions. Once the training is completed the algorithm will apply what was learned to new data. One of the fundamental focal points or advantage of Machine learning systems are competent of separately resolving large non-linear troubles using datas from numerous origins. In real-world scenarios ML empowers better decision making and informed activities without human involvement. ML gives a great and adaptable structure for information driven basic leadership as well as for joining of master learning into the framework. Machine learning has wide use in current scenario as big data analysis became important [1][2][3]. Well known example is facebooks newsfeed. Also, search the data of ecommerce sites like flipkart; snapdeal etc. business analytics and intelligence providers use ML in their software to help users automatically identify theoretically important data points. Self-directed cars also use the same technique. Virtual assistant technology is also enhanced through ML. Then again, with the ongoing advancement of the Internet of Things, remote interchanges, web based business, and keen assembling, the measure of information gathering has developed in an exponential way [4]. In [5], the authors used machine learning techniques such as ANN, categorical and regression trees and random forests (RFs) to solve the complexity of forecasting the pre-planting threat of Stagonospora nodorum blotch (SNB) in winter wheat. They have created hazard assessment models that can be useful in deciding on disease control earlier to gardening. In [6], the authors indicated pollution reduction and cost savings can be achieved by an automated decision procedure based on the Bayesian structure for the identification of weeds in corn cultures. In [7] authors exhibited that field information, for example, soil dampness, climate, water system attributes, and resulting yield could be intertwined by means of ML procedures to give automated suggestions to water system. Machine learning can be classified as supervised, unsupervised and reinforced type, where in supervised learning we have predefined outputs while in unsupervised learning we don't have such things. The reinforced learning works on reward system. Supervised leaning can be divided into two types, first classification in which Support vector machine, naïve bayes, nearest neighbors, discriminate analysis algorithms comes and second regression in which linear regression , decision trees, neural networks, ensemble methods and SVR, GPR are included. under unsupervised learning the category of algorithm is clustering in which fuzzy, hidden marcov model, Gaussian mixture etc comes. Reinforced learning is basically zone of deep learning which contains model iterating over several attempts to complete a procedure Steps that deliver great results are remunerated and steps that create undesired results are punished until the algorithms grasp the optimal process. Merits and Demerits of some of them have been listed by authors in [8]. Despite the major recent improvements in Machine Learning and the successful application in numerous areas. Machine Learning method has some primary restrictions when use idealistically in a solely data-driven manner. The exactness of the guesses and their uncertainties created by the Machine Learning techniques robustly depend on the data quality, model representativeness and the reliance between the target and input variables in the composed data's. Data with high noise levels, erroneous data, incidence of outliers and biases in data and scarce datasets can greatly reduce intellectual power models. An appropriate explanation of the machine learning model, such as the GP covariance function, the artificial neural network architecture and the SVR parameters, is also vital for achieving the best performance. To overcome these limitations, many schemes can be used, such as the integration of expert data into the covariance function, the detection of emissions, the transfer of learning and the choice of the model by computerized cross-checking. The purpose of this review is to express the competence of various machine learning methods to effectively solve these various but closely related tasks. An overview of current research in various fields of industry is presented, in which various methods of machine learning are used. Some technical features of the machine learning methods used in the studies are discussed.

2. Supervised Learning

Supervised learning has very wide range of applications. Supervised leaning have an input variable (Q) and the output variable (P) uses an algorithm to study the mapping function from the input to the output.

$$P= f(Q) \tag{1}$$

The goal is to approximate the display function so well that when you have new input data (Q), you can define the output variable (P) for the data. Controlled learning in the context of ML is a type of scheme in which both input and output are provided. Input or output data are labelled for providing is learning basis for processing future data. Supervised learning generates algorithm with known data to support future judgments. Supervised learning system are mostly related with retrieval based ML. In [9], the authors applied SVM-based classifier with methods of content intellectual analysis and prototypes of the semantic dialect to detect untrue reviews facilitated on Amazon.com. In [10], the authors tested controlled learning strategies, such as SVM, decision tree and logistic regression, to distinguish fraud from spontaneous online communication. Supervised learning problems can be clusters in classification and regression problems. The problem with classification is that the output variable is a class such as red or blue, a syndrome, or a no syndrome. The regression problem is that the output variable is an actual value, for example, dollars or weights [10] [11] [12].

2.1 Classification Algorithms

Classification is technique of supervised learning in which the desired output is already defined and comparison of different features of several commodities is carried out. SVM, Discriminant Analysis,

Naïve Bayes and Nearest Neighbours are the techniques which come under this algorithm. The application of this technique is very wide which is shown in table 1.

TABLE I. Classification Techniques Applications

Ref No.	Technique Used	Objective	Outcomes
[13]	SVM	Selective bipolar disorder from major depression using Whole-brain useful connectivity-a feature choice analysis With SVM-FOBA algorithm	The classification accuracy between bipolar disorder (BD) and major depression disease (MDD) up to 88% with leaving-one-out cruise authentication.
[14]	SVM	Support vector machine created differentiation between destructive and chronic periodontitis using microbial summaries	An SVC classifier using a section of 40 bacterial species was able to distinguish between PH, AgP in fresh individuals and ChP.
[15]	SVM	Failure exam of machinery bearings using the Variational mode of Decomposition and support vector machine as classifier.	Outcomes were obtained by Variational Mode Decomposition (VMD) and SVM process combinely are far better over normal process using SVM.
[16]	Discriminant Analysis	Fluid Dynamic Prototypes for Bhattacharyya-Based Discriminant Analysis	Proper prototypical reduction of prototypes to the optimal response to the events.
[17]	Discriminant Analysis	Sparse Exponential Discriminant Study and Its Uses to Fault Analysis	Sparse Exponential Discriminant Analysis (SEDA) algorithm can separate the faulty variables and simplify the discriminant prototype by removal of variables with tiny implication.
[18]	Discriminant Analysis	Improving the electrodiagnostic precision in Guillain-Barre disease subtypes: Criteria sets and sparse linear discriminant investigation	The error rates, the reference diagnosis, were: 15.3% for scarce LDA, 30% for our criteria, 45% for Rajabally and 48% for Hadden. LDA has presented the highest analytical precision.
[19]	Naïve Bayes	Planning a machine learning created software risk Ideal evaluation using the Bayes Naïve	Based on this ideal framework, we will analyze all aspects of risk and improve the risk assessment process.

		algorithm.	
[20]	Naïve Bayes	Human Heart Infection Prediction System Using Data Mining Techniques	The person's risk level is classified using extraction classification methods, such as naive Bayes, improved k-media etc.
[21]	. Nearest Neighbour	A nearest neighbour approximation of the remaining Variance	Problem of approximating the smallest achievable mean-squared error in regression function approximation solved
[22]	Nearest Neighour	Opposite k Nearest Neighbor Search over Paths	Path planning and capacity estimation. Has been done properly using this technique

2.2 Regression Algorithms

Regression is most commonly used method for defining connection between a dependent variable and one and new illustrative variable. This method is also known as pattern recognition technique. This method works on the problem where the output variable is real value. Linear Regression, SVR, GPR, Ensemble Methods, Decision Trees and Neural Networks are the techniques which come under this algorithm. The various applications of this technique is shown in table 2.

TABLE II. REGRESSION TECHNIQUES APPLICATIONS

Ref No.	Technique Used	Objective	Outcomes
[23]	Linear regression GLM	Prediction of Bitcoin Price using Machine Learning	Calculate the Bitcoin price exactly taking into consideration several parameters that bother the Bitcoin value.
[24]	Linear regression GLM	Citation Count Analysis for Papers with Preprints	Fitted regression model estimates that papers submitted to arXiv before acceptance, on average, tend to have 65% more citations in the following year compared to papers submitted after.
[25]	Linear Regression	A study of the User end workend application: Transitions from Freemium a Premium	The fitted regression model estimates that papers submitted to arXiv before acceptance, on average, tend to have 65% more citations in the following year compared to papers submitted after.
[26]	GPR AND SVR	A local perspective on the precision of machine learning predictions of tourism claim based on data features	The entropy and scatter display a negative relationship by precision, while the result of other data features on estimate accuracy is highly helpless on the prediction horizon.
[27]	SVR AND GPR	Calculate Twitter uses socioeconomic properties with networks and language test	The best performance model using graph-based functions (Graph) achieves an accuracy of 50.44% in the work classification. In the income statement, the MAE 9,048 is and the Pearson correlation is 0.63.
[28]	Ensemble methods	An collective learning scheme for a	As a outcome, about 2% of the novel Characteristics were nominated to construct

		4-way organisation of Alzheimer's disease and mild cognitive injury	a new space of functions, which can attain the last four-way classification by an accuracy of 54.38% in the test data over a classified grouping, upper than several another methods in judgement.
[29]	Ensemble methods	Healthy ensemble learning outline for day-ahead anticipating of household based energy consumption	The outcomes show a significant development in the simplification capacity, as well as the relief of numerous unstable calculation difficulties. The outcomes also deliver information on the capacity of the advised set prototype to produce better forecast performance with limited data, which shows the validity of the knowledge character of the set in the future prototype.
[30]	Decision tree	A Nonlinear Decision Tree based Arrangement Method to Predict the Parkinson's syndrome using Dissimilar Feature Sets of Speech Data	In this a new method by comparing presentation measures with dissimilar sets of features such as unique feature sets, as well as feature reduction technique based on principal component analysis to select feature sets. Here a nonlinear ordering approach to associate presentation indicators. We create an accuracy of 96.83% using accidental forestry classifiers using PCA-based feature sets. This examination will help clinicians distinguish the PD collection from the strong group based on the speech data.
[31]	Decision tree	Landslide weakness plotting using J48 Decision Tree with AdaBoost, Catching and Rotation Forestry ensembles in the Guangchang zone (China)	The result showed that all representations of avalanche material have high presentations (AUC> 0.8). However, the Yankees and the model forest rotation has the highest capacity calculations (AUC = 0.855), followed by the Yankees and AdaBoost (0.850), packaging (0.839) and Yankees (0.814) respectively. Thus, the result is that authenticates JDT to spin the forest is a model of best improved in this study and can be measured as a talented methods to map the sensitivity of the material in the case is similar to a good accuracy.

3. Unsupervised Learning

An Unsupervised learning has only input data (P) and the corresponding output variable. The goal of this kind of learning is to model an underscore structure or data distribution to learn more about the data. For example, there is no need to tag data to detect or classify spam and genuine reviewers over a website comments in unsupervised learning. Unsupervised learning does not need to be trained with result data. On the other hand, deep learning approach which is an iterative approach is used to review data and arrive at conclusions. Unsupervised learning algorithm is very useful for solving complex processing task such as image recognition and speech to text conversion. The main reason of using this algorithm is that it requires massive amount of training data like Big data. The application of unsupervised learning is self-driving car, facial recognitions and robots etc. In [32] authors employed unsupervised text mining strategies to distinguish basic attributes of countless insiders. In [33] authors developed an inclusive scoring method to calculate the degree of overall spam behavior of each reviewer. Unsupervised learning problems can be gathered as clustering and association problems. Clustering problems are basically where the inherent groupings in the data are discovered.

3.1 Clustering Algorithm

Clustering methods are not directly prognostic in nature; they are mostly descriptive and reveal an integral configuration that can be hidden in the data. Unlike regression, the application of the clustering method does not require a hypothesis. In a sense, clustering algorithms create hypotheses, revealing how data is combined together with group-specific functions. There are several different clustering algorithms, each of which has its advantages and disadvantages [34]. The most common algorithms are hierarchical clustering, k-means clustering and variable-shift clustering. Hierarchical and medium-term clustering gives more accurate and advanced results, but large data sets can reduce quality and productivity. K means, fuzzy, hierarchical, Gaussian mixture, neural network and hidden Markov model are methods that fall under this algorithm. Various applications of this method are shown in Table 3.

TABLE III. CLUSTERING ALGORITHM APPLICATIONS

Ref No.	Technique Used	Objective	Outcomes
[35]	K-means Clustering	Spatial K-means Clustering of HF Noise Trends in Southern California Waters	Results indicate elevated noise levels throughout the high frequency (HF) band with minor spatial variation.
[36]	*K*-Means Clustering	Examination of Card Sorting Data using K-Means and Multidimensional Scaling algorithms.	Analysis of the results indicates that the current approach can overcome the limitations of BMM, by allowing the card to be prepared again when the group formed.
[37]	Fuzzy C-means	Improved fuzzy C-means clustering algorithm use in medical image division	The results are shown as being NMFCM consequently less sensitive to noise than FCM. Conflict results NMFCM, FCM and FCM NMFCM show that the proposed work is very good and gives higher than the accuracy of segmentation and FCM FCM. Finally, the wrong root is the average error indication and mark the parameters of noise and experimental testing the contents of small and large and each interruption in image quality in the area well.
[38]	Fuzzy c-means algorithms	Methods for cholesterol and targets that can be gigantic GPCR seven helixel based cluster ghost and fuzzy c-organ algorithm	G-protein coupled receptors (GPCRs) plays a vital role for all type of complex signalling and inter cellular communication in membrane bilayer with secure connection. It is proved that an interaction between cholesterol and GPCR are tightly regulates with each other, using different parameters.
[39]	Hierarchical clustering	Nesterov's flattening technique and minimizing for	We detect that parameter selection is the conclusive factor in terms of accuracy and speed of junction of our proposed

		hierarchical differences of convex roles clustering	algorithms. The presentation of the future algorithms highly depends on the primary values set to the consequence and smoothing parameters, λ0 and μ0; and their particular growing/decaying factors σ1 and σ2.
[40]	Hierarchical clustering	Hierarchical Separating of the Output Space in Multi-label Data	The experimental results of experiments carried out on paper to convince Homer can restore the presentation techniques when applied to MLC. However, it is concerned about the way to honor the clustering algorithm optimal, because it is a part that affects the ability of the algorithm to the top. Especially for the work session, the results are positive can show that the number is more suited to chatting on the multi-stage large fire.
[41]	Gaussian mixture models	GGRaSP: R-technique to select a representative genome using a mixed model of Gauss	Creation of geometric, bio informational and graphic skills R, GGRaSP and accompanying R-script offers a single and customizable platform for launching several analyzes to create a subset of representative genomes. The user can specify clustering parameters and severity levels for ranking genomes, thereby allowing both generalized high throughput and more data to be used.
[42]	Gaussian mixture model	Anomaly detection and localization of video through Gaussian mixing Autoencoder Variational Full convolution	In the training stage, image patches of normal samples for each stream are extracted as input to train a Gaussian Mixture Fully Convolutional Variational Autoencoder (GMFC-VAE) that learns a Gaussian Mixture Model (GMM). In the testing stage, the conditional probabilities of each component of a Gaussian Mixture of test patches are obtained by employing the GMFC-VAE for each stream. Here a sample energy based method for predicting an appearance and motion anomaly score.
[43]	Neural network	Automatic identification of shock able and non-shock able life-threatening ventricular arrhythmias using CNN	Here left precision, sensitivity and specificity of 93.18%, 95.32% and 91.04%. Respectively Pintonanna above suggests that arrhythmias that can threaten life, potentially threatening your life can be detected directly and, therefore, increases the disclosure of which is the support provided by NGN or AED.
[44]	Neural network	NN adaptive fuzzy controller for robot constraints by learning impedance	Dynamics are certainly in the robot's work online, use the opportunity to exercise diffuse structure NN. To test the success of the proposed approach, there are four things that are considered: 1) control with control-free ban free space; 2) organize all countries ban the free space; 3) control the output limited to confined space; and 4) manage all

			state restrictions on confined space
[45]	Hidden markov model	Website finger printing attack on secrecy networks based on profile hidden markov model	The results of the research show that this approach has a high accuracy of the modern method of site classification, and beyond the identification of traditional web pages are separated. In addition, the study considered two types of hyperlink transition from state to trial. The result indicates that the method can be used to attack the environmental website is more applied, but the methods are only penetrate in a web page is the last one separated and ignore the discovery of the open potential Finally, two that could be willing to be identified to protect the intended methods
[46]	Hidden markov model	Facial Expression Recognition using Moments Invariants and Modified Hidden Markov Model	Experimental result shows that Proposed System achieves better than normal HMM and has the overall accuracy of 84% using JAFFE database.
[47]	Hidden markov model	Sentiment Analysis on the Online Reviews Based on Hidden Markov Model	By this device, it is possible to take consideration of the webpage layout from Amazon Japan's product review page under the human webpage reading behaviors, 2dHMM has shown the highest precision and f1 score. Such an extensive design of analysis model to higher dimension HMM is applicable not only to Amazon Japan, but also more to general commercial Web pages.

4. Conclusion

In this review paper total 47 paper are reviewed of different engineering discipline. Effectiveness in various applications is shown. This paper shows the advantages of Machine learning over other techniques in the market. Firstly an introduction about the machine learning is given than further the division of it in different categories is explained such as supervised and unsupervised learning algorithms and application of these algorithms with their further division is shown in the tables. It is found that most widely used techniques among the classifier algorithms are SVM and NN techniques SVM used in medical industry to find differentiation between destructive and chronic periodontitis using microbial summaries, NN used in civil industry in estimating road capacity etc. Similarly, linear regression and ensemble technique in regression and ANN and HMM in clustering techniques. All the algorithms such as supervised or unsupervised have different merits and demerits over each other when we talk about the applications for example if anyone wants to do weather forecasting than regression techniques will be very effective than other techniques like clustering and classification similarly if we talk about induction motor fault diagnosis classification techniques will be very effective like NN. Presently with the advancement in the machine learning techniques presently 2D/3D Convolutional Neural Network which comes under deep learning strategies becoming very effective and reliable because of removal of the feature extraction and calculation step.

References

[1] M. W. Libbrecht, & W.S. Noble. Machine learning applications in genetics and genomics. Nature Reviews Genetics, (2015), 16(6), 321.

[2] K, Kourou, T. P. Exarchos, K. P. Exarchos, M. V. Karamouzis, & D. I. Fotiadis. Machine learning applications in cancer prognosis and prediction. Computational and structural biotechnology journal, (2015), 13, 8-17.

[3] N. M. Nasrabadi. Pattern recognition and machine learning. Journal of electronic imaging, (2007), 16(4), 049901.

[4] Y. Lei, F. Jia, J. Lin, S. Xing, S.X. Ding, An intelligent fault diagnosis method using unsupervised feature learning towards mechanical big data, IEEE Trans. Ind. Electron. 63 (5) (2016) 3137–3147.

[5] L.K. Mehra, C, Cowger, K. Gross, P.S. Ojiambo. Predicting pre-planting risk of stagonospora nodorum blotch in winter wheat using machine learning models. Front. Plant Sci. (2016). 7, 390.

[6] A. Tellaeche, X.P. BurgosArtizzu, G. Pajares, A. Ribeiro. A vision-based classifier in precision agriculture combining bayes and support vector machines. In: IEEE International Symposium on Intelligent Signal Processing, (2007) pp. 1–6.

[7] A. Goldstein, L. Fink, A. Meitin, S. Bohadana, O. Lutenberg, G. Ravid, Applying machine learning on sensor data for irrigation recommendations: revealing the agronomist's tacit knowledge. Prec. Agric. (2017)

[8] I. Ali, F. Greifeneder, J. Stamenkovic, M. Neumann, C. Notarnicola, Review of Machine Learning Approaches for Biomass and Soil Moisture Retrievals from Remote Sensing Data. Remote Sens. (2015) 7, 15841

[9] R.Y.K. Lau, S.Y. Liao, R.C.W. Kwok, K. Xu, Y. Xia, and Y Li,. Text mining and probabilistic language modeling for online review spam detection. Transactions on Management Information Systems, ACM, 2, 4, (2011) 1–30.

[10] S.M. Ho, J.T. Hancock, C. Booth, and X. Liu, Computer-mediated deception: Strategies revealed by language-action cues in spontaneous communication. Journal of Management Information Systems, 33, 2 (2016), 393–420.

[11] B.Longstaff, S. Reddy, & D. Estrin. Improving activity classification for health applications on mobile devices using active and semi-supervised learning. In Pervasive Computing Technologies for Healthcare (PervasiveHealth), 2010 4th International Conference on-NO PERMISSIONS (2010) (pp. 1-7). IEEE.

[12] M. W. Libbrecht, & W. S. Noble,. Machine learning applications in genetics and genomics. Nature Reviews Genetics, (2015) 16(6), 321.

[13] N. F. Jie, E. A. Osuch, M. H. Zhu, M. Wammes, X. Y. Ma, T. Z. Jiang, ... & V. D. Calhoun. Discriminating bipolar disorder from major depression using whole-brain functional connectivity: A feature selection analysis with SVM-FoBA algorithm. Journal of Signal Processing Systems, (2018) 90(2), 259-271.

[14] M. Feres, Y. Louzoun, S. Haber, M. Faveri, L. C. Figueiredo, & L. Levin. Support vector machine-based differentiation between aggressive and chronic periodontitis using microbial profiles. International dental journal, (2018) 68(1), 39-46.

[15] K. R. Krishna, & K. I. Ramachandran,. Machinery Bearing Fault Diagnosis Using Variational Mode Decomposition and Support Vector Machine as a Classifier. In IOP Conference Series: Materials Science and Engineering (2018) (Vol. 310, No. 1, p. 012076). IOP Publishing.

[16] Y. K. Noh, J. Hamm, F. C. Park, B. T. Zhang, & D. D. Lee. Fluid Dynamic Models for Bhattacharyya-Based Discriminant Analysis. IEEE transactions on pattern analysis and machine intelligence, (2018) 40(1), 92-105.

[17] W. Yu, & C. Zhao. Sparse exponential discriminant analysis and its application to fault diagnosis. IEEE Transactions on Industrial Electronics, (2018) 65(7), 5931-5940.

[18] A. Uncini, L. Ippoliti, N. Shahrizaila, Y. Sekiguchi, & S. Kuwabara. Optimizing the electrodiagnostic accuracy in Guillain-Barré syndrome subtypes: Criteria sets and sparse linear discriminant analysis. Clinical Neurophysiology, (2017) 128(7), 1176-1183.

[19] K. Suresh, & R Dillibabu. Designing a Machine Learning Based Software Risk Assessment Model Using Naïve Bayes Algorithm. (2018).

[20] J. Thomas, & R. T. Princy, (March). Human heart disease prediction system using data mining techniques. In Circuit, Power and Computing Technologies (ICCPCT), 2016 International Conference on (2016) (pp. 1-5). IEEE.

[21] L. Devroye, L. Györfi, G. Lugosi, & H. Walk,. A nearest neighbor estimate of the residual variance. Electronic Journal of Statistics, (2018) 12(1), 1752-1778.

[22] S. Wang, Z. Bao, J. S. Culpepper, T. Sellis, & G.Cong. Reverse $ k $ Nearest Neighbor Search over Trajectories. IEEE Transactions on Knowledge and Data Engineering, (2018), 30(4), 757-771.

[23] S. Velankar, S. Valecha, & S. Maji, (February). Bitcoin price prediction using machine learning. In Advanced Communication Technology (ICACT), 2018 20th International Conference on (2018), (pp. 144-147). IEEE.

[24] S. Feldman, K. Lo, & W. Ammar. Citation Count Analysis for Papers with Preprints. arXiv preprint arXiv:1805.05238. (2018)

[25] C., Mulligan, C. V. Cruz, D. Healy, D. Murphy, M. Hall, Q. Nelson, & S. Caton, A Study of App User Behaviours: Transitions from Freemium to Premium. In International Conference on HCI in Business, Government, and Organizations (2018, July). (pp. 396-412). Springer, Cham.

[26] O. Claveria, E. Monte, & S. Torra,. A regional perspective on the accuracy of machine learning forecasts of tourism demand based on data characteristics (No. 201805). University of Barcelona, Research Institute of Applied Economics. (2018)

[27] N. Aletras, & B. P. Chamberlain,. Predicting Twitter User Socioeconomic Attributes with Network and Language Information. arXiv preprint arXiv:1804.04095. (2018)

[28] D. Yao, V. D. Calhoun, Z. Fu, Y. Du, & J. Sui. An ensemble learning system for a 4-way classification of Alzheimer's disease and mild cognitive impairment. Journal of neuroscience methods, (2018) 302, 75-81.

[29] M. H. Alobaidi, F. Chebana, & M. A. Meguid. Robust ensemble learning framework for day- ahead forecasting of household based energy consumption. Applied Energy, (2018) 212, 997-1012

[30] S. Aich, K. Younga, K. L. Hui, A. A. Al-Absi, & M. Sain,. A nonlinear decision tree based classification approach to predict the Parkinson's disease using different feature sets of voice data. In Advanced Communication Technology (ICACT), 2018 20th International Conference on (2018, February) (pp. 638-642). IEEE.

[31] H. Hong, J. Liu, D. T. Bui, B. Pradhan, T. D. Acharya, B. T. Pham, ... & B. B. Ahmad. Landslide susceptibility mapping using J48 Decision Tree with AdaBoost, Bagging and Rotation Forest ensembles in the Guangchang area (China). Catena, (2018) 163, 399-413.

[32] N. Liang, D.P. Biros, and A. Luse, An empirical validation of malicious insider characteristics. Journal of Management Information Systems, 33, 2 (2016), 361–392.

[33] E.P. Lim, V.A. Nguyen, N. Jindal, B. Liu, and H.W. Lauw, Detecting product review spammers using rating behaviors. Proceedings of the ACM International Conference on Information and Knowledge Management, 19, (2010), 939–948.

[34] CM. Bishop. Pattern recognition and machine learning. In: Information science and statistics.Springer: New York, (2006) xx,738 p.

[35] K. Buchanan, D. Gaytan, L. Xu, , C. Dilay, & D. Hilton,. Spatial K-means clustering of HF noise trends in Southern California waters. In URSI National Radio Science Meeting (USNC-URSI NRSM), 2018 United States National Committee of (2018, January) (pp. 1-2). IEEE.

[36] S. Paea, & R. Baird. Information Architecture (IA): Using Multidimensional Scaling (MDS) and K-Means Clustering Algorithm for Analysis of Card Sorting Data. Journal of Usability Studies, (2018) 13(3).

[37] K. M. Aljebory, & T. S. Mohammed, Modified fuzzy C-means clustering algorithm application in medical image segmentation. JEA Journal of Electrical Engineering, (2018) 2(1), 1-9

[38] R. Tripathy, D. Mishra, V. B. Konkimalla, & R. K. Nayak. A computational approach for mining cholesterol and their potential target against GPCR seven helices based on spectral clustering and fuzzy c-means algorithms. Journal of Intelligent & Fuzzy Systems, (Preprint), (2018) 1-10.
[39] N. M. Nam, W. Geremew, S. Reynolds, & T. Tran. Nesterov's smoothing technique and minimizing differences of convex functions for hierarchical clustering. Optimization Letters, (2018) 1-19.
[40] Y. Papanikolaou, G. Tsoumakas, & I. Katakis. Hierarchical partitioning of the output space in multi-label data. Data & Knowledge Engineering. (2018)
[41] T. H.Clarke, L. M. Brinkac, G. Sutton, & D. E. Fouts. GGRaSP: A R-package for selecting representative genomes using Gaussian mixture models. Bioinformatics, (2018) 1, 3.
[42] Y. Fan, G. Wen, D. Li, S. Qiu, & M. D. Levine. Video Anomaly Detection and Localization via Gaussian Mixture Fully Convolutional Variational Autoencoder. arXiv preprint arXiv:1805.11223. (2018)
[43] U. R. Acharya, H. Fujita, S. L. Oh, U. Raghavendra, J. H. Tan, M. Adam, ... & Y. Hagiwara. Automated identification of shockable and non-shockable life-threatening ventricular arrhythmias using convolutional neural network. Future Generation Computer Systems, (2018), 79, 952-959.
[44] W. He, & Y. Dong. Adaptive fuzzy neural network control for a constrained robot using impedance learning. IEEE transactions on neural networks and learning systems, (2018), 29(4), 1174-1186.
[45] Z. Zhuo, Y. Zhang, Z. L. Zhang, X. Zhang, & J. Zhang. Website Fingerprinting Attack on Anonymity Networks Based on Profile Hidden Markov Model. IEEE Transactions on Information Forensics and Security, (2018) 13(5), 1081-1095.
[46] M. Rahul, N. Kohli, & R. Agarwal. Facial Expression Recognition using Moments Invariants and Modified Hidden Markov Model. International Journal of Applied Engineering Research, (2018) 13(8), 6081-6088.
[47] X. Zhao, & Y. Ohsawa, Sentiment Analysis on the Online Reviews Based on Hidden Markov Model. Journal of Advances in Information Technology Vol, (2018) 9(2).

IOP Conf. Series: Materials Science and Engineering **594** (2019) 012024 doi:10.1088/1757-899X/594/1/012024

Classification of Human Bones Using Deep Convolutional Neural Network

Nitesh Pradhan[1,*], Vijaypal Singh Dhaka[2] and Himanshu Chaudhary[3]

[1]Department of Computer Science and Engineering, Manipal University Jaipur, Jaipur-303007

[2]Department of Computer and Communication Engineering, Manipal University Jaipur, Jaipur-303007

[3]Department of Electronics and Communication Engineering, Manipal University Jaipur, Jaipur-303007

Email: *nitesh.pradhan@jaipur.manipal.edu

Abstract. In human body, there are total 206 types of different bones. Each bone has its own importance. It is very important to correctly identify human bone and then suggest treatment. To classify the human bones, we will use Musculoskeletal Radiographs (MURA) dataset. MURA dataset is one of the largest public radiographic image datasets. MURA dataset contains total 40,005 x-ray images of 14,052 patients, in which 36,808 images use as a training set and rest 3197 images use the testing set. These all images belong to seven different categories of bones such as finger, elbow, hand, forearm, humerus, wrist and shoulder. This paper aims to present a novel classification method using a deep convolutional neural network (DCNN). Dataset is freely available at https://stanfordmlgroup.github.io/competitions/mura.

1. Introduction

A machine can be learned in two different ways either as supervised learning or unsupervised learning. In supervised learning, already we have some labeled data through which we train our machine and perform classification/prediction based on training [7, 8]. In unsupervised learning, we have unlabeled data. In our case, we used supervised learning to train our machine.

Classification is a concept of supervised machine learning in which the computer learns from the labeled data [1] we input and then uses this learning to classify new data given to it. Classification is used to predict the category of data we provide [3]. This classification can be done on binary input meaning 2 categories as well as multiple categories. An example of this would be a classic titanic question. In which we the computer learns from the given data and then tries to predict whether a person would have died or survived the Titanic crash based on the independent variables including age, gender, class etc. The classification has many applications ranging from medicine to marketing [4].

The convolutional neural network plays an important role in case of 2D/3D object prediction/classification [11]. The deep convolutional neural network also used to classify the biological images [1]. In the medical field, some task such as brain tumor segmentation, skin lesion segmentation etc can be easily performed by deep convolutional neural network [4, 8].

2. Dataset

For the classification of human bones, we require some x-ray images of the different body part. For that region, we used Musculoskeletal Radiographs (MURA) dataset. This MURA dataset contains seven different categories of human bones belonging to Elbow, Hand, Wrist, Shoulder, Finger, Humorous and Forearm. Each category of bone has some x-ray images with respect to the number of patients shown in below table.

Table 1: Classification of x-ray images in different categories

Study	Total No. of Patients	Total x-ray Images
ELBOW Bone	1863	5689
HAND Bone	2104	6312
WRIST Bone	3474	10502
SHOULDER Bone	2867	8571
FINGER Bone	2031	6293
HUMERUS Bone	719	1202
FOREARM Bone	994	1436
Total	**14052**	**40005**

2.1. *Training and Testing Set*

To classify/prediction the category of bone, we require training and testing dataset [5]. So MURA set also contain some training and testing set in the ratio of 90:10. It means if we have total x-ray images 40,005 then 36,808 images treated as training set and remaining 3,197 x-ray images treated as test set images. Same as x-ray images, patients are also divided into training set and test set shown in table 2.

Table 2: Division of dataset based on training and testing set

	Training Set		Test Set	
Study	**No. of Patients**	**No. of x-ray images**	**No. of Patients**	**No. of x-ray images**
ELBOW Bone	1711	5233	152	456
HAND Bone	1945	5835	159	477
WRIST Bone	3267	9881	207	621
SHOULDER Bone	2694	8052	173	519
FINGER Bone	1865	5795	166	498
HUMERUS Bone	587	887	132	315
FOREARM Bone	865	1125	129	311
Total	**12934**	**36808**	**1118**	**3197**

3. Model Architecture and Training

When we talk about classification, then classification is dividing into two different categories named as the binary classifier and categorical classifier. Binary classifier, classify given data in two class/category only such as yes and no or like and dislike etc. Categorical classifier, classify the given data into more than two classes/categories. Deep convolutional neural network (DCNN) model plays an important role in the case of classification/prediction [10,11].

3.1 *Deep Convolutional Neural Network*

The concept behind the deep convolutional neural network is to make the computer learn things like our brain does. It too has a neural network with neurons exactly like our brain. These neurons have different weights depending upon that neurons significance in classifying an object [6]. These weights get updated as our computer learns from the given data. These weights are then passed through an activation function and the output is given. An example of this would be a dog or cat classification problem. In which the machine will take thousands of images of dogs and cats and learn from them what features both have that differentiates them from one another and then predict whether the given photo is of a cat or a dog. DCNN is widely used on almost every image related problem because of its amazing accuracy that blows its competition out of water [9].

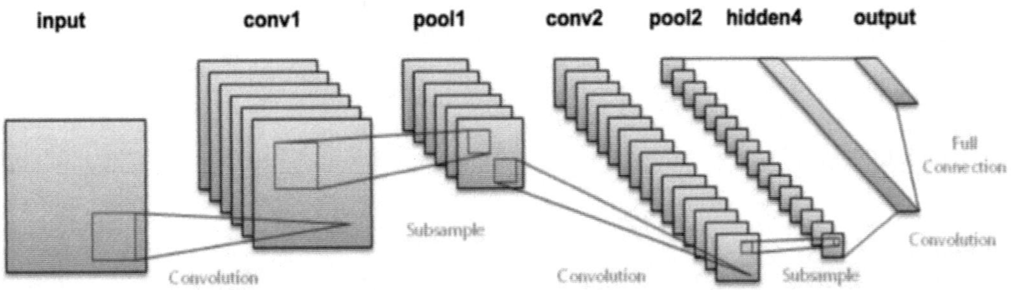

Figure 1: Deep Convolutional Neural Network [3]

In the deep convolutional neural network, mainly three layers are used namely convolutional layer, pooling layer and flatten layer [2]. Basically, all these three layers are used to extract useful information from an image. The convolutional layer is treated as a filter that is applied to the image and get the feature map. This feature map is the reduce size (in term of dimension) from the original image. Now on this feature map, we can apply a different type of pooling such as max pooling, min pooling and avg pooling etc based on your application. So by applying to pool on feature map we get pooled feature map (i. e. more reduce size in term of dimension). After applying the pooling layer, perform flattering on the pooled feature map. This flattening is the important feature of an image and treated as input layer neuron of an artificial neural network.

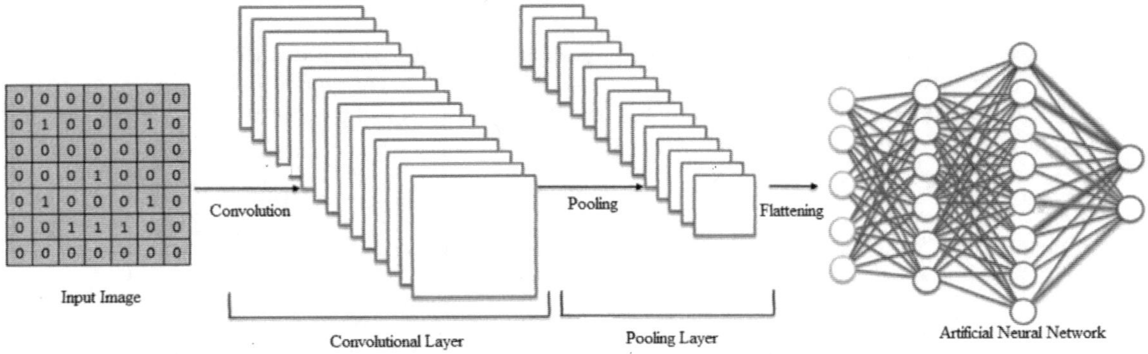

Figure 2: Work flow of deep convolutional neural network

As mentioned above, MURA dataset contains total 40,005 x-ray images of 14,052 patients, in which 36,808 images use as a training set and rest 3197 images use as a testing set. This MURA dataset contains seven different category of human bones belong to Elbow, Hand, Wrist, Shoulder, Finger, Humorous and Forearm. Our task is to identify the class/category of a particular image.

To classification, the model architecture uses four convolutional layers, four pooling layers, one flattened layer, and two hidden layers with dropout rate 0.2. The complete architecture of DCNN is described in table 3.

Table 3: model architecture of DCNN

Layer (type)	**Output Shape**	**Param #**
conv2d_4 (Conv2D)	(None, 62, 62, 32)	896
max_pooling2d_4 (MaxPooling2	(None, 31, 31, 32)	0

conv2d_5 (Conv2D)	(None, 29, 29, 32)	9248
max_pooling2d_5 (MaxPooling2	(None, 14, 14, 32)	0
conv2d_6 (Conv2D)	(None, 12, 12, 64)	18496
max_pooling2d_6 (MaxPooling2	(None, 6, 6, 64)	0
conv2d_7 (Conv2D)	(None, 4, 4, 128)	73856
max_pooling2d_7 (MaxPooling2	(None, 2, 2, 128)	0
flatten_2 (Flatten)	(None, 512)	0
dense_3 (Dense)	(None, 128)	65664
dropout_2 (Dropout)	(None, 128)	0
dense_4 (Dense)	(None, 128)	16512
dropout_3 (Dropout)	(None, 128)	0
dense_5 (Dense)	(None, 7)	903

4. Result

When we train our network by 36808 x-ray images with different epochs then we got the different accuracy of the training set. After training, we test our trained network by 3197 x-ray images and found good accuracy. The accuracy of training and testing set with respect to multiple epochs are mentioned in table 4.

Table 4: Result analysis based on different epochs

No. of epoch	Training Accuracy (in %)	Testing Accuracy (in %)
3	86.23	85.11
7	91.62	90.40
12	93.37	91.24
18	94.23	91.37

As we can see in table 4, in 18 epochs we got 94.17% accuracy it means if we give an image of bone to the machine, 94.17% chance that the machine will predict the correct class/category of bone.

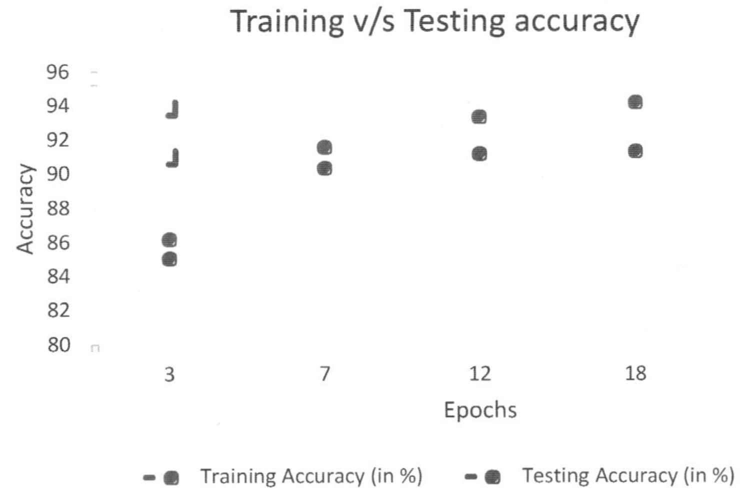

Figure 3: Result analysis

5. Conclusion

We successfully applied deep convolutional neural network for classification of human bone with four convolutional layers, four pooling layers, one flattened layer, and two hidden layers with dropout rate 0.2. We run this model architecture for multiple epochs and got the best accuracy 94.23% for the training set and 91.37% for the test set.

In this paper, we used only seven different classes/categories of bones such as Elbow, Hand, Wrist, Shoulder, Finger, Humorous and Forearm. In the further study, the number of classes/categories can be increased to predict/classify the different types of human bones.

6. Acknowledgements

We would like to acknowledge the Stanford Program for Artificial Intelligence in Medicine and Imaging for clinical dataset infrastructure support.

References

[1] Carlos, A., AndréLuis, D., Fábio, H., André, C., and Leon *"Deep learning for biological image classification"* Expert Systems with Applications 2017, pp. 114–122

[2] Grewal, Monika, Srivastava, Muktabh Mayank, Kumar, Pulkit, and Varadarajan, Srikrishna *"Radnet: Radiologist level accuracy using deep learning for hemorrhage detection in ct scans"* 2017, arXiv preprint arXiv:1710.04934.

[3] A. Rajkomar, S. Lingam, A. G. Taylor, M. Blum, and J. Mongan, *''Highthroughput classification of radiographs using deep convolutional neural networks''* J. Digit. Image 2017 vol. 30, no. 1, pp. 95–101.

[4] Sergio, P., Adriano, P., and Carlos, A *"Brain Tumor Segmentation using Convolutional Neural Networks in MRI Images"* IEEE Transactions on Medical Imaging 2016, pp. 1240-1251.

[5] H.-C. Shin et al., *"Deep convolutional neural networks for computer-aided detection: CNN architectures dataset characteristics and transfer learning" IEEE Trans. Med. Image May 2016,* vol. 35, no. 5, pp. 1285-1298.

[6] Nima, T., Jae, Y., Suryakanth, R., Todd, H., and Michael, B *"Convolutional Neural Networks for Medical Image Analysis: Fine Tuning or Full Training"* IEEE Transactions on Medical Imaging 2016, pp. 1299-1312.

[7] Criminisi, A *"Machine learning for medical images analysis"* Medical Image Analysis 2016, pp. 91–93.

[8] Jafari, M.H., Karimi, N., Nasr-Esfahani, E., Samavil, S., Soroushmehr, S.M.R., Ward, K. and Najarian, K *"Skin Lesion Segmentation in Clinical Images Using Deep Learning"* 23rd International Conference on Pattern Recognition (ICPR) 2016, pp. 337-342.

[9] W. Shen, M. Zhou, F. Yang, C. Yang, J. Tian, *"Multi-scale convolutional neural networks for lung nodule classification"*, in Information Processing in Medical Imaging June 2015 vol. 24, pp. 588-599.

[10] A. Krizhevsky, I. Sutskever, G. E. Hinton, *"ImageNet classification with deep convolutional neural networks", Proc. Adv. Neural Inf. Process. Syst 2012,* pp. 1097-1105.
[11] R. Socher, B. Huval, B. Bath, C. D. Manning, and A. Y. Ng, *''Convolutional-recursive deep learning for 3D object classification''* in Proc. Adv. Neural Inf. Process. Syst 2012, pp. 656–664.

IOP Conf. Series: Materials Science and Engineering **594** (2019) 012025 doi:10.1088/1757-899X/594/1/012025

Development and performance analysis of conformal UWB wearable antenna under various bending radii

Narbada Prasad Gupta[1], Mithilesh Kumar[2], Ranjan Maheshwari[2]

[1]Professor,Department of Electronics & Comm. Engg., ABES Engineering College, Ghaziabad
[2]Professor,RTU, Kota, Rajasthan

narbada.gupta@abes.ac.in

Abstract. Antennas for off-body wearable applications needs to be flexible and its electrical characteristics should be stable in different body postures, different bending or stretching environments. Therefore, the designed antenna should be such that, even if the antenna is bent frequently, it should operate properly. In this paper, a conformal UWB antenna is reported which has been optimized to be conformal to any body shape and size. The proposed antenna has been designed on RT Duroid substrate with dielectric constant of 2.2 and loss tangent 0.0009. The thickness of the material is 0.127 mm. Due to low thickness; it can easily be bent and mounted on curved surfaces. The size of the antenna is 35 × 31 mm^2. To accomplish the design process, two different bending radii have been considered to bend the proposed antenna. The two bending radii of 25 mm and 50 mm have been chosen; one corresponding to the arm of a healthy child of around 5 years age and another corresponding to a healthy person of age about 35 years with medium built. The simulated radiation efficiency and total antenna efficiency are more than 60% for the bending at 25 mm and 50 mm radii in the UWB frequency range from 3.1 GHz to 10.6 GHz. The gain of the proposed antenna for both the bending radii is changing from 1.4 dB to 5.3 dB. The return loss of the proposed antenna is obtained for 25 mm and 50 mm bending radii of curvatures and it is below 10 dB in UWB range. A prototype of proposed antenna has been fabricated and tested on Rohde & Schwarz 40 GHz VNA. Measured results are in close agreement with simulation results.

1. Introduction

Ultra-Wideband (UWB) radio is a potentially revolutionary approach to wireless communication. Recently, the demand for short-range, very high data rate, and video transmission devices has tremendously increased within the wireless user community. Research efforts in this area have exaggerated since the year 2002 when the United States Federal Communication Commission (FCC) allocated the frequency range of 3.1-10.6 GHz for unlicensed UWB indoor wireless communication [1]. A wide variety of antennas has been reported by many authors for off-body wearable applications [2]-[6]. There are two primary requirements for antennas in off-body communication. First, the antenna should be insensitive to the proximity of the body; and second, the antenna should have radiation properties like gain, directivity, etc., which always remain same for different body postures. The impedance matching of the antenna is naturally affected by the proximity of the body. A number of techniques are available for adaptively matching antenna impedance. It is important to note that if antenna is to be used for off-body wearable applications, it must be flexible and its electrical characteristics should be stable in different body postures, different bending or stretching environments [7]. Wearable antennas should have flexible substrates for their practical implementation. Therefore, the designed antenna should be such that, even if the antenna is bent frequently, it should operate properly. The variation in the dimensions due to bending or stretching affects the electromagnetic properties of the antenna [8]. Major effects are as follows:

- Changes in the length of antenna detune its operating frequency.
- Changes in the substrate thickness change the operating frequency.

- Changes in the substrate thickness also affect input impedance bandwidth.

In this paper, the flexible UWB antenna reported in [9] has been optimized to be conformal to any body shape. Design details are provided in [9] along with its parametric results. To accomplish the design process of this conformal UWB antenna, two different bending radii have been considered to bend the flexible UWB antenna reported in [9]. The two bending radii of 25 mm and 50 mm have been chosen; one corresponding to the arm of a healthy child of around 5 years age and another corresponding to a healthy person of age about 35 years with medium built. Simulation results of the conformal UWB antenna obtained from CST EM Simulator v.12 for the two bending radii have been compared with results of the flexible UWB antenna keeping it flat. The return loss of conformal UWB antenna bent at two different radii has been measured on Agilent E5071C 20 GHz VNA after design of the prototype. Section-II provides design and parametric simulation results of the proposed antenna. Section-III provides the detail of prototype designed and measured results. Also, measurement results are compiled in this section. Concluding remarks on the comparison of simulated and measured results are provided in section –IV.

2. Design of conformal UWB wearable antenna

Usually, wearable antennas for off-body applications are worn on arms, biceps, and thigh etc. All mentioned body parts are in general cylindrical in shape. So, two cylinders of radius 25 mm and 50 mm have been considered for the design of conformal antennas. These antennas are conformed to the surface of the corresponding cylinder to obtain a particular bending radius. Figure 1 shows the proposed antenna design bent at two radii 25 mm and 50 mm.

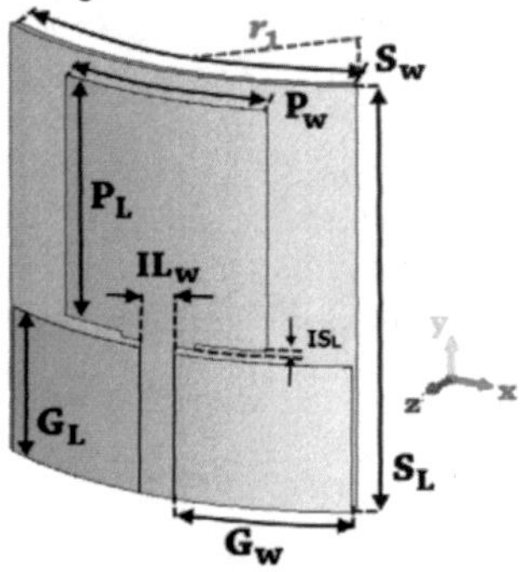

(a) Bending radius 25 mm

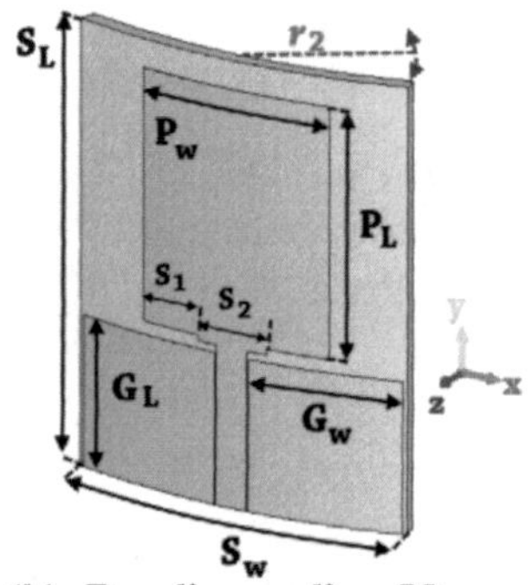

(b) Bending radius 50 mm

Figure 1 Structure of conformal UWB antenna for bending radius of r_1=25 mm and r_2=50 mm

The material used to model the antenna is RT Duroid with relative permittivity 2.2 and loss tangent 0.0009. The thickness of the material is 0.127 mm.

2.1. *Parametric study and optimization*

The conformal antenna shows a different response when it is bent. In the parametric study, 25 mm bending radius has been considered. Figure 2 represents the parametric results of ground length (G_L) varied from 10.36 mm to 10.56 mm in three steps of 0.1 mm. It is shown that increasing the ground length increases upper frequency to some extent but no significant change is observed in lower frequency. Parametric results for the step length (IS_L) are provided in Figure 3. The step length was varied between 0.6 mm to 1.4 mm in three steps of 0.4 mm. It was observed that if step length is varied, matching of the antenna in the lower as well as the upper frequency band is affected. More effect has been seen in upper frequency compared with lower frequency. Return loss is greatly increased in the middle UWB frequency range if step length is made larger.

IOP Conf. Series: Materials Science and Engineering **594** (2019) 012025 doi:10.1088/1757-899X/594/1/012025

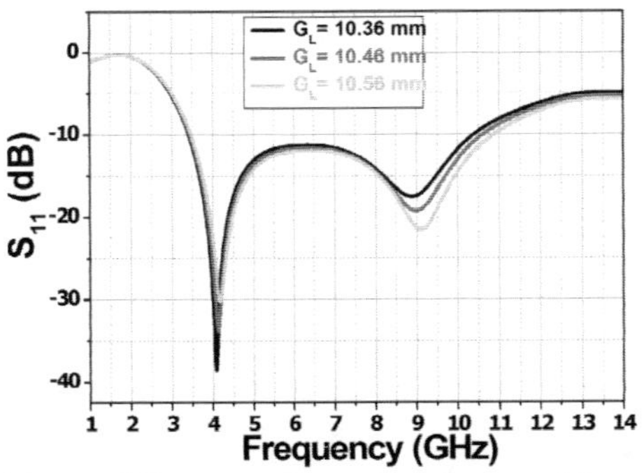

Figure 2 Parametric study of ground length (G_L)

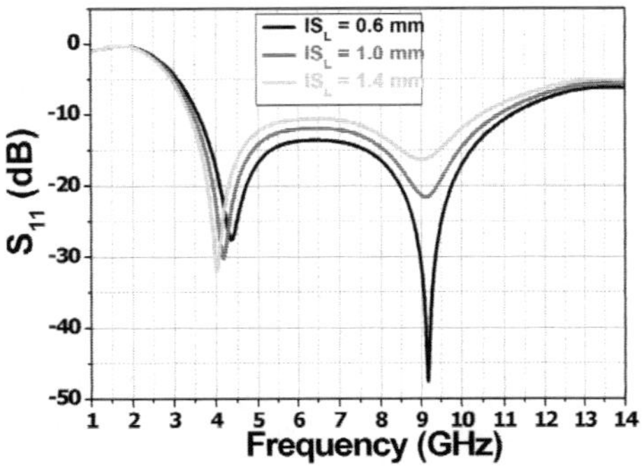

Figure 3 Parametric study of input step length (IS_L)

Similarly, the step width which is again a very important parameter while finalizing the parameter value of the antenna has been varied from 4.5 mm to 7.5 mm in four steps of 1 mm as shown in Figure 4. Performance of the antenna is greatly affected in the middle of UWB range when step width is altered. It also affects the impedance bandwidth. In fact, when step width is reduced to 4.5 mm, the return loss is above 10 dB in the middle UWB frequency range and entire operating band is disturbed at the step width of 4.5 mm. Parametric results for patch length (P_L) varied from 12.7 mm to 16.7 mm in five steps of 1 mm are shown in

Figure 5.

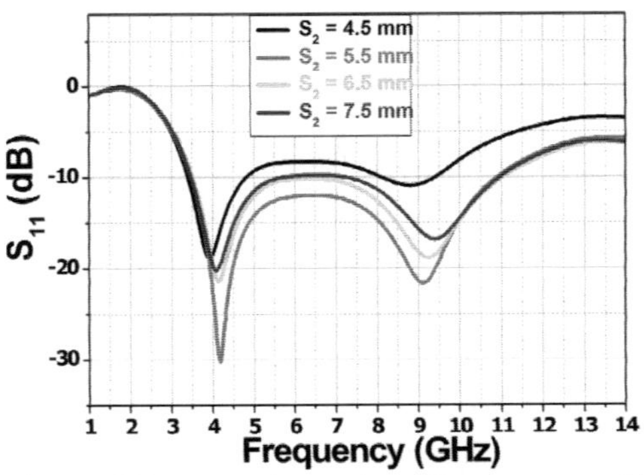

Figure 4 Parametric study of Input step width (IS_w)

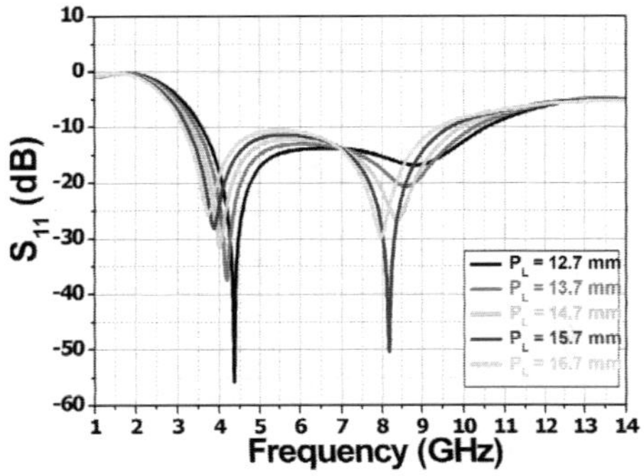

Figure 5 Parametric study of patch length (P_L)

It was observed that if patch length is changed, the upper, as well as lower frequency, are changed, which affect the impedance bandwidth of the conformal UWB antenna. The return loss comes closer to the 10-dB mark in the middle frequency band of around 6 GHz. It is evident that the proposed conformal UWB antenna works perfectly for patch length of 13.7 mm as the return loss bandwidth is 3.5 GHz to 10.5 GHz. So, the optimum value of the patch length is taken as 13.7 mm. Figure 6 represents the parametric results of the patch width (P_w), which was varied from 13.9 mm to 17.9 mm in five steps of 1 mm. Increasing the patch width results in decrease of upper frequency. At the same time, return loss in the middle UWB range of frequency is improved but impedance bandwidth is reduced significantly. Change in the patch width doesn't affect the lower frequency much but a very small decrease is observed in lower frequency also. The optimized value of patch width is taken as 14.9 mm.

IOP Conf. Series: Materials Science and Engineering **594** (2019) 012025 doi:10.1088/1757-899X/594/1/012025

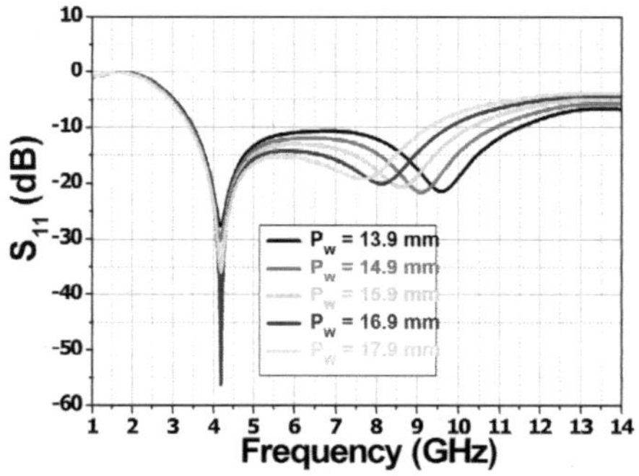

Figure 6 Parametric study of patch width (P_W)

Figure 7 and Figure 8 compares the return loss of conformal UWB antenna respectively bent at 25 mm and 50 mm radius with planar flexible UWB antenna reported in [9].

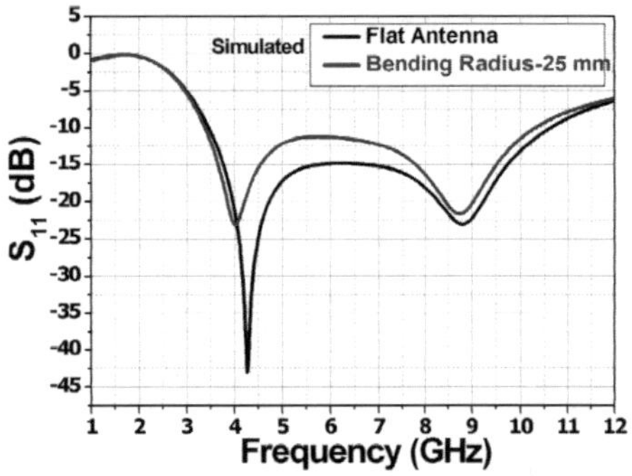

Figure 7 Return loss of flat antenna [9] and proposed antenna bent at 25 mm radius

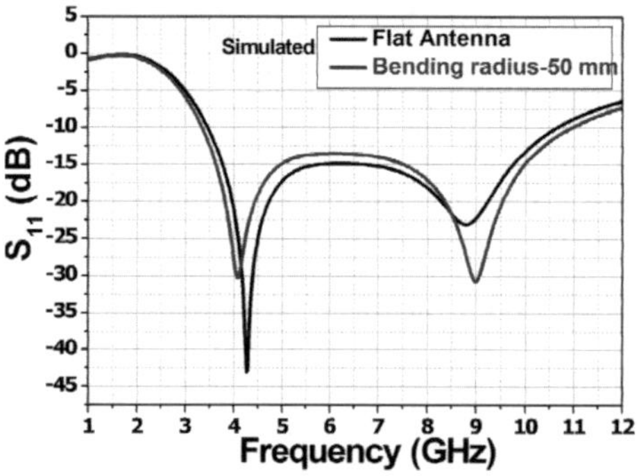

Figure 8 Return loss of flat antenna [9] and proposed antenna bent at 50 mm radius

Results show that planar antenna and conformal antenna have almost similar performance in the UWB range. There is only a little variation in the higher frequency as shown in Figure 7 and Figure 8. There is no much variation in the lower frequency due to bending of the conformal antenna. The return loss remains almost at the same level for the two bending radii. The parameters optimized for planar flexible UWB antenna as well as conformal UWB antenna are shown in Table 1.

Table 1 Optimized parameters of conformal UWB antenna

S. No.	Parameter Description	Dimension (mm)
1	Ground length (G_L)	10.56
2	Ground width (G_w)	11.66
3	Input line width (IL_w)	2.48
4	Step length (IS_L)	1.00
5	Step width (S_2)	5.50
6	Substrate length (S_L)	31.00
7	Substrate width (S_w)	26.00
8	Patch length (P_L)	13.70
9	Patch width (P_w)	14.90

IOP Conf. Series: Materials Science and Engineering **594** (2019) 012025 doi:10.1088/1757-899X/594/1/012025

After the antenna parameters were optimized for the proposed conformal antenna, these parameters were tested on the flexible UWB antenna proposed by [9] to confirm the design parameters.

Figure 9 shows return loss characteristics of the flexible UWB antenna keeping it flat and conformal UWB antennas with bending radius of 25 mm and 50 mm. There is very little variation observed in the lower and upper frequencies. But the change is not significant. Still, the antenna is valid for UWB range between 3.2 GHz to 10.6 GHz.

Figure 10 shows the antenna efficiency and radiation efficiency of the conformal UWB antenna bent at 25 mm radius.

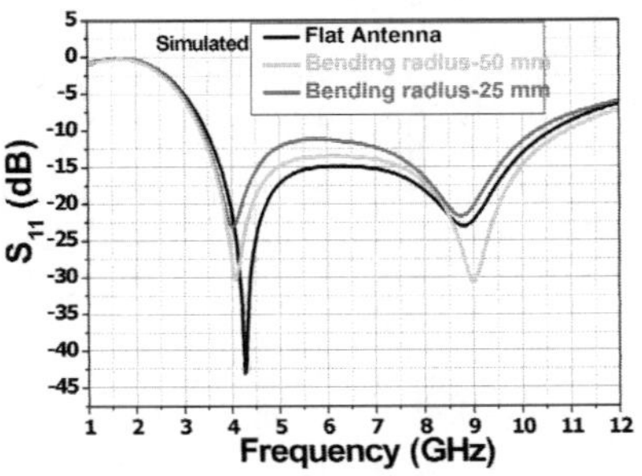

Figure 9 Return loss characteristics of flexible UWB antenna and conformal UWB antennas

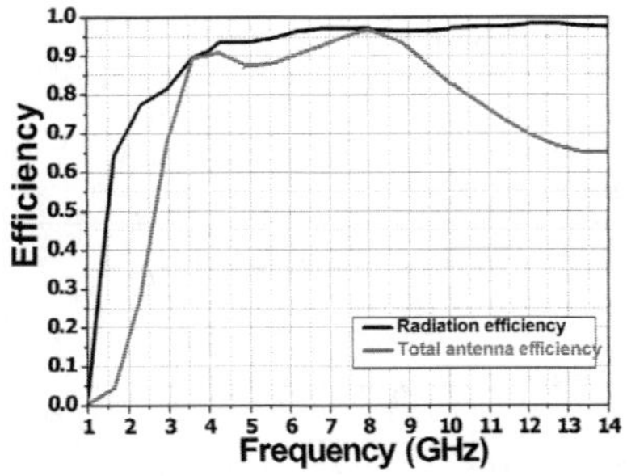

Figure 10 Efficiency of conformal UWB antenna for 25 mm bending radius

The radiation efficiency and total antenna efficiencies are more than 60% in the UWB range. Total antenna efficiency is shown to be ranging from 60% to 97.4% and radiation efficiency is ranging from 80% to 98.5% in the UWB frequency range from 3.1 GHz to 10.6 GHz. Figure 11, Figure 12, and Figure 13 show the far field radiation patterns of conformal UWB antenna bent at the radius of 25 mm at three different frequencies of 4GHz, 8 GHz and 10 GHz. Table 2 provides all the radiation properties of the proposed antenna bent at 25 mm radius.

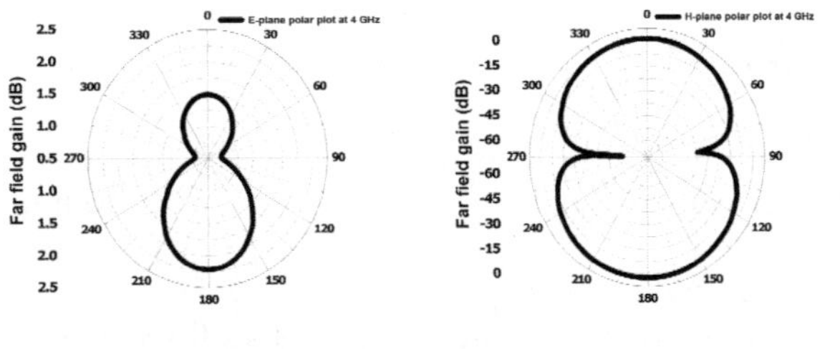

(a) E-plane pattern (b) H-plane pattern

Figure 11 E-plane and H-plane radiation patterns at 4 GHz for 25 mm bending radius

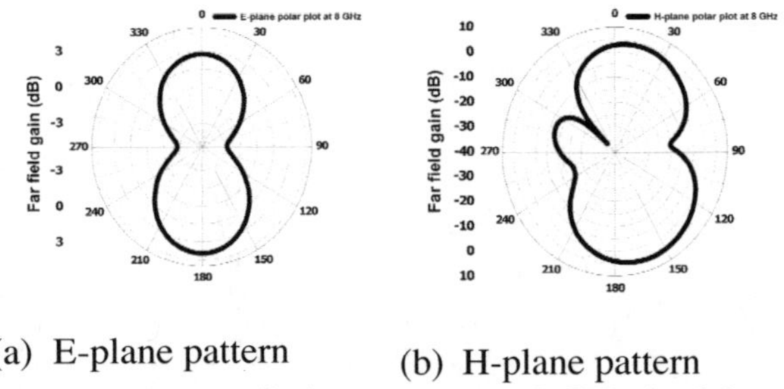

(a) E-plane pattern (b) H-plane pattern

Figure 12 E-plane and H-plane radiation patterns at 8 GHz for 25 mm bending radius

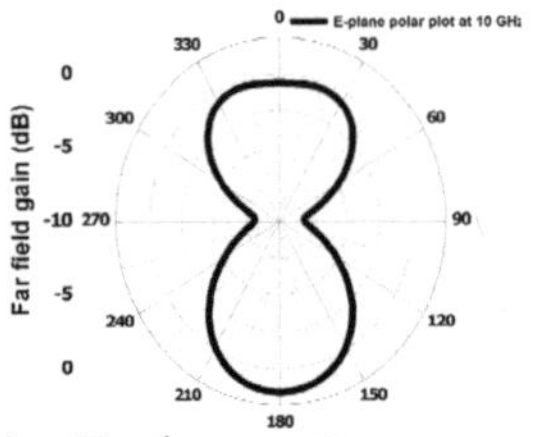

(a) E-plane pattern

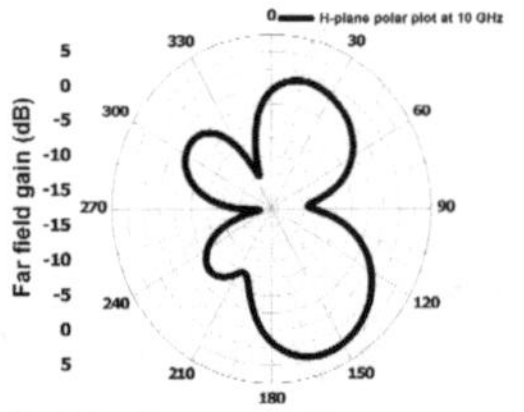

(b) H-plane pattern

Figure 13 E-plane and H-plane radiation patterns at 10 GHz for 25 mm bending radius

Table 2 Radiation properties of proposed antenna bent at 25 mm radius

S. No.	Frequency (GHz)	Main lobe direction	Main lobe magnitude (dB)	Beam width	Gain (dB)
1	4	180°	2.2	89.2°	2.219
2	8	180°	3.9	92.6°	5.34
3	10	180°	1.6	77.2°	4.93

Figure 14 depict the gain of the conformal UWB antenna for the bending radius of 25 mm. Figure 15 represents the radiation efficiency and total antenna efficiency of conformal UWB antenna bent at 50 mm radius. It is shown that the radiation efficiency varies between 86.1-97% and total antenna efficiency is varying between 71.8-94.3% in the UWB range from 3.1 GHz to 10.6 GHz.

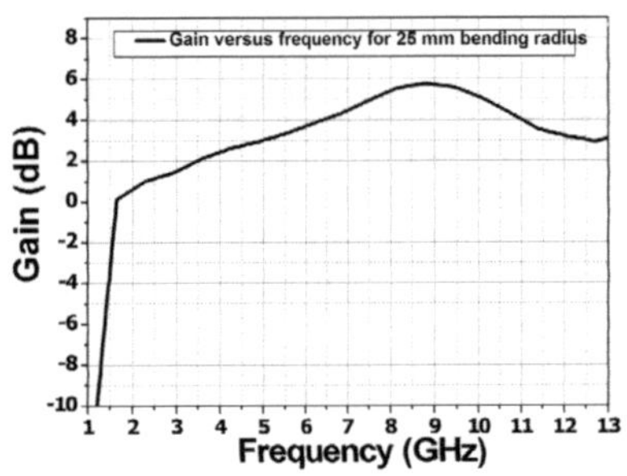

Figure 14 Gain of conformal UWB antenna bent at 25 mm radius

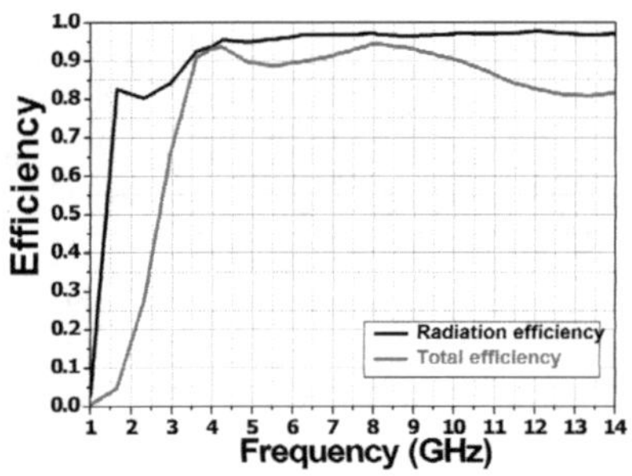

Figure 15 Efficiency of conformal UWB antenna for 50 mm bending radius

Figure 16, Figure 17, and Figure 18 depict E-plane and H-plane far field radiation patterns of conformal antenna bent at 50 mm radius. Table 3 provides all the radiation properties of the proposed antenna bent at 50 mm radius. Variation in the gain of conformal UWB antenna bent at 50 mm radius is shown in Figure 19. The gain is varying between 1.6 dB to 5.3 dB in UWB frequency range from 3.1 GHz to 10.6 GHz.

Table 3 Radiation properties of proposed antenna bent at 50 mm radius

S. No	Frequency (GHz)	Main lobe direction	Main lobe magnitude (dB)	Beam width	Gain (dB)
1	4	180°	2.6	82.2°	2.17
2	8	180°	2.6	94.9°	5.1
3	10	180°	1.0	94.5°	4.2

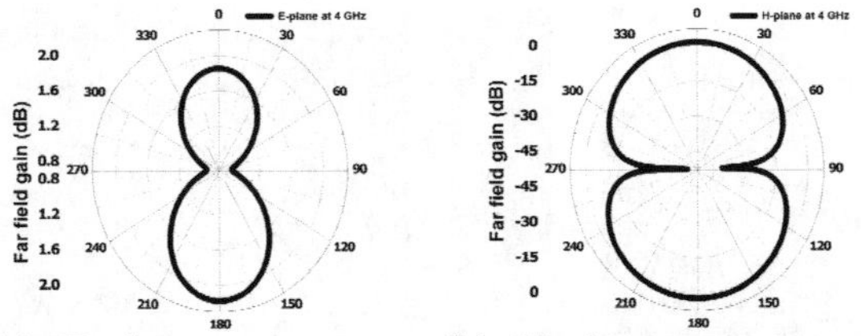

(a) E-plane pattern (b) H-plane pattern

Figure 16 E-plane and H-plane radiation patterns at 4 GHz for 50 mm bending radius

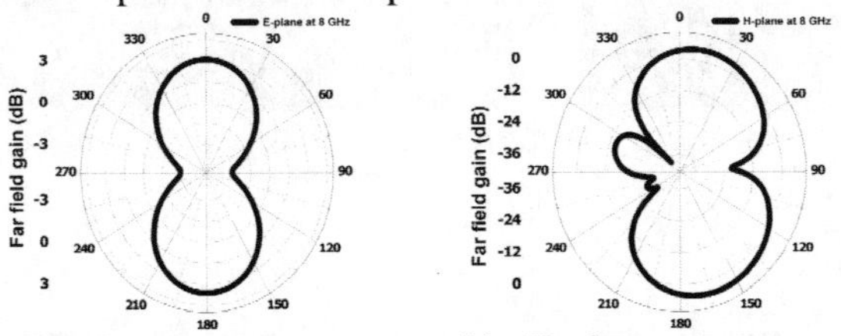

(a) E-plane pattern (b) H-plane pattern

Figure 17 E-plane and H-plane radiation patterns at 8 GHz for 50 mm bending radius

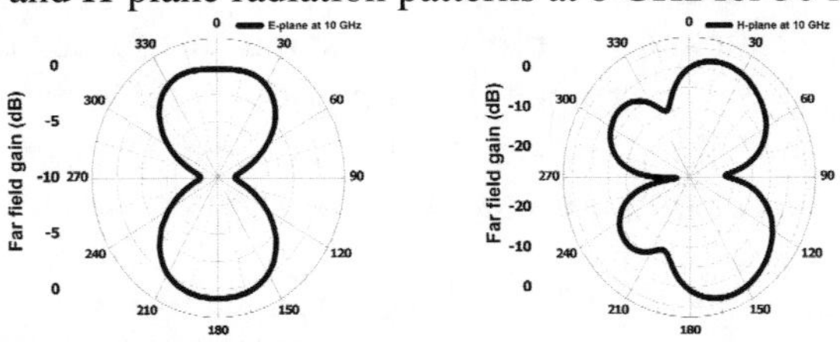

(a) E-plane pattern (b) H-plane pattern

Figure 18 E-plane and H-plane radiation patterns at 10 GHz for 50 mm bending radius

The proposed conformal UWB antenna exhibits the UWB properties and as far as simulations results are concerned it is suitable for conformal applications. Next section represents the measurement results for conformal UWB antenna, which was bent at two different radii. All measurements have been done on Agilent E5071C 20 GHz VNA.

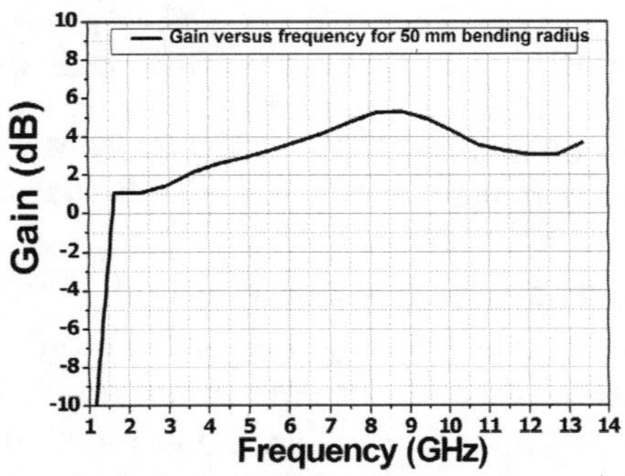

Figure 19 Gain of conformal UWB antenna bent at 50 mm radius

3. Measurement results

Figure 20 shows the measurement setup used for the measurement of return loss at bending radius of 50 mm. A flexible cardboard has been used to provide 50 mm bending radius. The antenna has been rolled on the cardboard to measure the return loss as shown in the Figure 20.

IOP Conf. Series: Materials Science and Engineering **594** (2019) 012025 doi:10.1088/1757-899X/594/1/012025

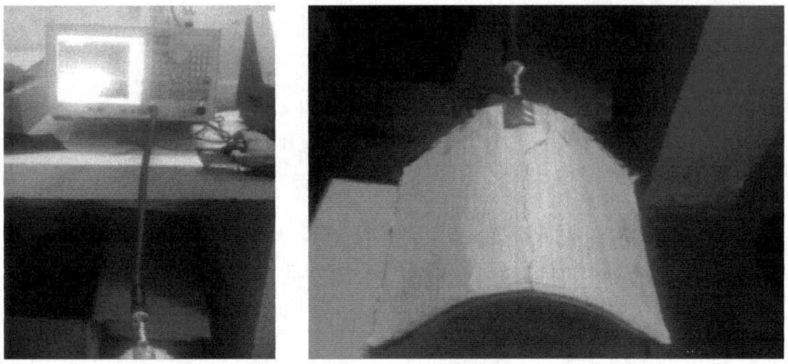

Figure 20 Measurement setup for conformal UWB antenna (bending radius- 50 mm)

Figure 21 compares the measured return loss of flexible UWB antenna in two different conditions when it was kept flat and when it was bent at the radius of 50 mm. Both the measured results closely matched. Almost the same response is observed in both the conditions. The measured lower and upper frequency is 2.9 GHz and 10 GHz respectively, for the conformal UWB antenna which is bent at the radius of 50 mm. The return loss is more around the frequency 8.4 GHz compared to the return loss measured for the flexible antenna in flat condition.

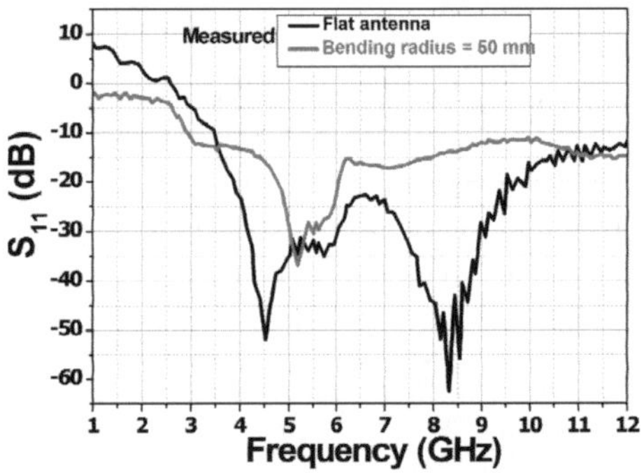

Figure 21 Measured return loss of conformal UWB antenna bent at 50 mm radius

Figure 22 Measurement setup for conformal UWB antenna (bending radius- 25 mm)

The shift in resonance frequency and deterioration in the return loss may be due to the use of cardboard, which was not present during the simulation. Also, the inaccurate 50 mm bending radius may contribute to the deterioration in the return loss. Figure 22 and Figure 23 respectively represent measurement setup to test the conformal antenna for 25 mm bending radius and measured return loss. It is clear from the graph shown in Figure 23 that the proposed conformal UWB antenna exhibits good conformal characteristics in the UWB range from 3 GHz to 10 GHz. The effect of bending the antenna at two different radii does not affect antenna performance much.

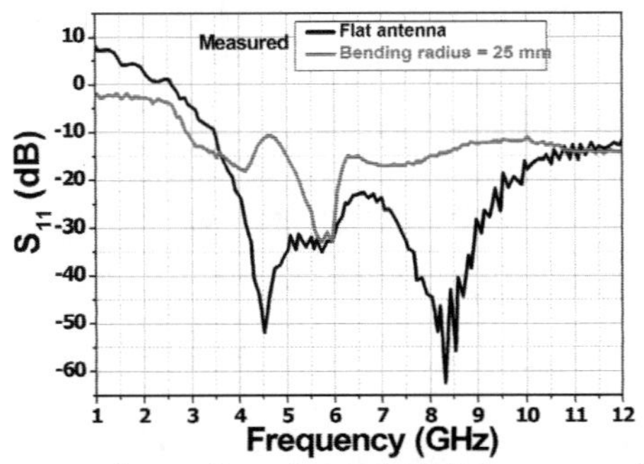

Figure 23 Measured return loss of conformal UWB antenna bent at 25 mm radius

There is only a little frequency shift observed in the result but it doesn't affect its UWB range. There is small variation observed in the measured results. This variation in the result may be due to cardboard in which the antenna has been wrapped for conformity test. The cardboard itself may have certain dielectric properties which certainly affect the frequency of operation. The material used to fix the antenna over the cardboard may also affect its performance. So, considering all these conditions, the return loss of the antenna is still maintained below 10 dB for UWB range, which provides the flexibility to neglect these effects. Therefore, the proposed conformal UWB antenna is perfectly working for UWB wearable applications and can easily be conformed to any bent surface.

4. Summary and conclusions

A conformal UWB antenna designed on RT Duroid flexible substrate with 0.127 mm thickness and relative permittivity of 2.2 has been tested for 25 mm and 50 mm bending radii. The size of the antenna is 35 × 31 mm^2. The radius of 25 mm corresponds to the arm of a medium built healthy child around 5 years age and 50 mm corresponds to a medium built healthy person of age about 35 years. Measured results are in close agreement with simulation results.

The simulated radiation efficiency and total antenna efficiency, gain, lower and upper frequencies are compiled in Table 4

Table 4 Simulated parameters for bending radius 25 mm and 50 mm

Bending Radius	**25 mm**	**50 mm**
Radiation Efficiency	80 to 98.5	86.1-97
Total antenna efficiency (%)	60 to 97.4	71.8 - 94.3
Gain (dB)	1.4 to 5.3	1.6 to 5.3
Lower frequency	2.9 GHz	2.9 GHz
Upper frequency	10 GHz	10 GHz

As far as return loss of the conformal UWB antenna is concerned, an equivalent response is obtained for the different radii of curvatures as shown in Figure 24. Return loss is below 10 dB for the UWB range for both the bending radii 25 mm and 50 mm.

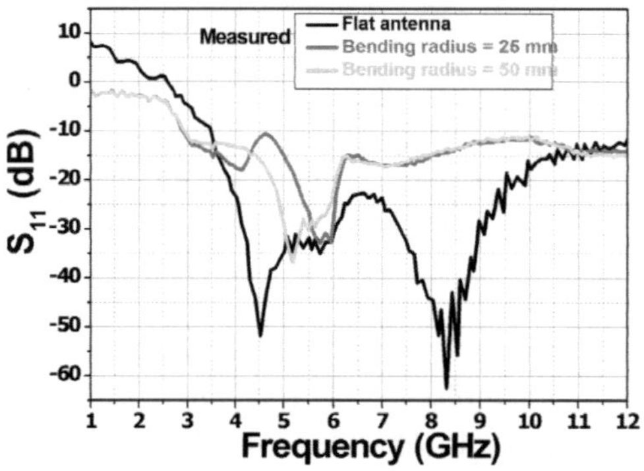

Figure 24 Comparison of flexible UWB antenna (planar) and conformal antennas

It was observed that the impedance bandwidth is between 3 GHz to 10 GHz for both the bending radii. There is only a little shift in frequency which is seen when the antenna is bent. Also, the return loss is more around the frequency of 8 GHz. Due to less thickness and flexibility of the substrate, the proposed antenna is easy to conform to any shape and radius. So, it can easily be employed for UWB wearable applications.

References

[1] First Report and Order, Revision of part 15 of the commission's rules regarding ultra-wideband transmission systems FCC, FCC 02-48, 2002.

[2] P. S. Hall and Y. Hao (edit), "Antennas and propagation for body-centric wireless communications," Artech House, 2006.

[3] Hans G. Schantz, "Three centuries of UWB antenna development," in proceedings of IEEE International Conference on Ultra-Wideband, ICUWB-2012, 2012, pp. 506-516.

[4] N. P. Gupta, M. Kumar, and R. Maheshwari "Advancement in ultra wide band antennas for wearable applications" International Journal of Scientific & Engineering Research, vol. 4, Issue 8, pp. 341-348 Aug. 2013..

[5] J. Liang, "Antenna study and design for ultra wideband communication applications," PhD thesis, University of London, UK, 2006.

[6] N. P. Gupta, M. Kumar, and R. Maheshwari "Design and verification of UWB antenna for wearable applications," in proceedings of National Conference NCASC-2014 Ghaziabad, India, Feb. 2014, pp. 124-129.

[7] S. Cheng et al., "Foldable and Stretchable Liquid Metal Planar Inverted Cone Antenna," IEEE Transactions on Antennas and Propagation, vol. 57, no. 12, pp. 3765-3771, Dec. 2009.

[8] Tommi Tuovinen et al., "Effect of the antenna-human body distance on the antenna matching in UWB WBAN applications," in proceedings of 7th International Symposium on Medical Information and Communication Technology (ISMICT), Tokyo, Japan, 2013, pp. 193-197.

[9] N. P. Gupta and M. Kumar "Improvement of Radiation Performance of UWB Antenna through slot insertion method" in proceedings of IEEE International Conference CSNT-2015, Gwalior, India, April 2015 pp.83-87.

[10] T. Tuovinen et al., "Antenna close to tissue: avoiding radiation pattern minima with an anisotropic substrate" IEEE Antennas and Wireless Propagation Letters, vol. 13, pp. 1680-1683, 2014.

IOP Conf. Series: Materials Science and Engineering **594** (2019) 012025 doi:10.1088/1757-899X/594/1/012025

[11] D. M. Pozar, "Microwave Engineering," 3rd ed. John Wiley & Sons Ltd. New York, USA, 1998.
[12] C.A. Balanis, "Antenna Theory and Analysis," 2nd ed., Wiley, New York, 1997.
[13] M. Kumar, A. Basu, and S. K. Koul, "UWB printed slot antenna with improved performance in time and frequency domains," Progress in Electromagnetic Research, vol. 18, pp. 197-210, 2011.
[14] Aliakbar Dastranj and Habibollah Abiri, "Bandwidth enhancement of printed E-shaped slot antennas fed by CPW and microstrip line," IEEE Transactions on Antennas and Propagation, vol. 58, no. 4, pp. 1402-1407, April 2010.
[15] Mithilesh Kumar et al., "Frequency and time domain characterization of a novel hexagonal shaped microstrip antenna for ultra wideband applications," IETE Journal of Research vol. 57, Issue-5, pp. 453-460, Sept. 2011.
[16] R. Garg, P. Bhartia, I. Bahl, and A. Ittipiboon, "Microstrip antenna design handbook", Artech House, Boston, 2001.

IOP Conf. Series: Materials Science and Engineering **594** (2019) 012026 doi:10.1088/1757-899X/594/1/012026

Optimization of Organic Waste Collection for Generation of Bio Gas using IoT Techniques

S S Manglorkar, *A O Sharma, D S Verma and S B Rane

Department of Mechanical Engineering, Sardar Patel College of Engineering, Mumbai, India.

E-mail: *9abhilashasharma999@gmail.com

Abstract: India is facing some major challenges in waste collection, segregation, transportation, treatment and disposal. Hence, there is a need to spread awareness about waste disposal and define better waste management techniques in India. The purpose of this paper is to utilize Internet of Things for waste management and optimized collection of wet (biodegradable) waste to produce bio-gas. It focuses on improving the current waste management system by using ultra-sonic sensors, Arduino-UNO and EC-GSM. Collection and segregation of waste in an optimized way is necessary because wet waste can be useful in producing bio-gas which in turn can be used to produce clean electricity and natural gas. Lastly, the paper also gives emphasis on cost analysis of this method to calculate revenue generation.

Index-Terms: EC-GSM, Ultrasonic Sensor, Arduino UNO, IOT, Bio-gas, Bio-CNG, wet waste.

1. Introduction

Most human activities generate waste and these wastes are generally collected and disposed of in dumping grounds. This can affect the environment and pose risks to human health [1]. Rapid urban and industrial growth in the country has resulted in overburdening of urban infrastructure services, including municipal solid waste management (MSWM) services. Solid waste management is one of the biggest difficulties faced by the civic bodies of the country today, other than supply of water, roads, electricity, education and public sanitation. Suitable facilities for disposal of MSW are not available, and the available facilities are not utilized properly due to poor management. Moreover, the increasing amount of waste generated daily in metropolitan cities is making its management more difficult. Along with household waste, industrialization is leading to generation of loads of industrial waste that may be hazardous in nature. Unscientific disposal of this waste causes an immense impact on environment and human health [2]. For the safety of life and environment, we need to device waste management methods that are economical, efficient and reliable in nature.

IOP Conf. Series: Materials Science and Engineering **594** (2019) 012026 doi:10.1088/1757-899X/594/1/012026

Internet and its applications have become an integral part of today's human lifestyle. It has become an essential tool in every aspect. Various fields like industrial automation, transportation, etc. are experiencing rapid growth with the application of IoT. Yet, not much has been achieved in the field of waste management using IoT. This paper aims to improve the waste management system of the country by implementation of IoT. The technology and its applications have been discussed briefly. All the equipment used in our day to day life can be controlled and monitored using IoT. Majority of the technology developed in this paper makes use of sensors. Sensors are deployed everywhere, and they convert raw physical data into digital signals and transmit them to its control center. EC-GSM and Arduino-UNO help in collecting data from the biogas plants and also in sending messages via internet to the waste collection trucks, where the message indicates the nearest available plant for waste disposal in a particular zone of a city. This system will therefore help in reducing the unused biodegradable waste in urban cities.

2. Literature Survey

Our plan focuses on implementation of IoT for optimized waste collection and utilization of wet waste generated in urban households for production of biogas. The following papers depict the advancements in the field of waste management using IoT:

1. Waste management Issues in India by Dr. Raveesh Agarwal, Mona Chaudhary & Jayveer Singh. It gives an idea of current practices related to the various waste management initiatives taken in India for human well-being.
2. Sustainable Waste Management Issues in India by Shikha Saxena, R.K. Srivastava and A.B. Samaddar. It highlights the possible improvement for sustainable solid waste management considering the technological, institutional and financial factors. It also gives a generalized statistics about types of waste in various states of India.
3. IoT Based Smart Garbage and Waste Collection Bin by S.S. Navghan, M.S. Killedar and Dr. V.M. Rohokale. This paper gave a basic idea about the type of sensors, micro-controllers and routers which can be used for segregation.
4. Automation of Smart Waste Management Using IoT To Support "SWACHH BHARAT ABHIYAN" – A Practical Approach by Bharadwaj B., M. Kumudha, Gowri Chandra N., Chaithra G. This paper gives an insight of how wet and dry waste can be segregated and how IoT can help in giving information about the level of garbage in a particular apartment.
5. IoT Based Solid Waste Management System by Abhay Shankar Bharadwaj, Rainer Rego and Anirban Chowdhury. It gave information about usage of LoRA technology for segregation of wet and dry waste and in a way helped to us identify which could be a better option for the system.

3. Motivation for Research

Current systems in India cannot cope with the volumes of waste generated by the increasing urban population, and this has an impact on the environment and public health. The usual practice followed is to lift solid waste from the point of generation and hauling it into distant places known as dumping grounds or landfill sites [3]. Over 160,000 metric tons of municipal solid waste is generated daily in the country out of which approximately 91,152 tons is collected and approximately 25,884 tons gets treated.

The government of India is promoting setting up of lakhs of biogas plants throughout the country. Central subsidy is being extended for setting up of biogas plants. Unfortunately, around 4.5% of the inspected Biogas plants are non-functional. The main reasons for this as explained by the government are:

1. Construction & Maintenance defects.
2. Lack of raw material.
3. Problems related to equipment (burners, pipeline and fittings).

Moreover, the biogas plants being setup focus on utilization of agricultural and animal wastes as raw material. As per statistics, almost 65-70 % of the waste in households consists of bio-degradable waste like food, garden waste, etc. Out of this, almost 80% biodegradable waste goes untreated because there is no segregation. There are no fixed government policies for utilization of organic waste generated in urban households to produce biogas. Thousands of tons of organic waste generated in the major cities ends up being disposed at dumping sites. If this waste is systematically segregated and used as a raw material in biogas plants, it can generate enormous amounts of biogas. This will solve the problems of waste disposal and availability of renewable energy sources. But it remains a big challenge for India to manage such huge amounts of waste. Current waste management systems are more focused towards the traditional method of managing waste. The motive of our project is to develop an efficient waste management system using IOT techniques.

4. Research Methodology

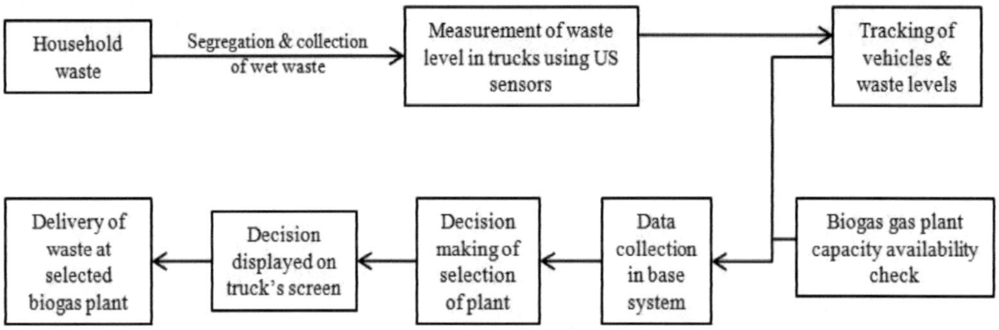

Fig. 1: Process Description Flowchart

5. Process Description

Currently, most of the cities use centralized way of waste management. For cities like Mumbai, centralized way of managing waste becomes hectic, complex and costly. Application of a decentralized system would be more efficient and effective in such cities. It focuses on division of city in several zones and effectively collecting, segregating and managing waste of each zone separately. Our aim is to provide a decentralized, efficient and simple waste management system using IOT techniques.

Critical to the implementation of such a system is segregation of waste into wet and dry waste. The best way to segregate waste is to segregate at the source itself. Hence, pertaining to 'Swachh Bharat Abhiyan', people can be motivated to perform segregation of household waste into wet and dry waste by giving some considerable allowances. The enormous amount of wet waste generated in cities can be used to generate huge amount of biogas which can be sold as fuel and thus generate revenue. This generated revenue will help in providing the desired allowance to people.

5.1. Application of IoT

After segregation of waste, optimized collection will begin. The level of wet waste in collection trucks will be measured by ultrasonic sensors embedded in them. Arduino-UNO will be used for programming the stored data. EC-GSM will transmit this data to the Municipality Base Center. Simultaneously, data from each bio-gas plant will be manually fed in the web-page and this data will be directly transmitted to the database via internet to the base system. GPS will be used to track the position of trucks. Once the

IOP Conf. Series: Materials Science and Engineering **594** (2019) 012026 doi:10.1088/1757-899X/594/1/012026

ultrasonic sensors sense that the truck is completely filled with waste, and after receiving the location of the trucks and capacity information of the plant, the center will assign the nearest available biogas plant to the truck. We have divided our system into device and management layers [4] as explained below.

5.1.1. Device layer: The device layer here consists of the vehicle which is carrying waste with the embedded hardware. Ultrasonic sensors in the vehicle will sense the garbage level in the vehicle. These sensors interfaced with Arduino will sense the garbage level and ensure that the vehicle is filled up to its total capacity and share information with Arduino. The Arduino will use the information received from the sensor and check if it is filled to the maximum capacity of the vehicle. The Global positioning system will also be connected to the vehicle so that the location of the vehicle can be tracked.

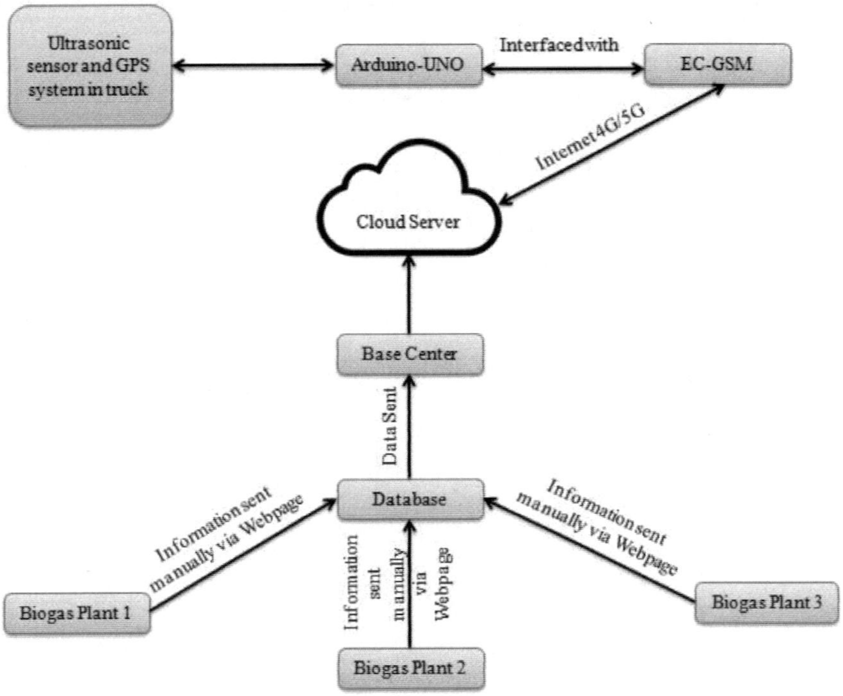

Fig. 2: IOT Network Flowchart

5.1.2. Device Transmission Layer: The Transmission device used is GSM module interfaced with Arduino-UNO. This device will be placed in each vehicle and programming will be done in such a way that data taken from the sensor will get stored and manipulated. This manipulated data or garbage level will be sent to the Municipality Base Centre by transmitting device i.e. GSM module interfaced with Arduino.

The data from the bio-gas plant about the waste required for generating electricity or natural gas is stored manually in the web page. This stored data is directly transmitted to the database via internet to the Municipality Base Center.

5.1.3. Device Communication Layer: The communication device which is used is the EC-GSM module. Data stored in the micro-controller will be sent to EC-GSM Transmitter pin in the vehicle. This data will be delivered to the EC-GSM receiver pin at the base center through the internet.

5.1.4. Management Layer: The single Arduino based module which is at the base center will compare the data received from the vehicle (level of wet waste in the vehicle) and the bio-gas plant (how much more wet waste is required to operate the plant) and the appropriate location will be assigned to the vehicle operator. If there is a similar requirement by all the bio-gas plants, the GPS will track the vehicle and assign the nearest available location.

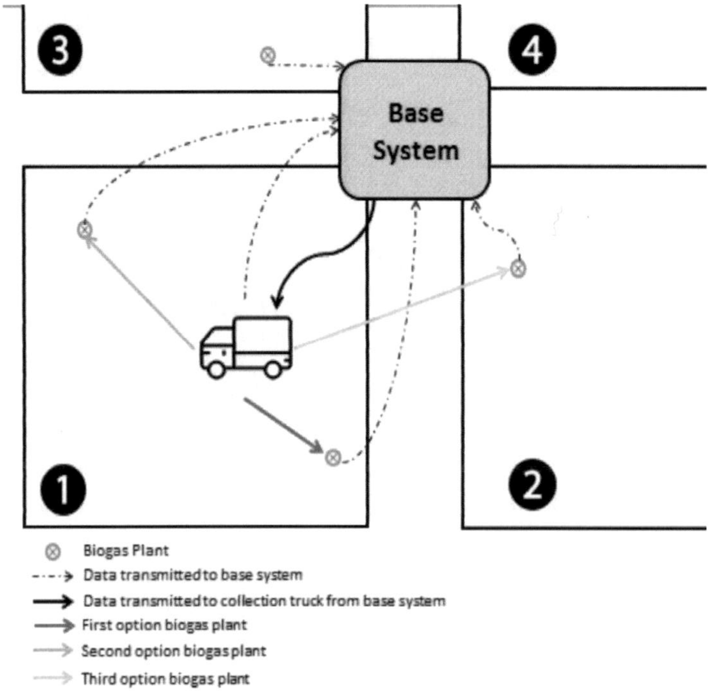

Figure 3: City Structure

6. Cost Analysis

Basic cost calculations were done to calculate the revenue generated from the process for the city of Mumbai. Over 70% of household waste is organic in nature and can be used to generate biogas. This biogas can be sold back to the households in the form of bio-CNG which can be used as fuel for cooking and other purposes. This biogas generated can also be used to generate electricity (this form of utilization is not considered in the cost analysis).

Essentials	Value	Unit
	Revenue Calculation	
Waste generated per day	7,500.00	Metric ton
Wet waste generated per day	5,250.00	
Waste collection efficiency	60	%
Wet waste collected per day	3,150.00	Metric ton
Biogas generated per ton of wet waste	100.00	m^3
Biogas produced per day	31,500.00	m^3
Density of biogas	1.15	Kg/m^3
Weight of biogas produce per day	36,2250.00	kg
Cost of bio CNG	35.00	Rs./kg
Revenue generated per day	12678750.00	Rs
	Fixed Cost Calculation	
Cost of 25 ton capacity biogas plant	6,000,000.00	Rs.
Land cost per plant (0.25acre)	20,000,000.00	Rs.
Total fixed cost per plant	26,000,000.00	Rs..
Total no of bio gas plant required	126	nos
Total fixed cost	3,276,000,000.00	Rs.
	Variable Cost Calculation	
Water cost rate	4.80	Rs./metric ton
Water required per day	3,150.00	Metric ton
Water cost per day	15,120.00	Rs.
	Break Even Analysis	
Days achieve breakeven	256.69	Days

The analysis shows that even at the collection rate of 60%, we would be able to generate enough revenue to break even in less than one year. This revenue will also help provide incentives to people to motivate them to segregate waste.
Moreover, waste water from households can be used in these plants, reducing the amount of water required. The bi-product of the process (fermented slurry) can be sold as fertilizer, thus providing an alternative revenue source.

7.Discussion

7.1. Advantages:
Firstly, this method focuses on motivating people to segregate waste at home by providing incentives. This will solve the problem of waste segregation. Secondly, the major chunk of urban household waste, i.e. wet waste, which is getting disposed of in dumping grounds will be used to generate biogas which will ultimately help in revenue generation. Thirdly, this system will ensure that vehicle moves at full capacity and thus save the cost of transportation by avoiding extra trips. Finally, implementation of IoT in the waste management system makes it more reliable and efficient. The better management and decision making achieved with the help of IoT integration creates scope for infinite future improvements in the waste management system for cities.

7.2. Future Scope
Fabrication of an IoT based product that segregates waste at home. This will promote waste segregation which is a critical success factor for the whole system. Complete automatization of the system without any human interference. The data of biogas plant capacity can be sensed and transmitted automatically to the base system. Human help will then be required for monitoring purposes only.

IOP Conf. Series: Materials Science and Engineering **594** (2019) 012026 doi:10.1088/1757-899X/594/1/012026

7.3. Limitations

Availability of good and continuous network connectivity is necessary. Any kind of barrier in communication can hamper the smooth functioning of the system. The success of this system has a high dependency on the fact whether people will segregate waste at home or not. Provision of incentives will promote waste segregation.

References

[1]. Zhu, D., Asnani, P. U., Zurbrugg, C., Anapolsky, S., & Mani, S. K. (2007). *Improving municipal solid waste management in India: A sourcebook for policymakers and practitioners*. The World Bank.

[2]. Gupta, S., Mohan, K., Prasad, R., Gupta, S., & Kansal, A. (1998). Solid waste management in India: options and opportunities. *Resources, conservation and recycling*, *24*(2), 137-154.

[3]. Agarwal, R., Chaudhary, M., & Singh, J. (2015). Waste Management Initiatives in India for human well-being. *European Scientific Journal, ESJ*, *11*(10).

[4]. Bharadwaj, A. S., Rego, R., & Chowdhury, A. (2016, December). IoT based solid waste management system: A conceptual approach with an architectural solution as a smart city application. In *2016 IEEE Annual India Conference (INDICON)* (pp. 1-6). IEEE.

IOP Conf. Series: Materials Science and Engineering **594** (2019) 012027 doi:10.1088/1757-899X/594/1/012027

A Relative Survey on Handover Techniques in Mobility Management

Pramod Kumar P* and Sagar K

Department of CSE, Osmania University, Hyderabad, Telangana, India
Department of CSE, Chaitanya Bharathi Institute of Technology, Hyderabad, Telangana, India
Email: * pramodpoladi111@gmail.com

Abstract. Innovation is advancing at a speedier pace and necessity for web organizations at whatever point wherever is required. To impact this vision ever to better organizations open to a customer, effective mobility managing procedures must be composed. Two key intends to execute mobility administrations are Handover and area administration. Therefore, we need to have an efficient mobility management system among heterogeneous wireless networks where several wireless networks can interoperate to provide users with good QoS. In this paper, distinctive issues related to mobility administration in the coming period of frameworks are highlighted and besides throws a light on the uncertain zones to ensure proficient mobility management.

Index Terms— Location Management, Mobile Node, Handover, Mobility Management, Mobility Models.

1. Introduction

An enormous number of termini are connected over the Internet worldwide with a mutual aim of being 24 by 7 service. This has incited improvement in all areas, business, enlightening, transportation et cetera. Now a days the network technologies are speedily continuing forward in the advancement pathway, various access administration systems have been proposed to address the issues of predictable accessibility. The organization of mentioned frameworks is the prime point of view as proposed frameworks requires more resource allocation [1]. The Figure.1: Mobility administration in various heterogeneous wireless networks shows the varied setup of systems through mobility administration structure [8].

2. Related Work

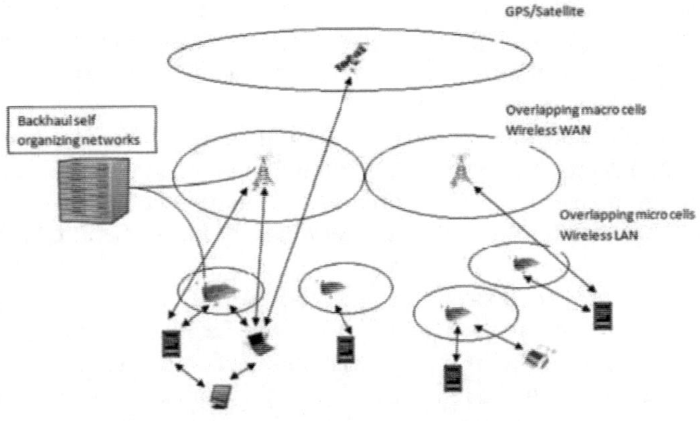

Fig. 1. Mobility in Heterogenous Network

The below territory incorporates the earlier work done by the diverse proud analysts. An expansive examination beginning from QoS, administration of mobility and administration of handover has been finished. Systems watching out for the concerns winning explored in mentioned areas.

The investigators in [4] depicted the multihoming highlight of Stream Control Transmission Protocol (SCTP), however, it's wastefulness in helping to move towards center points from this time forward examined mSCTP which connects with the vehicle layer portability. The Cross-Layer Route Method in SCTP with MIPv6 has experienced age [11]. Changed the basic Binding Update system and joined the exchanges between the SCTP and MIPv6.

Analysts in [6] proposed an innovative quality-careful flexible Concurrent Multipath Transfer Quality Arrangement (CMT-QA) that employs SCTP for File Transfer Protocol (FTP) for the transmission of data and steady multimedia in remote heterogeneous systems. CMT-QA screens and examinations often every way's data dealing with limit and settles on data movement alteration decisions to pick the qualified routes for simultaneous data exchange.

The analysts in [12] proposed an overhauled Fast handovers for proxy mobile IPv6 (FPMIPv6) strategy to improve the vertical handover (VHO) process by using low data rate and enhanced synchronization of upheld divide and Terminal Node (TN) trading, that realizes an essentially of reduced package delay.

The analysts in [3] exposed a particular research concerns and arrangements on various traits of 4G remote structures of present and earlier. This is one of such a issue shows explores brings about different parts of 4G remote, for example, booking, resource allotment, subjective and pleasing exchanges, multicast administrations and extension and organizing of little cells.

The analysts in [11] have set up a course to direct the system in which various features like the imperativeness of the hub, area, transmission limit of the package are analyzed. Along these lines, when the bundles ought to be sent, it examines each such feature and moves up the course with the effective hub that has one ricochet check with the help of SCTP transport convention. In the process of transmission of groups some imperativeness from the hub gets decreased. Hence, there is a more need for stop up, as the breaking point of the hub is diminished. By implementing the SCTP, this package hardship rate & endways grant is decreased and Quality of Service is improved.

The analysts in [10] inspected the best in class multihoming systems using SCTP. An entire diagram of changes has conveyed all possible essential research zones, specifically: handover administration, Concurrent Multipath Transfer (CMT), and cross-layer works out.

The analysts in [9] included diverse concerns and encounters related for mobility administration of 4G – heterogeneous systems. Elaborated the concerns with the arrangement of handover administration like hailing overhead, QoS, control need, flexibility, immovable quality and generosity.

The analysts explored late headways in area administration. Study of several mechanisms for handover administration among heterogeneous structures have been obtainable. Analyzed techniques among system handovers in distributing and organizing the convention layer in which the handovers happens. Displayed open issues for mobility administration in future remote systems.

The analysts explored late headways in area administration. Survey of several methods for Handover administration between heterogeneous structures has been obtainable. Analyzed techniques for between system Handovers in distributing organizes as demonstrated by the convention layer in which the Handovers happens. Displayed open issues for mobility administration in future remote systems.

3. Mobility Management

Mobility Management oversees the following supporter's area, steering, endorsement, and check and dealing with Handovers. Mobility between the versatile terminals is considered into two essential creates:

1. Inter-framework(InterSystem)
2. Intra-framework(Intra System)

Intra-framework mobility is the improvement of the portable terminals in the middle of the cells of a similar system. It can in like way be alluded to as Horizontal Handover. Intra-framework mobility administration methods depend upon close system interfaces and conventions. Between framework, wandering implies moving between various systems which change in measures and innovation.

Mobility administration in remote systems is dealt with by its two key approaches: area administration and Handover administration [8]. Location Management is associated with finding and following the location and improvement of a portable hub & invigorating the versatile terminals and the location information is stored in the databases like Home Location Register (HLR) and Visitor Location Register (VLR) while handover administration attentions on the exchange of the association beginning with the base station then onto the following in the middle of the advancement of the portable hub to keep up data cognizance. Hence, the issues and troubles related to mobility administration has to consider for all of these frameworks and the concerns must be elaborated. The concerns in area administration aren't relevant to directing and subsequently, isn't convention subordinate, while Handover computations require steering and resource administration and are liable to organize conventions. The below shown diagram depicts segments connected with mobility administration [3].

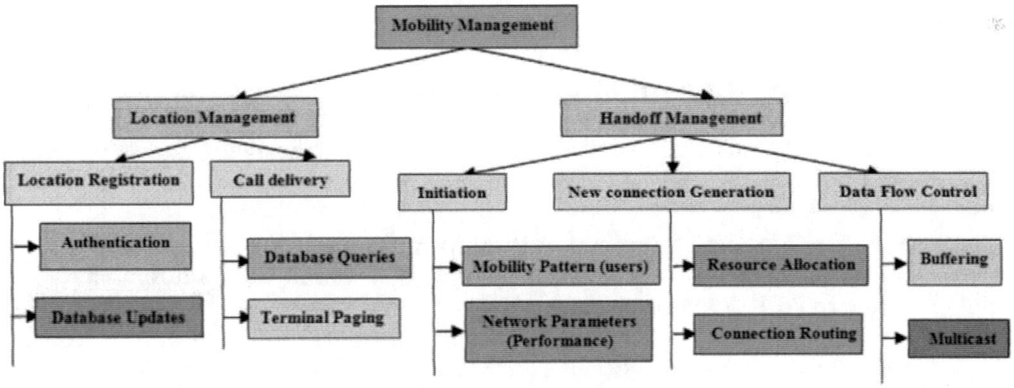

Fig. 2. Mobility Administration

3.1 Location Management: The Location Management allows a framework to monitor the current locations of Mobile Node (MN) in the middle of its course of improvement at various systems. It joins two critical assignments. The underlying procedure is the updation in area enlistment, where MN teaches the framework at standard interms of time to invigorate the appropriate location repositories with it's present location information. The next one is about call movement, in which the structure chooses the present area of the MN in light of the information present in the framework repositories when the MN correspondence is begun. The latter is known as paging, in which the messages are sent to each telephones/subnets abiding in enlistment zone of the tracked mobile terminal. For the intersystem nomadic, the arrangement of the location administration strategies are inline with challenges:

3.2 QoS improvement by decreasing signals: Latency is one of the important basic constraints included bythe specialists in light of the fact that to decrease inertia is the apex essential. It is required to pass on administrations to the area of MN and revive with the least delay. Hailing issue is also another measure to access to the related repositories that are to be chopped down. Thusly, the adaptive research work can be started and appropriated proposed methods can be used chop down the multifaceted thought of territory strengthening & selections.

3.3 Control on enrollment of MN in heterogeneous networks: If the MN goes into the non-coverage area then it winds up its services, then it's challenging to pick that with which the MN must accomplish enrollment. Now and again the wrong enlistment may incite greater dormancy and manufactures overhead.

4. Handover

Handover/Handoff is a method of changing the point of connection from one cell to another cell. A Handover strategy happens when a versatile customer moves past the radio extent of its passage centers to another [2]. In the midst of this, some latency is a direct result of which the customer can't send or get development for a couple of minutes even there are various neighboring access centers in a remote circumstance. Therefore, handover parts are to be used with the end goal of that which they can pick the best & perfect Access Point (AP) to acknowledge determined administrations [3].

In this way, unique strategies are proposed and get some information about are being improved handover techniques. Truth be told, even experts utilize the cushy strategy for thinking of neural frameworks and to pick the best base station or AP. With the occurrence of the new frameworks advances and change in this adaptable correspondence below is the couple of unmistakable issues that need to be thought for a capable handover.

- **Handling of Inter-area and intra space correspondence**: In the front-line time it is basic to analyze the execution of a convention with the objective that it can be all around perceived what convention that can works better in intra area & between space conditions.

- **Need for Fast and predictable Handover**: There is a need for the handover that should be smart when required to give a continuation of administrations to realize divide and call dropping or blocking.

- **Need for directing viability**: As the handover is liable to the conventions so relevant conventions, it must be needed to outset issues, for instance, steering delay, triangular directing issue, et cetera e.g. MIP has given a course streamlining method by getting rid of the triangular steering issue.

- **Enhanced QoS**: There are some new strategies has been offered to pass on Quality of Service, yet a couple of issues are guided by remote innovation which ruins in its execution. Some of these issues are limited information transmission, low confident quality, and deferment.

- **Mobility between getting to signs of progress**: requires a passageway innovation which offers more information transmission with small deferral, less package hardship & more security. Be that as it may, it is attempting to fulfill each one of these requirements together, so a customer needs to keep up an administration require and suitably pick the passageway innovation [9].

- **Convention parameters optimization**: One of the approaches like fuzzy basis can be associated with parameter enhancement and a regularly expanding number of credits adding to organize execution ought to be researched [8].

- **Optimization of cross-layer:** There should be a suitable co-arrangement among the system layer & a data interface layer of handover process. Adaptive systems which are proposed should emphasis on the most ideal approach to developing fitting investment between different layers.

4.1 Mobility management model in heterogenous networks (General)

Mobile Node and Corresponding Node: Portable hub naturally winding hub and that can make a change in it's motivation of association beginning with one center hub then next to the following, it may be either flexible in communication among all the portable hub and home network.

Home Network (HN) and Foreign Network (FN): Here the home network at first doles out IP convey to the convention center and to which the flexible center point is incessantly passing on. Remote framework to which the versatile center point is beginning at now related amidst its course of headway. The outside framework converses with the versatile center point through a concise Care of Address (CoA).

Fig. 3. Mobility Management-General Framework

Home Address and Care of Address (CoA): Place of residence is constantly unalterable deliver doled out to the movable hub through its home system. CoA – is having capability to allot and deliver doled out to portable hub by the non-local system until the duration center abides in the outside system.

Home Agent (HA) and Foreign Agent (FA): Home agent master assistances in the association of a versatile hub when it changes to outside system.

IOP Conf. Series: Materials Science and Engineering **594** (2019) 012027 doi:10.1088/1757-899X/594/1/012027

Remote mobility administrator: it is a switch outwardly organize which causes in getting to the web and along these lines tolerating datagrams passed on to CoA. The going with figure exhibits the basic part connected with general mobility system as analyzed in [5].

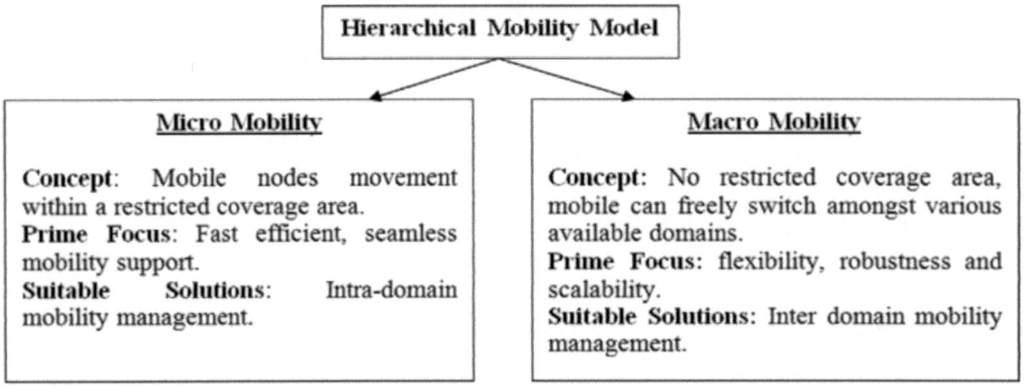

4.2 A Model - Mobility Management for hierarchical:

It exhibits the space-based portability thought which gives a good control over portability to the extent execution and flexibility especially for the hubs which are from time to time moving. Figure 4 clears up the more broad reason for the disconnection between large scale and miniaturized scale mobility designs [5].

Features of Hierarchical Mobility Management:

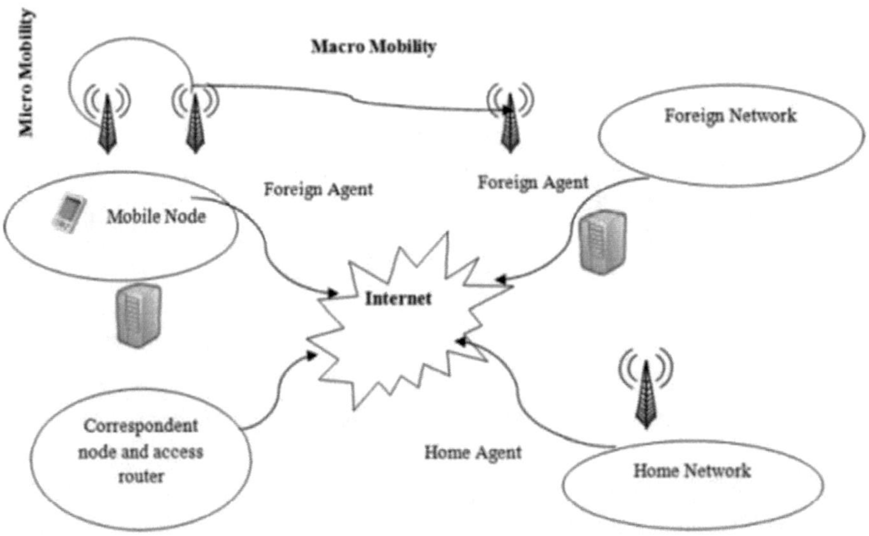

Fig. 4. Hierarchical Mobility Management Framework

IOP Conf. Series: Materials Science and Engineering **594** (2019) 012027 doi:10.1088/1757-899X/594/1/012027

5. Proposed Work

Observing monster advancement in the modernized and handy devices, administrations being given and the mobility outlines, the deliberated concerns, and troubles exhibit the necessity for the learning and examination of more factors that may be measured and participated in existing mobility system with the objective that the customer hint at change QoS. From this time forward, the optimization of cross-layer coordination ought to be functioned in further process, with the objective that subject on the area, the individual layer that can subsidize to mobility administration in handover process.

6. Conclusion

As a consistently expanding number of administrations are being offered to the customers over portable systems. The customer want has swung to be all the all the more requesting and need to be driven. This paper to total things up discusses the fundamental locales in a field of mobility administration in remote systems. Doubtlessly various specialists are finding their enthusiasm here yet in the meantime various issues still exist and these issues are a direct result of issues which are controlled in remote innovation i.e data transmission, unwavering quality, inertness, and postponement. In any case, as we are drifting towards 4G and higher generations, heterogeneous systems can put a conclusion to these issues by proposing more degree data rates, and transfer speed. In this way, a proficient mobility system ought to be made which is more versatile, savvy and administration center.

References

[1] M. Kottkamp, "LTE-Advanced Technology Introduction," Rohde & Schwarz, 2010.

[2] "5G : A Technology Vision", www.huawei.com/ilink/en/download/HW_314849.

[3] M. Monemian, P. Khadivi, and M. Palhang, "Analytical model of failure in LTE networks," IEEE, 2009, pp. 821-825.

[4] S. Barre, C. Paasch, and O. Bonaventure. Multipath TCP: from theory to practice. In Proceedings of the 10th international IFIP TC 6 conference on Networking - Volume Part I, NETWORKING'11, pages 444–457. Springer-Verlag, 2011.

[5] V. Stencel, A. Muller, and P. Frank, "LTE Advanced-A further evolutionary step for Next Generation Mobile Networks," IEEE, 2010.
[6] Y. H. Nam, L. Liu, and Y. Wang, "Cooperative communication technologies for LTE-advanced," IEEE, 2010, pp. 5610-5613.

[7] B. Jiang, Y. Cai, and D. Towsley. On the resource utilization and traffic distribution of multipath transmission control. Perform. Eval.,68(11):1175–1192, Nov. 2011.

[8] B. Karakaya, H. Arslan, and H.A. Cirpan, "An adaptive channel interpolator based on Kalman filter for LTE uplink in high Doppler spread environments," EURASIP Journal on Wireless Communications and Networking 2009:7.

[9] G. A. Abed, M. Ismail, and K. Jumari, "Traffic Modeling of LTE Mobile Broadband Network Based on NS-2 Simulator," Computational Intelligence, Communication Systems and Networks (CICSyN), 2011 Third International Conference on,2011, pp. 120-125.

[10] A. Balasubramanian, R. Mahajan, and A. Venkataramani. Augmenting mobile 3g using wifi. In Proceedings of the 8th international conference on Mobile systems, applications, and services, MobiSys '10, pages 209–222. ACM, 2010.

[11] A. Ford, C. Raiciu, H. M., and O. Bonaventure. TCP extensions for multipath operation with multiple addresses. Draft ietfmptcpmultiaddressed- 07,2012.

[12] A. Ghosh, J. Zhang, G. Andrews, and R. Muhamed, "Fundamentals of LTE," Prentice Hall, 2010.

IOP Conf. Series: Materials Science and Engineering **594** (2019) 012028 doi:10.1088/1757-899X/594/1/012028

A Compact Multiband Rectangular Microstrip Antenna for UWB Applications

Sandeep Kumar Singh[1,*], Tripurari Sharan[2], Arvind Kumar Singh[2]

[1]ECE Department, SET, Sharda University, India
[2]N. E. R. I. S.T., Nirjuli, India
E-mail: *hit.ssk@gmail.com

Abstract. This manuscript develops the design of rectangular patch antenna with loaded on defected ground structure (DGS), which covers the entire range of ultra-wideband (UWB). The area of proposed antenna 33×35 mm^2 and is printed on FR-4 plane substrate, its permittivity (ε_r) 4.2 and loss tangent (tanδ) is 0.02 at entire frequency range. Proposed structure renders wider impedance bandwidth extended from 2 GHz to 10.7 GHz at < -10 dB return loss (VSWR< 2) with multi-bands. The UWB antenna suitable for UMTS 2.1 GHz, WLAN 3.5 GHz, WiMAX 5.8 GHz, and X-band 7.5 GHz applications. The parameters of antenna such as loss of return are reduced, and width of band, as well as gain in VSWR, is improved to an acceptable limit with reasonable radiation pattern by means of CST V.17 EM simulator. The results from simulation and experiment results are almost similar to acceptable limit.

1. Introduction

The increasing expansion of the wireless network and several far-reaching types of wireless applications like GSM/UMTS/ DCS/PCS/IMT, Bluetooth, ISM, Wi-MAX, and WLAN etc. raises the demand to cover a wide frequency range and design an antenna with broadband. In wireless communication, the patch antennas have played an important role and to be used as multi-band applications due to its smart profile, and lesser power consumption capabilities. The size reductions of antenna gain and impedance bandwidth enhancement are the major approaches for practical applications of patch antennas. The different shapes can be applied developed a patch antenna, that has been used for achieving as much as wide impedance bandwidth and gain of antenna. Satellite and wireless communication often require antenna with conveniently small size, ease of fabrication and capable of operating multi-band of frequencies [1– 6].

In 2002, a US Federal Communications Commission (FCC) permitted a (3.1 - 10.6) GHz unlicensed radio spectrum of band designated by used as an ultra-wideband (UWB). UWB is generally applied because of its lower cost, lower power spectral density and low capacity of channel [7]. The design of UWB patch antennas compared to narrowband antennas has more complex and challenging due to the large bandwidth and linearity requirements [8].

A simple and compact UWB defected ground structure (DGS) antenna is proposed and investigated for achieving the gain and bandwidth enhancements for multiband operation [9– 13]. The dimensions of patch antenna have a small and operate an entire frequency range 2.0 - 10.7 GHz and suitable for UMTS 2.1 GHz, WLAN 3.5 GHz, Wi-MAX 5.8 GHz, and X-band 7.5 GHz applications. The remainder of this paper is structured into three parts. The geometrical parameters of antenna are described in section 2. The analysis of results of UWB antenna is presented in part 3 and summarized in section 4.

IOP Conf. Series: Materials Science and Engineering **594** (2019) 012028 doi:10.1088/1757-899X/594/1/012028

Table1. Impedance bandwidth comparison of UWB with other antennas

Reference	Area $[mm^2]$	UWB (3.1 - 10.6 GHz) Notchfrequencies(f_N)
[5]	25×40	3-10.6 GHz (f_N=3.4, 6.3, and 8.9 GHz)
[6]	40×40	2.9-19.2 GHz (f_N=3.7, 5.7, and 9.5 GHz)
[7]	57×77	3-10 GHz (f_N=4 , 7, and 10 GHz)
[8]	42×24	2.3-2.5 GHz, 3.1-12 GHz (f_N=2.45 , 3.5 , 6.5 ,8.5 GHz)
[9]	60×60	2.67-13 GHz (f_N=5.12, 5.73, and 6.42 GHz)
[10]	40×35	2.62-15.45 GHz (f_N=3.4, 6.8, and 8.9 GHz)
Proposed	33×35	2.0 - 10.7 GHz (f_N=2.1,3.5 ,5.8, and 7.3 GHz)

2. Structure of Antenna

The structure and dimensions of the UWB multiband antenna is shown in Fig 1. The radiating patch of the antenna is loaded on FR-4 material. With the entire range of frequency, it's loss tangent (tanδ) is 1.2 and permittivity (ε_r) 4.4. The area (W_{sub}×L_{sub}) of the substrate is 33×35 mm^2 and thickness 1.6 mm respectively. By etching a radiating patch (W_p×L_p) 15.7×15.8 mm^2 on the substrate plane and connected to a 50Ω feed line (W_f= 2.8 mm) on same plane. The UWB antenna and its radiating tuning stub are tuned empirically to find the performance of the antenna in terms of enhanced bandwidth impedance with stable radiation pattern to entire operating bandwidth. The parameters related to design UWB antenna are mentioned in Table 2. To optimize the design parameters for the multiband application, simulation is carried out.

Table 2. Optimal unit value of the antenna shown in Fig. 1

Parameters	Lsub	Wsub	Lp	Wp	Lf	Wf	Lslot
Unit (mm)	35	33	15.8	15.7	10.25	2.8	24.75

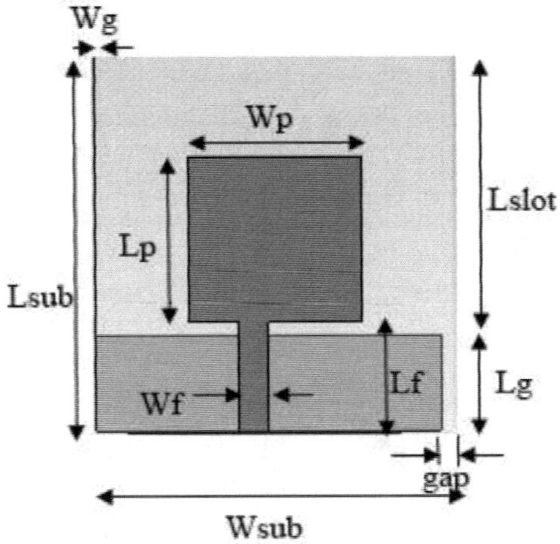

Fig. 1. Configuration of the UWB antenna

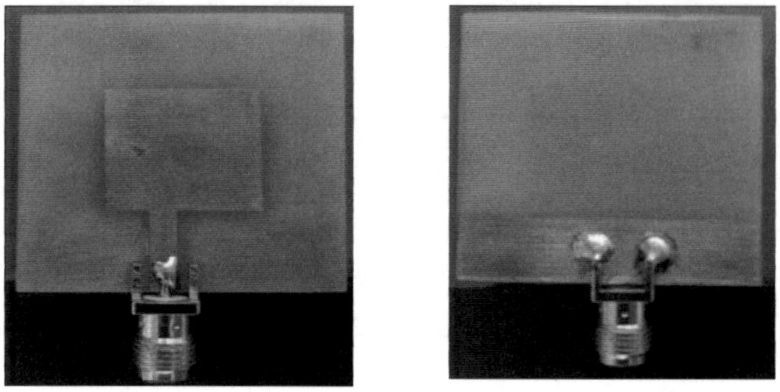

(i) (ii)

Fig. 2. Fabricated antenna (i) Front, (ii) Backimage

3. Result Discussions

The simulated performances of antenna have been used (CST V.17 EM) simulator, and experimental results of the proposed design had measured with vector network analyzer. The different structural parametric results are carried out and presented, after analysis, which is inevitable for any UWB antennas. The optimization results are best performance characteristic of the antenna to suit for bandwidth impedance, VSWR, gain and radiation pattern demonstrated in Fig. 3. The optimal unit values of the antenna structure are given in Table 2.

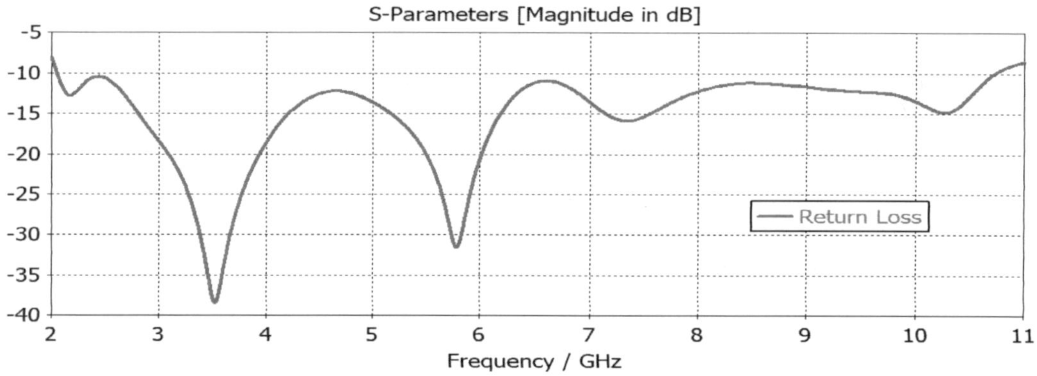

(i)

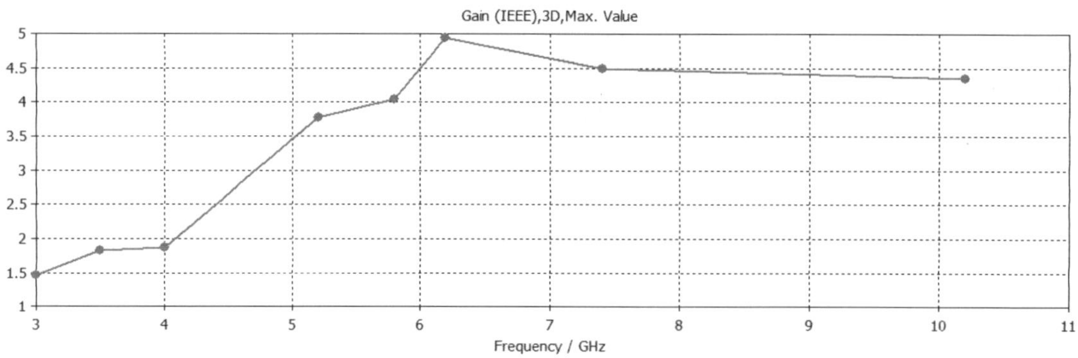

(ii)

IOP Conf. Series: Materials Science and Engineering **594** (2019) 012028 doi:10.1088/1757-899X/594/1/012028

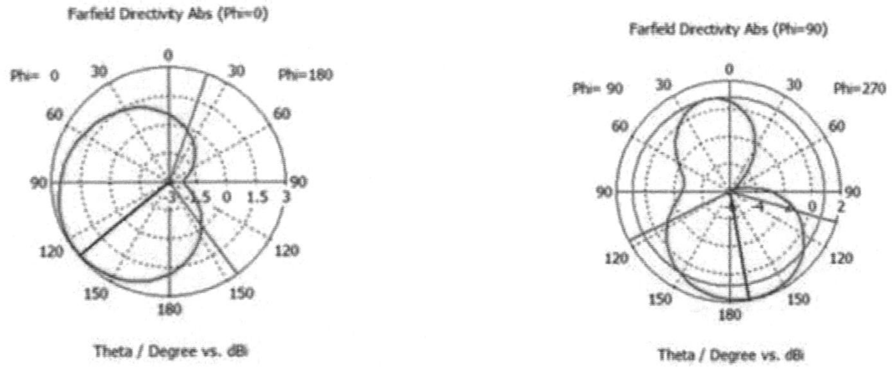

(iii)

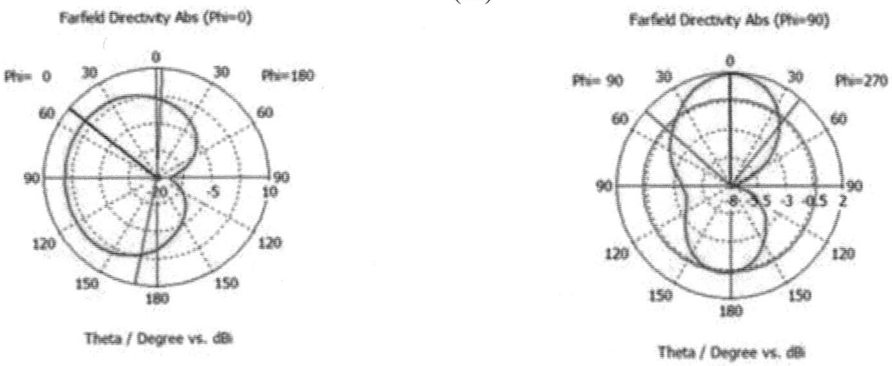

(iv)

Fig. 3. Simulated characteristics of proposed antenna
(i) Reflection coefficient, S_{11}<-10 dB,(ii) Gain of antenna
(iii) Radiation pattern E/H at 3.5 GHz,(iv) Radiation pattern E/H at 5.8 GHz

3.1. *Parametric Results Analysis*

The antenna simulations process is carried out for the parametric geometry studies and conducted by to investigate the design technique and optimization methods. For the same geometry, only one parameter is all the owed to change during any one period while remaining parameters remained same. A parametric analysis of dimensions such as Wg, Wp, Lp, Wf and gap is carried out.

3.1.1. *Impact of Wg variation*

A width of ground strip (Wg) shown in Fig. 1. The width of ground strip is varied from 0 mm to 2 mm. The deviation of notch frequency is pragmatic in term of reflection coefficient. When Wg = 0 mm, the optimum desired result is achieved, and loss of return is stable at a low and high frequency of operation presented in Fig. 4.

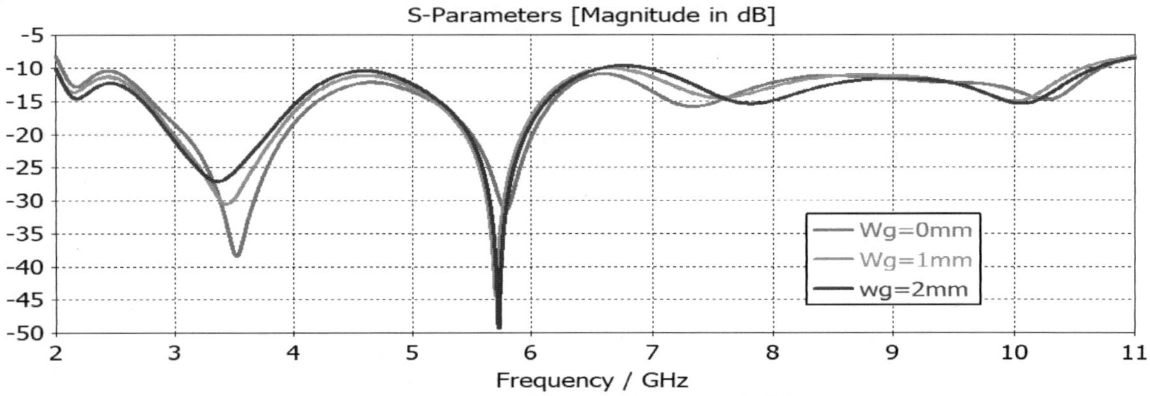

Fig. 4. Reflection coefficient with variable Wg

3.1.2. Impact of Wp variation

The parametric variations of radiating patch width (Wp) are done to study effect on reflection coefficient characteristic of proposed antenna. Reflection coefficient notch frequency is observed, when Wp =14.8 mm, 15.8 mm and 16.8 mm respectively are shown in Fig. 5. We can see that the optimized result has been observed at Wp = 15.8 mm with large bandwidth and good reflection coefficient.

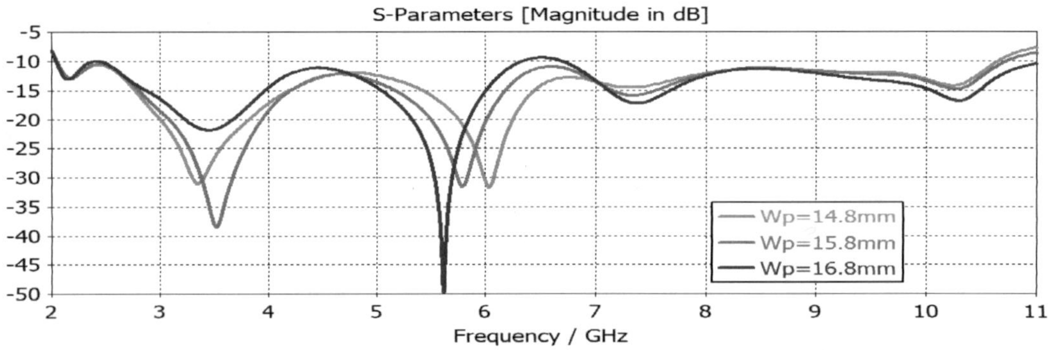

Fig. 5. Reflection coefficient with variable Wp

3.1.3. Variation Effect of Parameter (Lp)

As it can be analyzed in Figure 6, which shown that at the radiating patch length (Lp) from 14.8 mm to 16.8 mm, we achieve a satisfied UWB characteristic of proposed antenna and good reflection coefficient. The optimum desired result has been observed at Lp =15.7 mm

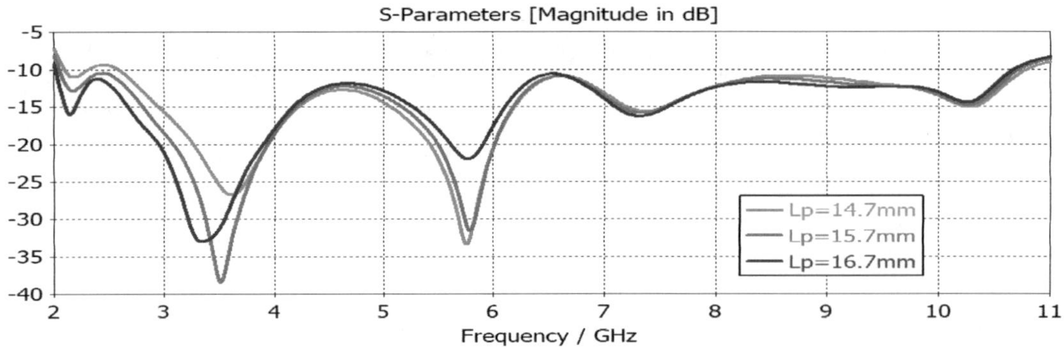

Fig. 6. Reflection coefficient with variable Lp

IOP Conf. Series: Materials Science and Engineering **594** (2019) 012028 doi:10.1088/1757-899X/594/1/012028

3.1.4. Impact of Feed Line Width Parameter (W_f) Variation

The feed line width is varies from 2.7 mm to 2.9 mm as shown in Fig. 7. When Wf =2.8 mm, the reflection coefficient achieved desired level as compared to others feed line width. But when Wf is 2.7 mm and 2.9 mm, very minor changes in notch frequency.

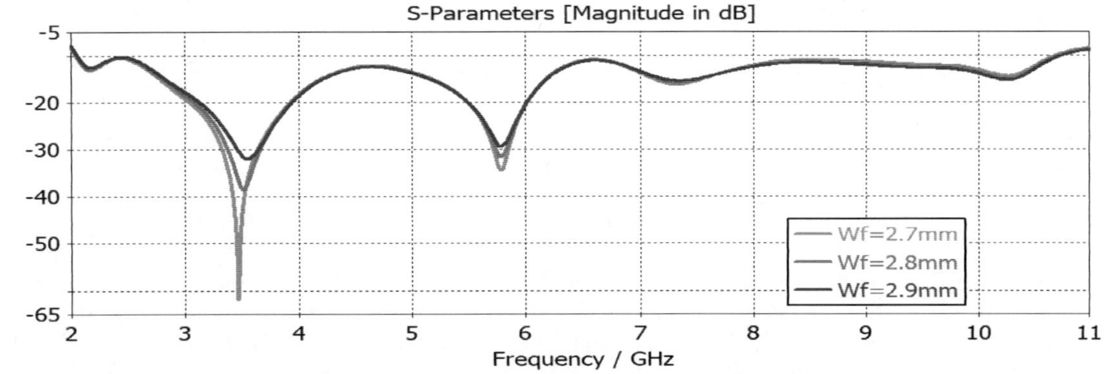

Fig. 7. Reflection coefficient with variable Wf

3.1.5. Impact of Ground Gap Variation

The impact of variation in 'ground gap' on the coefficient of reflection and performance of UWB is presented in Fig. 8. If gap 0.2, 1.2, and 2.2 mm, achieved a stable multi-deep notch frequency bands of wireless communication.

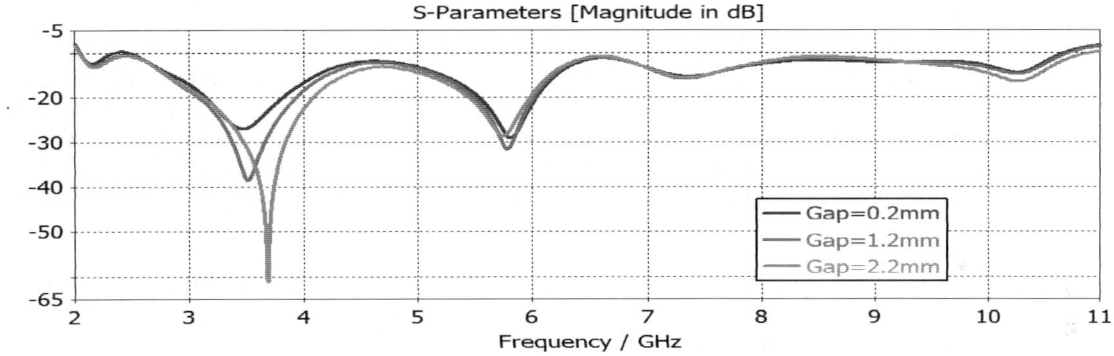

Fig. 8. Reflection coefficient with variable gap

4. Experimental Results and Discussion

A vector network analyzer (N5230A) has been applied to find characteristics performance of the developed design of antenna in terms of bandwidth impedance and VSWR. Fig. 9 presents the deliberated and simulated results such as return loss and VSWR of a compact multiband rectangular micro-strip antenna. The return loss is measured and plotted to indicate that it covers UWB of 8.7 GHz (2 to 10.7 GHz). Good agreements between the simulated and measured results are achieved as shown in Fig. 9.

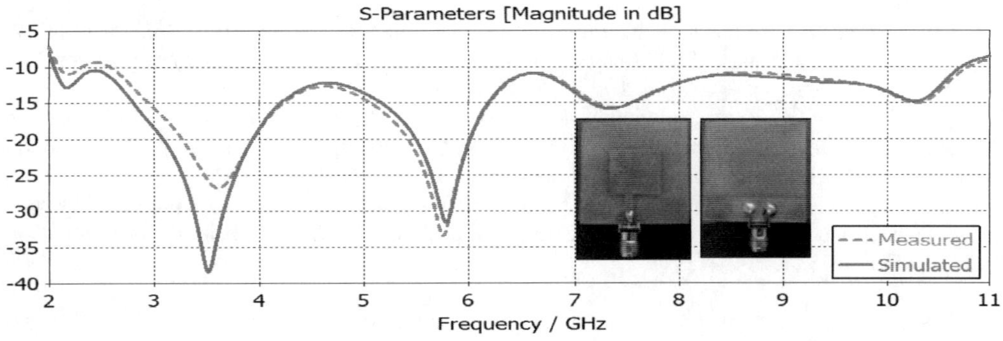

(i)

IOP Conf. Series: Materials Science and Engineering **594** (2019) 012028 doi:10.1088/1757-899X/594/1/012028

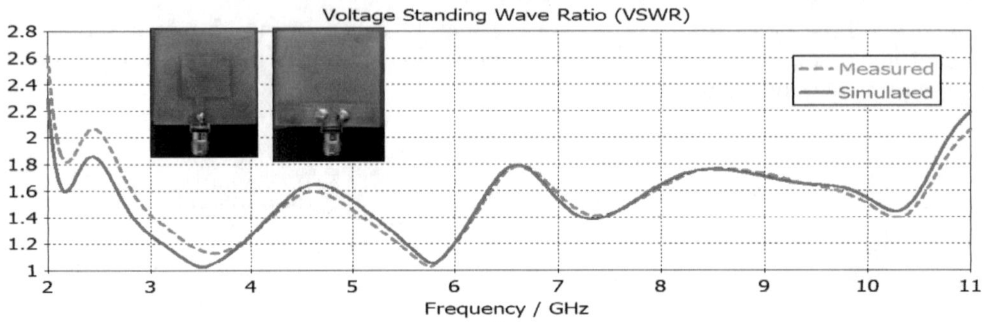

(ii)

Fig. 9. Simulated result compared with measured, (i) Reflection coefficient,(ii) VSWR

5. Conclusion

The proposed UWB antenna have covered the bandwidth 8.7 GHz from 2.0 GHz to 10.7 GHz in which it shown four distinct notch frequency, first 2.1GHz, second 3.5GHz, third 5.8GHz and fourth 7.3GHz respectively, which covers frequency bands of UMTS (1.9 - 2.17) GHz, Wi-MAX (3 - 4) GHz, WLAN (5.2 - 6.5) GHz and X-band (7.1 - 7.9) GHz respectively with a return loss < -10 dB and VSWR < 2 at the specified operating bands. The antenna also showed desired flat gain (close to 5dB), better directivity and good radiation patterns in the desired frequency bands all possible because of the proposed geometry. This type of antenna might prove to be a useful structure for modern wireless communication systems, including devices like- UMTS, WLAN, Wi-MAX and lower band of UWB communication systems.

References

[1] F. Fereidoony, S. Chamaani, and Seyed A. Mirtaheri, "UWB Monopole Antenna with Stable Radiation Pattern and Low Transient Distortion", *IEEE Antennas and Wireless propagation Letters*, **Vol. 10, 2011**

[2] Debdeep Sarkar, Vaibhav Srivastava and Kushmanda Saurav, "A Compact Microstrip-Fed Triple Band Notched UWB Monopole Antenna", *IEEE Antennas and Wireless Propagation Letters*, **Vol.13, 2014**

[3] Ahmed S. Elkorany, Ghidaa T. Ahmed, and Demyana A. Saleeb "Double Rotated ESSRs" *Advanced Electromagnetics*,
Vol. 7, Jan 2018

[4] Hari Shankar Mewara, Deepak Jhanwar, Mahendra M. Sharma, Jitendra K. Deegwal, "A Novel Hammer- Shaped UWB Antenna with Triple Notched-Band for Rejecting RLS, WLAN and XSCS bands", *Advanced Electromagnetics*, **Vol. 6, No. 4, October 2017.**

[5] F. M. Alnahwi, K. M. Abdulhasan, and Naz E. Islam, "An Ultra Wideband to Dual-Band Switchable Antenna Design for Wireless Communication Applications", *IEEE Antenna and Wireless Propagation Letters*,**Vol. 14, 2015**

[6] A. Boutejdar and W. Abd Ellatif, "A Novel Compact UWB Monopole Antenna with Enhanced Bandwidth Using Triangular Defected Microstrip Structure and Stepped Cut Technique", *Microwave Optical Technological Letters,***Vol. 58, 2016**

[7] Federal Communications Commission, Washington, DC, USA, Revision of Part 15 of the commission's rules regarding ultra-wide-band transmission systems First Report FCC 02. **Vol. 48, *Tech. Rep*. 2002.**

[8] S. K. Mishra, R. Kumar Gupta, and J. Mukherjee, "A Compact Dual-Band Fork-Shaped Monopole Antenna for Bluetooth and UWB Applications", *IEEE Antennas and Wireless propagation Letters*, **Vol. 10, 2011**

[9] J. Pourahmadazar, Ch. Ghobadi, J. Nourinia, N. Felegari, and HamedShirzad, "Broadband CPW-Fed Circularly Polarized Square Slot Antenna With Inverted-L Strips for UWB Applications", *IEEE Antennas and Wireless propagation Letters*, **Vol. 10, 2011**

[10] E. S. Angelopoulos, A. Z. Anastopoulos, D. I. Kaklamani, A. A. Alexandridis, F. Lazarakis, and K. Dangakis, "Circular and Elliptical CPW-Fed Slot and Microstrip-Fed Antennas for Ultra Wideband Applications", *IEEE Antennas and Wireless propagation Letters*, **Vol. 5, 2006**

[11] Kai Da Xu, Y. H. Zhang, Ronald J. Spiegel, Y. Fan, William T. Joines, Qing Huo Liu, "Design of a Stub- loaded Ring- Resonator Slot for Antenna Application", *IEEE Trans Antennas Propagation* **Vol. 63, No. 2, Feb 2015**

IOP Conf. Series: Materials Science and Engineering **594** (2019) 012028 doi:10.1088/1757-899X/594/1/012028

[12] J. Y. Sze, and C. C. Chang, "Axial-Ratio Bandwidth Enhancement of Asymmetric CPW-Fed Circularly Polarized Square Slot Antenna", *Electronics Letters,* **Vol. 44, No 18, Aug. 2008**

[13] Sajjad Ur Rahman and Majeed A, "Design and System Characterization of Ultra Wideband Antennas with Multiple Band-Rejections",***IEEE Access*** **2017**

Analysis and design of solar photovoltaic systems and its applications by using different MPPT techniques

Pramod Kumar[1] **and Ashish Shrivastava**[2]

[1]Department of Electrical and Electronics Engineering Skyline Institute of Engineering and Technology, Greater Noida, India

[2] SM IEEE, Department of Electrical Engineering Manipal University Jaipur, India
E-mail: kumarpramod2002@gmail.com

Abstract- In this paper a 40KW solar photovoltaic design is presented and displaying the output curve with different parameters i.e. temperatures and irradiations. The different MPPT techniques are used and make comparison among these techniques by using solar PV modeling to get better results using in matlab simulation tool boxes. This paper is useful for solar PV modeling and power electronics converter design. The perturb and observe and incremental conductance technique is compared to get maximum output voltage, current and power of photovoltaic array and using feedback PI controller to regulate the output of different power electronic converters i.e. buck-boost and cuk converters. The comparative analysis confirms the best method by studying the performance.

Keywords – PV model, circuit, MPPT techniques, modeling, Buckboost and Cuk converters, simulation results.

1. Introduction

In this era, the world is facing so many challenges i.e. pollution, population density. More population demands more facilities and more pollution deploy the epidemics. To acknowledge these problems, the research work is going on to rectify this problem. All population needs un-interrupted electrical energy, which is mostly being produced by conventional power plants i.e. hydro, gas and coal fired plants etc. These plants are producing more pollution and high cost of electrical energy. But people needs a clean and green energy with less cost, which increased the focus towards renewable energy, which is being obtained by renewable energy sources i.e. sun, wind, geothermal etc. The more work is being done on solar energy.

The solar energy is a clean, green and environment friendly energy. The solar photovoltaic (PV) cell are exposed in the sun and the sun energy is converted into electrical energy. For getting desired output from the solar, the photovoltaic (PV) cells are connected in series and parallel. The series-parallel connected photovoltaic cells make an array, which will increase the efficiency of the photovoltaic array output.

To increase the efficiency of solar photovoltaic system, the maximum power point techniques (MPPT) are used. These techniques also produced the desired output of the PV array. The maximum power point (MPP) trackers are used to extract the maximum power point voltage, current and maximum power from solar PV systems. From years, different type of MPPT techniques are used i.e. perturb and observation (P & O), incremental conductance (INC) method, open circuit voltage method and short circuit current method. In this paper, perturb and observation and incremental conductance techniques are used to get MPP voltage (Vmpp), MPP current (Impp) and maximum power (Pmax). The output of solar photovoltaic array is feeding to DC-DC converters to get a regulated output of converters and results are compared. Similarly, when PV array is connected to the DC-DC converters, the output of converters are regulated by feedback controllers i.e. PI controller and results are compared. Analyze all the results and find the best one.

2. Modeling of PV cells, Module and Array

2.1. *Single PV cell circuit Model*

The fig(1) shows a single PV cell circuit model. This is an equivalent PV circuit model [1], the I-V characteristics are described by mathematical equations.

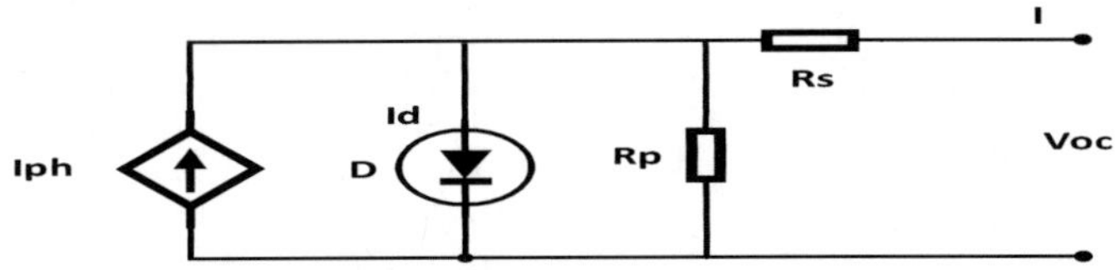

Fig. 1. Single PV cell model

$$I = I_{ph} - I_d - \frac{V_{oc} + IR_s}{R_p} \quad (1)$$

$$I = I_{ph} - I_0[\exp\left(\frac{V_{oc} + IR_s}{V_t}\right) - \frac{V_{oc} + IR_s}{R_p} \quad (2)$$

$$V_t = \frac{a\,k\,t\,N_s}{q} \quad (3)$$

2.2. *Analytical Modeling of PV array*

In this paper, there has been developed a design of 40 KW solar PV system. The I-V characteristics of solar PV cell depends on solar insolation, solar temperature, weather conditions and PV cell contact resistances i.e R_s and R_p. The variation in solar photovoltaic current (I_{ph}) is depending on the variation of solar insolation, atmospheric conditions and solar temperature. But for getting a best I-V curve fitting, the suitable value of Rs and Rp [1] should be choosen. The photovoltaic current equation for solar PV cell is:

$$I_{ph} = \{I_{scn} + K_i\,(T - T_r)\frac{G}{G_r} \quad (4)$$

Where I_{scn} is the reference short circuit current by solar light generated in Ampere at T_r = 25°C (reference temperature) and G_r = 1000 W/m^2 (reference solar insolation), $\Delta T = T - T_r$ (°K), G = actual solar insolation on the surface of solar PV cell and K_i = short circuit temperature coefficient (A/°K).

In solar PV cell, the diode reverse saturation current (Io) is depending on temperature [1] and the modified [1] equation is:

$$I_o = \frac{I_{scn} + K_i * \Delta T}{\exp\left[\frac{V_{oc} + K_v * \Delta T}{V_t}\right] - 1} \quad (5)$$

A single PV cell produces approximately 3 W and 0.6 V. If we achieve more power from PV cell then there will be some cells in series and some parallel [2]. If N_s are the no. of cells connected in series and N_p are the no. of cells connected in parallel then the equation (2) will be written as:

IOP Conf. Series: Materials Science and Engineering **594** (2019) 012029 doi:10.1088/1757-899X/594/1/012029

$$I = N_p I_{ph} - N_p I_o [\exp\left(\frac{V_{oc} + IR_s}{V_t}\right) - \frac{N_p V_{oc} + IR_s N_s}{N_s R_p} \quad (6)$$

The design [2] of an array can be made by connecting modules in series and parallel. In this paper, there are 14 PV modules are connected in series and 22 PV modules are connected in parallel and fig(2) shows the PV modeling of an array. The parameters observed of the model are shown in table 1. The actual output parameters of the PV array are shown in table 2.

TABLE 1

Specifications	Array output
Short circuit current (Isc)	180.4 A
Open circuit voltage (Voc)	302.4 V
Voltage at MPP	252 V
Current at MPP	163 A
Power at MPP	41.076 KW
No. of series cells (Ns)	36

TABLE 2

Specifications	Array output
Short circuit current (Isc)	180.4 A
Open circuit voltage (Voc)	302.4 V
Voltage at MPP	240.2 V
Current at MPP	161.4 A
Power at MPP	38.768 KW
Efficienciy (η)	95 %

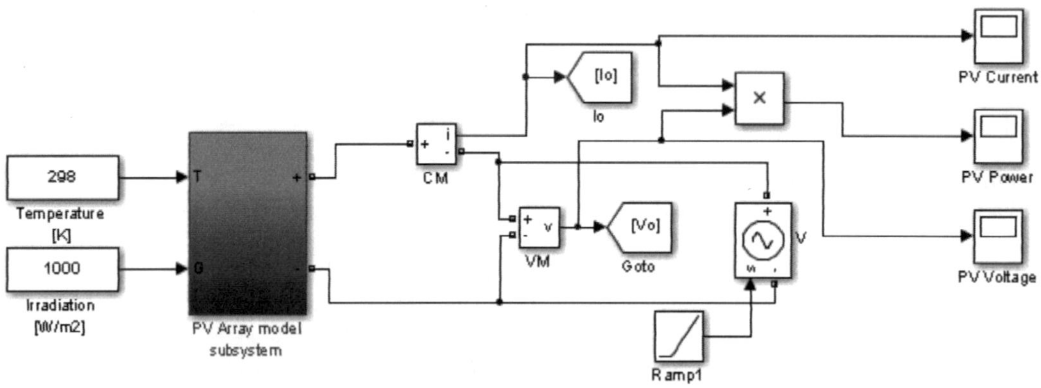

Fig. 2. Solar PV modeling matlab circuit

In this solar PV matlab cicuit, it has seen that at different insolation, the current is reduced at decreasing insolation and I-V and P-V characteristics are shown in fig(3) and fig(4).

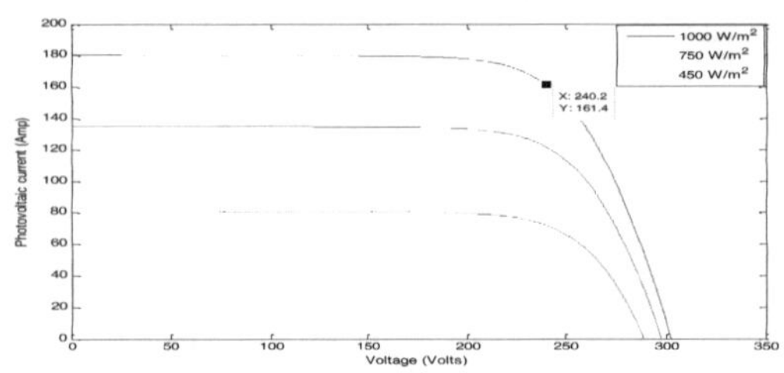

Fig. 3. I-V characteristics of array

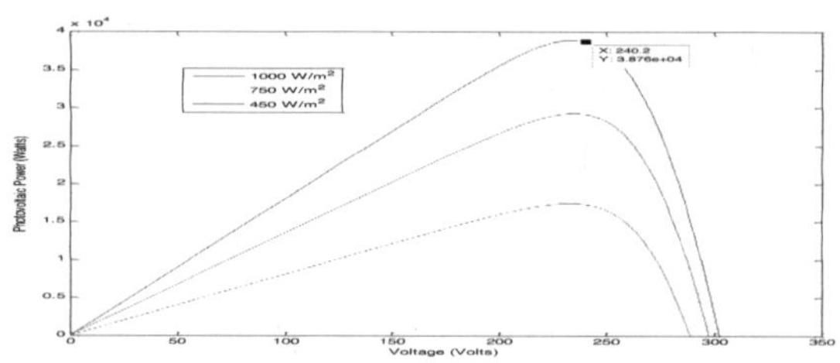

Fig. 4. P-V characteristics of array

IOP Conf. Series: Materials Science and Engineering **594** (2019) 012029 doi:10.1088/1757-899X/594/1/012029

This is also observed that at different temperature i.e. if temperature is reduced then the PV array voltage as well as power is reduced. The I-V and P-V characteristics are shown in fig(5) and fig(6) below.

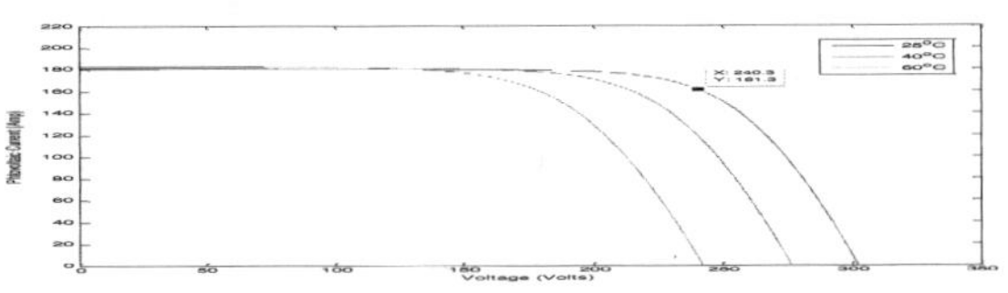

Fig. 5. I-V characteristics of PV array

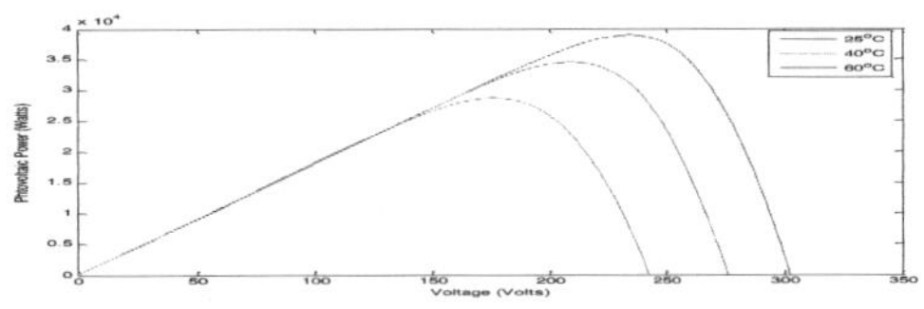

Fig. 6. P-V characteristics of PV array

3. Design of DC-DC converters

3.1. *Buckboost converter*

The buckboost converter is fed with unregulated supply and it will produce a regulated output. There are different weather conditions and if a solar array exposed in the sun, then it will face these variable weather conditions and reaches to the surface of solar PV array panel. Therefore, the output of solar PV array is variable. This variable output of PV array is feeding to the buck-boost converter [3], then it will produce a constant output. The buck-boost converter circuit is shown in fig(7) and it displayed the inverted output. The switching frequency should be high to reduce the switching losses and design for continuous conduction mode [4]. In this circuit 40 Khz switching frequency is used.

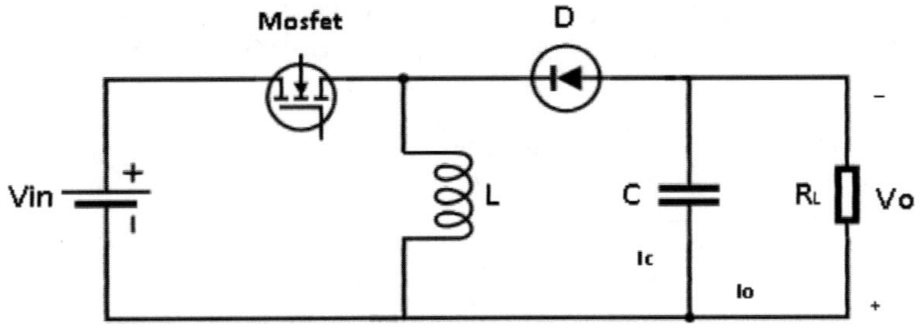

Fig. 7. Buck-boost DC converter

The value of L and C is declared by the equations as:

$$L = \frac{V_{in} * D}{\Delta I * f_s} \quad (7) \qquad C = \frac{I_o * D}{f_s * \Delta V_c} \quad (8)$$

3.2. *Cuk converter*

This converter gives smooth and constant output either supply is regulated or unregulated. The output of cuk converter is inverted and it is designed for continuous conduction mode [4] as shown in fig(8), the switching frequency is used 40 Khz for avoiding swiching losses. The values of inductors and capacitors [5] can be determined as:

$$L_1 = \frac{V_{in} * D_{max}}{\Delta I_1 * f_s} \quad (9) \qquad L_2 = \frac{V_{in} * D_{max}}{\Delta I_2 * f_s} \quad (10)$$

$$C_1 = \frac{I_{in}(1 - D_{max})}{\Delta V_{c1} * f_s} \quad (11) \qquad C_2 = \frac{\Delta I_2}{8 * \Delta V_{c2} * f_s} \quad (12)$$

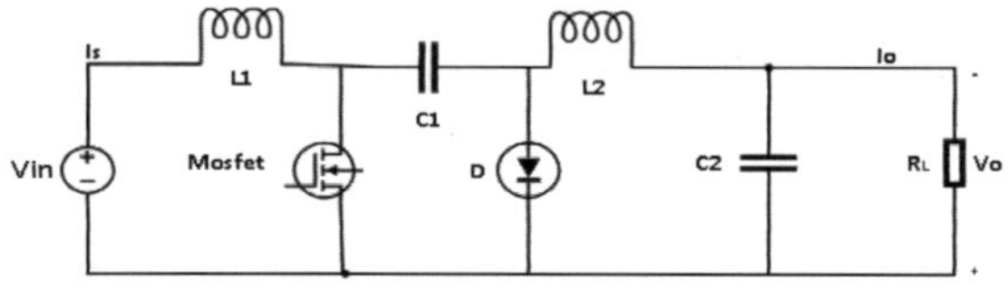

Fig. 8. Cuk DC converter

4. MPPT Techniques

The different MPPT techniques are discussed below:

4.1. *Perturb and Observation Method*

The Perturb and observation (P and O) technique is the hill climbing technique [6], it can extract the maximum voltage, maximum current and it available maximum power from a PV array system. The temperature and insolation are always changing [6], in this changing weather conditions, it can extract maximum power point voltage (V_{mpp}), maximum power point current (I_{mpp}) and available maximum power (Pmax). The perturb and observation method can extract the maximum power and fed to a DC-DC converter i.e. buck-boost and cuk converter [7] to get a constant output voltage, current and power. A flowchart is shown in fig(9) for P and O method [7], in which power increases upto desired maximum value and after maximum value the perturbation start reverses.

It is simple and easy algorithm to extract maximum electrical energy [7] from a photovoltaic array irrespective of weather parameters. The voltage and current is sensed and fed to P and O MPPT, which is producing a duty cycle. If the duty cycle increases, power will be increased or vice-versa. This algorithm always keeps the power maximum. This duty cycle is fed to a DC-DC buckboost and cuk converters, which will control the MOSFET switching and produces desired output power.

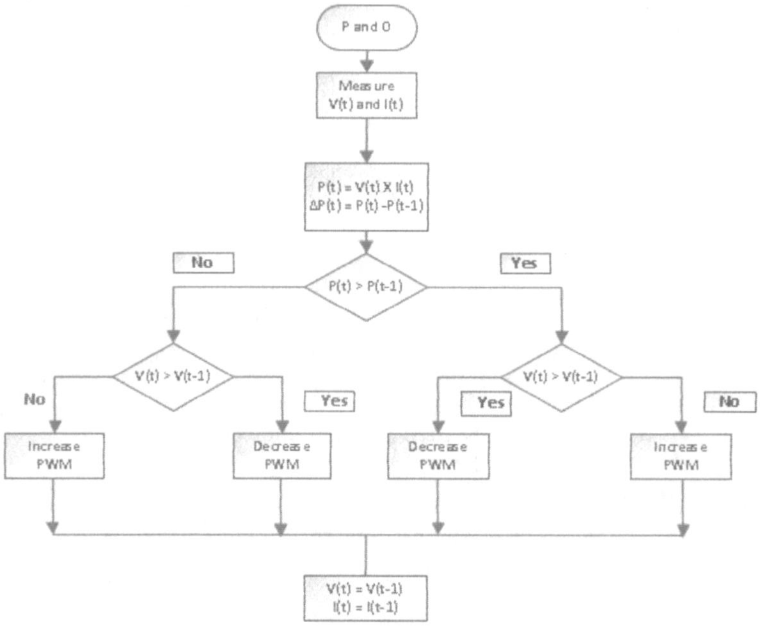

Fig. 9. Perturb and Observation algorithm

4.2. *Incremental Conductance Method*

This is a hill climbing technique, which provide quick response because of rapid variation in irradiation and temperature [8]. If, we see the power output and derivative of power as

P = VxI

dP/dV = I + VdI/dV

at MPP , the dP/dV = 0 and I/V + dI/dV = 0 at MPP

I/V + dI/dV > 0 on the left side of MPP

I/V + dI/dV < 0 on the right side of MPP

At given MPP as shown in above graph and in above derivation, the sum of instantaneous conductance (I/V) and Incremental conductance (dI/dV) [10] is zero at MPP. The sum is negative, when voltage increases from maximum value, at that time the INC [9] algorithm is shown in fig(10), decrease the duty cycle. If sum is positive, when voltage decreases from maximum value, at that time the INC algorithm increase the duty cycle [10]. This algorithm gives fast response and make the output power stable at different weather conditions.

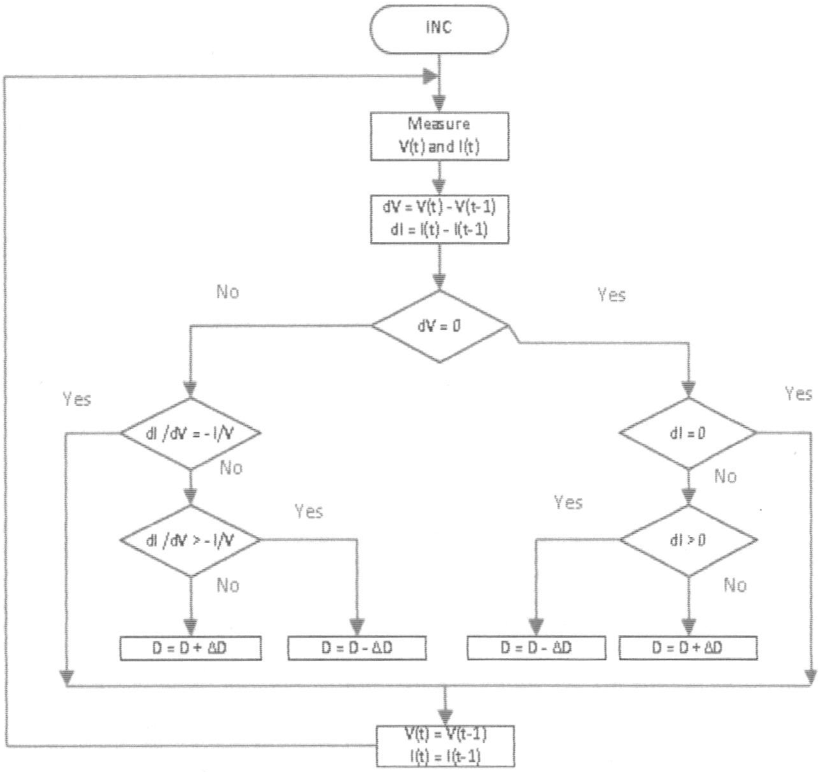

Figure 10. Incremental conductance algorithm

5. Performance of PV array with matlab simulation

5.1. *Maximum power point tracking based simulink*

The different dynamic MPPT techniques i.e. perturb and observation and incremental conductance are discussed above. The perturb and observation algorithm [11] provide oscillation and slow tracking as compared to incremental conductance algorithm[12]. When the output power decreased the tracker starts perturbation and set the duty cycle at maximum power tracking point. A matlab simulation circuit of buckboost converter with P and O is shown in fig.(11) and its associated waveform of input I_{in}, V_{in}, P_{in} and output I_o, V_o, P_o is shown in fig. (12), (13), (14) with constant temperature (25° C) and insolation (1000 W/m^2).

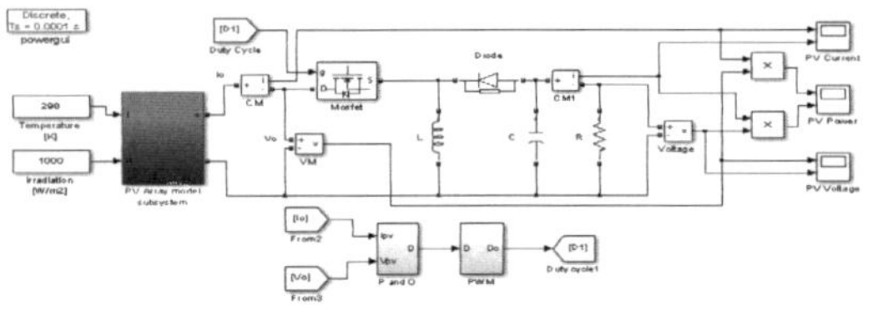

Figure 11. Buckboost with P and O simulink

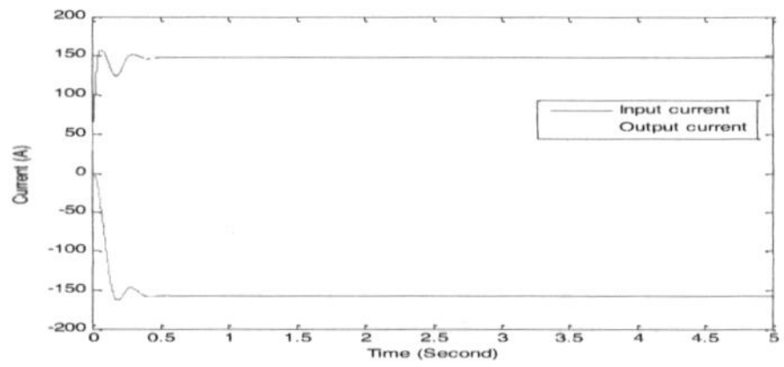

Figure 12. Input and output currents

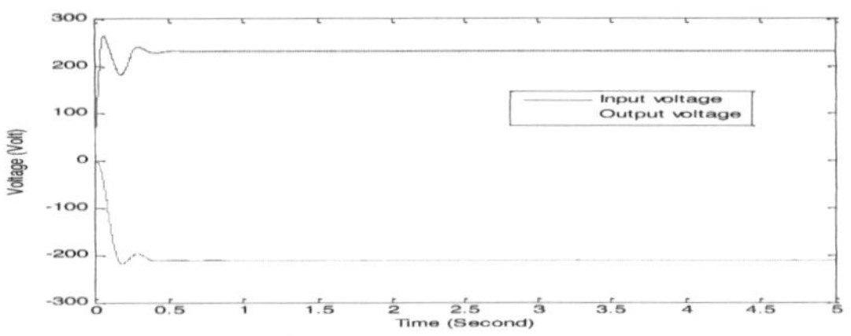

Figure 13. Input and output voltages

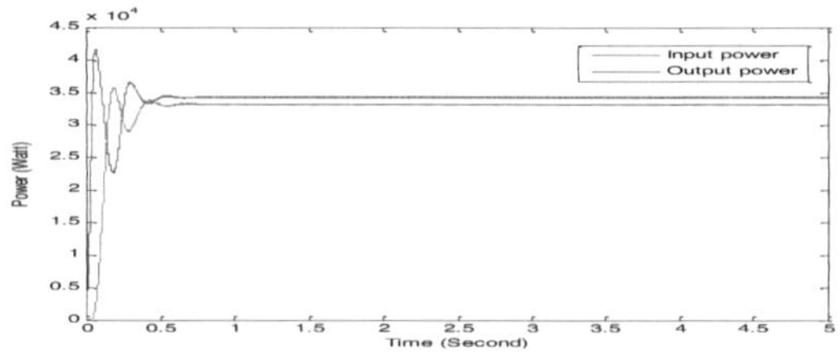

Figure 14. Input and output power

Now, a buckboost converter is tuned with Incremental algorithm as shown in fig.(15) and its associated waveform of input I_{in}, V_{in}, P_{in} and output I_o, V_o, P_o is shown in fig. (16), (17), (18) with constant temperature (25° C) and insolation (1000 W/m^2). There are some variation in outputs.

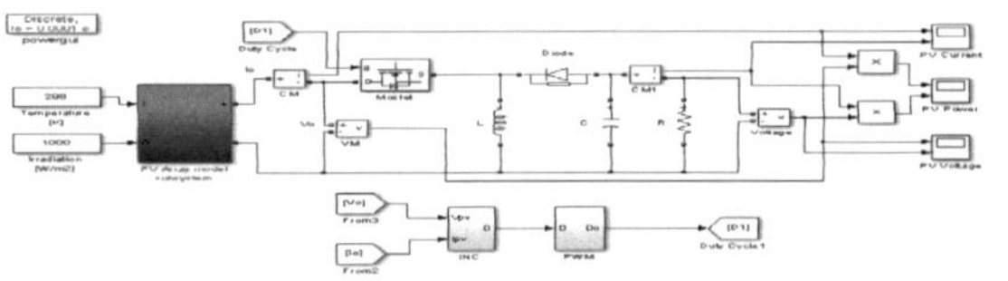

Figure 15. Buckboost with INC matlab simulink

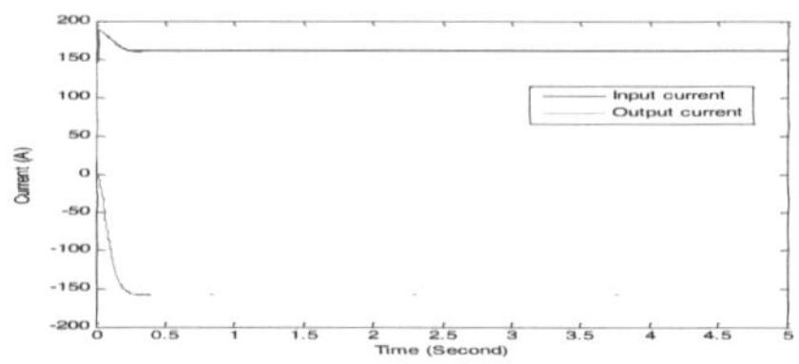

Figure 16. Input and Output currents

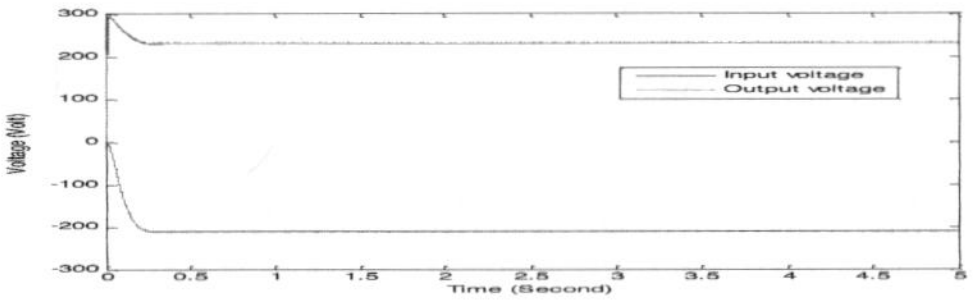

Figure 17. Input and Output voltages

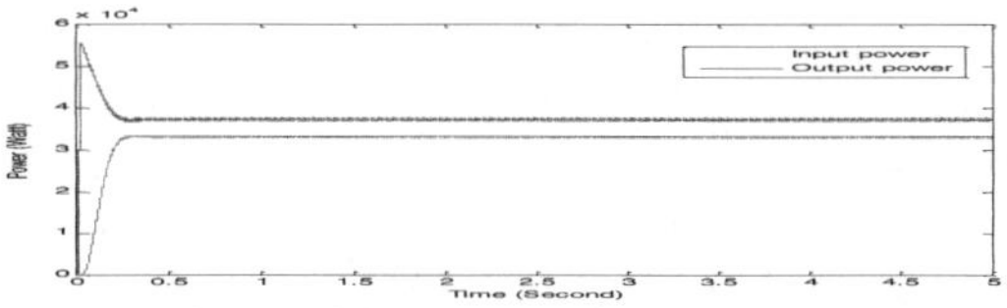

Figure 18. Input and Output powers

In a similar manner, there is a cuk converter tuning circuit with P and O as shown in fig.(19) and its associated waveform of input I_{in}, V_{in}, P_{in} and output I_o, V_o, P_o is shown in fig. (20), (21), (22) with constant temperature (25° C) and insolation (1000 W/m^2).

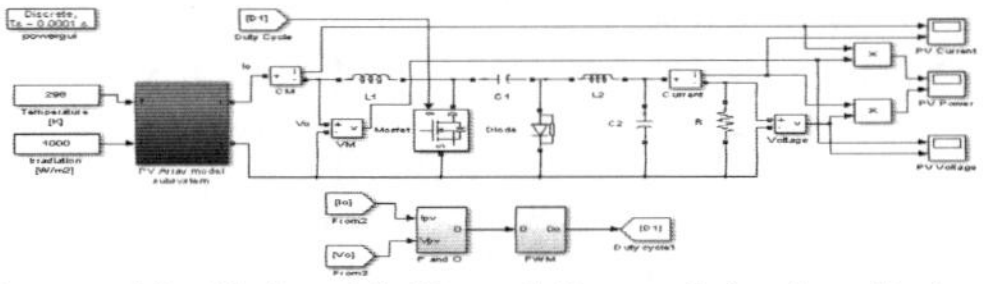

Figure 19. Cuk with P and O matlab simulink

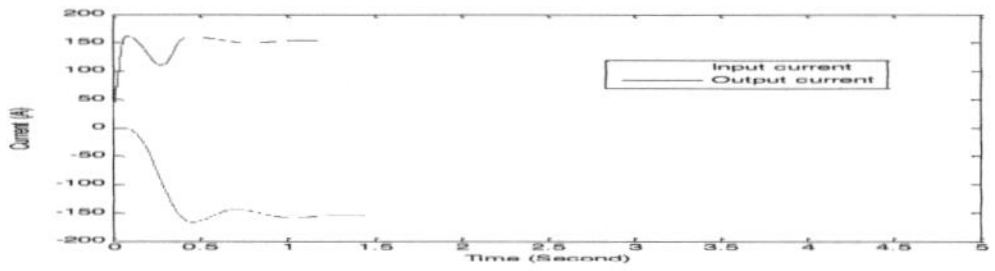

Figure 20. Input and Output currents

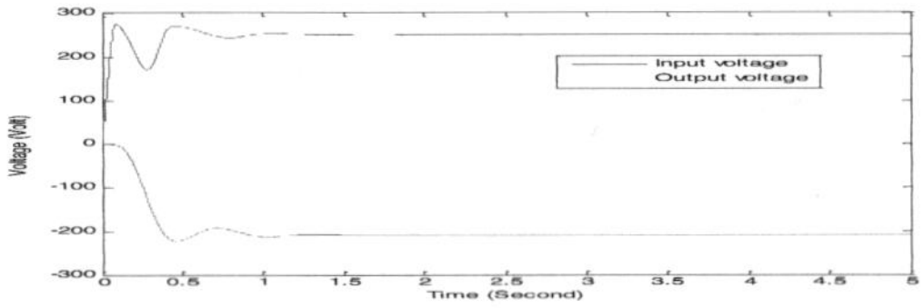

Figure 21. Input and output voltages

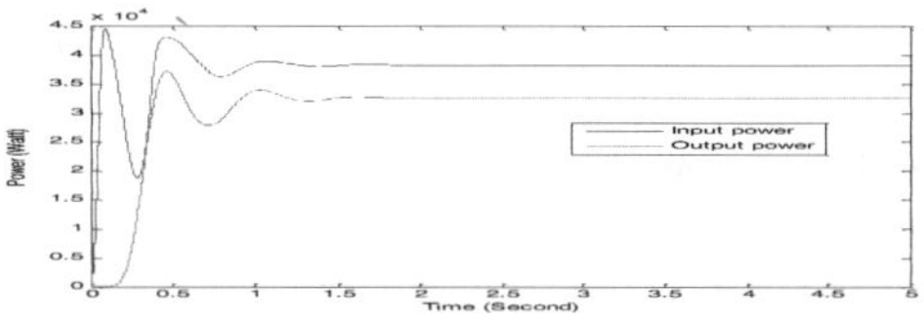

Figure 22. Input and Output powers

In a similar manner, there is a cuk converter tuning circuit with Incremental conductance algorithm as shown in fig.(23) and its associated waveform of input I_{in}, V_{in}, P_{in} and output I_o, V_o, P_o is shown in fig. (24), (25), (26) with constant temperature (25° C) and insolation (1000 W/m^2).

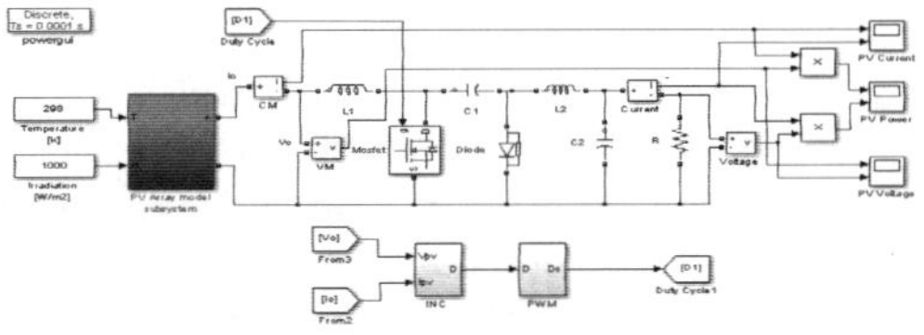

Figure 23. Cuk with INC matlab simulink

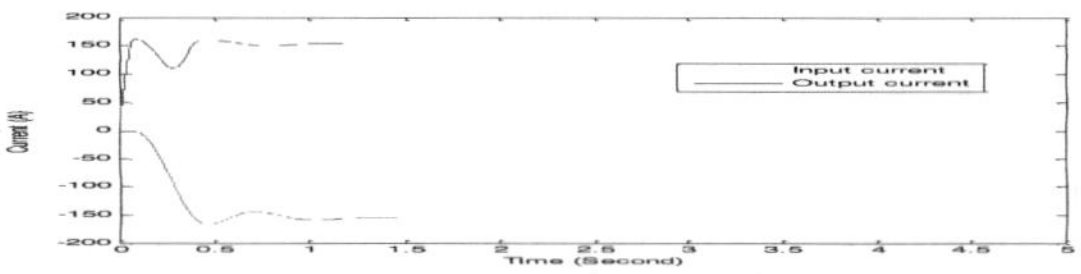

Figure 24. Input and Output currents

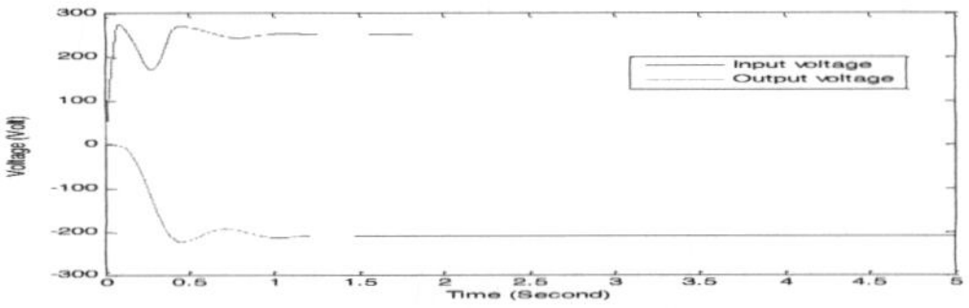

Figure 25. Input and Output voltages

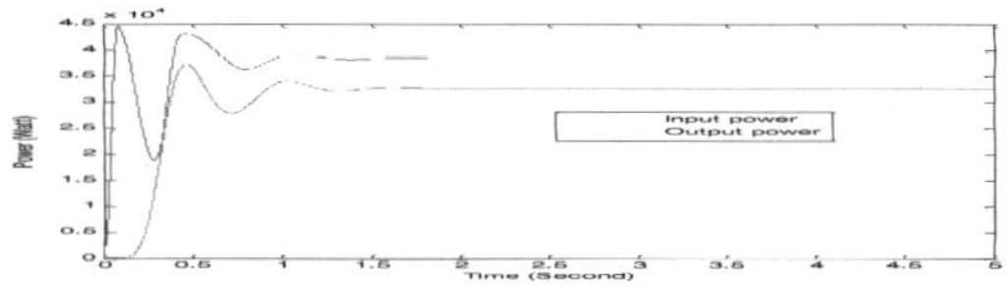

Figure 26. Input and Output powers

5.2. *PI controller based simulink*

There is buckboost converter as shown in fig.(27) tuned with feedback PI controller to get smooth outputs. The output of PI controller [13] can be adjusted by varying the value of Kp and Ki. The design motive of this controller [14] to produce a desired output. But there are transients in outputs as shown in waveform of input and output currents, voltage and power is shown in fig. (28), (29), (30) with constant temperature (25° C) and insolation (1000 W/m^2).

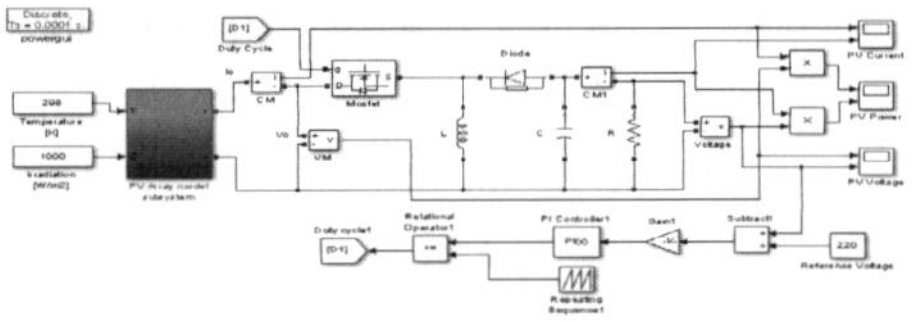

Figure 27. Buckboost with PI matlab simulink

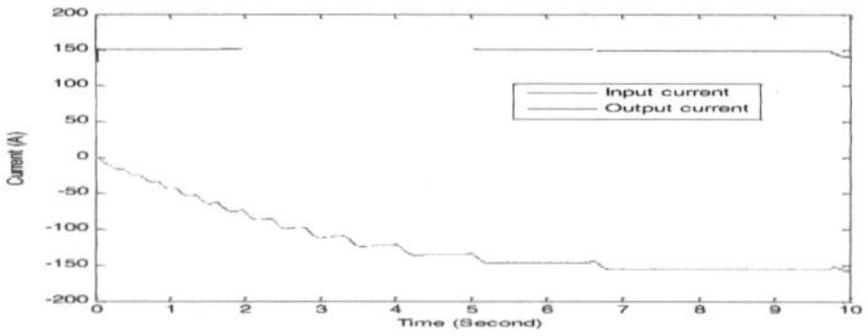

Figure 28. Input and Output currents

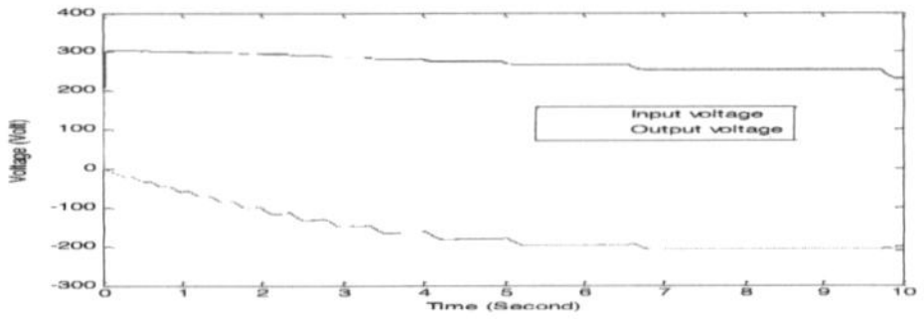

Figure 29. Input and Output voltages

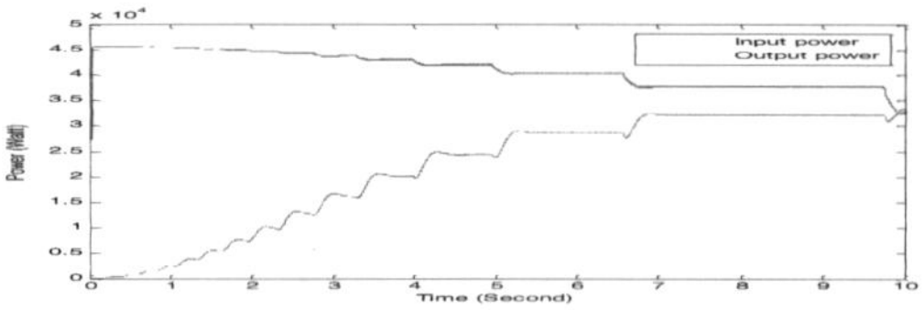

Figure 30. Input and Output powers

The cuk converter produce a DC output voltage, if there is transition in input of converter then output of converter is varied[15]. To eliminate this variation of output, a feedback control is desired. The output of converter is compared with reference input [16] and producing an error. This error is fed to PI controller, which will produce a signal for PWM by adjusting the value of Kp and Ki. The PWM actuate the MOSFET to produce a constant output of converter.

There is a cuk converter tuning with PI controller as shown in fig.(31) and its associated waveform of input I_{in}, V_{in}, P_{in} and output I_o, V_o, P_o is shown in fig. (32), (33), (34) with constant temperature (25° C) and insolation (1000 W/m^2).

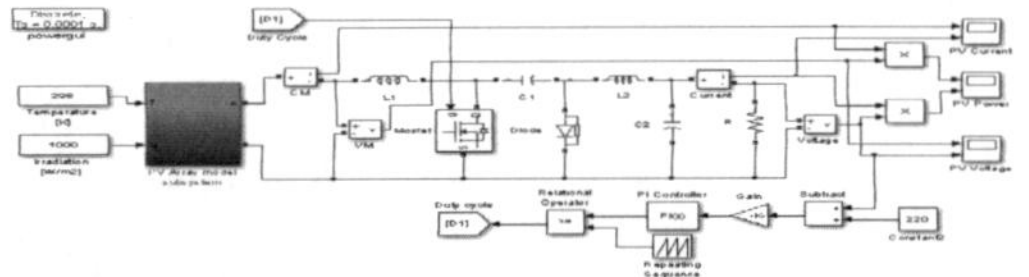

Figure 31. Cuk with PI matlab simulink

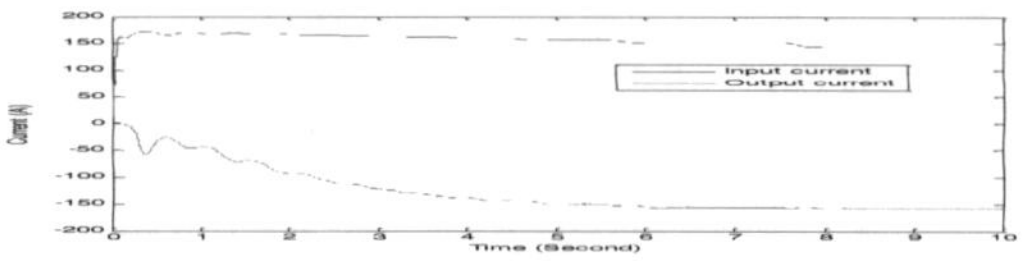

Figure 32. Input and Output currents

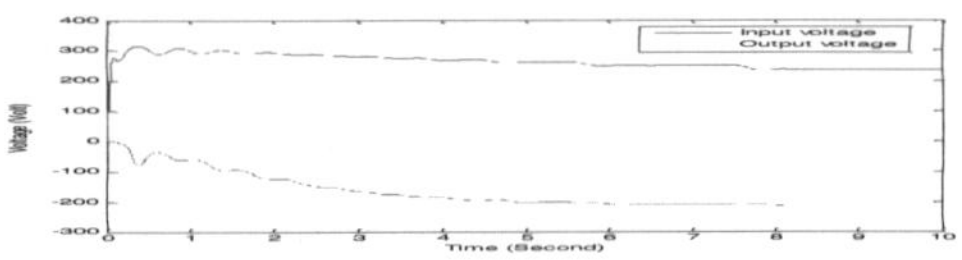

Figure 33. Input and Output voltage

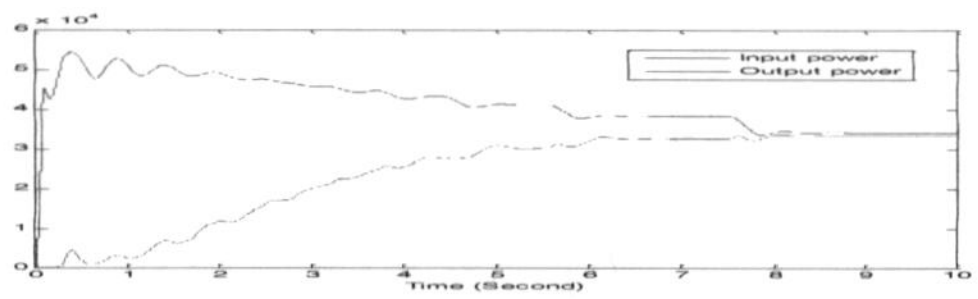

Figure 34. Input and Output powers

IOP Conf. Series: Materials Science and Engineering **594** (2019) 012029 doi:10.1088/1757-899X/594/1/012029

6. Results and Discussion

In this paper, The PV array output voltage 240.2 V and current 161.4 A at MPP and array is design 14 module in series and 22 module in parallel. This array output is fed to buckboost and cuk converters. These converters are designed with 220 V and 163 A output and switching frequency is 40 Khz. The converters are tuned with P and O and Incremental conductance MPPT and feedback PI controllers are used to get the desired output of converters. But due to transient in temperature and insolation the output of converters is not steady. The output is in pulse form, there are more overshoot and more settling time of outputs. The output parameters are shown in table 3.

TABLE 3

Type of circuits	1000 W/m^2, 25^0 C					
	V_{in} (Volt)	I_{in} (A)	P_{in} (KW)	V_o (Volt)	I_o (A)	P_o (KW)
Buckboost with P & O	231.5	147.9	34.2	210.9	157.4	33.2
Buckboost with INC	230.5	160.7	37	210.3	157	33.02
Cuk with P & O	250.8	152.6	38.3	209	156	32.6
Cuk with INC	250.8	152.6	38.3	209	156	32.6
Buckboost with PI	253	149	37.7	207.8	157	32.6
Cuk with PI	235.4	145.3	34.2	211.4	157.8	33.4

The above table 3 shows the variation in output V_o, I_o and P_o due to transient in output of solar photovoltaic array. The cuk converter performs better than buckboost converter.

7. Conclusion

In this paper, a 40 KW photovoltaic array is designed and deals with different control techniques i.e. P and O, incremental conductance and PI controller. For getting desired performance, the output of solar photovoltaic array is feeding to buckboost and cuk converters. The inductor and capacitor of converters are designed with 10% current ripple and 1% in voltage ripple.

The PV array output voltage is set 240V at MPP and converters output voltage is set at 220V. The measured values are verified with proposed control techniques. The output voltage reaches the desired value, but not steady. There are some pulses in output due to transient in PV array output.

It is ensure that any variation in weather parameters i.e. temperature, insolation will be considered by the control techniques and produced a bounded output.

References

[1]. M.G. Villalva, J.R. Gazoli, E.Ruppert F. 2009 University of Campinas- UNICAMP, *Brazilian Journal of power electronics* Modeling and circuit based simulation of photovoltaic arrays.

[2]. S. Sumathi et al 2015 Green energy technology, *Springer International Publishing Switzerland* Solar PV and wind energy conversion systems.

[3]. Vijaylakshmi S. and Sree Renga Raja T. 2014 *J Electr Eng Technol* Time domain Based Digital Controller for Buck-Boost Converter.

[4]. Priyadarshni, Dr. Shantharam Rai 2012 *International Journal of Science and Research (IJSR)* Design, Modeling and Simulation of a PID Controller for Buck-Boost and Cuk Converter.

[5]. R. Sriranjani, A. Shree Bharthi and S. Jaylalitha 2013 India EEE Department, EIE Department, School of EEE, SASTRA University, Thanjavur, Tamilnadu, *Research Journal of Applied Science, Engineering and Technology* Design of Cuk Converter Powered by PV

IOP Conf. Series: Materials Science and Engineering **594** (2019) 012029 doi:10.1088/1757-899X/594/1/012029

array.
[6]. Jirada Gosumbonggot march,2016 Thailand *International Electrical Engineering Conference,* Maximum Power Point Tracking Method using Perturb and Observe Algorithm for small scale DC voltage converter.
[7]. Bikram Das 1,Anindita Jamatia1, Abanishwar Chakraborti,Prabir Rn.Kasari and Manik Bhowmik July 2012, India *International Journal of Advances in Engineering & Technology,* New perturb and observe MPPT algorithm and its validation using data from PV module Vol- 4 pp. 579-591.
[8]. Mohammed A. Elgendy, Bashar Zahawi, David J. Atkinson Janaury 2013 *IEEE Transactions on Sustainable Energy* Assessment of the incremental conductance maximum power point tracking algorithm Vol.4, No.1.
[9]. Jitendra Kasera, Vinod Kumar, Dr. R.R. Joshi, Jai Kumar Maherchandani Journal of Electrical Engineering Design of Grid Connected Photovoltaic System Employing Increamental Conductance MPPT Algotithm.
[10]. Fangrui Liu, Shanxu Duan, Fei Liu, Bangyin Liu, and Yong Kang July, 2008 *IEEE Transactions on Industrial Electronics* A Variable Step Size INC MPPT Method for PV Systems Vol. 55, No. 7.
[11]. Ahmed M. Atallah, Almoataz Y. Abdelaziz, and Raihan S. Jumaah February 2014 *Emerging Trends in Electrical, Electronics & Instrumentation Engineering: An international Journal* Implementation of Perturb and Observe MPPT of PV System with direct control method using buck and buckboost converters Vol. 1, No. 1.
[12]. R.Durga Devi, K.Deepa 2018 *International Journal of Pure and Applied Mathematics* Comparative Analysis of MPPT methods for cuk converter based Solar System vol. 118, No. 22. Minh Quan Duong, Gabriela Nicoleta Sava, Marco Mussetta 2017 *IEEE excellence Research GrantsProgram, UPB –GEX* Design and Simulation of PI-Type Control for the Buck Boost Converter.
[13]. Minh Quan Duong, Gabriela Nicoleta Sava, Marco Mussetta 2017 *IEEE excellence Research Grants Program, UPB –GEX* Design and Simulation of PI-Type Control for the Buck Boost Converter.
[14]. Ersan KABALC, Goksel GOKKUS, Alper GORGUN June 2015 *ECAI-International Conference–7th Edition Electronics,Computers and Artificial Intelligence*, Bucharest, ROMÂNIA-27 Design and Implementation of a PI-MPPT Based Buck-Boost Converter.
[15]. Srushti R.Chafle, Uttam B. Vaidya, Z.J.Khan "Design of Cuk July 2013 *International Journal of Innovative Research in Electrical, Electronics, Instrumentation and Control Engineering* Design of Cuk Converter with MPPT Technique vol. 1,4.
[16]. P.Nagalakshmi Kanna, Dr.B.Meenakshi 2015 *International Conference on Circuit, Power and Computing Technologies* Analysis and Design of DC- DC/AC Non Isolated Cuk Converter using Sliding Mode Controller.

IOP Conf. Series: Materials Science and Engineering **594** (2019) 012030 doi:10.1088/1757-899X/594/1/012030

Cost benefit modeling of AB_2X_4 (A= Cd; B= Ga; X= S, Se) solar photovoltaic (PV) materials

Pancham Kumar[1,2] Amit Soni[2] and Jagrati Sahariya[3]
School of Electrical Skills, Bhartiya Skill Development University, Jaipur, Rajasthan 302042, India
[2]Department of EEE, Manipal University Jaipur, Jaipur, Rajasthan 303007, India
[3]Department of Physics, Manipal University Jaipur, Jaipur, Rajasthan 303007, India
E-mail: pancham066@gmail.com

Abstract
Present paper focus towards implementation and commercialization of emerging solar PV technology with cost effective manner. Energy band gap, absorption coefficients and module implementation cost for 1KW solar power plant for ternary AB_2X_4 (A= Cd; B= Ga; X= S, Se) solar photovoltaic (PV) system is investigated here. All the material related calculations are carried out using density functional theory [1] and cost related calculations by using mathematical expressions. By performing a comparative cost study between proposed materials based solar photovoltaic module and crystalline based solar photovoltaic module we concluded that present material will be cost effectively commercialized as compare to Crystalline Si.

Index Terms: Solar PV, Cost, DFT

1. Introduction

With increasing gap between global energy supply and demand due to limited conventional energy resources, it is mandatory to harness maximum energy from renewable energy resources. To full fill above requirements researchers must pay their attention towards developments of emerging renewable energy harnessing techniques. It is found that out of all available renewable energy sources solar energy is worldwide authorized and having great potential to fulfill the gap between energy supply and demand. In this present paper our focus is towards investigation of such emerging solar collector materials. Ternary chalcogenides materials AB_2X_4 (A= Cd; B= Ga; X= S, Se) show excellent attractional properties as a suitable solar collector materials. Full potential linearized augmented plane wave (FP-LAPW) based density functional theory (DFT) approach have been implemented here to study electronic and optical behavior of proposed materials [1]. At last a cost comparative analysis between present material based solar PV module and Crystalline Si based PV module are done and on the basis of that we concluded about suitability and economical benefit of present ternary chalcogenides AB_2X_4 (A= Cd; B= Ga; X= S, Se) materials.

Regarding previous literatures available, Maestro et al.2016 [2] presents the optimization of three solar PV compounds to provide the sufficient electricity as per requirement of typical household (10 kWh/day). A systematic study has been carried out by Chandel et al.2014 [2] to achieve energy demand (2.5 MW capacity solar PV plant) for garment zone in Jaipur. Pavlovic et al. 2013 [3] has been studied about generation of electricity from 1 MW capacity of solar PV plant in Serbia. Dale et al.2013 [4] has analyzed the meta-analysis result related to wind, PV and CSP like renewable energy technologies. Srivastava et al.2013 [5] explore their idea on the impact of solar PV technology and their upcoming effects on Indian economy. Pandey et al.2012 [6] has been studied the role of solar PV technology in the development of Rajasthan as well as for western India. The potential requirement for solar PV recycling policies is analyzed by Donald et al.2010 [7] for major types of solar PV materials.

IOP Conf. Series: Materials Science and Engineering **594** (2019) 012030 doi:10.1088/1757-899X/594/1/012030

Kalowekamo et al.2009 [8] evaluates manufacturing costs of OSC that lies between \$48.8 and \$138.9/m^2. Muneer et al.2005 [9] has been studied about how the electricity obtained from solar PV systems become major part of the future energy solution. The economic viability of a stand-alone solar photovoltaic system has been analyzed by Kolhe et al.2009 [10]. Prasanna et al. [11] have been presented the financial feasibility of 100 KW roof top solar PV power based on life cycle cost of energy. The basic problem behind the costs of PV systems has been studied by Smested et al. [12].

2. Theoretical Methodology

The electronic band gap and absorption coefficient spectra are obtain by full-potential linearized augmented plane wave (FP-LAPW) as embodied in WIEN2k software package. mBJ exchange potential are used to investigate the required calculations [13] by using density functional theory (DFT). The cost of AB_2X_4 (A= Cd; B= Ga; X= S, Se) solar photovoltaic (PV) materials are expressed on per-unit-area basis, whereas the same for solar PV module are based upon cost per watt that obtained under peak solar radiation.

3. Result and Discussion

The energy band gap value calculated from FP-LAPW investigation is presented in Table 1. Both the compounds showing direct band gap nature. The integrated area under the absorption curve near solar visible spectrum from 1-6eV is presented in table 2.

Efficiency of proposed chalcopyrite's materials are around 10% for S based and 20% for Se based compounds [19, 20] as indicated in table 2. For future comparison also 15% efficiency [21] of crystalline silicon is taken. The cost investigation has been organized by calculating the manufacturing amount of $CdGa_2X_4$(X= S, Se) materials per square meter area. We use bottom-up cost estimation for the materials, overhead, and labor costs.
The PV modules cost (\$/Wp) has been obtained by dividing the fabrication cost per square meter by the output of the equal space, in this case 1000 Wp/m^2 times efficiency. The reported amount in \$/Wp because it carries efficiency, cell and module yield, and it empower the opinion to be similar with another solar PV technologies. For initiation of cost calculation first we start with elemental composition calculation [14] of proposed $CdGa_2X_4$(X= S, Se) compounds. The elemental composition calculated value along with density (gm/cm^3) [15] and cost (\$/100gm) [16] of each element which are taken from various online resources are mention in table 3.

Table 1 Band gap (in eV) for $CdGa_2S_4$ and $CdGa_2Se_4$ compounds

Method	Approach	$CdGa_2S_4$	$CdGa_2Se_4$
FP-LAPW	DFT-mBJ	3.22	2.36

Table 2. Integrated absorption coefficients (in the energy range 1-6eV) along with efficiency

Material	IAC FP-LAPW	η(%)
$CdGa_2S_4$	77.99	10[20]
$CdGa_2Se_4$	135.81	20[20]
Crystalline Si	_	15[21]

Table 3 $CdGa_2S_4$ and $CdGa_2Se_4$ elemental composition along with density (gm/cm^3) and cost ($/100gm)

$CdGa_2S_4$			$CdGa_2Se_4$		
Elemental Composition	**Density (gm/cm^3)**	**Cost ($/100gm)**	**Elemental Composition**	**Density (gm/cm^3)**	**Cost ($/100gm)**
Cd (29.57%)	8.65	46	Cd (19.8%)	8.65	46
Ga (36.69%)	5.907	220	Ga (24.56%)	5.907	220
S (33.74%)	2.07	24	Se(55.64%)	4.79	61

Now first we start with cost calculation of each element of $CdGa_2S_4$ compound according to their elemental composition percentage and after that combined all individual obtained elemental cost to get $CdGa_2S_4$ compound cost.

Since 29.57% of Cd in 8.65gm/cm^3 = (29.57*8.65)/100 = 2.55gm/cm^3

As from table 3 the cost of 100gm/cm^3 Cd is $46

Hence 2.55gm/cm^3 Cd cost = ($46* 2.55)/100 = $1.173

Similarly

Since 36.69% of Ga in 5.907gm/cm^3 = (36.69*5.907)/100 = 2.16gm/cm^3

As from table 3 the cost of 100gm/cm^3 Ga is $220

Hence 2.16gm/cm^3 Ga cost = ($220* 2.16)/100 = $4.752

Similarly

Since 33.74% of S in 2.07gm/cm^3 = (33.74*2.07)/100 = 0.69gm/cm^3

As from table 3 the cost of 100gm/cm^3 S is $24

Hence 0.69gm/cm^3 S cost = ($24* 0.69)/100 = $0.165

Hence total cost ($/100gm) of $CdGa_2S_4$ compound = ($1.173+ $4.752+$0.165) = $6.09

Similarly we calculate cost of $CdGa_2X_4$(X= S, Se) compounds as indicated in table4.

Now once the proposed material cost is calculated we are are in position to calculate over all module cost of solar photovoltaic system. We know very well that bulk semiconductor thickness lies between 200- 300 micron [17] hence an assumption is made for current proposed material thickness is 200 microns. Therefore, semiconductor material quantity required is 200g/m^2 and the cost of 200gm proposed materials are shown in table 4.

After the bulk material cost estimation, the calculation of the photovoltaic module cost ($/m^2) is needed. The module cost estimation includes semiconductor material cost, electrical interconnections, glass substrate, protective cover, packaging material cost, absorbing dye, catalyst, electrolyte and labor cost [8], [12]. Kalowekamo et al.2009 [8] is taken as cost reference from literature for electrical interconnections, glass substrate, protective cover, packaging material cost, absorbing dye, catalyst, electrolyte cost and for labor cost reference taken are Smestad et al. 2008 [12]. Including all the cost components we estimate the PV module cost ($/m^2) as indicated in table 5.

Table 4 Material Cost of $CdGa_2X_4$(X= S, Se) compounds

Compound	**$CdGa_2S_4$**	**$CdGa_2Se_4$**
Cost ($/100gm)	6.09	5.529
Cost ($/200gm)	12.18	11.18

Table 5 PV module Cost ($\$/m^2$) of $CdGa_2X_4$(X= S, Se) compounds

Cost Component	**$CdGa_2S_4$-Cost($\$/m^2$)**	**$CdGa_2Se_4$-Cost($\$/m^2$)**
Semiconductor Compound	12.18	11.18
Electrical Contact & Interconnection	6.84	6.84
Substrate	45.50	45.50
Protective cover	45.58	45.58
Packaging Material	4.35	4.35
Other(absorbing dye, Catalyst, Electrolyte)	14.50	14.50
Labor	5	5
Total	**133.95**	**133.95**

Table 6 $/Wp and $/KWp cost of PV module

Compounds	**Cost $/Wp**	**Cost $/KWp**
$CdGa_2S_4$	1.343	1343
$CdGa_2Se_4$	0.669	669
Crystalline	0.7589[18]	758.90[18]

The PV module cost per square meter area is calculated, and further it is required for dollar per watt and dollar per kilo watt cost calculation of PV module.

For $CdGa_2S_4$ compound $/Wp cost

$$\$/Wp = \frac{\$/m^2}{\eta \times 1000W_P/m^2} = \frac{133.95\$/m^2}{0.10 \times 1000W_P/m^2} = 1.339\$/W_P$$

Here η indicates efficiency of the module.

Hence $/KWp cost of $CdGa_2S_4$ compound = 1000* 1.339$/Wp = 1339 $/KWp

Similarly calculate $/Wp and $/KWp cost of all the proposed compounds along with crystalline PV module are indicated in table 6. From table 6 it is clear that Se based PV modules require less cost compare to other modules.

4. Conclusion

The energy band gap and absorption coefficient spectra of ternary chalcogenides materials AB_2X_4 (A= Cd; B= Ga; X= S, Se) using FP-LAPW methods are computed here. The present calculation shows that $CdGa_2Se_4$ poses ideal band gap value for maximum absorption of solar incoming radiation. From the cost point of view $CdGa_2Se_4$ is comparatively less expensive and more efficient compare to other compounds.

References

[1] P. Blaha, K. Schwarz, G.K.H. Madsen, D. Kvasnicka, J. Luitz, Wien2K Code, An augmented plane wave plus local orbitals program for calculating crystal properties. Vienna University of Technology, Vienna, Austria (2014).

[2] C.J.S Maestro, R. Dufo-Lopez and J.B. Agustin, "Analysis of photovoltaic self-consumption systems, energies article, 25, 2016.

IOP Conf. Series: Materials Science and Engineering **594** (2019) 012030 doi:10.1088/1757-899X/594/1/012030

[3] M. Chandel, G.D. Agrawal, S. Mathur, A. Mathur, "Techno-economic analysis of solar photovoltaic power plant for garment zone of Jaipur city," Thermal Engineering, vol. 2, pp. 1-7, (2014).
[4] T. Pavlovic D. Milosavljevic, I. Radonjic, L. Pantic, A. Radivojevic, M. Pavlovic, "Possibility of electricity generation using PV solar plants in Serbia," Renewable and Sustainable Energy Reviews, vol. 20, pp. 201-218, (2013).
[5] M. Dale, "A comparative analysis of energy costs of photovoltaic, solar thermal, and wind electricity generation technologies," Appl. Sci., vol. 3, pp. 325-337, (2013).
[6] S. Pandey, V.S. Singh, N. Gangwar, M.M. Vijayvergia, C. Prakashe, D. Pandey, "Determinants of success for promoting solar energy in Rajasthan, India," Renewable and Sustainable Energy Reviews, vol. 16, pp. 35933598, (2012).
[7] N.C. McDonald, J.M. Pearce, "Producer responsibility and recycling solar photovoltaic modules," Energy Policy, vol. 38, pp. 7041-7047, (2010).
[8] J. Kalowekamo, E. Baker, "Estimating the manufacturing cost of purely organic solar cells," Solar Energy, vol. 83, pp. 1224-1231, (2009).
[9] T. Muneer, M. Asif, S. Munawwar, "Sustainable production of solar electricity with particular reference to the Indian economy," Renewable and Sustainable Energy Reviews, vol. 9, pp. 444-473, (2005).
[10] M. Kolhea, S. Kolhea, J.C. Joshi, "Economic viability of stand-alone solar photovoltaic system in comparison with diesel-powered system for India," Energy Economics, vol. 24, pp. 155-165, (2002).
[11] M. G. Prasanna, S. M. Sameer, G. Hemavathi, "Financial Analysis of Solar Photovoltaic Power plant in India," Journal of Economics and Finance, PP 09-15.
[12] G.P. Smestad, "The Basic Economics of Photovoltaics," Optical Society of America, (2008).
[13] F. Tran, P. Blaha, Phys. Rev. Lett. 102, 226401-1–226401-4, (2009).
[14] G.L. Gilbert, "Education percentage composition and empirical formula—A new view," Journal of chemical education, vol. 75, pp. 851, (1998).
[15] http://periodictable.com/Properties/A/Density.al.html.
[16] https://en.wikipedia.org/wiki/Prices_of_elements_and_their_compounds.
[17] https://en.wikipedia.org/wiki/Wafer_(electronics).
[18] http://www.eai.in/ref/ae/sol/rooftop/cost.
[19] Siebentritt, S. (2011): What limits the efficiency of chalcopyrite solar cells, Solar Energy Materials & Solar Cells, 95: 1471-1476.
[20] W. Shockley and H. J. Queisser "Detailed Balance Limit of Efficiency of pn Junction Solar Cells", J. Appl. Phys. 32, 510 1961.
[21] Saga, T. (2010): Advances in crystalline silicon solar cell technology for industrial mass production, NPG Asia Materials, 2: 96-102.

IOP Conf. Series: Materials Science and Engineering **594** (2019) 012031 doi:10.1088/1757-899X/594/1/012031

Comparative Study of Different Configuration Techniques to Address the Outcome of Partial Shading Conditions on Solar Photovoltaic System

Mona Sharma[1*], Smita Pareek[2] and Kulwant Singh[1]

[1]Dept. of ECE, Manipal University Jaipur, Rajasthan, India

[2]Dept. of ECE, BKBIET Pilani, Rajasthan, India

E-mail: * mona.sharma30@gmail.com

Abstract:
As per energy and power productivity of a solar photovoltaic (SPV) system is evaluated, Partial Shading Conditions (PSC) play a significant role. Under PSC, the panels of a SPV modules receive various levels of solar irradiance, as a result power generation of a SPV system diminishes, and these losses in SPV panels may be decrease by changing the configuration of the panels in array/module. The panels may be configure in many different configuration such as Series(S), Parallel (P), Series Parallel (SP), Total Cross Tied (TCT), Bridge Linked (BL), and Honeycomb (HC) to improve the output energy and power efficiency. This work is aimed to present all the configuration that already been presented in literature and result of PSC on SPV systems are referenced and analyzed. There are four configuration 4×4 array of solar photovoltaic panels to be discussed in this paper. Four configuration are Series parallel (SP), total cross tied (TCT), bridge linked (BL), and honey comb (HC). Four simulated models were performed to decide to effect of shadow with 10 shading patterns. The simulated results shows as a power against voltage (PV) curves of 4×4 array of SPV under PSC for above mentioned configuration. The PV curve of the proposed configuration is enhanced the power efficiency and the minute and exact finding of global maximum peak is simpler. The test of the simulated modules and its theoretical results ensure the effective implementation of this techniques in hardware setup also. It is confront that this work will be a reference of useful and important information for researchers in Solar panel area. The work presented in this papers envisages to be a source.

Keywords: photovoltaic cells, power enhancement, partial shading, global peaks, series parallel (SP), total cross tied (TCT) bridge link (BL) and honey comb (HC)

1. Introduction

While planning an inexhaustible renewable energy and power production, efficiency and productivity factor must be careful evaluated to confound endeavours and venture. Power production from SPV is one of the dependable and support source of power that utilizations daylight to create electrical energy and power [1, 2]. Sun oriented energy production has been accomplishing well known acknowledgment because of quick falling price of photovoltaic (PV) system and huge mechanical enhancements in the field sunlight of intensity converters. In the present situation, control age by PV clusters confront various difficulties, production intensity and efficiency of PV generators is seen to be incredibly impacted by the working conditions, for example, the natural temperature, sun's light levels, its topographical area and field issues [3]. The decrease of SPV system output is affected by variation in irradiance with respect to partial shading conditions The shading happens because of mists, flying creatures, nearby building, shaft, tree and residue. The deprivation of output power throughout the PSC depends on type and sequence of shade, position of cell/module within the array and aside from the corresponding of the shaded space of array [4-6]. The reduced efficiency of PV arrays

IOP Conf. Series: Materials Science and Engineering **594** (2019) 012031 doi:10.1088/1757-899X/594/1/012031

because of partial shading is a predominant barrier in the growth of SPV systems. This issue has been widely discussed in the literature and number of schemes has been introduced to mitigate the results of PSC. Various available methods given by several researchers are studied and compared in terms of their advantages and limitations [7]. Bypass Diode: In uniform irradiance condition the PV characteristics of SPV cell produce single peak power, but Superior photocurrent continues to conduct at the condition of non-uniform irradiance, unshaded cells because the current should be equal when solar cells are connected in series. A known solution for this problem is to attach bypass diodes in anti-parallel but in PSC, bypass diode starts conducting if any solar cell/module is reverse biased. This successively permits the total quantity of current to experience them and it prevent damage of solar cells. However, production of solar panel with bypass diodes is expensive and increase overall price of the system. Also in PSC, the small amount of energy may be very less but produced by shaded cells is totally lost. The total loss is remarkable where low-voltage is required. The non-linearity of PV characteristic is significant with condition by the addition of bypass diodes under PSC due the presence of multiple maxima during tracking. Out of many local maxima, there exit only single GMPP. Because of multiple peaks, existing MPPT techniques does not perform well. In literature, various techniques are reported to find out GMPP in between many local MPPs. These techniques differentiate each other by many factors such as their different accuracy, variable tracking speed, price, hardware necessities, range and kinds of device needed, quality etc [8].

Rest approaches to accord with losses due to partial shading are micro-converters, adaptive reconfigurations of solar array connections, AC modules, generation control circuits, multi-level, and battery equalizer energy recovery. Above mentioned approaches supply sensible results, as every module works at its MPP, but high price of those techniques build power production by PV less attractive and useful [9-10].

The reconfigurable PV array is additionally projected for PV modules to reduce the impact PSC however the price of reconfigurable array is additionally high, and it's appropriate for small PV systems having less number of PV modules [11-12]. Interconnections scheme of PV modules additionally influence to reduce mismatch losses as an outcome of partial shading [13-14]. It is given that impact of partial shading condition may also be decreased by connecting the modules in TCT (total cross tied), BL (bridge linked) [4] etc. shown in figure 1(a). Schemes instead of standard SP (series- parallel) scheme because of additional parallel connection. On the other hand, the SP interconnection has less number of wiring compared to TCT and BL connections and so the price of SP is a smaller as compared to others. Additionally these interconnection schemes are investigated just for little part shaded electrical phenomenon fields (PS-PV) in literature and also the generalized affiliation laws for giant PV fields are still under analysis. [15] This paper is aimed to generate a standard method for interconnection of shaded and unshaded modules, and also to reduce the influence of shadow as an outcome to improve power generation.

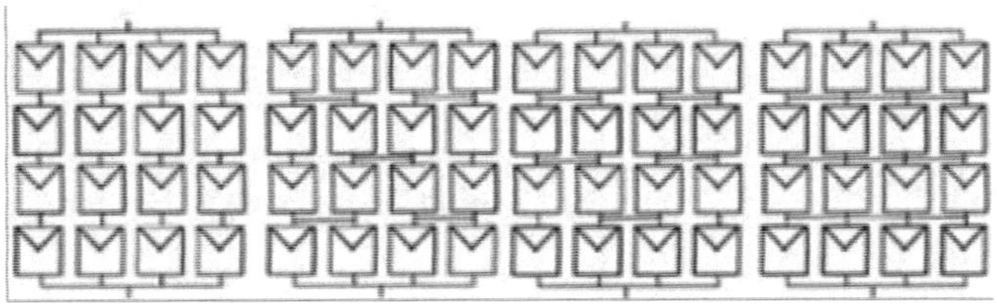

Fig. 1 (a) SP (b) BL (c) HC (d) TCT [18].

This paper is arranged as follows: In Section two, modelling of SPV module are discussed. Section three presents the investigation of 4×4 PV array under different PSC. Section four includes summary and conclusion.

IOP Conf. Series: Materials Science and Engineering **594** (2019) 012031 doi:10.1088/1757-899X/594/1/012031

2. System Configuration

2.1. Photovoltaic module

For simulation and implementation of photo voltaic cell, many models are used analyse performance. Every set up and module is an advancement over the previous one with few additional component and hardware. But complexity of module increases as advancement occurs due to attachment of extra hardware. Among all existed models, one-diode and two-diode models are commonly used [16]. The main demerits of other existing models are that they are not accurate and also require many parameters to model. In between two-diode model and one diode model, commonly one diode model is preferred to implement [17] because it requires only five parameters to implement and model solar cell and has sufficient accuracy. The single diode model has the merit of good flexibility to estimate modifications due to irradiance and temperature changes. This model is shown in figure 2.

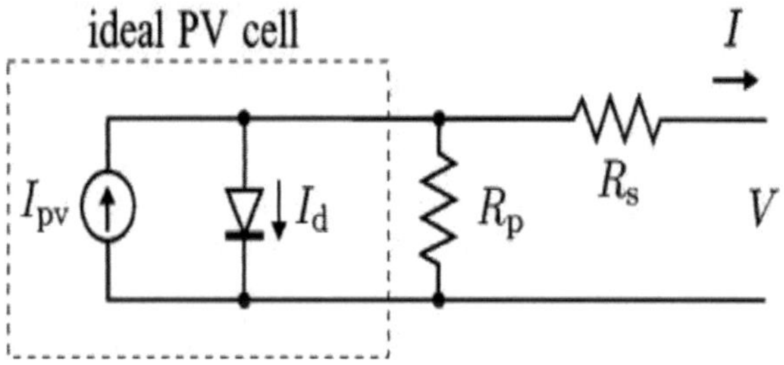

Fig. 2. Equivalent circuit of practical single PV device

I-V characteristic equation of an ideal PV cell is given by:

$$I = Ipv - Io\left[\exp\left(\frac{V + RsI}{Vta}\right) - 1\right] - \frac{V + RsI}{Rp} \tag{1}$$

Where I and Ipv are the total current and photovoltaic (PV) currents, respectivel.

Vt = NskT/q represents thermal voltage of the array where Ns is number of cells in series.

The photovoltaic and saturation current (Ipv & Io) can be given by:

$$Ipv = (Ipv,n + K1\Delta t)\frac{G}{Gn} \tag{2}$$

$$Io = \frac{Isc,n + K1\Delta t}{\exp\left(\frac{Voc,n + Kv\Delta t}{aVt}\right) - 1} \tag{3}$$

Where Δt =T- Tn (T and Tn represent the actual and nominal temperatures [in Kelvin], respectively), G stands for irradiation on the device surface in watts per square meters, Gn stands for nominal irradiation, and K1 and Kv stands current and voltage coefficient respectively.

IOP Conf. Series: Materials Science and Engineering **594** (2019) 012031 doi:10.1088/1757-899X/594/1/012031

2.2 Simulation

All simulations in this paper are done using MATLAB/ Simulink. Here simulation purpose, parameters are taken PV KC200GT, a typical 200W PV. The standard parameters are irradiance of 1000 W/m2, cell temperature of 25 LC and spectrum of 1.5 air mass as shown in table 1.

Table 1 Datasheet of Kyocera PV module.

Parameters	Values
Maximum Power (P_{max})	200 W (+10%/ 5%)
Maximum Power Voltage (V_{mpp})	26.3 V
Maximum Power Current (I_{mpp})	7.61 A
Open Circuit Voltage (V_{oc})	32.9 V
Short Circuit Current (I_{sc})	8.21 A
Max System Voltage	600 V
Temperature Coefficient of V_{oc}	1.23 $\times 10^{-1}$ V/LC
Temperature Coefficient of I_{sc}	3.18 $\times$ 10^{-3} A/LC
Number per Module	54

Four Shading patterns which are used for this research work on SP, TCT, HC and BL configuration in PSC are mentioned below:

Case I Single Row Shading: In this case we consider two different level of irradiations to evaluate the performance of existed system. Group one (top three rows) receive 1000W/m^2 of irradiation and group 2(bottom row) receive 200W/m^2 of irradiation as shown in figure 3(a).

Case II Double Row Shading: In this group one (top two rows) receive 1000W/m^2 of irradiation and group 2 (bottom two row) receive 200W/m^2 of irradiation as shown in figure 3(b).

Case III Oblique Shading: In this group one receive 1000W/m^2 of irradiation and group 2 receive 200W/m^2 of irradiation as shown in figure 3(c).

Case IV Quarter Array Shading: In this case Group one receive 1000W/m^2 of irradiation and group 2 receive 200W/m^2 of irradiation as shown in figure 3(d).

IOP Conf. Series: Materials Science and Engineering **594** (2019) 012031 doi:10.1088/1757-899X/594/1/012031

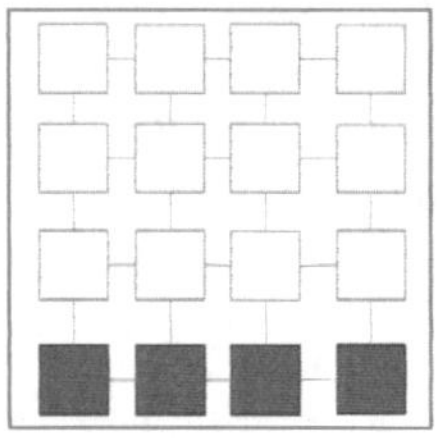

Figure 3(a) Single Row

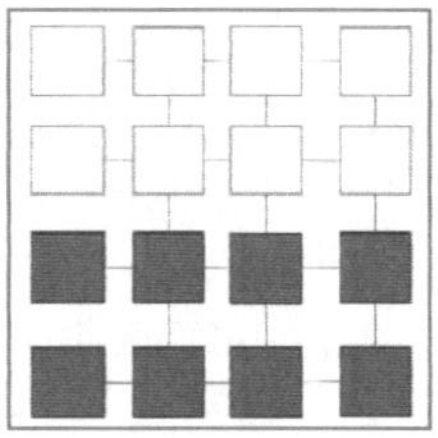

Figure 3(b) Double Row

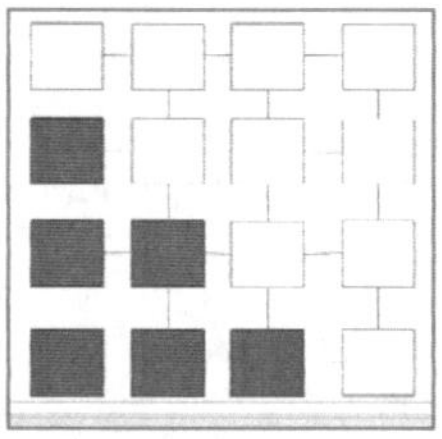

Figure 3(c) Oblique

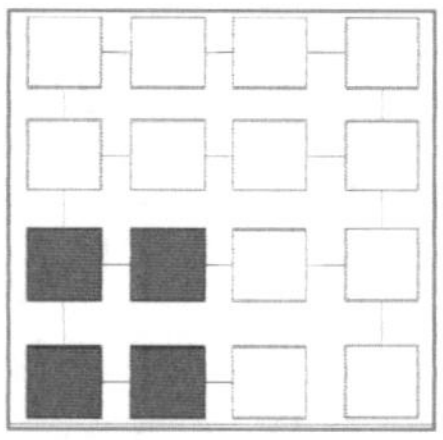

Figure 3(d) Quarter Array

2.3 Interconnection Schemes

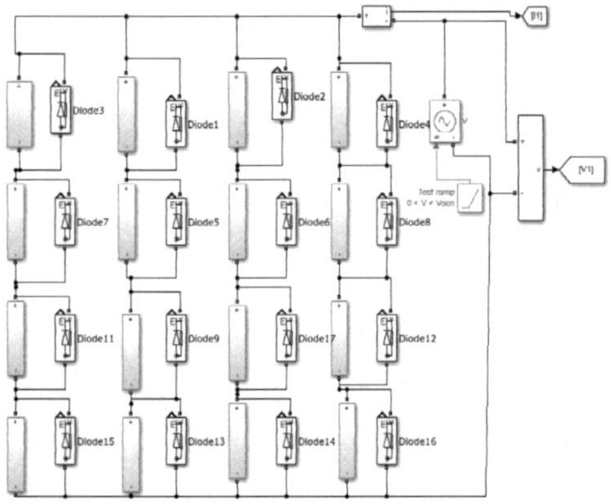

Fig 4(a) Series parallel

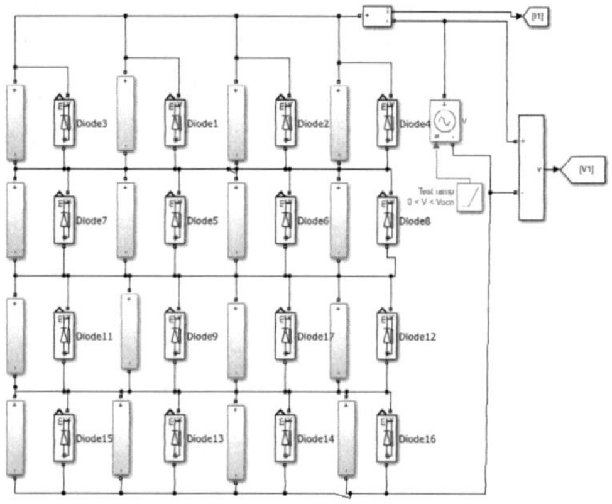

Fig 4(b) Total Cross Tied

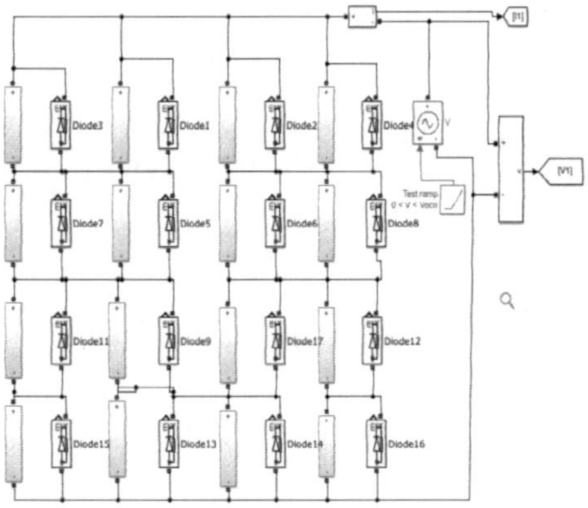

Fig 4(c) Bridge Link

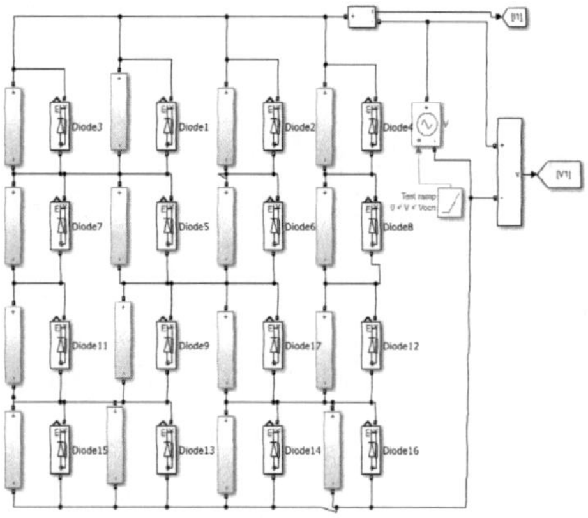

Fig 4(d) Honey Comb

IOP Conf. Series: Materials Science and Engineering **594** (2019) 012031 doi:10.1088/1757-899X/594/1/012031

3. Results and Discussion

To find out the impact and results of shadow on different configuration like SP, TCT, HC and BL topologies, four simulations were performed. Figure 5 shows the results as a PV curve of different shading scenario for SP, TCT, BL and HC topologies

Table 2 shows electrical parameters (Im, Vm and Pmax) of SP, TCT, BL and HC configuration on PSC with various shading scenario. It is concluded from the results that, there exist a relationship between shading patterns and corresponding power generated. In case when number of shaded modules increased corresponding power losses is high as a result the power production is small.

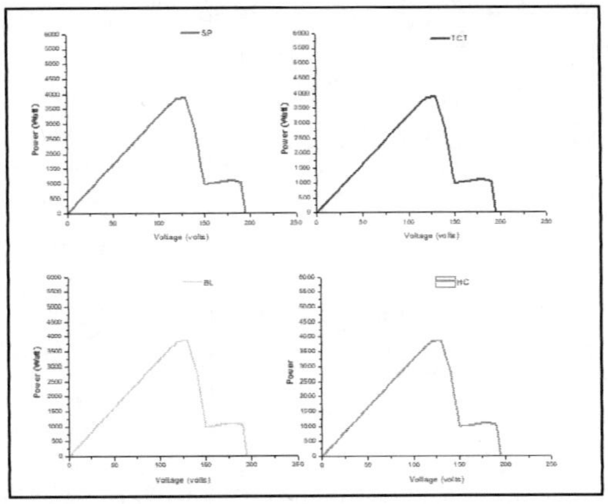

Figure 5(a) PV Characteristics Single Row

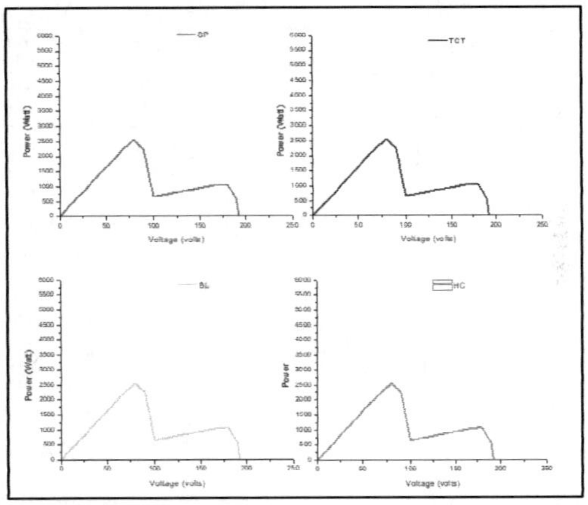

Figure 5(b) PV Characteristics Double Row

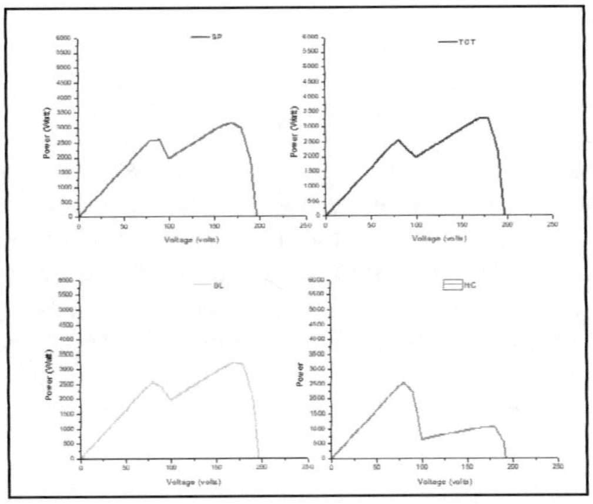

Figure 5(c) PV Characteristics Quarter Array

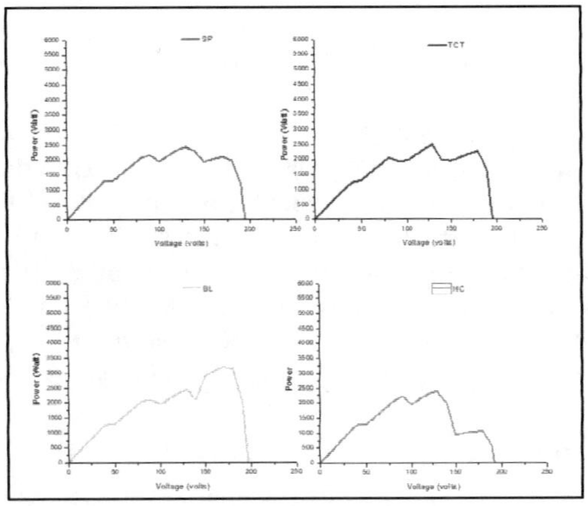

Figure 5(d) PV Characteristics Oblique

Table 2 Electrical parameters of SP,TCT,BL and HC Configuration

Case	Topology	P_m(kW)	V_m(V)	Im	Best
Uniform Irradiance	SP	5.2634	170	30.9	SP, TCT, BL and HC
	TCT	5.2634	170	30.9	
	BL	5.2634	170	30.9	
	HC	5.2634	170	30.9	
Single Row	SP	3.563	110	32.3	SP, TCT, BL and HC
	TCT	3.563	110	32.3	
	BL	3.563	110	32.3	
	HC	3.563	110	32.3	
Double Row	SP	2.5464	80	31.83	SP, TCT, BL and HC
	TCT	2.5464	80	31.83	
	BL	2.5464	80	31.83	
	HC	2.5464	80	31.83	
Oblique	SP	2.225	90	24.12	TCT
	TCT	2.497	130	19.3	
	BL	2.4343	130	19.3	
	HC	2.2255	90	24.72	
Quarter Array	SP	2.599	90	28.77	SP
	TCT	2.546	80	31.83	
	BL	2.5612	80	32.01	
	HC	2.5465	80	31.83	

3.1 Observations

- All topology produces same power with uniform irradiance and when single and double rows are fully shaded
- TCT topology gives best results and it produces high power and current when shadow is oblique.
- SP topology gives best and it produces high power and current when shadow is Quarter Array

4. Summary and Conclusion

An analysis and review of SP, TCT BL and HC configuration under PSC is presented and evaluated in this paper. It is concluded from the results that the power produced depended on type of partial shading and also number of shaded modules within array. Based on result and analysis, it can be seen that TCT configuration is showing good performance as compare to SP configuration when shadow is oblique. This occurs due to high connections of cable which provides more current path to prevent the reduction of current similarly, SP interconnection produces high power and current as compared to TCT configuration in case of quarter array shadow. The results of Bridge link and Honey comb lies between SP and TCT interconnection schemes. Future research work may include study of optimum configuration scheme which can be more appropriate for large PV fields and with frequent change in shading scenario.

References

[1]. Anssi M. aki, Student Member, IEEE, and Seppo Valkealahti, Member, IEEE "Power Losses in Long String and Parallel Connected Short Strings of Series-Connected Silicon-Based Photovoltaic Modules Due to Partial Shading Conditions", IEEE Transactions on Energy Conversion, Vol. 27, No. 1, March 2012, pp 173-183.

[2]. M.Arun Bhaskar, B.Vidya, R.Madhumitha, S.Priyadharcini, K. Jayanthi, G.R.Malarkodi" A Simple PV Array modeling Using MATLAB" 2011 IEEE pp 122-127
[3]. Ramabadran, Ramaprabha, and Badrilal Mathur. "Effect of shading on series and parallel connected solar PV modules." Modern Applied Science 3.10 (2009): 32
[4]. Rudi Darussalam1, Rakhmad Indra Pramana1, Ahmad Rajani1 "Experimental Investigation of Serial Parallel and Total-Cross-Tied Configuration Photovoltaic under Partial Shading Conditions" IEEE ICSSEA Conferences 2017, pp 140-144.
[5]. Hiren Patel and Vivek Agarwal, Senior Member, IEEE "MATLAB-Based Modeling to Study the Effects of Partial Shading on PV Array Characteristics" IEEE Transactions on Energy Conversion, Vol. 23, No. 1, March 2008 pp 302-310.
[6]. K. Ishaque, Z. Salam, H. Taheri, Accurate MATLAB Simulink PV System Simulator Based on a Two-Diode Model, Journal of Power Electronics, 11 (2011)
[7]. Drif, M., et al. "A new estimation method of irradiance on a partially shaded PV generator in grid-connected photovoltaic systems." Renewable Energy 33.9 (2008): 2048-2056.
[8]. Mutoh, Nobuyoshi, Masahiro Ohno, and Takayoshi Inoue. "A method for MPPT control while searching for parameters corresponding to weather conditions for PV generation systems." IEEE Transactions on industrial electronics 53.4 (2006): 1055-1065
[9]. Bidram, A., Davoudi, A., Balog, R.S., 2012. Control and circuit techniques to mitigate partial shading effects in photovoltaic arrays. IEEE J. Photovolt. 2 (4), 532–546.
[10]. Busquets-Monge, S., Rocabert, J., Rodríguez, P., Alepuz, S., Bordonau, J., 2008. Multilevel diode-clamped converter for photovoltaic generators with independent voltage control of each solar array. IEEE Trans. Industr. Electron. 55 (7), 2713–2723
[11]. Patel Hiren and Vivek Agarwal, "MATLAB-based modelling to study the effects of partial shading on PV array characteristics," IEEE Transactions on Energy Conversion, vol. 23, no. 1 , pp. 302-310, 2008.
[12]. R. Ramaprabha and B. L. Mathur, "A comprehensive review and analysis of solar photovoltaic array configurations under partial shaded conditions," International Journal of Photo energy, vol. 2012, 2012.
[13]. Wang Yaw-Juen and Po-Chun Hsu, "An investigation on partial shading of PV modules with different connection configurations of PV cells," Energy, vol. 36.5, pp. 3069-3078, 2011.
[14]. D. Nguyen and B. Lehman, "An adaptive solar photovoltaic array using model-based reconfiguration algorithm," IEEE Trans. Ind. Electron., vol. 55, no. 7, pp. 2644–2654, Jul. 2008
[15]. Velasco-Quesada, et al., "Electrical PV array reconfiguration strategy for energy extraction improvement in grid connected systems," IEEE Trans. Ind. Electron., vol. 56, no. 11, pp. 4319–4331, Nov. 2009
[16]. Ishaque, K., Salam, Z., Taheri, H., 2011. Modeling and simulation of photovoltaic (PV) system during partial shading based on a two-diode model. Simul. Model. Pract. Theory 19 (7), 1613–1626.
[17]. Villalva, M.G., Gazoli, J.R., Ruppert Filho, E., 2009. Comprehensive approach to modeling and simulation of photovoltaic arrays. IEEE Trans. Power Electron. 24 (5), 1198–1208.
[18]. Vijayalekshmy S, Bindu G R, and S Rama Iyer, "Estimation of Power Losses in Photovoltaic Array Configurations under Passing Cloud Conditions" Proceedings of the World Congress on Engineering 2014 Vol I, WCE 2014, July 2 - 4, 2014, pp 270-275.

IOP Conf. Series: Materials Science and Engineering **594** (2019) 012032 doi:10.1088/1757-899X/594/1/012032

Grid Connected-Photovoltaic System (GC-PVS): Issues and Challenges

Bhuwan Pratap Singh[1,*], Sunil Kumar Goyal[1] and Shahbaz Ahmed Siddiqui[2]

[1]Department of Electrical Engineering, Manipal University Jaipur, Jaipur (India)
[2]Department of Mechatronics Engineering, Manipal University Jaipur, Jaipur (India)

Email: [*]halobhuwan@gmail.com

Abstract. Smart grid is the key technology for an effective utilization of the Renewable Energy Sources (RES). The utilization of RES for the generation of electricity is increasingly gaining interest of researchers during the last decades. The main reason behind this is global incentivization, the increasing price of petroleum products, climate issues and deregulations in the energy market. As the Government of India, (MNRE i.e. Ministry of New & Renewable Energy) is targeting to generate 20000 MW power through grid-connected solar PVS by the year 2022 therefore, the main focus in this paper has been presented on power generation through grid-connected PVS. The emerging smart grid technology has enabled the grid-connected PVS as an evolving process in today's world for electrical power generation. However, apart from so many advantages, there are several issues and challenges associated with the integration of PVS to the electric utility grid hence, the investigation to find out available possible solutions to overcome these issues becomes essential in order to enhance the performance of grid-connected PVS. The most severe constraint associated with this emerging technology is it high penetration level. If during low load conditions there is some mismatch found between the real power output and the load profile characteristics of PVS then it may result into large reverse power flow, high power losses or severe voltage violation. In this paper, several issues and challenges associated with the integration of solar PVS with the electric utility grid are presented.

1. Introduction

Utilization of RES for electrical power generation is increasing drastically due to depletion of Conventional Energy Sources (CES) and associated environmental problems. The integration of RES into the current electric utility grid is become a challenging task because most of the RES are variable and uncertain in nature, they lead to several technical and environmental issues while integrating. In order to maintain a reliable as well as economically effective power supply, new efforts need to be considered in order to manage the energy networks, integration of RES to the current electric utility grid, demand management as well as for a variety of some other technical and cost-effective aspects of decentralized energy markets.

Recently, utilization of RES in smart grid system has been continuously increasing. Numbers of programs have already been implemented in different parts of the world; most of them are in those countries which either developed or developing countries. Various studies show that these already adopted technologies can provide reliable electricity service comparatively at low cost. To feed the electrical energy demand of the world, RES offer excellent options that enable the impact on the environment, social and economic factors in to consideration.

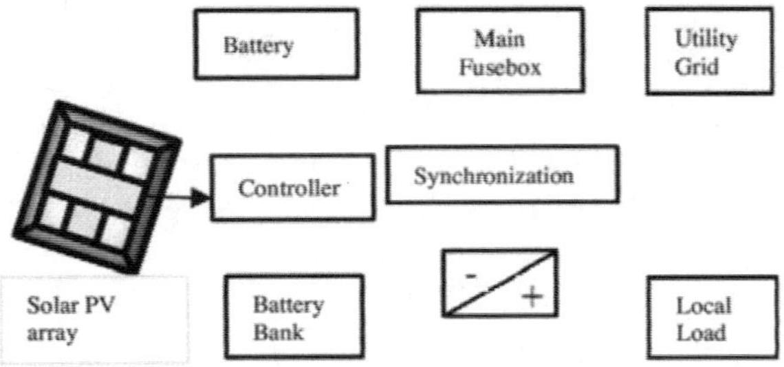

Fig. 1. Grid-connected hybrid power system (PV/Battery)

Further, during the last decades, the use of RES has increased by the development for RES in grid-connected technology. The interest of industries and academia is attracting increasingly towards Grid-connected solar PVS primarily encouraged by the capability to deliver a substitute to the power generation by conventional fossil fuels. It helps to achieve the energy demands which are increasing day by day and to limit the environment pollution due to fossil fuel emissions. The Grid-connected solar PVS configurations are growing consistently in the solar market over the last 20 years by 20–25% PA and that is why they are gaining interest.

The PVS is an array of PV cells which are combined with the controlling device, power electronic equipment and some additional related accessories which convert solar energy received from the sun directly into the electrical energy [1-3]. The output received from the PV cell is subjected to the solar insolation, position of PV modules and temperature of PV cell. The operation of PVS can be either in integration with the existing electric utility grid for peak shaving during the extreme load conditions or as a stand-alone unit to supply demand of a specific load. The configuration and connection of PVS to the existing electric utility grid can be either through net-metering or through feed-in. As the operation of grid-connected PVS is in parallel with the connected utility grid so they do not require any energy storage system. Regardless of the fact that there are some variability and uncertainties always associated with the traditional electrical power system, it is still in the process to know how to tolerate with the diversity and the requirements of high power quality in case of DER sources particularly PVS. Out of the several concerns in the integration of PVS with the electrical power system, the major one is the great trend of a divergence between load profile characteristics and the output of the PVS, which might cause severe violation of voltage regulation, high transmission losses and, during low load conditions a large amount of reverse power may flow. This inconsistency might be the effect of unpredictable cloud movements and sometimes hardware failures [4-6].The grid-connected solar PVS (GCPVS) might be building integrated (BIPVS) or terrestrial (TPVS). TPVS could be located in desert area, tidal area, and saline-alkali land [7].

This paper ultimately presents a comprehensive overview of assembly, benefits, issues and challenges, and the control strategies adopted to overcome the challenges associated with grid-connected photovoltaic systems (PVS). In this paper we have reviewed over 50 research publications including journal papers, conference papers, research articles and magazines. We have also presented the recent trends and developments in the adopted control strategies to overcome the various issues and challenges associated with grid-integration.

2. Benefits of Grid-Connected PVS

A survey has been conducted by World Energy Council in 2007 on several available Energy Resources and according to the survey it has been found that the energy received from sunlight is directly converted into the electrical energy only in the photovoltaic energy conversion process. It has also been studies that there is no interference of heat engine involved in this process. The design of PVS devices are very robust and simple so they possess biggest advantages on a very little maintenance cost. They are capable to produce power outputs ranging from microwatts to megawatts as stand-alone systems. In this way, they are utilized tremendously in variety of applications e.g. for water pumping, interchanges, remote structures, invert osmosis plants, sun oriented home frameworks, space vehicles, satellites and for even control plants on megawatt scale. As a result, the interest in the application of PVS is expanding each year. Solar energy is almost everywhere available and need not to be imported from different regions of the country or across the world as a result it reduces the environmental impacts associated with transportation and also reduces the dependence of consumers or power system engineers on the imported fuels. The PVS are modular; their installation is not complicated, compatible to the environmental conditions, no fuel consumption, pollution free with a long life-span and minimum operation and maintenance costs. Among all grid-connected RES, the grid-connected PVS are coming as the top leading source for the electricity generation and meeting demand. Regardless of abundant benefits of using grid-connected PVS, it suffers from high capital requirements for installation therefore; governments of several countries have taken a step to encouraging the customers for the installation of PVS at their respective locations by offering support in the form of feed-in-tariff (FIT) and net-metering (NEM) [8].

Unlike isolated photovoltaic systems, the operation of the PVS connected to the grid is parallel to the electric utility grid, so they do not need any kind of storage devices to store electrical energy. The PVS connected to the grid help to overcome the issue of greenhouse gas emissions by displacing the required electricity by local connected load and providing supplemental power to the electric utility grid. This is because when generated electricity is more than the demand of local connected load, the grid-connected PVS fed power back into the grid. As such during maximum solar irradiance (peak solar hours), few plants for conventional power generation are required. Additionally, it has also been studied that the transmission and distribution (T&D) losses are also reduced in grid-connected PVS [9]. Therefore, in modern power system, applications of PVS or Distributed Generation (DG) connected to the grid help to meet the load requirements of the consumers and applications of DG technologies are more efficient and reliable to supply additional power to the grid on the appropriate quality and continuity.

The PVS can be of any size depending upon the energy requirements. The small scale solar power plants up to 2 KW of capacity can be mounted on the unused space on rooftops of small buildings like homes, shops etc. In case of large scale solar power plants, the advantage of unused space on rooftops of large buildings can be taken e.g. universities, industries or unused agriculture land. Further, the owner of PVS can move or enlarge his/her system with changes in energy needs. If the energy usage and financial resources grow then owners can add modules or move the whole system to some other relevant place. In comparison with the operating and maintenance costs of other renewable energy based systems, the solar PVS are considered to be low, almost negligible. The PVS can sometimes produce more than needed electrical energy especially during the hot summer months; this excess amount of energy can either be stored in a storage device called battery or sometimes fed directly back into the locally connected electric utility grid if the PVS is grid-connected. The main advantages of connecting the PVS to the local electric utility grid is its simplicity, reducing electricity bills of the owner and low operating and maintenance costs. Further it also reduces the consumption of conventional energy sources to produce electricity since a part of required energy is met by feeding back the excess generated power by the grid-connected PVS. Table 1 represents the technical, environmental and economic benefits of using grid-connected PVS [10, 11-14].

3. Challenges and Impacts of PVS on Grid

Since the overall costs of owing grid connected PVS are decreasing these days therefore, the rate of adoption of this technology at residential, industrial, commercial and utility scale is increasing. Apart from many of the benefits associated with the adoption of grid connected PVS, such as its low operating and maintenance costs and long lifespan of around 25-30 years also, environmental benefits to overcome the massive use of fossil-fuels, this technology have its own set of issues and challenges. Various research scholars and industry researchers are working on this and have suggested the rapid rise in adoption of this technology could produce huge stress on the electrical utility grid. Xin et al. stated that if the infrastructure is not technically proper then a large number of grid-connected PVs will have a negative impact on power network of electrical utility grid [15]. Zarina et al. in [16] explained that if the grid-connected PVS are not dispatchable then it will become difficult to control the effects of PVs on the grid. Omran et al. in [17] stated that in the upcoming stage, many independent system operators and utilities may start implementing more rigorous rules on the application of PVS in grid integration mode because of its high penetration level and unpredictable output power. All of these conclusions made on the basis that the function of grid-connected PVS is inherent in nature. It is well known fact that the power output generated by PVS decreases with the decrease in solar irradiation. Also, the stand alone grid-connected PVSs are unable to contribute adequately when demand increases mainly at the time of sunset which is generally the time of peak demand. During this time period, the electrical utility grid import generation from conventional power plant in order to meet the surge in demand. In addition, Wang et al. argued in [18] that energy storage systems are essential to smooth the operation of grid-connected PVS to provide more reliable and stable system for adequate capacity. Walawalkara et al. in [19] stated that the applications of large scale storage have its own barriers in technical as well as marketing areas that often rise the capital costs in comparison to conventional peak power generation.

Table 1. Benefits of grid-connected photovoltaic systems (PVS)

Technical Benefits					**Economic Benefits**	**Environmental Benefits**
Improvement in Reliability	Improvement in Voltage profile/quality	Reduction in Line/energy losses	Improvement in Security	Benefits in Operations		
1. Power system reliability could be improved 2. Capacity release could be reduced 3. Generation diversity could be improved 4. Reduction in peak power	1. Improvement in voltage quality 2. Improvement in voltage profile 3. Reduction in voltage flicker 4. Better regulation and voltage support	1. Line losses are reduced 2. Control of reactive power is better	1. Security of the critical loads could be enhanced 2. Security risks to the grid could be reduced 3. Power utilities security is improved 4. Reduced impacts of cyber-attacks 5. Reduced Vulnerability of terrorist attacks	1. Facility of ancillary services 2. Productivity is increased 3. Installation is quick and easy 4. Maintenance and operation is easy 5. Reserved requirement is reduced 6. Improvement in total efficiency and infrastructure resilience	1. Reduced operation and maintenance cost for a long time 2. Reduction in cost associated with losses 3. No fuel cost	1. Effects of land use are reduced 2. Improvement in health 3. Reduction in green house gas emission pollutants 4. Reduction in emission of CO_2

3.1 Impacts on voltage stability.

On the other hand if we talk about technical issues then most important is the voltage instability is generally caused by the grid-connected PVS because of its high penetration level, several studies have been carried out about this issue out of which Yan et al in [20] and Thomson et al. in [21] both have concluded that the voltage at the distribution level is mainly affected by the cloud transient as the action of tap changing transformers is not as quick as required to maintain the desired voltage level at distribution level.

Apart from cloud coverage, the second very most important factor to consider is the distance of feeder, according to the studies in [22] the performance of the system at distribution end is mostly affected by the voltage level at the feeder which ultimately depends on the size and location of PVS.

IOP Conf. Series: Materials Science and Engineering **594** (2019) 012032 doi:10.1088/1757-899X/594/1/012032

Therefore, the optimum location is desirable for the PVS at the distribution level in order to achieve reduced power losses and voltage stability.

The study about voltage stability affected by large scale PVS has not been carried out broadly till date, however, Omran et al. in [23] and Wang et al. in [24] have investigated the effects of large scale PVS on voltage profile which are installed at different sites and connected to the grid. Their studies show that when the power generated by the PVS crosses the desired limit of 0.2-10 MW, the voltage profile at the feeder varies as a parabolic trend. The overall impedance of the PVS including impedance of the transmission line decides the peak of this curve. These studies do not consider the output power variation of the PVS during the day but they have considered that the PVS supplies maximum possible power output. However, if we talk about low voltage networks, Rylander et al. in [25] investigated the response of two feeders on PVS at different penetration levels, they have found that the response on the penetration level of PVS is unique in each feeder case and it is depends on the power factor of the inverter, size of PV and its location characteristics. Pompodakis et al. in [26] presented that in low voltage networks, the reactive power regulation in the inverters can modify the overvoltage. The impacts of penetration level on the traditional voltage regulation devices and voltage profile has been investigated by Liu et al. in [27] and reported that the capacitor banks connected to support the voltage across the feeder can be replaced by the overvoltage at high penetration level (30-50%) in the secondary circuits with PV inverters. Shahnia et al. proved in [28] that PVS may responsible for the variation in voltage profile depending on their location, size, and the interconnected load. They have also presented that the impact of PVS on voltage imbalance installed at rooftops will be more severe at the end of the feeder as compared to the starting point of the feeder in case of low voltage distribution network.

Therefore, today, the most important challenge is to investigate how the voltage profile is affected by the large scale PVS connected to the grid due to cloud coverage, solar irradiation and temperature as there are no deep studies have been carried out on how the environmental effects affects the performance of the grid in grid-connected large scale PVS.

3.2 Impacts on frequency stability.

The frequency of generation in current electrical system depends upon the rotating speed of the active synchronous generator. The operation of these synchronous generators is controlled by the difference between demand and generated electrical power of each machine governs by the acceleration or deceleration to increase or to decrease. To control the frequency in the electrical system, most important role is played by the inertial response of the generators. The supply of kinetic energy can limit the initial rate of frequency, if the demand and generation imbalance occurs in the electrical system. Seneviratne et al. and Rahmann et al. in [29-30] argued that a big difference in the operation compared with the conventional power plants can be made by the involvement of inverter and PV array in the case of large scale PVS. They said that in the electric utility system, the reduced number of rotating machines and the power variability during the day time could be responsible for the several problems from the frequency point of view. Yan et al. carried out a study on Australian network in [31], in this network a wind and solar power plant were integrated. They concluded that the frequency regulation problems in the electrical utility system are caused by the low inertia and variation in PVS. They have considered a small PVS integrated with electrical utility grid, the grid has a net installed capacity of 500 MW. Abdlrahem et al. investigated in [32] that how the interconnected system of 200 MW made up of four large scale PVS of 50 MW each in an electrical utility system affects the frequency. The study is carried out to see the response of electrical system to the increased and decreased radiations on each PVS. They conducted the analysis with constant conventional power generation and made conclusions that the frequency stability is not drastically affected by the penetration level of PV however, on the basis of simulation results they commented that the frequency stability of the electrical power system is better in a grid-connected PVS. Also, the frequency oscillations damped out faster with an increasing level of PV penetration. On the other hand another investigation based on statistical evaluation of the European continental electrical utility system in [33] says that the high penetration level of PV affects the frequency stability.

The detailed studies and analysis about this phenomenon are still not achieved till date however, it is important and necessary to investigate that how the lack of inertia in PVS affects the electric utility grid as the large scale PVS have been installed and in operation throughout the world. The switching from conventional power plant to grid-connected PVS requires that the synchronous generators must be re-designed to accommodate the variation in power output of solar PVS in order to maintain the connected electrical utility system stable. The investigation carried out by Rahmann et al. in [34] could be considered as initial step to understand that how the lack of inertia creates problem in the current electrical utility system. Moreover, it is suggested that a power reserve is not at all required to control the frequency stability if the large scale PVS works at MPP. Today, the most important challenge associated with the large scale PVS is to overcome with the primary and secondary control of frequency which is generally affected by the lack of inertia. However, the challenge is not limited up to large scale PVS but also for conventional power plants.

3.3 Impacts on active power regulation.

The peak of active power varies non-linearly depending on the solar irradiance. To track this point called as MPP, many researchers have developed and implemented various tracking algorithms, some most widely used are named as Particle Swarm Optimization (PSO), Perturb and Observe (P & O), Hill Climbing HC), incremental conductance (IncCond), artificial neural network, fuzzy logic control. There are many more as they are explained in [35]. Their time of response, the time delay to notice the variation in solar radiation and the required number of sensors are considered as their limitations. However, under these techniques, the large scale PVS will always be operating at MPP in any solar radiation. Sometimes, large and quick active power fluctuations are experienced by the electrical power system due to cloud transients. This response in result affects the power balance of the grid as well as economic dispatch. In [36] the authors have analyzed the impact of the PVS connected to the grid on the active power dispatch during the day, in the PVS which is installed at the distribution network, the PV production due to variation in solar radiation shows higher ramp rates more than 2.5 MW/min.

Therefore, for active power control: ramp rate control and power curtailment are the two types of grid codes which are necessary to be implemented. As per the literature, it would be necessary to continuously track the power required by the Transmission System Operator (TSO) instead to ensure that the PVS to be work at MPP all the day. The MPP of the PVS must be less than the overall power handling capacity of the large scale grid-connected PVS otherwise the installation of a storage system is required.

3.4 Impacts on reactive power regulation.

The PQ control is not available with the recent advancements of PV inverters which are implemented at the distribution end as inverter itself can perform it however this control is also not required by IEEE1547. The MPPT technique implemented in normal operation of PVS limits the performance of the system as it does not allow the complete control of the reactive power during the day. The PV inverter connected in large scale PVS are required not only to control the active power but to control the reactive power as well due to the increased development of grid-connected PVS [37]. The limitations of temperature, dc solar radiation and dc voltage are not reflected by the developers of PV inverter in their capability area.

Thus, to understand the limitations of the inverters connected in PVS considering the variation in temperature, dc solar radiation and dc voltage is a big challenge for today.

3.5 Impacts on equipment loading and power losses.

The power flows as a function of current through the transformers, lines and loads therefore, in the electric utility grid the power losses are generally ohmic in nature. Power losses and equipment loading both are related to each other as they vary as a function of rms value of the current. An installation scheme is presented by Weckx et al. in [38] in which central and local converters control the two balancing inverters, which results in the reduction of distribution losses in the electric utility grid. The absolute values measured for the local voltage are considered in local controller whereas the line current which controls the voltage drop over the line is considered in central controller. The grid losses in the distribution network were reduced by 10%. Also, the high nodal voltages are noticed as a result of high PV penetration systems which may lead to rise the feeder section loading and no-load distribution transformer losses.

Moreover, the resultant increment or decrement in power losses by the integration of PVS with the electric utility grid depend on location and size of PVS on the grid. In [39] authors reported that non-optimal size or location of a PVS may lead to rise in grid power losses. To find out the optimal size and location of the PVS for the distribution network, various methods have been presented in the existing literature which are employed to minimize the system losses. Additionally, in [40] authors have done an exhaustive review on the methods used to optimize the performance of the current electric utility grid interconnected with the PVS.

3.6 Impacts on voltage regulation devices.

An innovative attempt is made in [41] to show that how the voltage regulation devices are affected by the high penetration level of PVS in the distribution system. Garret et al. in [41] have experimentally obtained the switching operations performed by the voltage regulation device when integrated with a relatively large PVS, 90 operations were observed by the authors. Hariri et al. in [42] performed simulation using QSTS (Quasi-Static Time Series) to show that the traditional system control is an essential requirement because the operations of voltage regulator might increase whenever the PV inverter system is permitted to participate in voltage regulation operations. The on load tap changers (OLTC) and operation of the voltage regulator is affected by the PVS as a result the increased number of operation and runaway issues arises. Hence, Agalgaonkar et al. in [43] developed a new control methodology by taking reactive power supply from the inverter system. With this strategy authors have alleviated the negative impact of the PVS. Investigation has also achieved the voltage control when the PVS and OLTC operated at set points which are required for optimum control.

Sandia National Laboratory recently conducted a study on the impact of grid-connected PVS on the operation of conventional voltage regulation devices. The verities of impact have been noticed on various devices such as line voltage regulators, OLTC and switched capacitor banks and it was found that all depend on operational settings of the device, location and level of load. It has been noticed that the operations of the devices are reduced for the first capacitor adjoining to the substation as well as for the OLTC and increased for the rest of the capacitors.

3.7 Impacts on Harmonics.

According to the previous studies, it has been found that one of the harmonics delivery source in to the electric utility grid including inter-harmonics may be the PVS connected to the grid, which is also found as one of the reason for worst power quality [44-45]. In the latest research, to study and investigate the impact of grid-connected PVS inverters on harmonics, many attempts have been made. It has been found that since, the interaction between different controllers such as the current controllers and the dc link voltage as well as the switching behavior of the power devices are non-linear in nature hence, PV inverters are generally responsible for the introduction of harmonics [46-47]. Moreover, the issues related to inter-harmonic in PVS are hardly observed. However, the conclusion given in some of the recent articles shows that inverters connected in PVS may responsible for the introduction of inter-harmonics as well in the electric utility grid currents, resulting flickering and overloading.

In the most terrible conditions, inter-harmonics may result in triggering the protection circuit, and as a result isolates the PVS unintentionally from the electric utility grid [48]. Impacts of such actions comprise challenges from the system stability point of view and substantial energy losses in case of large-scale grid-connected PVS. Sangwongwanich et al. in [48] has defined the concept of inter-harmonics as “frequency components with non-integer times of the fundamental frequency”. In [49], it has been investigated experimentally that MPPT device which is used to track MPP for the PVS to operate at maximum power output sometimes imposes variations in the dc power and as a results produce inter-harmonics. During the investigation, a significant amount of inter-harmonics has been measured from PV inverters during the steady-state operation of MPPT controllers. Particularly, it is recommended that "the inter-harmonic frequency spectrum is somehow correlated with the MPPT frequency".

A brief discussion has been presented in [50] on the inter-harmonic emission mechanisms through various simulations, an in-depth analysis is still not achieved about this mechanism and has not been validated experimentally either.

4. Conclusion

The PVS in integration with the electric utility grid is one of the most emerging tool in the field of DERs. Some most important issues of PVS that affects the performance of the connected grid are discussed which are available in the literature and the challenges to overcome these issues are also highlighted. The environmental issues like cloud transients and solar irradiation leads to challenging issues regarding instability in voltage and frequency profile. The Power electronic devices are found as feasible solution to minimize the problems associated with the variability and uncertainty in the grid-integration. Further, to overcome with the variation in power generated by the PVS, MPPT controllers and energy storage devices could be employed. However, the involvement of PV inverters in the PVS also leads to challenging issue regarding harmonics. Moreover, the up gradation in stability of the system by incorporating some new materials and storage equipment may employed to enhance the performance of the grid-connected PVS. Therefore, the developments in grid-connected PVS are still needed that incorporates technical and financial aspects. This would be useful to assess the balance of the electricity price for grid-connected PVS.

References

[1] K. Gabrovska, A. Wagner, and N. Mihailov, "Software system for simulation of electric power processes in photovoltaic-hybrid system," ACM, New York, pp. 1–7, 2004.
[2] R. Shah, N. Mithulananthan, and R. Bansal, "Damping performance analysis of battery energy storage system, ultracapacitor and shunt capacitor with large-scale photovoltaic plants," Applied energy, vol. 96, pp. 235–244, 2012.
[3] R. Shah, N. Mithulananthan, R. Bansal, and V. Ramachandaramurthy, "A review of key power system stability challenges for large-scale pv integration," Renewable and Sustainable Energy Reviews, vol. 41, pp. 1423–1436, 2015.
[4] M. Begovic, A. Pregelj, A. Rohatgi, and D. Novosel, "Impact of renewable distributed generation on power systems," in hicss, p. 2008, IEEE, 2001.
[5] J. Smith et al., "Stochastic analysis to determine feeder hosting capacity for distributed solar pv," Electric Power Research Inst., Palo Alto, CA, Tech. Rep, vol. 1026640, 2012.
[6] W. A. Omran, M. Kazerani, and M. Salama, "Investigation of methods for reduction of power fluctuations generated from large gridconnected photovoltaic systems," IEEE Transactions on Energy Conversion, vol. 26, no. 1, pp. 318–327, 2011.
[7] J. Wid´en, E. W¨ackelg°ard, J. Paatero, and P. Lund, "Impacts of distributed photovoltaics on network voltages: Stochastic simulations of three swedish low-voltage distribution grids," Electric power systems research, vol. 80, no. 12, pp. 1562–1571, 2010.
[8] R. Bakhshi and J. Sadeh, "Economic evaluation of grid–connected photovoltaic systems viability under a new dynamic feed–in tariff scheme: A case study in iran," Renewable Energy, vol. 119, pp. 354–364, 2018.
[9] U.S. Energy Information Agency. United States electricity profile. Summary statistics for supply and disposition of electricity, URL ⟨http://www.eia.gov/electricity/state/unitedstates/⟩ 1990–2012; 2012.
[10] T. Adefarati and R. Bansal, "Integration of renewable distributed generators into the distribution system: a review," IET Renewable Power Generation, vol. 10, no. 7, pp. 873–884, 2016.
[11] A. Chowdhury, S. K. Agarwal, and D. O. Koval, "Reliability modelling of distributed generation in conventional distribution systems planning and analysis," in Industry Applications Conference, 2002. 37th IAS Annual Meeting. Conference Record of the, vol. 2, pp. 1089–1094, IEEE, 2002.
[12] K. Balamurugan, D. Srinivasan, and T. Reindl, "Impact of distributed generation on power distribution systems," Energy Procedia, vol. 25, pp. 93–100, 2012.
[13] H. A. Gil and G. Joos, "Models for quantifying the economic benefits of distributed generation," IEEE

Transactions on power systems, vol. 23, no. 2, pp. 327–335, 2008.
[14] B. Delfino, "Modeling of the integration of distributed generation into the electrical system," in Power Engineering Society Summer Meeting, 2002 IEEE, vol. 1, pp. 170–175, IEEE, 2002.
[15] H. Xin, Z. Lu, Y. Liu, and D. Gan, "A center-free control strategy for the coordination of multiple photovoltaic generators," IEEE Transactions on Smart Grid, vol. 5, no. 3, pp. 1262–1269, 2014.
[16] P. Zarina, S. Mishra, and P. Sekhar, "Deriving inertial response from a non-inertial pv system for frequency regulation," in Power Electronics, Drives and Energy Systems (PEDES), 2012 IEEE International Conference on, pp. 1–5, IEEE, 2012.
[17] W. A. Omran, M. Kazerani, and M. Salama, "Investigation of methods for reduction of power fluctuations generated from large grid connected photovoltaic systems," IEEE Transactions on Energy Conversion, vol. 26, no. 1, pp. 318–327, 2011.
[18] H. Wang and X. Bai, "Adequacy assessment of generating systems incorporating wind, PV and energy storage," in Innovative Smart Grid Technologies-Asia (ISGT Asia), 2012 IEEE, pp. 1–6, IEEE, 2012.
[19] R. Walawalkar, J. Apt, and R. Mancini, "Economics of electric energy storage for energy arbitrage and regulation in New York," Energy Policy, vol. 35, no. 4, pp. 2558–2568, 2007.
[20] R. Yan and T. K. Saha, "Investigation of voltage stability for residential customers due to high photovoltaic penetrations," IEEE transactions on power systems, vol. 27, no. 2, pp. 651–662, 2012.
[21] M. Thomson and D. Infield, "Impact of widespread photovoltaics generation on distribution systems," IET Renewable Power Generation, vol. 1, no. 1, pp. 33–40, 2007.
[22] W. Pattaraprakorn, P. Bhasaputra, and J. Pattanasirichotigul, "Impacts of PV power plants on distribution grid for voltage stability and economic values," in Electrical Engineering/Electronics, Computer, Telecommunications and Information Technology (ECTI-CON), 2015 12th International Conference on, pp. 1–6, IEEE, 2015.
[23] W. A. Omran, M. Kazerani, and M. M. Salama, "A clustering-based method for quantifying the effects of large on-grid pv systems," IEEE Transactions on Power Delivery, vol. 25, no. 4, pp. 2617–2625, 2010.
[24] W. Yi-Bo, W. Chun-Sheng, L. Hua, and X. Hong-Hua, "Study on impacts of large-scale photovoltaic power station on power grid voltage profile," in Electric Utility Deregulation and Restructuring and Power Technologies, 2008. DRPT 2008. Third International Conference on, pp. 2575–2579, IEEE, 2008.
[25] M. Rylander, J. Smith, D. Lewis, and S. Steffel, "Voltage impacts from distributed photovoltaics on two distribution feeders," in Power and Energy Society General Meeting (PES), 2013 IEEE, pp. 1–5, IEEE, 2013.
[26] E. E. Pompodakis, I. A. Drougakis, I. S. Lelis, and M. C. Alexiadis, "Photovoltaic systems in low-voltage networks and overvoltage correction with reactive power control," IET Renewable Power Generation, vol. 10, no. 3, pp. 410–417, 2016.
[27] Y. Liu, J. Bebic, B. Kroposki, J. De Bedout, and W. Ren, "Distribution system voltage performance analysis for high-penetration pv," in Energy 2030 Conference, 2008. ENERGY 2008. IEEE, pp. 1–8, IEEE, 2008.
[28] F. Shahnia, R. Majumder, A. Ghosh, G. Ledwich, and F. Zare, "Sensitivity analysis of voltage imbalance in distribution networks with rooftop PVs," in Power and Energy Society General Meeting, 2010 IEEE, pp. 1–8, IEEE, 2010.
[29] N. Miller, K. Clark, and M. Shao, "Frequency responsive wind plant controls: Impacts on grid performance," in Power and Energy Society General Meeting, 2011 IEEE, pp. 1–8, IEEE, 2011.
[30] C. Seneviratne and C. Ozansoy, "Frequency response due to a large generator loss with the increasing penetration of wind/pv generation–a literature review," Renewable and Sustainable Energy Reviews, vol.57, pp. 659–668, 2016.
[31] R. Yan, T. K. Saha, N. Modi, N.-A. Masood, and M. Mosadeghy, "The combined effects of high penetration of wind and pv on power system frequency response," Applied Energy, vol. 145, pp. 320–330, 2015.
[32] A. Abdlrahem, G. K. Venayagamoorthy, and K. A. Corzine, Frequency stability and control of a power system with large pv plants using pmu information," in North American Power Symposium (NAPS), 2013, pp. 1–6, IEEE, 2013.
[33] Y. Wang, V. Silva, and M. Lopez-Botet-Zulueta, "Impact of high penetration of variable renewable generation on frequency dynamics in the continental europe interconnected system," IET Renewable Power Generation, vol. 10, no. 1, pp. 10–16, 2016.
[34] C. Rahmann and A. Castillo, "Fast frequency response capability of photovoltaic power plants: The necessity of new grid requirements and definitions," Energies, vol. 7, no. 10, pp. 6306–6322, 2014.
[35] A. Cabrera-Tobar, E. Bullich-Massagu´e, M. Arag¨u´es-Pe˜nalba, and O. Gomis-Bellmunt, "Review of advanced grid requirements for the integration of large scale photovoltaic power plants in the transmission

IOP Conf. Series: Materials Science and Engineering **594** (2019) 012032 doi:10.1088/1757-899X/594/1/012032

system," Renewable and Sustainable Energy Reviews, vol. 62, pp. 971–987, 2016.
[36] R. Guerrero-Lemus, B. Gonz´alez-D´ıaz, G. R´ıos, and R. N. Dib, "Study of the new spanish legislation applied to an insular system that has achieved grid parity on pv and wind energy," Renewable and Sustainable Energy Reviews, vol. 49, pp. 426–436, 2015.
[37] A. Ellis, R. Nelson, E. Von Engeln, R. Walling, J. MacDowell, L. Casey, E. Seymour, W. Peter, C. Barker, B. Kirby, et al., "Reactive powerperformance requirements for wind and solar plants," in Power and Energy Society General Meeting, 2012 IEEE, pp. 1–8, IEEE, 2012.
[38] S. Weckx, C. Gonzalez, and J. Driesen, "Reducing grid losses and voltageunbalance with pv inverters," in PES General Meeting— Conference & Exposition, 2014 IEEE, pp. 1–5, IEEE, 2014.
[39] I. El-Samahy and E. El-Saadany, "The effect of dg on power quality in a deregulated environment," in Power Engineering Society General Meeting, 2005. IEEE, pp. 2969–2976, IEEE, 2005.
[40] W.-S. Tan, M. Y. Hassan, M. S. Majid, and H. A. Rahman, "Optimal distributed renewable generation planning: A review of different approaches," Renewable and Sustainable Energy Reviews, vol. 18, pp. 626–645, 2013.
[41] D. Garrett and S. Jeter, "A photovoltaic voltage regulation impact investigation technique," IEEE Trans. Power Electronics;(United States), vol. 4, no. 1, 1989.
[42] A. Hariri, M. O. Faruque, R. Soman, and R. Meeker, "Impacts and interactions of voltage regulators on distribution networks with high PV penetration," in North American Power Symposium (NAPS), 2015, pp. 1–6, IEEE, 2015.
[43] Y. P. Agalgaonkar, B. C. Pal, and R. A. Jabr, "Distribution voltage control considering the impact of pv generation on tap changers and autonomous regulators," IEEE Transactions on Power Systems, vol. 29, no. 1, pp. 182–192, 2014.
[44] D. G. Infield, P. Onions, A. D. Simmons, and G. A. Smith, "Power quality from multiple grid-connected single-phase inverters," IEEE Transactions on Power Delivery, vol. 19, no. 4, pp. 1983–1989, 2004.
[45] J. H. Enslin and P. J. Heskes, "Harmonic interaction between a large number of distributed power inverters and the distribution network," IEEE transactions on power electronics, vol. 19, no. 6, pp. 1586–1593, 2004.
[46] S.-G. Jeong and M.-H. Park, "The analysis and compensation of dead-time effects in pwm inverters," IEEE Transactions on Industrial Electronics, vol. 38, no. 2, pp. 108–114, 1991.
[47] Y. Du, D. D.-C. Lu, G. James, and D. J. Cornforth, "Modeling and analysis of current harmonic distortion from grid connected PV inverters under different operating conditions," Solar Energy, vol. 94, pp. 182–194, 2013.
[48] Y. Du, D. D.-C. Lu, G. James, and D. J. Cornforth, "Modeling and analysis of current harmonic distortion from grid connected pv inverters under different operating conditions," Solar Energy, vol. 94, pp. 182–194, 2013.
[49] R. Langella, A. Testa, S. Djokic, J. Meyer, and M. Klatt, "On the interharmonic emission of pv inverters under different operating conditions," in Harmonics and Quality of Power (ICHQP), 2016 17th International Conference on, pp. 733–738, IEEE, 2016.
[50] A. Sangwongwanich, Y. Yang, D. Sera, and F. Blaabjerg, "Interharmonics from grid-connected pv systems: Mechanism and mitigation," in Future Energy Electronics Conference and ECCE Asia (IFEEC 2017- ECCE Asia), 2017 IEEE 3rd International, pp. 722–727, IEEE, 2017.

IOP Conf. Series: Materials Science and Engineering **594** (2019) 012033 doi:10.1088/1757-899X/594/1/012033

Hybrid human powered power generating unit by using multiple transducers

Avanish Gautam Singh[1] and B Ravi Subrahmanyam[2]

[1,2]Department of Electrical and Electronics Engineering, GCET (Dr. A.P. J. AKTU) Greater Noida, Uttar Pradesh-201306 India
E-mail: avanish.g.s@ieee.org

Abstract. This paper provides concept of capturing the different kinds of energy conversions that can be in form of motion, physical force, heat energy and solar energy and converting all form of energy into useful electrical energy. It can provide endless supply of energy to electronic device also has attracted large number of researchers and has bought a growing amount of attention to energy harvesting. The hybrid human powered power generating unit is an electronic device, purpose of proposed model is to supply electricity for charging devices when the user is out of reach to the charger or charging port. This device basically transforms many different kind of energies to electricity (direct current) example piezoelectric plate produces electrical energy when mechanical force is applied on it and another we have Peltier plate (thermal electric generator) which transforms temperature differences into electrical energy. The proposed model is very much portable, so any individual easily can carry with them. This device does not need any external power supply to charge it, because this device made up of four energy inducing elements or can say transducers and one buck booster which step up the output voltage of the energy produced or induced by the generating elements assembled with a super-capacitor. This device is very useful in the situation where the person doesn't have power source around and need to charge any electronic device. This device can replace your heavy power bank with this light weighted device. Anyone can easily use this electronic device simply by squeezing it and shaking horizontally, these all physical movements induces electrical energy and got stored in super capacitor and further consume by the charging devices.

1. Introduction

Electronic wireless communication devices play a vital role in today's era. These devices majorly consist of microprocessor, sensors, transducers, screen, etc. [1], these kinds of elements connected in a systematic pattern to run the device for performing a specific task. To run any electronics device, it needs a specified or rated quantity of power supply (generally rated 3.7V) and required current according to device load demand [2]. The above literature shows that battery is one of the basic components of any device and if the battery dies or the state of charge become "0" unfortunately, the device will die and it will stop working completely, this situation leads any individual barring from intercellular communication, world wide web or internet and many other important applications which may cause anyone heavy loses. So it is necessary for electronics or smart electronics devices to work uninterruptedly. The uninterrupted supply from the battery is little difficult task because it is always possible that there might occur few situations when battery's state of charge is near to zero and the person is out of reach to the power supply, currently most of the people carries power bank, this may be the liable solution for the above stated problem, but there are also some demerits of power bank like power bank needs to charged again and again, so with time power bank's capacity degrades. Another demerit is that it is heavy in weight. And yet another demerit is that power bank can only supply limited capacity of charge once it is fully charged. So this paper provides smart solution for the electronic device's batteries, which encounters all the above stated problems. The proposed model is of hybrid human power generating unit using multiple transducers, the device consists of five kinds of energy converting elements like Piezoelectric Plates, Peltier Module (thermal electric generators),

IOP Conf. Series: Materials Science and Engineering **594** (2019) 012033 doi:10.1088/1757-899X/594/1/012033

Solar Panel (5w), small DC generator (dynamo) and a coil with permanent magnet. These elements are connected to a boost converter (XL6009-module) which is further connected to super-capacitor and then to IC-78S05 and at the output, a constant 5-voltage and current not more than 2A is obtained.

Here, we investigate a complete design and construction for an electronic circuit that will be used for charging electronic devices. The paper is organized in four different sections. The first section consists of introduction. The second section consists of description of all the elements present in the proposed model. Third section consists of working and the last and forth section consist of results and discussion.

2. Technical Studies

This section includes different units of the device explained individually along with their fundamental mathematical output expressions.

2.1. Piezoelectric Plates

The piezoelectric plates are used in the proposed model in a particular pattern as shown in Fig. 1. The piezoelectric means the electricity generated with the help of pressure, the Quartz crystal is one of the example of natural piezoelectric crystal, where the Rochelle salt, lithium sulphate, dipotassium tartrate are the example of manmade crystal. Piezoelectric is a kind of energy converting element which converts pressure and vibration into electric energy. The piezoelectric effect is due to the presence of electric dipole moments in solid materials. The induced ions on crystal lattice sites with different kinds of charge surroundings like BaTi03 and PZTs or may be directly carried with molecular groups. The dipole density can easily be calculated for crystal by adding the dipoles per volume of crystallographic unit cell [3]. As dipoles are vector quantity, dipole density is vector field. The dipoles those are nearer to each other having the tendency to align in a regions known as Weiss domains. The domains are generally having the random orientations. But with the help of poling (process by which a strong electric field is applied across material, at elevated temperature. And not all piezoelectric materials can be polarized. The different piezoelectric material has different piezoelectric constant like PZT-5H has D31 of -274 PM/V that means if you applied a Voltage of 1V it will produce a charge of 1 Pica Meter. Negative sign depends upon its direction [4]. Output voltage of piezoelectric crystal can be calculated by using formula-

$$V = P \times g \times t \tag{1}$$

where **P** is the pressure applied in N/(sq.m)
g is the sensitivity of the material
t is the thickness of the material
V is output voltages like quartz crystal

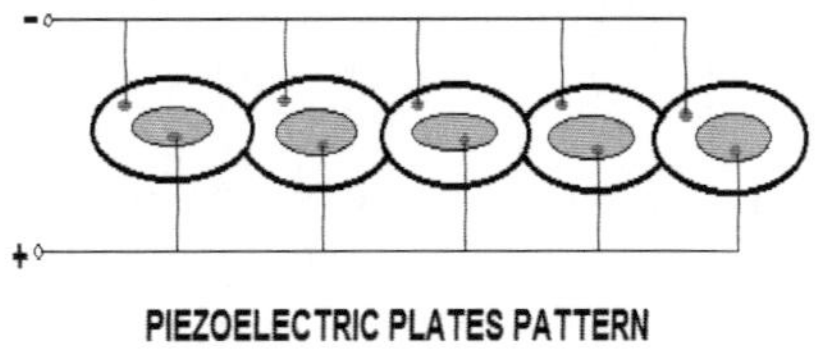

Fig. 1. Piezoelectric plates connected in parallel and placed in fixed-pattern.

2.2. Dynamo

The dynamo is a small electric generator. It works on the principle that if the coil is rotating in the presence of electric field then the electricity induced in the coil, which is proportional to the flux density and number of turns. So that it converts the mechanical rotations into direct electric current. Dynamos were the first large scale generators available for transferring power for industrial complex power supply. Regardless of size, dynamo generators obey ohms' law(V=IR), where V is voltage, I is current and R is resistance respectively. For this reason, lap-wound armatures used in dc generators require several pairs of poles and brushes. Wave Winding is used in dc generators employed in high-voltage applications. So the proposed model consists of lap winding dc dynamo generator. The e.m.f. output equation of dynamo generator is-

$$E = \alpha PNZ / (60\ A) \tag{2}$$

Let, **α** = flux/pole in Wb (weber)
Z = Total number of conductors.
P = Total number of generator poles.
A = Total number of armature poles in generator.
N = Rotational speed of armature in revolution per minutes(rpm).
E = E.M.F. induced in parallel path.

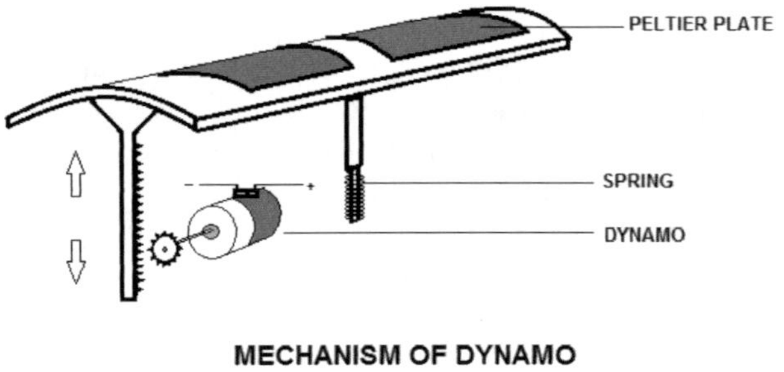

Fig. 2. Dynamo is physically coupled with teeth of the movable part in the device.

2.3. Photovoltaic panel

Photovoltaic effect, phenomenon used to produce electricity from the photos or light energy from the sun. There are some diodes (photosensitive) which are sensitive to the sunlight and make electrons move, when the beam of light falls on it. Several such diodes combined as a cell and several such cells put together called modules. Generally, the modules are made up of wafer based crystalline silicon or thin-film cells, and are connected in series (electrically) with one-another, so to obtain the desired output. The output of the panel is in the form of electric current and given below is a formula regarding-

$$E_g = S_a * Y_\% * R_{avg} * pr \tag{3}$$

E_g is the Energy generated (kWh)
S_a represents Total area of solar panel (in m^2)
$Y_\%$ represents Yield (%) of solar panel
R_{avg} is Average radiation on panels per annum (without shadings on panel)
pr represents Performance ratio, coefficient for losses (range between 0.5 and 0.9, default value = 0.75)
$Y_\%$ is the yield of solar panel given by the ratio of electrical power (in KW) to the area of one solar panel.

2.4. Linear Alternator

According to Faraday's law change in magnetic flux of a coil will cause e.m.f. to be induced in the coil, i.e. there should be a relative movement between the coil and the magnet in order to change the magnetic flux linkage of the coil. Proposed model deals with a moving magnet within a coil of conducting wire (as shown in Fig. 3), hence flux linkage between the magnet and the coil will change causing e.m.f. induction within the coil. The e.m.f. induced can be given according to Faraday's law as,

$$\text{E.m.f.} = -N(\Delta(BA))/\Delta t \tag{4}$$

Where E.m.f. denotes the voltage induced, N denotes the number of turns in the coil, B is the magnetic field of the coil, A is the Area and t is time.

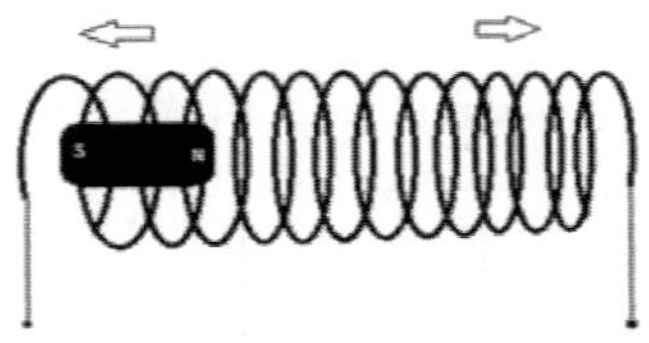

Fig. 3. Linear Alternator

As shown in Fig. 3 the magnet is kept in between the coil (kept inside a hollow non-ferrous and non-conducting cylindrical tube and the coil is wound around it [5]). As soon as the magnet is made to move within the coil, e.m.f. induces in the coil.

2.5. Peltier Plate

Peltier Plate (also called as thermoelectric generator (TEG), seebeck generator), is a solid state device that converts heat energy directly into electrical energy through a phenomenon called the seebeck effect [6]. Peltier Plates are used to convert electrical energy into a temperature difference between both sides of plates and vice-versa.

The relation between heat and current can be given through the formula below,

$$Q = P \times I \times t \tag{5}$$

Where **Q** denotes the amount of heat, **P** is the Peltier coefficient (depends on the material used), **I** is the current and t is time.

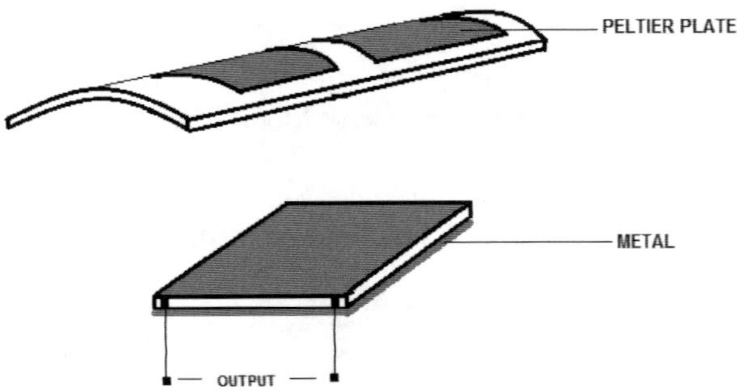

Fig. 4. Peltier Plates

Peltier Plates are used in the proposed model at the top of the device as shown in the Fig. 4.

2.6. IC 78S05

IC 78S05 belongs to family of linear voltage regulator integrated circuits i.e. 78XX, and is used as a voltage regulator to regulate the output voltage to 5V, 2A. In the proposed model the IC has been used at output end, so as to obtain the desired output of 5V, 2A. The circuit diagram of the model has been shown in Fig. 5,

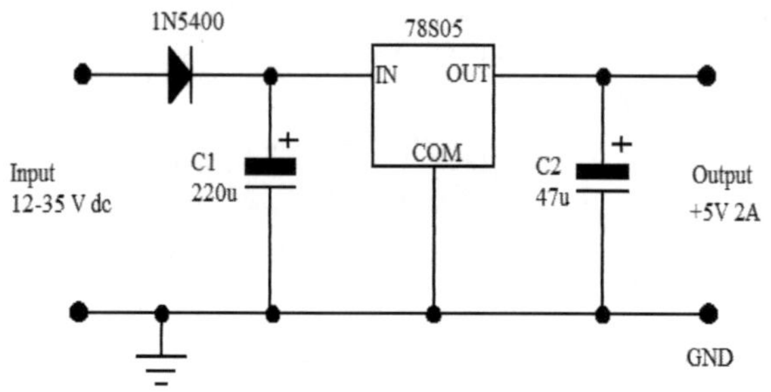

Fig. 5. Circuit Diagram of IC 78S05.

2.7. XL6009 Module

XL6009 is used as a boost converter in the model. Boost converter is a step up DC-DC power converter used to step up the voltage from its input side to its output side. In the proposed model the module has been used after all the power producing transducers to obtain desired output.

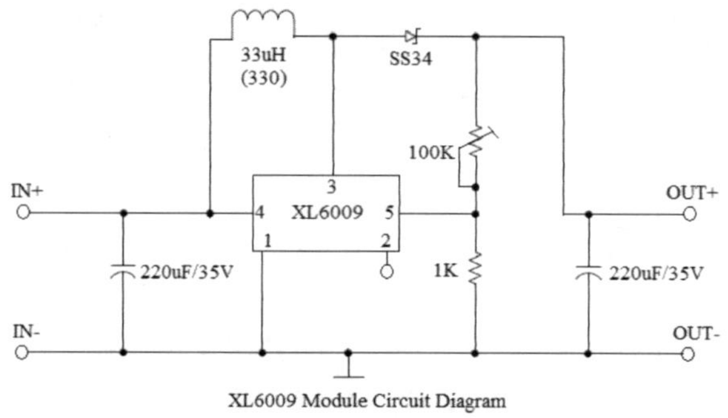

Fig. 6. Circuit Diagram of Module XL6009.

3. Methodology

This section consists of working of proposed model, explanation of flow chart and physical appearance of model.

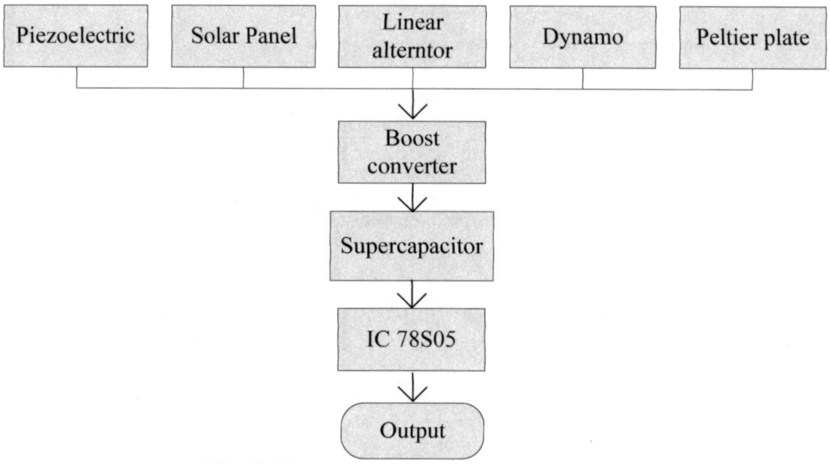

Fig. 7. Flow chart of proposed model.

3.1. Flow chart

Fig. 7 shows flow chart, where the proposed model consists of five power generating elements those are connected in parallel and combined output of these five elements provide input to the boost converter, boost converter step ups the input and the output energy get stored in the super capacitor. The stored energy is made to release through IC-78S05, which provides the desirable output for the electronic devices.

3.2. Circuit diagram and Outer model

Fig. 8 shows circuit diagram of each and every element present in the proposed model including the required diodes and bridge rectifiers with labeling.

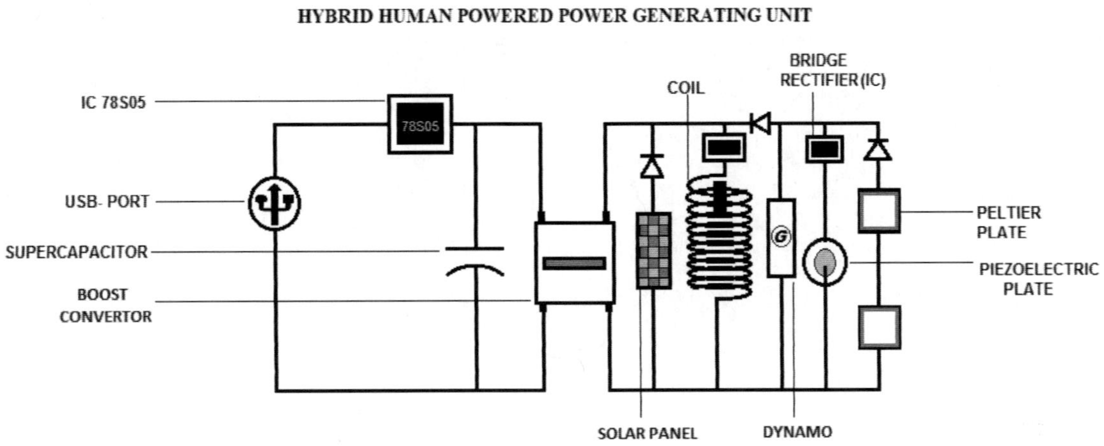

Fig. 8. Circuit Diagram

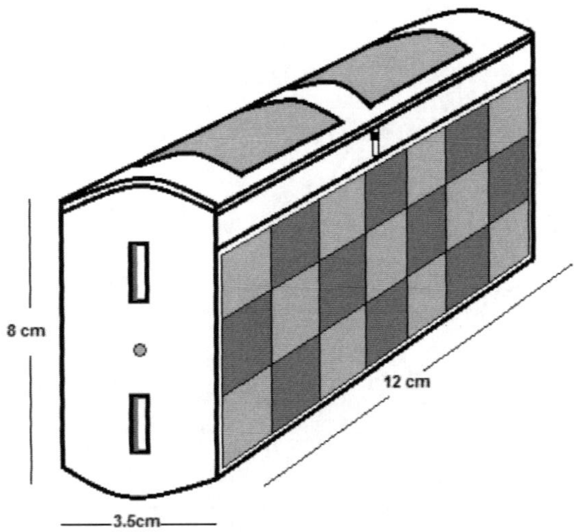

Fig. 9. outer appearance of proposed model.

4. Conclusion

Portable wireless electronic devices generally consist of a battery for required power, and thus counter the problems associated. The device proposed in this paper thus provides with respective solutions. Hybrid human powered power generating unit consists of five different transducers used to convert different forms (heat, solar, rectilinear motion and pressure) of physical energy into electrical energy. The output of individual transducers is obtained as per the energy equations given, and thus output power obtained by the device is sum of the outputs of individual transducers. And thus obtaining required output power for charging wireless electronic devices.

IOP Conf. Series: Materials Science and Engineering **594** (2019) 012033 doi:10.1088/1757-899X/594/1/012033

References

[1] Carrol A and Heiser G 2010 An analysis of power consumption in a smartphone (Berkeley, CA, USA) In proceedings of the 2010 USENIX annual technical conference, USENIX ATC'10, page 21,.

[2] Rahimi A.M and Emadi A July 2009 Active damping in dc-dc power electronic converters: A novel method to overcome the problems of constant power loads", IEEE Trans. Ind. Electron, volume 56, no. 5, pp.-1428-1439,.

[3] Birkholz M 1995 Crystal-field induced dipoles in heteropolar crystals–II physical significance. Archived from the original on 2016-10-30.

[4] Trolier-McKinstry S (2008) Chapter3: Crystal Chemistry of Piezoelectric Materials". In A. Safari; E.K. Akdogan. Piezoelectric and Acoustic Materials for Transducer Applications. New York: Springer. ISBN 978-0-387-76538-9.

[5] https://patents.google.com/patent/US5347186A/en

[6] Adroja, Nikunj Mr., Mehta B , Prof. Shruti Shah, Mr. Pratik; 1, March 2015 Review of thermo-electricity to improve energy quality, JETIR.

IOP Conf. Series: Materials Science and Engineering **594** (2019) 012034 doi:10.1088/1757-899X/594/1/012034

Automatic moving floor for smart rooms

B Ravi Subrahmanyam and A Ambikapathy

Department of Electrical and Electronics Engineering, GCET (Dr. A.P.J. AKTU)
Greater Noida, Uttar Pradesh-201306 India
E-mail: ravisubbu97@gmail.com

Abstract. This paper presents Automatic moving floor of a room for smart homes; invention is provided to solve the problems like moving of heavy furniture in homes, cleaning of floor everyday and to help the old people to roam around the room. The complete invention consists of two disconnected parts namely, CoS (Control System) box and the Tiles. Tiles arranged in the form of an array, having high torque motors internally and set of gears. There will be a false floor just beneath the main designed purposely for several instructions (like cleaning of tiles, moving of tiles, etc.) to work upon. The CoS box is a control system that listen, record, analyze the data given by the user and finds the best suitable solution form the array with given conditions. The CoS box contains a microphone and a touch panel to give instructions, then it provides a suitable way or path for the tiles to move from starting point to desired locations within the room. The instructions to the CoS box can also be given through an android application to have more user-friendly interface.

1. Introduction

Smart homes and home automation are now quite common words in technological domain. There are a number of ways in which inventers are doing at their best to make homes future ready. Even smart Airports with moving floors are available. Home Automation Systems have been designed and developed for mobile phones having Android platform to automate an 8 bit Bluetooth interfaced micro-controller which controls a number of home appliances like lights, fans, bulbs and many more using on/off relays [1]. Even a Home Automation system that employs the integration of multi-touch mobile devices, cloud networking, wireless communication, and power-line communication to provide the user with remote control of various lights and appliances within their home has been developed [2]. IoT (Internet of Things) technology has been used in homes in order to achieve smartness. In order with the above technologies GSM based home automation, Phone based home automation, ZigBee based home automation, Wireless control systems are also been proposed for making homes more user friendly [3].

Some inventions in home automation have also been made in order to help the old and disabled people. 'Voice Recognition Based Home Automation System for Paralyzed People' was an invention by Mukesh Kumar, Shimi S.L in which the design of the low cost voice recognition based home automation system has been made, for the physically challenged people suffering from quadriplegia or paraplegia (who cannot move their limbs but can speak and listen) to control the various home appliances just by the voice commands according to their need and comfort with the help of voice recognition module, Arduino UNO microcontroller, and relay circuit [4]. 'Automatic Fall Detection System using Sensing Floors', paper describes the work of detecting falls of old and elderly people in independent living apartments using accelerometer concealed under tiles [5]. Also a huge research has been done on floor cleaning of smart homes. 'Design and Development of Floor Cleaner Robot (Automatic and Manual)' in this paper written by Manreet Kaur, Preeti Abrol; they used a microcontroller AT89S52 for all hardware and software operations, including a RF module, IR sensor, LCD display and a 12V battery and developed a robot to clean the floors [6].

IOP Conf. Series: Materials Science and Engineering **594** (2019) 012034 doi:10.1088/1757-899X/594/1/012034

This paper is suggesting an idea in the domain of home automation, smart airport and smart hospitals. Here the concept of automatic moving floor has been discussed with the help of two sets of tiles (with false and actual tiles) and CoS box specially designed. The every individual tiles of whole floor of the room consists of two tiles namely false tile and actual tile. These combination of tiles carries pressure sensor, a set of gears and motors along with transmitter and receiver pair for effective communication with the CoS box. The CoS box is designed as an intermediate between the tiles and the user, it is embedded with voice recognition system and a touch panel for the instructions from the user.

2. Description

The invention concentrates on the movement of floor in a room for smart homes and next generation houses. The floor is divided into an array of 'tiles' which are controlled by an automatic system called 'CoS box'. There are two floors, the main floor and the false floor one below the other. There will be gap of one floor height between both the main floor and the false floor. Each floor consists of same number of tiles arranged with the precision.

Fig. 1 shows, the top view of the floor having the tiles arranged as shown. The black spots around each corner of the tile are the poles that are attached to the ground to have vertical movements of the tile.

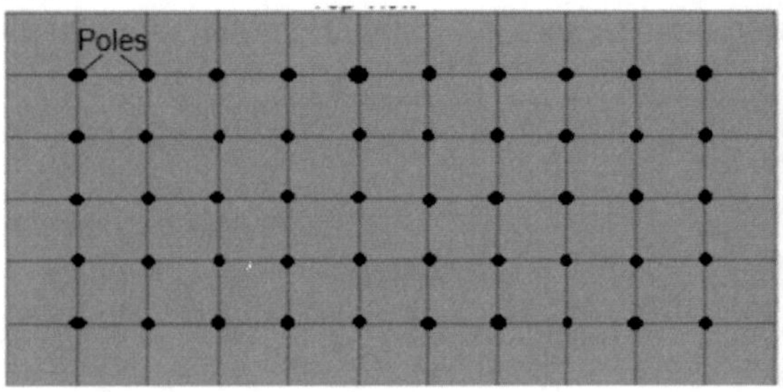

Fig. 1. Top View of Tiles in the Room

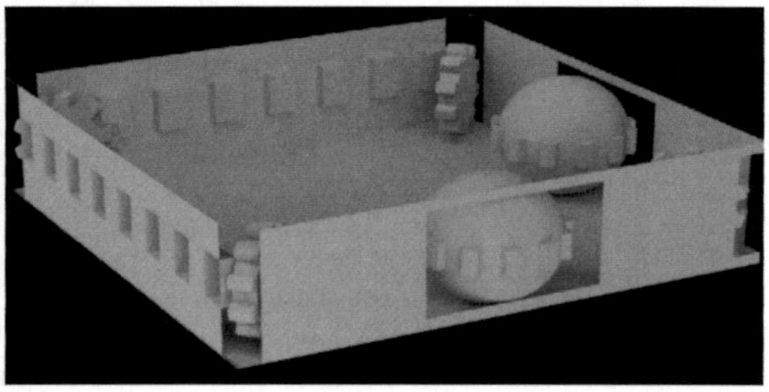

Fig. 2. Layout of Tile

2.1. Tiles

Tiles are designed with materials having strong strength that can withstand a pressure somewhat more than that of human and the furniture in the room. Also, each tile consists of pressure sensor, so to detect the presence of human or any object. Each tile consists of gears two big and four small gears (as shown in Fig. 2), that will help in the movement of tiles in any direction.

The two big gears that are held at the center of two adjacent faces are used for the movement along the same plane (i.e. Left, Right, Forward and Backward), whereas the four small gears are at the each of the four corners will be used for up and down movement (i.e. Along the z-axis), the other two adjacent sides of the tile are in the shape of teeth (extruded inwards).

All the gears are divided into two categories; one those are used to move in the same plane will be referred as Horizontal Gears hereafter; other those are used for up-down movement will be referred to as Vertical Gears hereafter.

There are small poles on each corner of the tile, these poles are attached to the ground, and height of each pole will be about three and a half times of height of each tile. These poles act as a support to the tiles at the time of up-down movement. The poles and the tiles are designed in a way that there will not be any missing of the gears (as shown in Fig. 3).

The tiles contains a programmed chip which is used to have communication with the CoS box, and to perform the specified tasks.

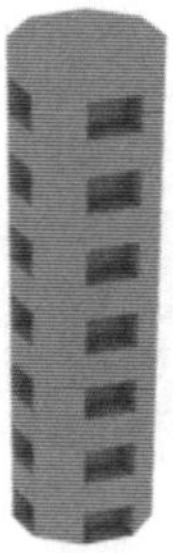

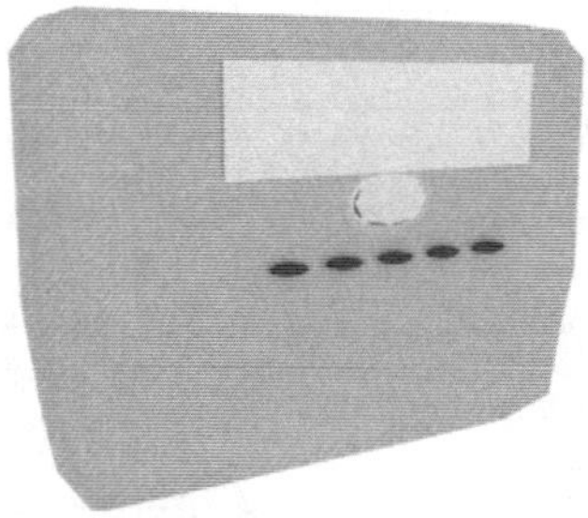

Fig. 3. Design and Shape of Pole **Fig. 4.** Pictorial representation of CoS box

2.2. CoS box

CoS box is a specially designed control system box that accepts, analyze and commands information as and when needed.

CoS box consists of a microphone, and a touch panel embedded for input from the user. Then it will process the information and come to a conclusion on the basis of input, and the respective instructions are then transmitted from the CoS box to each of the respected tile which is to be moved. The CoS box is connected to each and every tile of the room and have complete access to move those tiles as per the requirement.

The CoS box is also connected to an android application to have ease on assess with the commands. The data will be transmitted to the respective tiles, and the tiles move with proper and calculated time to avoid confusion within the tiles.

Fig. 4 shows a pictorial representation of CoS box, having a touch screen, along with micro-phone for input signals. Some buttons are also provided at the bottom for manual operation as and when needed.

2.3. Movement of tiles

Each tile consists of two big gears at the center of two adjacent sides i.e. horizontal gears; these gears are allowed to rotate in both the directions on a fixed axis with the help of high torque and low rpm motor, this will allow the tile to move on all the four directions on the same plane

The four small gears on the corners of the tile i.e. the vertical gears are held to move the tile up and down.

When the tile has to move in the same plane except the two main responsible horizontal gears; the vertical gears shift a little inwards, so to have a free movement without any inconvenience. Similarly, when the tiles have to move in up-down direction except the vertical corner gears; the remaining two horizontal gears shift a little inwards.

When gears of one category move the other set has been shifted a step back (inwards). When this happens, the gears are programmed to come to a reset position so next time overlapping will not cause a problem.

When the tile has to move in the forward direction (say north) then the gear on the on the east face will rotate in clockwise direction, and the gear on the east face of adjacent tile (left to the concerned tile) will rotate in anticlockwise direction, now with the effect of Newton's third law of motion the concerned tile will move forward (as shown in Fig. 5).

Fig. 6 shows an instant of up-down movement of the tile. Here all the vertical gears will move in the direction so as to achieve the desired task efficiently.

Fig. 5. Forward and backward movement, movement within same plane

Fig. 6. Up-Down Movement of tile

Let us consider a room with the arrangement of furniture as shown in Fig. 7. When a command from the user inputs as "*Take me to the door*" and the user stands on the tile with a star mark. As the CoS box takes the input and possess it with a conclusion that all the tiles with a dot mark have to move in the up and down direction only whereas tile with star mark and tiles beneath the floor will have to move in the same plane. So the CoS box will make the gear arrangement of all the necessary tiles as per their movement, hence to have an uninterrupted and smooth movement from bed to the door.

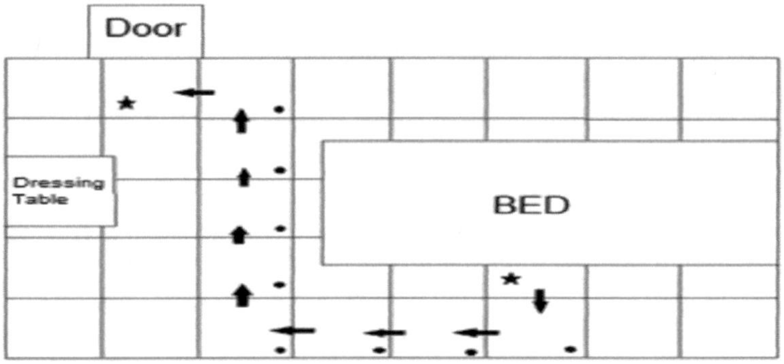

Fig. 7. Optimal path between the two positions, as found by the CoS box.

For an example as shown in Fig. 8; let a command says a tile 1 to move from position 1 to position 3 as shown in adjacent Fig.. Now tile1 will move forward to take the position of tile 2, tile 2 starts moving down to the place of tile 5, tile 5 shifts left to position 6 i.e. at the place of tile 6 and finally tile 6 move upwards to position 1, now tile 1 is at position 2. In the similar manner tile 1 move's one more block to left and all the adjacent tiles change their positions for tile 1 to reach its destination. The tiles 2 and 6 in first movement (i.e. when tile one move from position 1 to position 2) will move first and take their position in the gap between the two floors, so to have a smooth movement for tiles 1 and 5, then tiles 1 and 5 move in their respective directions to complete the path. The movement and timing of tiles will been calculated by the CoS box beforehand so to have no time lag or any confusion among the tiles and a continuous flow is achieved. The speed for the movement of the tiles is set calculated so that the person using should not experience any effect of moment of inertia and/or momentum.

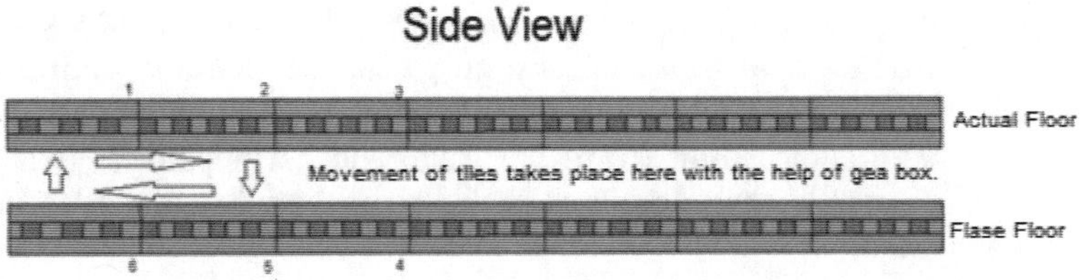

Fig. 8. Side view of the floor (movement of the tiles)

2.4. Charging of tiles Circuitry

Charging of tiles will take place through the concept of real time charging. All the tiles on both the floors will be connected to the supply. A battery is used to store the energy and will be utilized during the movement of tiles. And in steady state the energy used for the functions will be directly consumed from the supply.

The design of the tile will be such that when the two tiles will be in complete contact the circuit for charging of tiles will get complete and the charging of battery takes place. And when the tile is in moving state the contact breaks, the circuit will not be complete and the energy from the battery is used to feed the whole circuit.

2.5. Cleaning of tiles

The cleaning of tiles can be divided into two categories; temporary cleaning and complete cleaning.

In temporary cleaning the two of four vertical gears will move a little downwards so to form a little slope of the tile, and due to the effect of gravity the heavy dust particles will fall down from the tile, resulting to a temporary clean tile.

In complete cleaning of tile the tile will move to the location assigned for cleaning. In the false floor a corner, which is not as useful, will be made as cleaning centre for tiles

2.6. Replacement of tiles

When there is any problem in any of the tile it will be diagnosed by the CoS box and will be informed to the user about the problem. If after troubleshooting still the problem persists and requires a replacement, then on the command of user all the four standing gears set will move in a direction such that the tile will come out from the array of main floor. In the similar manner a new tile will be kept back in its position.

3. Overview of the entire operation(Flowchart)

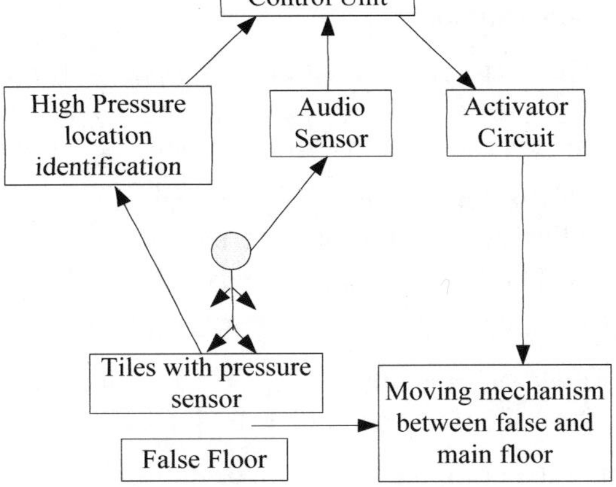

Fig. 9. Flow Chart

Fig. 9 shows a flow chart representing the entire process briefly. Here the block control unit refers to the CoS box of the system, the input to which is high pressure location identification data (as it has been connected to all the tiles), along with the voice command through the audio sensor (embedded within the CoS box to have audio input from the user), and as the output it activates the moving mechanism between false and the main floor (i.e. the gears between the tiles depending upon the instructions, will get activated).

4. Result and Conclusion

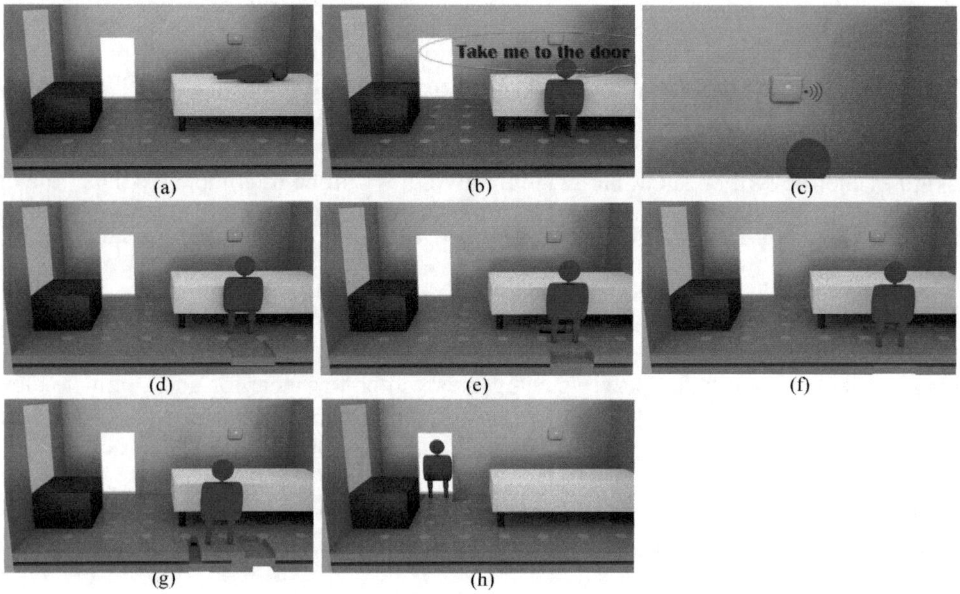

Fig. 10. Desired Outcome shown pictorially

Fig. 10 shows a pictorial representation of the desired output. The various steps in the flow are shown through pictures represented in increasing number format from (a) to (h). Where,

a) Represents that there is a man in the room who is sleeping at any instant.

b) Represents that after some time interval the man wants to go to the door, and he gives a voice input saying "Take me to the door"

c) CoS box listens to the instruction through the microphone embedded in it. And analyze it to come to the best optimum solution.

d) At first the tile placed in the forward direction of the path will move down. As well as the tile just beneath the target moving tile will move upwards. There is a space of one tile height between the two floors. So both the tiles will travel half way down at first.

e) As soon as both the tiles reach their half way, the remaining two tiles (i.e. the target tile on which the man is standing and the tile beneath the forward placed tile moving downwards.) will move in the horizontal direction as per the instructions given by the CoS box.

f) In this ways all the tiles move as of completing a vertical quadrilateral path.

g) Each tile is embedded with a transmitter-receiver pair to have a communication with the CoS box. In the similar manner the man reaches his destination.

h) Finally the man reaches the desired destination location.

IOP Conf. Series: Materials Science and Engineering **594** (2019) 012034 doi:10.1088/1757-899X/594/1/012034

Fig. 11 shows the schematic representation of the selection of the optimum path by the CoS box as soon as it receives the instruction. The various steps in the process flow are represented by the increasing number format from (a)-(e). Where,

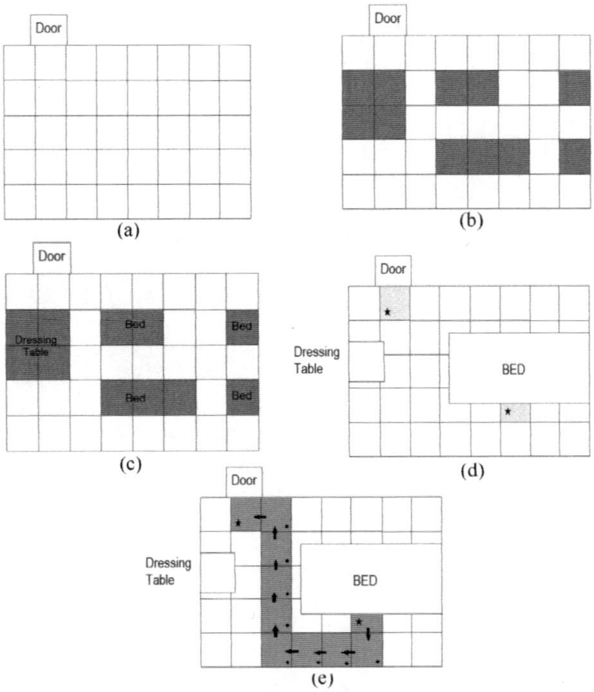

Fig. 11 Schematic diagram representing Selection of optimum path by the CoS box.

a) Represents the basic layout of the room an array of tiles.

b) Pressure sensors are embedded in each tile, so as to know if there is any object on any particular tile. Hence all the tiles encountering certain pressure are obtained.

c) As the position of all the furniture (i.e. bed and dressing table in this particular case) in the room is already defined in the CoS box at the time of installation.

d) So it will be easy for it to find out the starting and ending tiles for the movement.

e) CoS box finds the optimum path for the tiles to move.

References

[1] Shirisha Tadoju and Mahesh J; August-2015 Int. J. of Adv. Technol. and Innovative Research, ISSN 2348–2370 Vol.07,Issue.10; 'Bluetooth Remote Home Automation System using Android Application'; Pages:1815-18

[2] Dipti P Wale, Prof Patil S S and Dr. Anekar S V; May-2015 Int. J. of Innovative Technol., ISSN 2321-8665 Vol.03,Issue.01; 'Home Automation using Cloud Network and Mobile Devices'; Pages:0054-58

[3] Satish Palaniappan, Naveen Hariharan, Naren T Kesh, Vidhyalakshimi S and Angel Deborah S; April 2015 Int. J. of Comp. App. (0975 – 8887) Vol.116 – No. 11 'Home Automation Systems - A Study'; Pages:11-18

[4] Mukesh Kumar and Shimi S L; October 2015 Int. J. of Adv. Research in Electronics and Communication Engg. ISSN: 2278 – 909X, Vol.4, Issue 10, 'Voice Recognition Based Home

IOP Conf. Series: Materials Science and Engineering **594** (2019) 012034 doi:10.1088/1757-899X/594/1/012034

Automation System for Paralyzed People'; Pages:2508-15
[5] Mohamad Daher, Ahmad Diab, Maan El Badaoui El Najjar, François Charpillet and Mohamad Khalil; December 2016 Int. J. of Computing & Information Sciences Vol. 12, No. 1 'Automatic Fall Detection System using Sensing Floors' Pages 75 – 82 DOI: http://dx.doi.org/10.21700/ijcis.2016.110
[6] Manreet Kaur and Preeti Abrol; July 2014 Int. J. of Comp. App. (0975 – 8887) Volume 97–No.19 'Design and Development of Floor Cleaner Robot (Automatic and Manual)'; pages:32-38

IOP Conf. Series: Materials Science and Engineering **594** (2019) 012035 doi:10.1088/1757-899X/594/1/012035

A retrofit design of safety and stability mechanism for two wheelers

Nitin C M and Santosh Patil

Department of Mechanical Engineering, Manipal University Jaipur, 303007, Jaipur, India

nitinchandrashekarmuthe@muj.mainpal.edu, santosh.patil@jaipur.manipal.edu

Abstract. This paper presents an idea of a retrofit design of a mechanism which helps in achieving the safety and stability of the two wheeler vehicle. The aim of the mechanism is to avoid the two wheeler from fall/skid and also work as an attachment for physically challenged drivers. The mechanism is a hydraulic system clubbed with linkages, where an extra two small sized wheels are grounded to provide safety from fall and give stability to the two wheeler. The proposed retrofit mechanism would replace the double stand of the vehicle. These extra wheels move along with the vehicle, hence providing the required stability. This forms a replacement of the costly attachments which the physically challenged drivers add to their vehicles. The shock absorbers would be included accordingly in the mechanism. The probable improvements on deploying the said mechanism on a two wheeler will provide easy riding, as the vehicle self-stabilize itself when the small wheels are grounded. This helps the less abled riders to easily ride the vehicle with no worries of grounding their feet. A number of designs were developed, studied and three designs were shortlisted based on the feasibility of the mechanism. Finally, a better design is proposed for further improvement and implementation.

1. Introduction

In present day scenario, transportation has become integral part of human civilization. Road transport has always been the dominant mode of transport in India, both in terms of traffic share and in terms of contribution to the national economy. Hence, to meet the demands of road transportation a big network of road connectivity has been led and is still growing. The harmful part of the growing road network along with motorization and urbanization is the increase in road accidents and crash fatalities. Today, road traffic injuries are one of the leading causes of death, disabilities and hospitalization in the country imposing huge socio-economic costs. It has been observed from the survey that out of all road accidents, two wheelers take the major share with around 35% [1]. Many accidents are due to the negligence of the two wheeler drivers and others due to the skid of the vehicle or fall due to imbalance. Figure 1 shows the vehicle topple due to skid, fall due to imbalance, etc. [2]

Figure 1(a). Vehicle fall due to skid/due to imbalance [2]

Figure 1(b). Vehicle fall due to slight collision [2]

In another angle, for a physically disabled or a debilitated person, transportation is a major hindrance and so the mobility of physically disabled people is among the great concern of the human civilization. It is really very hard to realize the problems of a physically disabled person who has a limited mobility. In India, the contemporarily modified scooters used by the physically disabled people for transportation are not very ergonomic and have oversized attachments. The modification done to the scooter is heavy and also expensive. These attachments also reduce the efficiency and performance of the vehicle. Figure 2 shows a few conventionally modified scooters for disabled people.

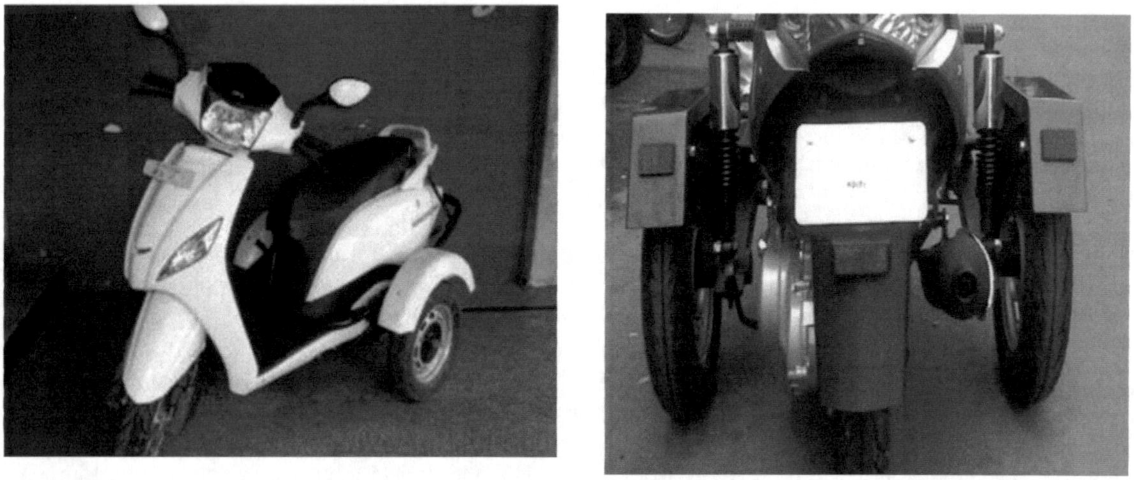

Figure 2. Conventionally modified scooters generally used in India [3]

It can be observed that the modified attachments are heavy and huge and also occupy more space. This becomes difficult for others while driving besides these vehicles. Hence to avoid this difficulty and also to prevent the two wheeler fall due to imbalance and skid, a retrofit solution needs to be developed. In this paper a feasible retrofit mechanism for scooters is proposed. This mechanism will behave like a supporting stability device which would not allow the vehicle to topple down. The main objective of a self-balanced two wheeler vehicle is to ensure safety of the rider and vehicle. This can be achieved if a retrofit stability mechanism is developed which can balance the vehicle.

2. Literature Review

Every year Government of India does survey on number of accidents (Road Transport and Highways report on Road Accidents in India), the survey of year 2015 and 2016 showed that around 29% and 36%, respectively were the accidents caused by two wheelers [4]. So, the safety and stability mechanisms and/or other safety measures need to be implemented in order to reduce the number of accidents and quantum of injuries.

Significant literature is available in which various designs for a modified scooter are presented. Few works concentrated on the two wheeler seating design changes. Neville [5] in his thesis studied the basic principles of two wheeler seating & positioning of kids and youth with physical disability. Based on his examination of various literatures the author complied the better designs and proposed best principles for seating of disabled kids and teenagers. An overview of the normal and abnormal seating postures was also examined. Chandore and Deshmukh [6] presented an extensive review on cushion foam and seating structures. In general, about the seating designs for a two wheelers. The work examined included, the finite element simulation of the vibration of cushion foam and the supportive structures of the seat, along with the human buttocks' soft tissues in the seated position. They examined and analysed the various studies of other researchers that might facilitate redesigning or modification, assessment and examination of the seats of two wheeler riders for greater comfort while riding.

The other major design study on two wheelers which concentrated on safety was carried out by Kumar et al. [7]. The author and team designed a three-wheeled vehicle for easy riding, for physically challenged people. The work introduced two design concepts named as Sholay and Chariot. Further, Sholay design was propagated as it proved to be a better design. Once the detailed design was performed in CATIA, the ergonomic issues of the model were carried out using Jack software. Finally, a working prototype model was developed.

Rumi et al. [8] proposed a design of a self-balancing two wheeler vehicle where they have used the gyroscopic principle. An android application was deployed, which measures the tilt angle and takes it as an input. This input (tilt angle) taken from the gyroscope is further sent to the receiver. Further, the Arduino controls the motor based on the received data, which further control the tilt direction of the two rotating disks of the self-balancing mechanism. This vehicle is designed to provide the safety that two wheeler vehicle does not have during an impact or at zero velocity.

There was a US patent entitled "Motorcycle safety device" [9]. In which a safety frame was designed where the rider could be saved from major injuries caused by the impact. The safety device or the exterior frame envelops the rider during impact and reduces the quantum of injuries.

To the best of the author's knowledge there has been no reporting on retrofit safety solution for two wheeler skid or fall due to imbalance. But a lift mechanism was observed in a "Scalevo" an electric wheel chair design [10]. In this design, two small sized wheels are drawn downwards so that the wheel chair is lifted up. This mechanism also helps in mounting the ram of the wheel chair, so that it can climb the stairs. In this paper a similar concept of extra two wheels are used for the stability of the two wheeler vehicle. The conceptual design of the stability mechanism has been generated, which can be attached such that two extra wheels (small sized) are grounded and the balance of the vehicle is achieved.

3. Modelling and Design of the Stability mechanism

In this section, we mainly discuss the designs of the stability mechanism which are proposed and would be fitted under the scooter for balancing of the vehicle. Various designs are generated and then their comparison is done on the basis of their attributes in order to propose the best design concept.

3.1. Design Type-1

In this design concept, a hydraulic system is clubbed with a linkage mechanism. As the cylinder extracts the linkage is moved downwards where the wheels attached to the links touch the ground. These wheels touching the ground provide the required stability to the vehicle. Figure 3 shows the conceptual design of the stability mechanism. This design concept focuses on basic functionality rather than the ergonomics and external factors induced on it. It features a spherical joint on each side above the wheels to provide better maneuverability in case of vehicle imbalance. The wheels are designed in such a way that a straight path be maintained during skidding and the speed can be reduced due to friction between wheels and road. Figure 4 shows the detailed drawings of the Design type-1. The major advantages offered by this design is easy to produce the components and simple assembly with economic installation.

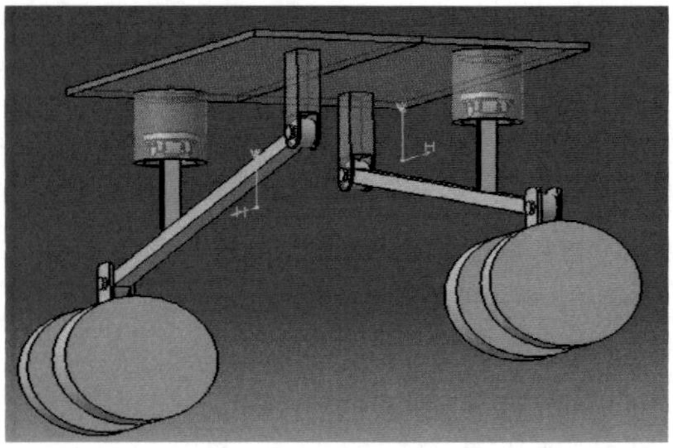

Figure 3.CAD Model of Proposed Design Type-1

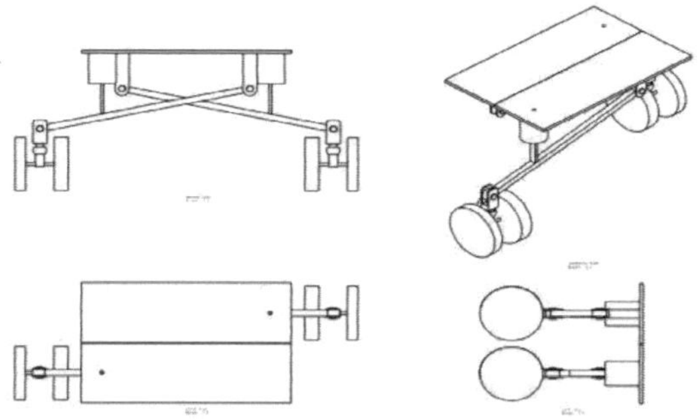

Figure 4. Drawings of Proposed Design Type-1

However, the disadvantages of this design include more technical aspects such as the torque would be generated on the wheels, as they engage with the road. Synchronization of both the pistons when extracting the wheels will be critical. If synchronization is not proper than imbalance of vehicle will be caused. Wear associated with the wheels' outer surfaces as the wheels engage during vehicle motion. Also, vibrational jerk would be induced when the mechanism is engaged while traveling at high speeds.

3.2. Design Type-2

This design was conceived with the aim of replacing the vertical actuation of the piston with horizontal actuation to increase the vertical usable area of the apparatus. This was achieved using a 3bar linkage to convert horizontal motion into rotational motion. The output link of the 3-bar linkage was connected to a collar which would slide over the rotary arm resulting in extraction or retraction of the arm, two hydraulic piston and cylinder arrangements were mounted on the collar to pivot the wheels about a joint on the arm (as shown in Figure 5).

Figure 5. CAD Model of Proposed Design Type-2

This design aims to save space while obtaining the same functionality offered by the previous design. It also addresses the flaws faced by the previous design. The pivoting half of the rotary arm would provide better control on uneven roads and absorb shocks caused due to impact. The use of a single cylinder also eliminates the synchronization issues faced by the previous design. Figure 6 shows the detailed drawing of Design type-2.

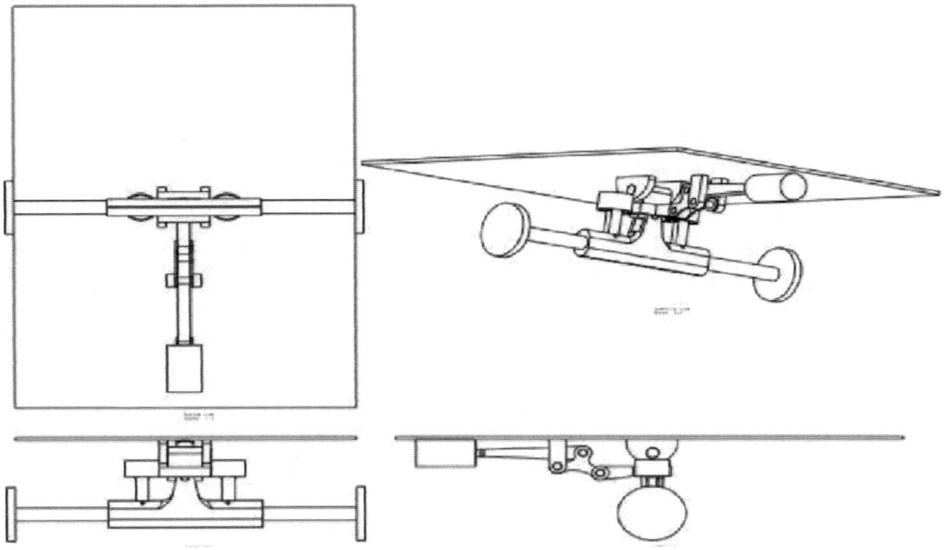

Figure 6. Drawings of Proposed Design type-2

Major advantages of this design are the increased features, better control and increased safety. The disadvantages of this design are higher cost, manufacturing of complex parts, difficulty in designing the 3-bar linkage and its installation.

3.3 Design type-3

In this design horizontal actuation was switched with vertical actuation yet again to simplify the design, reduce of cost and decrease the number of moving parts. This design can achieve a balance between the simplicity of the first design and functionality of the second design. This design only involves 4 moving elements i.e. arm, connecting rod, piston and wheels (as shown in Figure 7).

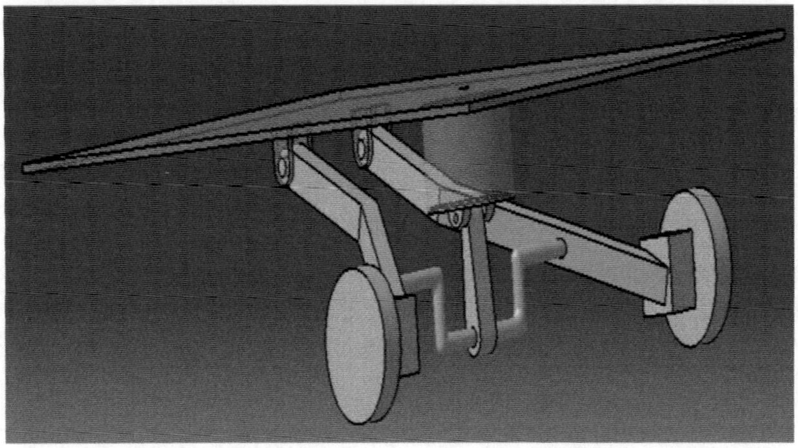

Figure 7. CAD Model of Proposed Design type-3

This design features complete retraction of the wheels up to the platform. It also replaces the double stand of the vehicle, as the wheels in extracted position would behave as the stand for the vehicle. This design achieves complete retraction by virtue of a U-shaped bar connected to both the arms and also connected to the piston via a connecting rod into which a shock absorber could be integrated. Figure 8 shows the detailed drawing of the Design type-3.

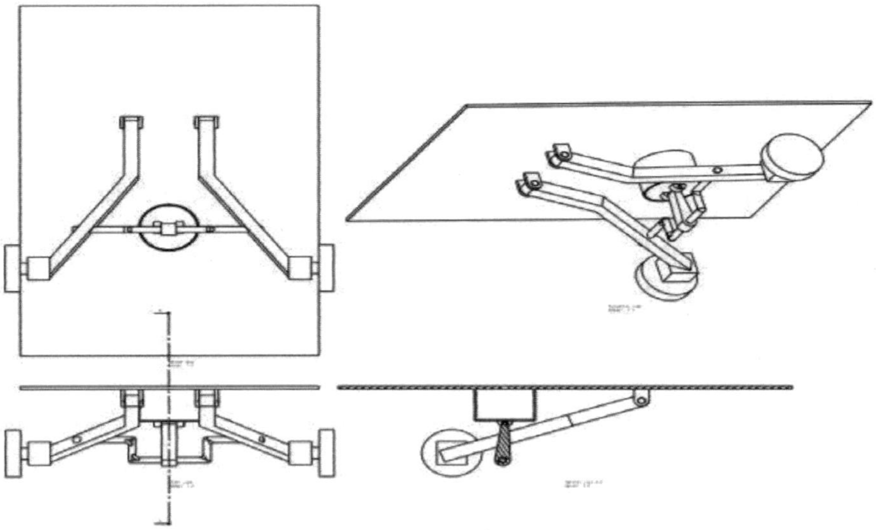

Figure 8. Drawings of Proposed Design type-3

Advantages of this design include easy installation, simple and low number of parts, lower costs, easily reproducible and no requirement of synchronization of the hydraulic system. This system currently has limited disadvantages related to its practicality and life.

Based on the comparison of all the three designs, Design type-3 proves to be the better and simpler to install. This system has been the better concept, as it provides the required functionality with less problems. Whereas, the practical applicability needs to be tested. The mechanism will be fitted below the vehicle thus the clearance between the vehicle and the road will be reduced. The efficiency of the vehicle would reduce a little, as extra weight is being added to the vehicle. Also, if the road conditions are not good and/or bumps are present, then the mechanism may face slight problems. However, these problems can be solved with proper shock absorbers and slow riding in typical conditions. Practical testing will be carried out further on the mechanism (Design type-3).

4. Conclusion

The proposed mechanism is basically on the concept to avoid the fall of a two wheeler vehicle and also form a replacement for the physically challenged drivers. The mechanism will help the two wheeler rider to ride the scooter without worrying about the stability of the vehicle. In was observe that Design type-3 provided better functionality and is easier to install compared to other proposed designs. However practical testing needs to be performed for further implementation of the mechanism. The idea of the retrofit stability mechanism is a new viewpoint and will provide greater advantage in reducing the quantum of injuries, fatalities in the present two wheelers.

References

[1] "Deadly afternoons: 35% of road accidents in India reported between 3pm and 9pm"[Online]. Available:*https://www.hindustantimes.com/india-news/deadly-afternoons-35-of-road-accidents-in-india-reported-between-3pm-and-9pm/story-FlxcGYsmGc62GLHjHtCuTL.html*

[2] "Motorcycle Accident: Statistics, Reasons, Common & Severe Injuries, Protective Measure, Role ofAttorney"[Online].Available:*https://www.epainassist.com/personal-injury/autoaccident/motorcycle-accident-injuries-statistics-role-ofattorney.*

[3] "Modification of Vehicles for Physically Handicapped." [Online]. Available: *http://bajeesons.blogspot.com/.*

[4] T. Dawda, "Two-wheelers and commercial vehicles see highest road accident fatalities," *AUTOCAR, Professional*, pp. 1–3, 2016.

[5] L. Neville, "The Fundamental Principles of Seating and Positioning in Children and Young People with Physical Disabilities," B.Sc project thesis, University of Ulster, 2005.

[6] A. S. Chandore and T. R. Deshmukh, "Design of Two Wheeler Seat: A Review," *Int. J. Pure Appl. Res. Eng. Technol.*, **vol. 2, no. 9**, pp. 450–458, 2014.

[7] S. B. Kumar, K. V. Banthia, and K. A. Ray, "Design of three wheeler vehicle for physically challenged people," *Sas Tech J.*, **vol. 12 (1)**, pp. 80–89, 2013.

[8] S. Mohibul, I. Rumi, M. F. Hossain, I. S. M. S. Islam, and K. Rahman, "System Design of Two Wheeler Self-balanced Vehicle," in *10th France-Japan/8th EuropeAsia Congress on Mecatronics (Mecatronics-2014 - Tokyo)*, 2014, pp. 331–336.

[9] J. J. Bouvier, "United States Patent 4227717- MOTORCYCLE SAFETY DEVICE," 1980.

[10] S. Smith, "Scalevo, An Electric Wheelchair, Can Climb Most Staircases And Hopes To Negate The Need For Ramps And In-Home Lifts," *Medical Daily*, 2015.

Acknowledgements

The authors would like to thank Mr. Dharmesh Yadav, faculty Automobile Dept., MUJ, for his useful contribution in preparing this paper and Dr. Ravi Kumar Sharma, faculty Mechanical Dept., MUJ, for proof reading.

IOP Conf. Series: Materials Science and Engineering **594** (2019) 012036 doi:10.1088/1757-899X/594/1/012036

Design of Read Circuitry for Nonlinear Smart Sensors

S Ananthi, Himanshu Chaudhary and Kulwant Singh

Electronics and Communication Engineering, Manipal University Jaipur, India

E-mail: ananthi232@gmail.com

Abstract. The purpose of this paper is comparative study, design and simulation of two stage, analog to digital convertor to convert the properly conditioned stepwise linearized analog output signal from a ratio metric NTC thermistor circuit. The design part consists of linearization of thermistor temperature- resistance characteristics in analog domain. A CMOS sample and hold circuit is used to bridge linearizing circuit in analog domain and the flash ADC circuit in digital domain. Ratio of the voltage is sensed rather than absolute voltage across the thermistor, as part of linearization. Further linearization is done in two stage flash ADC consisting of a reference voltage divider resistance, comparators, analog multiplexers and 8bit differential ADC. A graph is plotted between temperature and the corresponding resistance of NTC thermistor NTCLE100E10 [1], having a resistance of 10kΩ at 25°C. The graph is an exponential curve and it is linearized [2] around 50°C by drawing a curve between temperature and voltage ratio across a resistor connected in series to the thermistor. This curve is still nonlinear at extreme values of the considered range -20 °C to 100 °C. The full range is divided into four sub ranges and piecewise linearization is done simultaneously with analog to digital conversion. Analog and digital integrated circuits are generally low powered and the voltage limit is 5V. Thermistor is compatible in this range. A number of signal conditioning circuits are devised in the past. Here the linearizing steps involved in simple circuit elements and simple design. Signal conditioning of thermistor response has a number of design circuits already devised, still there is room for research to make it is small in size, low powered, high sensitivity and cost effective.

Index-Terms: Mathematical modelling, Nonlinear transfer function, piecewise linear analog signal, flash ADC

1. Introduction

Any sensor that senses a physical parameter produces an output electrical parameter. To convert this response of the sensor to a measurable level it must be amplified. To measure temperature four different types of sensors are commonly used [3]: infrared, thermocouple, RTD and thermistor. Kufre Esenowo et al [4] designed a thermal detector that can convert incident power into an electrical output and signal conditioning circuit &data acquisition elements for processing the chopped IR sensor signal. Rotating shutter is configured to supply the chopped signal. A multi sensor embedded system is developed [5] along with reconfigurable automation and communicating with host. It attempts to develop a generic platform to support hardware interface, sensors, communication and actuators. Results of developed prototype system with RTD temperature sensor, over a wireless connectivity for various distances are presented.

A smart CMOS sensor with a combined read out for temperature and humidity measurement [6] with low power consumption, high sensitivity, no post processing and small in size is designed. The design has two modes of operation and the sensor signal is converted into a digital code representing a frequency ratio between to oscillators. Lm35 is a calibrated readily available semiconductor sensor. [7]Shantanu et al. designed a signal conditioning circuit using Lm35. The designed system is divided into two parts, a personal unit and the base unit. A detailed design of biomedical sensors with the necessary signal conditioning circuit is presented.

A temperature sensor CMOS based, for the range -5ºC to 120ºC is designed and analyzed in [8]. It proposes a design with temperature range selection circuit, so that the thermistor linearizing circuit automatically switches to a corresponding linearity calibration loop and the method is implemented in TSMC 0.5μm CMOS high voltage mixed signal based polycide. A total of four states are described circuit. It is designed to extend sensing range through a temperature range auto selecting circuit. Output results demonstrate a high linearity with linearity error less than 1.4% is achieved. It resolves the problem that the temperature range of a single thermistor temperature sensor is too small.

An ANN based linearization of an NTC thermistor connected to an inverting operational amplifier is presented [9]. The nonlinearity of signal conditioning circuit due to temperature voltage characteristic of the thermistor is reduced with the help of ANN technique. A low cost linearizing circuit is developed [10] using inverting operational amplifier. The gain of the arrangement can be adjusted over a wide range by simply varying the feedback resistance. For a narrow span the linearity is ±0.5%.

This paper discusses with mathematical modeling of nonlinear response and its application to thermistor sensors. Analog signal is then converted into digital signal in two stages and finally by successive approximation register, an differential flash ADC converter having resolution of 8bit.The read circuitry for linearization of the signal from a thermistor is designed and simulated using Multisim12.0.

2. Mathematical modeling of thermistor

Thermistor is a Semiconductor material. It measures temperature through a corresponding change in its resistance. It has the highest sensitivity among all other temperature sensing elements, for a small change in temperature a large change in its resistance. In NTC type thermistor the resistance decreases as the temperature increases. But its response, change in its resistance with respect to change in temperature is highly nonlinear response. This nonlinearity of the sensing element causes low accuracy during measurement. Linearity is a desired property, because the sensitivity remains constant throughout the range since calibration can be done with fewer points. The characteristics of thermistor, temperature versus resistance can be modeled mathematically. The NTC thermistor NTCLE100E3103 is taken as example.

Referring to table 1, the thermistor has 10KΩ resistance at temperature 25°C. The temperature-resistance characteristics drawn in Fig.1, can be [11] expressed as

$$R_T = R_0 * \exp\left(\frac{\beta}{T} - \frac{\beta}{T_0}\right) \quad (1)$$

Where,
R_T = thermistor resistance at temperature T (K)
R_0 = thermistor resistance at temperature T_0(K)
T_0 = the reference temperature taken as (25+273) K
β = the material property, assumed to be constant for a given temperature range.

Taking natural logarithm on both sides and performing some mathematical rearrangement of the parameters, the equation (1) becomes,

$$T* \frac{T_0}{(T_0-T)} *[\ln \left(\frac{R_T}{R_0}\right)] = \beta \quad (2)$$

Table 1. Temperature and the corresponding resistance of NTC thermistor NTCLE100E3103

Temperature (°C)	Resistance of Thermistor $R_T(\Omega)$	$V_{out} = 5/(1+R_T/R_S)$ (Volts)
-20	96358	0.19
-15	72500	0.25
-10	55046	0.33
-5	42157	0.42
0	32554	0.53
5	25339	0.66
10	19872	0.81
15	15698	0.99
20	12488	1.18
25	10000	1.4
30	8059	1.62
35	6535	1.856
40	5330	2.1
45	4372	2.34
50	3605	2.58
55	2989	2.82
60	2490	3.04
65	2084	3.25
70	1753	3.44
75	1481	3.61
80	1256	3.77
85	1070	3.9
90	915.4	4.04
95	786	4.15
100	677.3	4.25

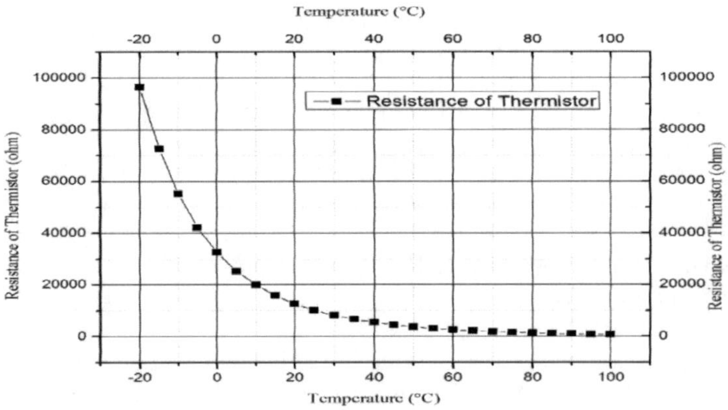

Fig. 1. Exponential relation between the parameters

Considering its mid temperature range resistance, R (at T=50°C) = 3605Ω, β can be calculated using equation (2) as

β = (298*323/ (298-323)) * ln (3605/10000) = 3928.17Ω. Thus the n-type thermistor resistance decreases exponentially as the temperature increases.

2.1. Thermistor Linearization

The exponential relation between thermistor resistance and the temperature can be linearized by connecting it in a DC circuit with a parallel or series resistor [12]. A simplest linearization circuit containing thermistor with a series resistor is configured This acts as a voltage divider of the DC source as shown in Fig. 2.

$$V_{out} = V_{in} * \frac{R_S}{R_T + R_S} \tag{3}$$

$$V_{out} = V_{in} * \frac{1}{1 + \frac{R_T}{R_S}} \tag{4}$$

where R_T = thermistor resistance at temperature T (Kelvin).

R_s = series resistor connected to thermistor for linearization. As the temperature increases further R_T << R_s, V_{out} have a value near ≈ V_{in}. If the temperature is very low R_T >> R_s, V_{out} is very low value ≈ 0. Hence (R_T / R_s) is a function of temperature. This can be verified from the table 1 as voltage across R_s.

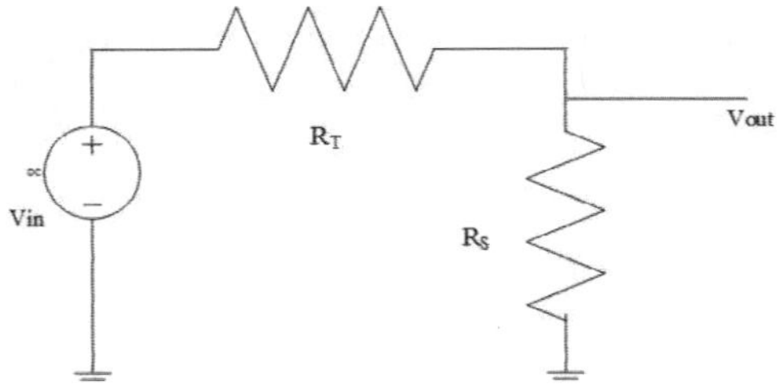

Fig. 2. Series voltage divider circuit

Substituting equation (1) in (4) for R_T and take second derivative & equate it to zero, gives point of infection.

$$R_s = R_0 * \left(\frac{\beta - 2T_1}{\beta + 2T_1}\right) * \exp\left(\frac{\beta}{T_1} - \frac{\beta}{T_0}\right) \tag{5}$$

where T_1 = temperature of inflection.

In this linearization process temperature range is selected from - 25°C to 100°C, T_1 =40°C, R_0 = 10KΩ, T_0 = 25°C and β = 3928.17K which gives R_s = 3856.9Ω. V_{in} is taken as 5V and using equation (3), a plot for temperature versus V_{out} is drawn. Compared with Fig. 1, the graph drawn in Fig. 3, linearity is very much improved around T= 40°C. But for wider range temperature measurement the nonlinearity at extremes in Fig. 3 is to be linearised. In this paper a method is proposed such that the nonlinearity is further improved in digital domain while performing A to D conversion.

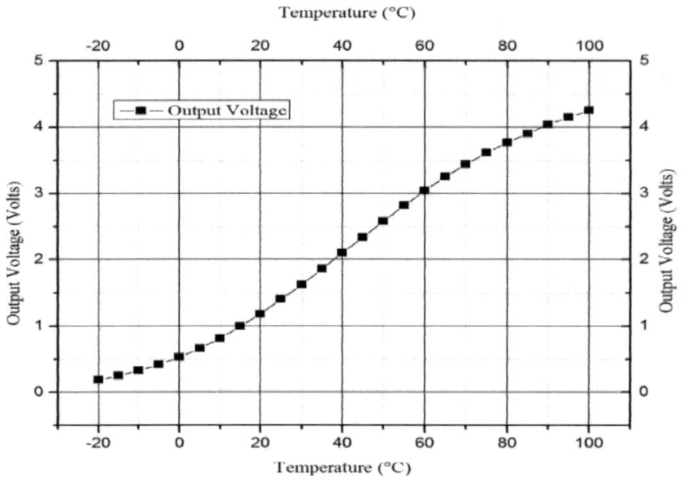

Fig. 3. Linearization is obtained near T = 40°C and showing nonlinearity at extremes.

2.2. Thermistor analog circuit

Output of the circuit is which is partially linearised is further processed by digital conditioning circuit by connecting through a unity gain op-Amp and a Sample and hold circuit [13] which is acting as a buffer for impedance matching as shown in Fig. 4.

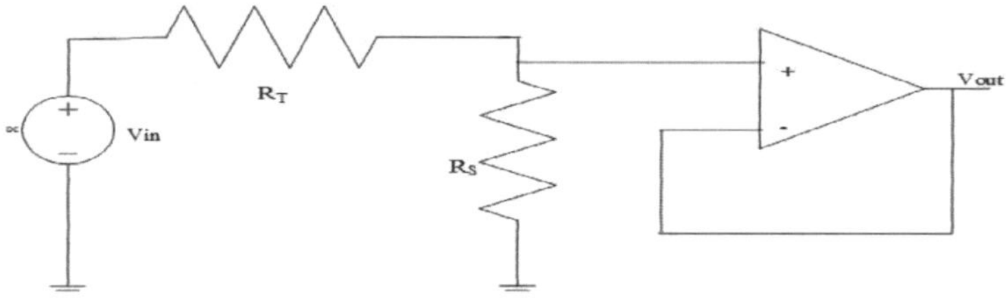

Fig. 4. Analog sensor circuit

3. Piecewise linearized design for thermistor sensor

While performing piecewise linearization the quantization steps is not linear. Generally they are based on two step conversion process [14]. First coarse conversion followed by the fine conversion. Coarse nonlinear conversion is done by flash ADC of fewer bits and the second step by differential eight bits flash ADC. The second conversion can also be done by a DAC, converting the coarse digital value into analog value, thereby restricting the input voltage for second conversion. But it need more area and consumes more power as explained in [15].

To improve the linearity and minimizing the quantization error nonlinear piecewise ADC can be used. The step size (for temperature range -20 to 100◦C) is divided into 4, but the corresponding voltage steps are not equal due to the nonlinear transfer function analog output versus temperature as shown in Fig. 3.

Reference voltage for first stage is taken as 5V. Four unequal resistors [16] are used to set the reference voltages for comparators. These voltages represent the range of each segment of the piecewise linear transfer function. In Fig. 5 the segments are represented by differently colored straight line segments. The voltages correspond to the temperatures obtained by dividing the total temperature range into four parts:-20-10, 10-40, 40-70 and 70-100◦C as shown in Table 2.

Table 2. Piecewise temperature range and the corresponding voltage range

Temperature range	(Centigrade)	Voltage (Volts)
Linear range 1	-20 to 10	0.192 to 0.813
Linear range 2	10 to 40	0.813 to 2.1
Linear range 3	40 to 70	2.1 to 3.44
Linear range 4	70 to 100	3.44 to 4.25

Below Fig. 5 shows the variation of voltage with respect to temperature divided into four segments, the four different segments are represented by different coloured straight line segments. Four different segments basically define a linear rise in voltage with the increase in temperature.

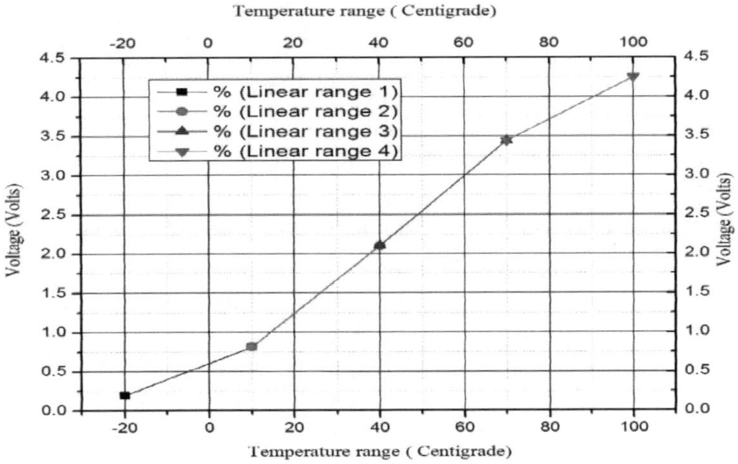

Fig. 5. Nonlinear Flash ADC transfer function.

4. The proposed method

Flash ADC converts analog into digital with highest speed. It has very simple circuits but the number of comparators required increase exponentially as the number of output binary bit increases. In this paper, two stage of analog to digital conversion is done and the number of comparators used is only two. The total range of the analog input voltage is divided into two parts in stage 1. Out of the four linear ranges as given in table.2, one particular range is chosen by stage 2, which is explained below.

In stage 1, the middle value of the total range is given as reference voltage to the negative input terminal of first comparator (CompA) and sensor output voltage after sampling and hold is given to positive input terminal of the comparator (CompA).

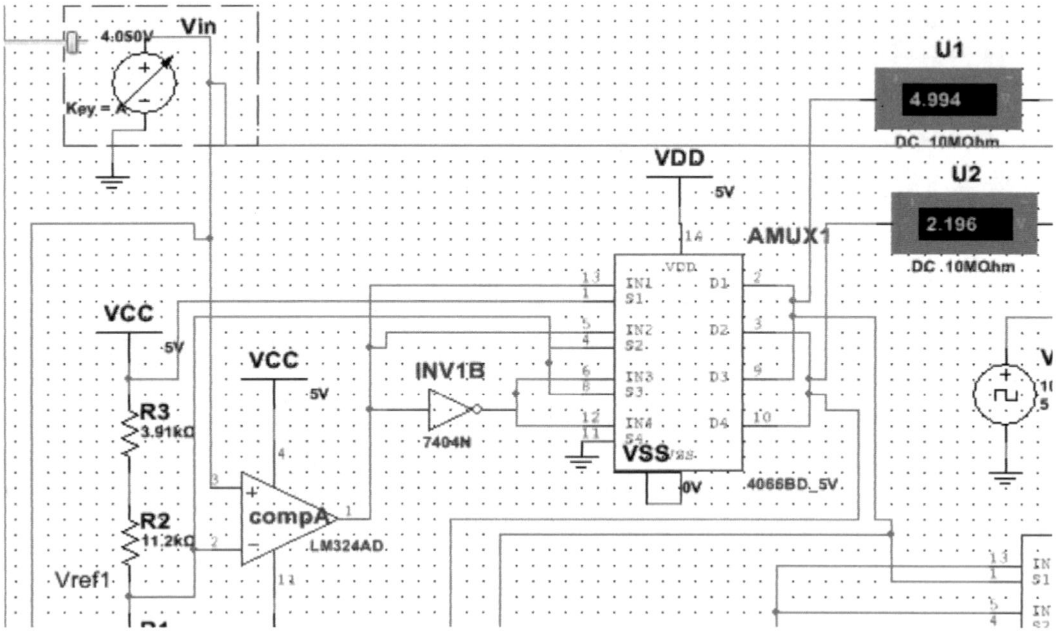

Fig. 6. Circuit connection between multiplexer and comparator of stage1

Referring to the table 2, the total range of the analog signal is from 0.192 to 4.25 Volts. Voltage divider network is designed [14] such that the middle value is 2.1Volts and it is given as Vref1 to compA. In Fig. 6, the input pins of Amux1 connections are as follows:

Pin1 is connected to 5Volt, pin4 and pin8 to V_{ref1} and pin 11 to ground. The compA compares V_{in} with V_{ref1}. CompA output is connected to Amux1 to control pin 13 and 5 & negated output is connected to control pin 6 and 12. Its output terminal pin 2 and 9 are interconnected (at a time only one goes high).Voltmeter U1 shows its reading. Its output terminal pin 3 and 10 are interconnected (at a time only one goes high).Voltmeter U2 shows its reading. If $V_{in} > V_{ref1}$, CompA output goes high. Amux1 Pin1 is connected to pin2 (internally since the control pin 13 goes high), the voltmeter U1 displays pin2 value. Amux1 Pin4 is connected to pin3 (internally since the control pin 5 goes high), the voltmeter U2 displays pin3 value.

If $V_{in} < V_{ref1}$, CompA output goes low. Amux1 Pin8 is connected to pin9 (internally since the control pin 6 goes high as a result of compA output negation), the voltmeter U1 displays pin9 value. Amux1 Pin11 is connected to pin10 (internally since the control pin 12 goes high as a result of compA output negation), the voltmeter U2 displays pin10 value. Comparator compB is connected to the node between the series resistors R4 and R5 (ref voltage for compB) the other terminal of R4 is connected to the interconnected junction between terminals pin 2 and 9 of Amux1. Other terminal of R5 is connected to the interconnected junction between terminals pin 3 and 10 of Amux1.

IOP Conf. Series: Materials Science and Engineering **594** (2019) 012036 doi:10.1088/1757-899X/594/1/012036

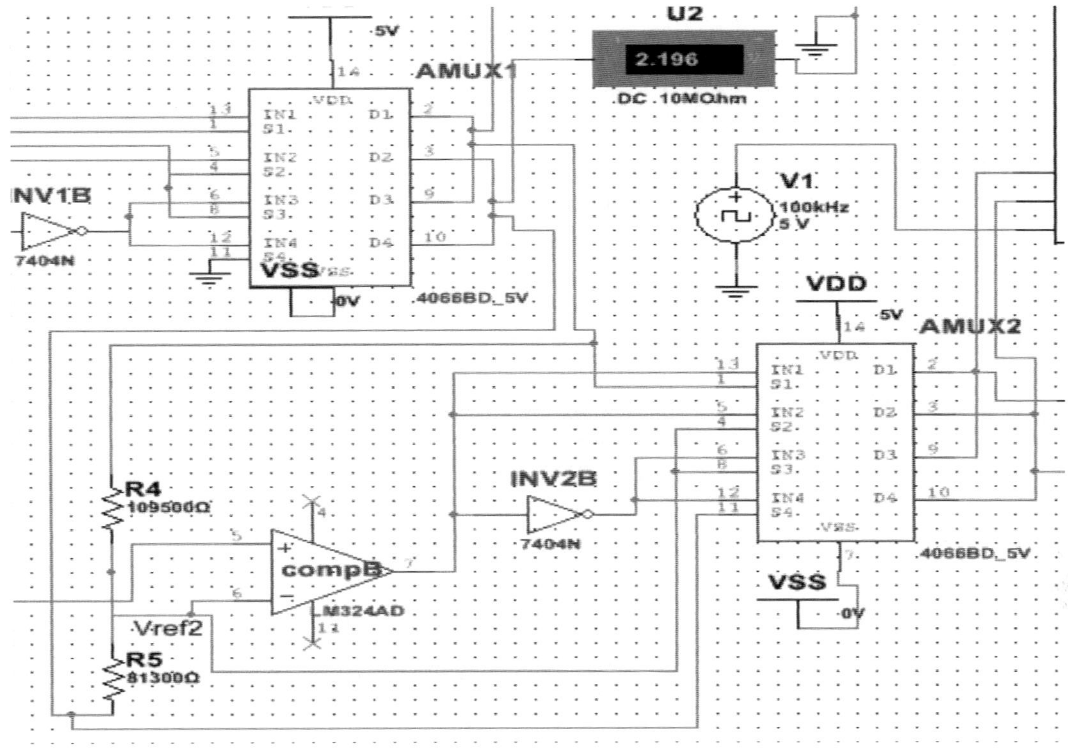

Fig. 7. Circuit connection between stage 1 and multiplexer of stage2

The comparator B in Fig. 7 compares V_{in} with V_{ref2}. The input pins of Amux2 connections are as follows: Pin1 is connected to interconnected junction between terminals pin 2 and 9 of Amux1. Pin4 and pin8 to V_{ref2} of compB and pin 11 connected to interconnected junction between terminals pin 3 and 10 of Amux1. Going by the same procedure as that of Amux1, The compB compares V_{in} with V_{ref2}. compB output is connected to Amux2 to control pin 13 and 5 & negated output is connected to control pin 6 and 12. Its output terminal pin 2 and 9 are interconnected (at a time only one goes high).Voltmeter U3 shows its reading. Its output terminal pin 3 and 10 are interconnected (at a time only one goes high).Voltmeter U4 shows its reading.

If $V_{in} > V_{ref2}$, compB output goes high. Amux2 Pin1 is connected to pin2, when control pin 13 goes high the voltmeter U3 displays pin2 value. Amux2 Pin4 is connected to pin3, when control pin 5 goes high the voltmeter U4 displays pin3 value. If $V_{in} < V_{ref2}$, compB output goes low. Amux2 Pin8 is connected to pin9, when control pin6 goes high the voltmeter U3 displays pin9 value. Amux2 Pin11 is connected to pin10, when control pin12 goes high the voltmeter U4 displays pin3 value.

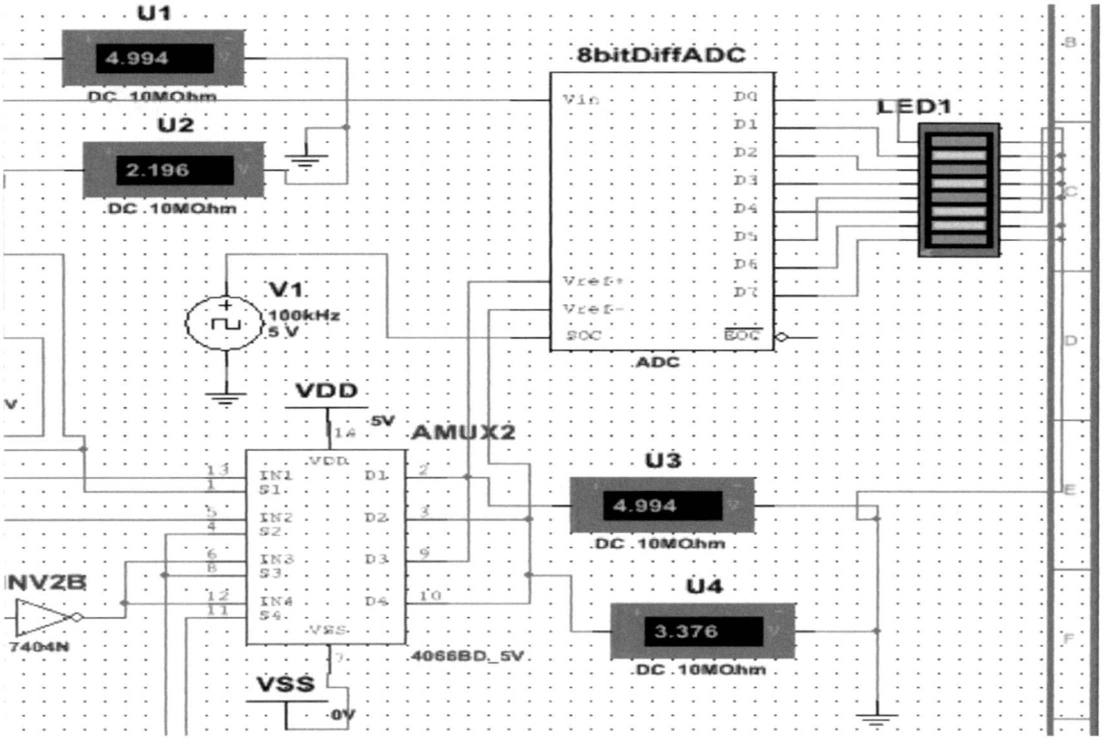

Fig. 8. Circuit connection between stage2 of proposed circuit and output device LED

In Fig. 8, the output of Amux2, either pin2 or pin9 is connected to V_{ref2}+ and either pin3 or 10 to V_{ref2}- pins of 8bit differential ADC converter. V_{in} is connected to the V_{in} pin of 8 bit differential ADC converter. The output of 8bit ADC is shown in 8bit bar LED1

5. Simulated Results

The proposed circuit is simulated using Modelsim12.0. A variable voltage source is used to represent the analog output voltage of thermistor analog circuit, V_{out}. Four such cases are discussed one from each piecewise voltage range (and thereby representing the corresponding temperature range). The digital value shown by the LED1 is converted into decimal and compared withV_{in}, the input variable voltage value. From these four different cases the input results are compared with the simulated results.

Case1: When input voltage V_{in} = 4.05 Volt

The displayed binary value is (64+32+8+2) = 106

U3 = voltmeter reading connected to V_{ref}+ pin of ADC =4.994Volt.

U4 = voltmeter reading connected to V_{ref}- pin of ADC =3.371Volt.

V_{in} as calculated from simulated result = (106/256)*(4.994 - 3.371) + 3.371 = 4.0459Volt.

Here, it can be seen that applied input voltage and simulated results are approximately equal.

IOP Conf. Series: Materials Science and Engineering **594** (2019) 012036 doi:10.1088/1757-899X/594/1/012036

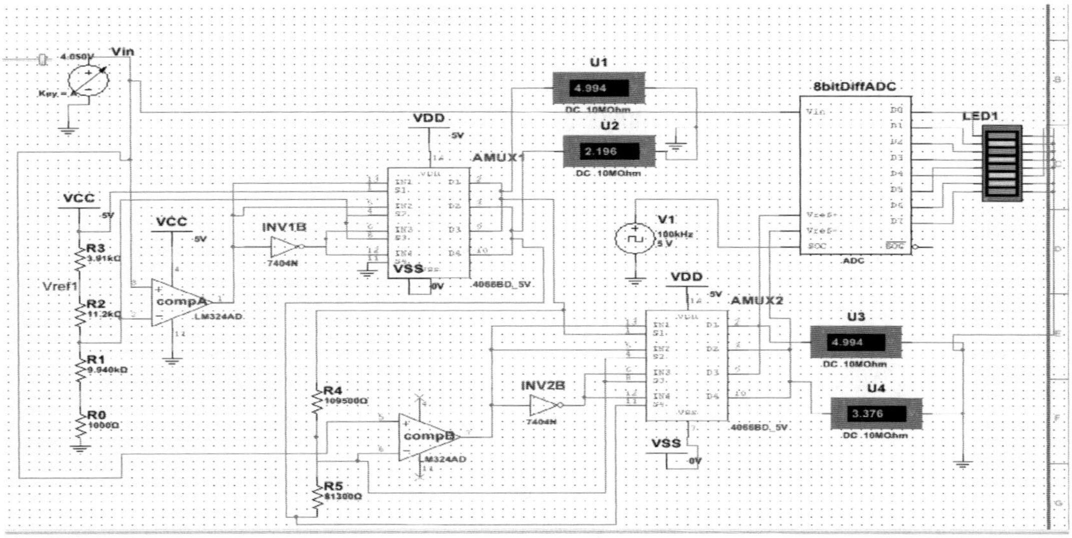

Fig. 9. Simulated result Case1, V_{in}= 4.05Volt

Case2: When input voltage V_{in} = 2.7 Volt
The displayed binary value is (64+32+8+4+1) = 109.
U3 = voltmeter reading connected to V_{ref}+ pin of ADC =3.371Volt
U4 = voltmeter reading connected to V_{ref}- pin of ADC =2.194Volt.
V_{in} as calculated from simulated result = (109/256)*(3.371 – 2.194) + 2.194 = 2.695Volt.
Here, it can be seen that applied input voltage and simulated results are approximately equal.

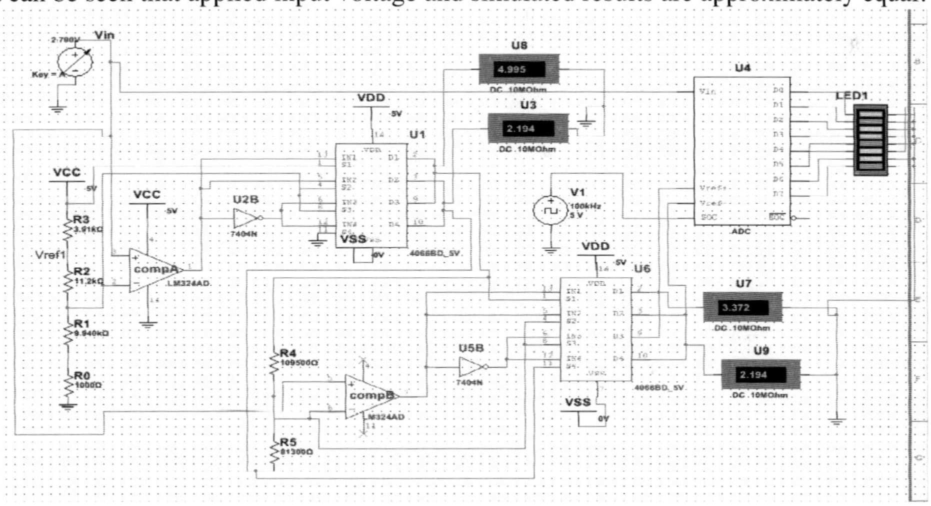

Fig. 10. Simulated result, Case2: V_{in} = 2.7 Volt

Case3: When input voltage V_{in} = 1.575 Volt
The displayed binary value is (128+16+8+4) = 156

IOP Conf. Series: Materials Science and Engineering **594** (2019) 012036 doi:10.1088/1757-899X/594/1/012036

U3 = voltmeter reading connected to $V_{ref}+$ pin of ADC =2.026Volt.
U4 = voltmeter reading connected to $V_{ref}-$ pin of ADC = 0.864Volt.
V_{in} as calculated from simulated result = (156/256)*(2.026 – 0.864) + 0.864 = 1.572Volt.
Here, it can be seen that applied input voltage and simulated results are approximately equal.

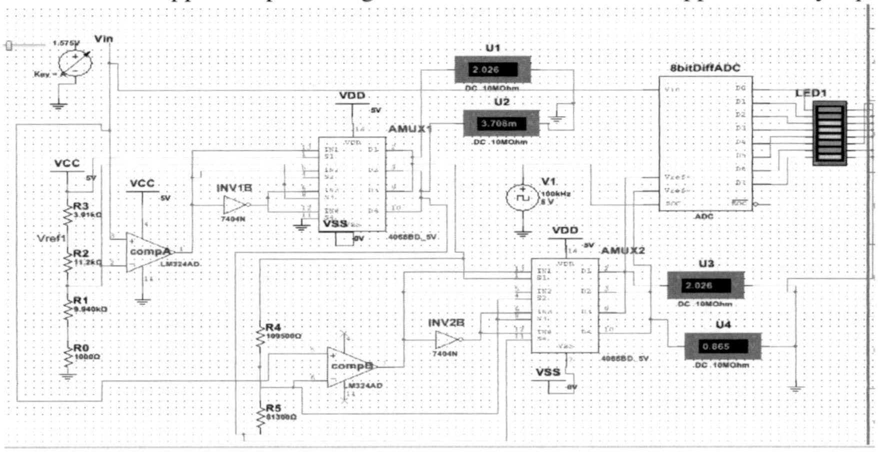

Fig. 11. Simulated result, Case3: V_{in} = 1.575 Volt

Case4: When input voltage V_{in} = 0.45 Volt
The displayed binary value is (128+4+1) = 133.
U3 = voltmeter reading connected to $V_{ref}+$ pin of ADC =2.026Volt.
U4 = voltmeter reading connected to $V_{ref}-$ pin of ADC = 0.864Volt.
V_{in} as calculated from simulated result = (133/256)*0.864 = 0.4478Volt.
Here, it can be seen that applied input voltage and simulated results are approximately equal.

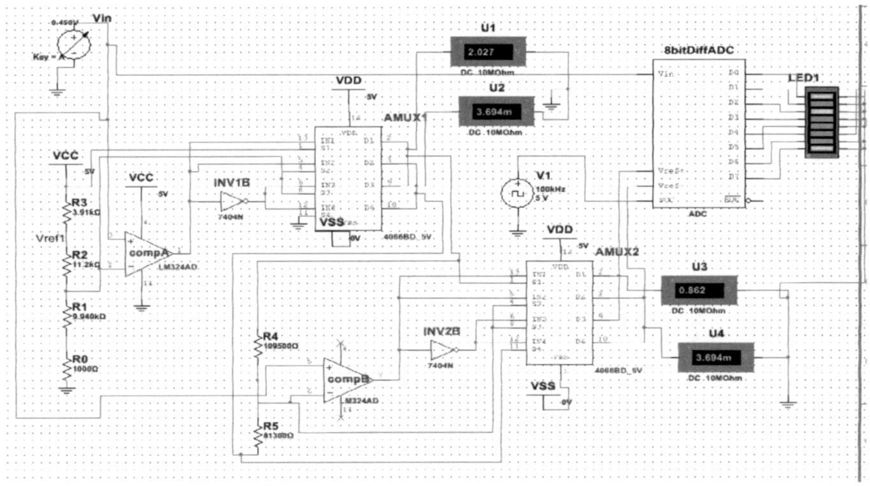

Fig. 12. Simulated result, Case 4: V_{in} = 0.45 Volt.

IOP Conf. Series: Materials Science and Engineering **594** (2019) 012036 doi:10.1088/1757-899X/594/1/012036

6. Error Calculation

For more input voltage values, the decimal equivalent of displayed binary values are calculated and the same is tabulated in table 3.

Table 3. Calculation of error between input voltage and measured value by the proposed circuit design

V_{in} from variable voltage source	Range shown by Multiplexer Amux1	Range shown by Multiplexer Amux2	Voltage calculated from binary display LED1	Error %
3.825	4.994, 2.196	4.994, 3.376	3.8247	$7.8*10^{-3}$
3.6	4.994, 2.196	4.994, 3.376	3.59	0.277
2.92	4.994, 2.196	3.371, 2.194	2.92	0
2.475	4.994, 2.196	3.371, 2.194	2.4744	0.0242
2.025	2.026, 0	2.026, 0.862	2.026	0.049
1.125	2.026, 0	2.026, 0.862	1.122	0.266
0.675	2.026, 0	0.862, 0	0.674	1.48
0.225	2.026, 0	0.862, 0	0.222	1.33

7. Conclusion

Partial linearization of sensors characteristics in analog domain is done by passive circuit elements. Using two stage PWL flash ADC having inverse transfer function to that of sensor has improved performance. By using the proper ADC transfer function, linearization of nonlinear sensor characteristic is possible in digital domain. This will eliminate further correction circuits to handle the nonlinearity problem. A to D conversion and linearization are done simultaneously in digital circuitry, by which it is possible to make more efficient and cheap smart sensors. Nonlinearity is removed at acquisition speed.

The input range is divided into four linear ranges and the range in which sensor voltage falls is selected by using two comparators and two analog multiplexers. Out of four ranges, according to the input value only one range is linearly converted into digital domain by 8bit flash ADC converter. The simulated result by Multisim12.0 shows negligible error in higher three ranges and of the order of 1% in the fourth lower most range.

References

[1] R. Brown and R. Orange, "*NTCLE100E3 Vishay BCcomponents NTC Thermistors , Radial Leaded* , Standard Precision NTCLE100E3," pp. 1–17.

[2] Thermistor Mathematics. (2011). Retrieved from http://mathscinotes.com/2011/07/thermistor mathematics/

[3] Trancã, D. C., Rosner, D., Tãtãroiu, R., Stegaru, S. C., Surpãţeanu, A., & Peišić, M. (2018, September). *Precision and linearity of analog temperature sensors for industrial IoT devices*. In 2018 17th RoEduNet Conference: Networking in Education and Research (RoEduNet) (pp. 1-6). IEEE.

[4] Jack, K. E., Etu, I., & Ukanide, V. N. (2016). *The Design of a Signal Conditioning and Acquisition Elements of a Chopped Broadband Radiation Pyrometer*. In Proceedings of the World Congress on Engineering and Computer Science (Vol. 1).

[5]Anand, C., Sadistap, S., Bindal, S., Botre, B. A., & Rao, K. S. N. (2010). *Wireless multi-sensor*

embedded system for Agro-industrial monitoring and control. International Journal on Advances in Networks and Services Volume 3, Number 1 & 2, 2010.

[6] Eder, C., Valente, V., Donaldson, N., & Demosthenous, A. (2014). A CMOS smart temperature and humidity sensor with combined readout. Sensors, 14(9), 17192-17211.

[7] Kodgirwar, S., Chandrachood, A., Vyas, S., & Kodgirwar, V. (2016). *Design of Signal Conditioning Circuit for Biomedical Sensors and Battery Monitoring Circuit for a Portable Communication System* American Journal Of Engineering Research (AJER), (6), 100–107.

[8] Wang, C. C., Hou, Z. Y., & You, J. C. (2018). *A high-precision CMOS temperature sensor with thermistor linear calibration in the (– 5 C, 120 C) temperature range*. Sensors, 18(7), 2165.

[9] Kumar, V. N., & Narayana, K. V. L. (2015). *Development of thermistor signal conditioning circuit using artificial neural networks*. IET Science, Measurement & Technology, 9(8), 955-961.

[10] Sarkar, A. R., Dey, D., & Munshi, S. (2013). *Linearization of NTC thermistor characteristic using op-amp based inverting amplifier*. IEEE Sensors Journal, 13(12), 4621-4626.

[11] Becker, J. A., Green, C. B., & Pearson, G. (1946). *Properties and uses of thermistors—Thermally sensitive resistors*. Electrical Engineering, 65(11), 711-725.

[12] Stankovic, S. B., & Kyriacou, P. A. (2011). *Comparison of thermistor linearization techniques for accurate temperature measurement in phase change materials*. In Journal of Physics: Conference Series (Vol. 307, No. 1, p. 012009). IOP Publishing.

[13] Rajput, A., Kanathe, S., & Bhopal, R. (2012). *Design of sample and hold circuit*. Int. J. Sci. and Research Publications, 2.

[14] Jovanović, J., & Denić, D. (2016). *A Cost-effective Method for Resolution Increase of the Two stage Piecewise Linear ADC Used for Sensor Linearization*. Measurement Science Review, 16(1), 28-34.

[15] Jovanović, J., Denić, D., & Jovanović, U. (2017). *An Improved Linearization Circuit Used for Optical Rotary Encoders*. Measurement Science Review, 17(5), 241-249.

[16] Lukić, J., & Denić, D. (2015). *A Novel Design Of An NTC Thermistor Linearization Circuit. Metrology and Measurement Systems*, 22(3), 351-362.

IOP Conf. Series: Materials Science and Engineering **594** (2019) 012037 doi:10.1088/1757-899X/594/1/012037

Neodymium magnetic shock absorber for two wheelers Automobiles

B Nandish*, K P Muthanna, M B Kaveriappa and P S Biddappa
Mechanical Engineering Department, Coorg Institute of Technology, Ponnampet, South Coorg, Karnataka, 571216, India.

Email: *nandishb77@gmail.com

Abstract: This paper presents a research work on magnetic suspension system of two wheelers Automobiles, which are usually depending on spring type, Hydraulic and Pneumatic suspension systems. In this proposed magnetic suspension system, two permanent magnets made of Neodymium material are placed inside the shock absorber cylinder such that both facing same pole. So they produce a repulsive magnetic flux force, when they come closer due to shocking load. This repulsive magnetic flux force is used as shock absorbing media and provides damping force. Proposed suspension system proves to be more efficient over other type of suspension systems, absorb more number of shocks with high accuracy, has no leakage problem unlike in Hydraulic and Pneumatic system. So, all these beneficial qualities make the magnetic suspension system to work efficiently with less maintenance cost and hence the Automobile.

1. Introduction

Shock absorber is a mechanical member, fitted on and held between wheel and body of the automobile. Its primary function is to absorb shocking force emerging from wheel of Automobile, when wheel moves over road irregular surfaces. Hence shocking force is prevented to reach body of vehicle and rider of vehicle feels comfort. Main types of suspension system are Mechanical spring type, Hydraulic type, Pneumatic type. In current days, magnetic suspensions system is used because of its many advantageous over other systems. Present work consists of design and analysis of magnetic suspension system, which is used in two wheeler automobiles.

2. Description of Neodymium magnet.

It is a permanent magnet formed with alloy of Neodymium, Iron and Boron. It is having $Nd_2Fe_{14}B$ tetragonal crystalline structure and is commercially available in many grades. Grade used in the present work of suspension system is N4518, easy availability and desirable properties are the reason to choose this grade. It's important properties are shown in Table 1.

Table 1.Important properties of Neodymium magnet of grade N4518.

Parameter and its unit.	Values
Remanence (T)	1-1.3
Co-ercivity (MA/m)	0.875-1.99
Relative permeability	1.05
Compressive strength (N/mm^2)	1110

3. Design of magnetic suspension system.

In the present work of suspension system, two permanent magnets made of Neodymium are placed end to end inside the shock absorber cylinder, such that both magnets face same pole and a considerable air gap is allowed between them. These magnets are mounted individually on ends of two piston rods of shock absorber. Among two piston rod, one is connected to wheel of vehicle, hence it move along with wheel, other piston rod is connected to body of vehicle and it is fixed. Since both magnets face same pole, a magnetic repulsive flux force is generated between them. This repulsive force is used as shock absorbing media. In the course of working of this suspension system, if the wheel of vehicle moves over road surface irregularities, wheel would move up and down, this in turn causes the piston rod which is connected to it to reciprocate linearly inside the shock absorber. Hence magnet on it moves closer to other magnet, which is on another piston rod. So a repulsive magnetic flux force is generated, this repulsive force shove the movable magnet away from fixed one and constrains action of moving piston rod. As a result, displacement force of wheel is prevented from reaching to body of Automobile. So rider on it feels a comfort drive. In addition to magnetic suspension, a helical steel spring is also provided inside shock absorber. This spring provides additional suspension support, absorb shocks and prevents magnets from direct collision on severe level of shocking load. Fig. 1. shows schematic diagram of magnetic shock absorber.

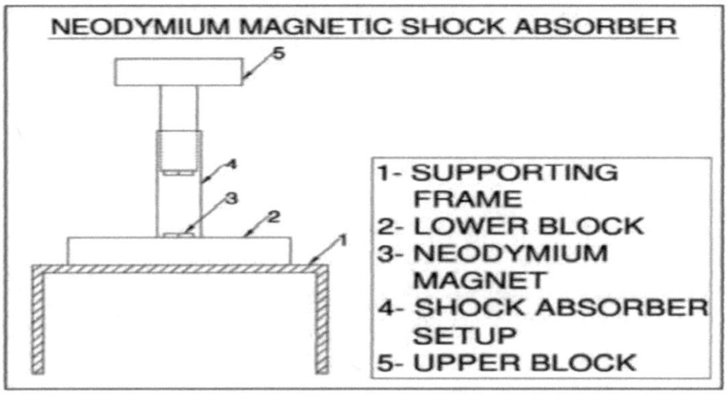

Fig. 1. Neodymium magnetic shock absorber.

3.1 Design of spring used in shock absorber.

Outer diameter and length of spring are selected based on geometrical requirement of shock absorber. Maximum load sustained by spring is found as follows.
Outer diameter of spring D_o =25mm, length of spring L=82mm, number of turns n =11, wire diameter of spring d=2.5mm,
Inner diameter D_i = D_o- d = 25-5 = 20mm,
Spring is made of steel and for it, shear stress (from design data book) =600N/mm^2.
Spring index D_O/d =25/2.5 = 10. For this spring index, from design data book, Wahl's correction factor of spring K=1.18.
Load carried by spring P=shear stress* K
Hence P =600*1.18 = 708 N
So, spring can bear load of 708 N, remaining load on shock absorber will be shared by flux power of magnet.

3.2 Design of magnets used in shock absorber

Considering a fraction weight of Automobile acting on suspension system (excluding weight of wheel, engine weight and etc. since they are not supported by suspension system) = 15 Kg=147 N and weight of person sitting on it = 60kg = 588 N.
Total Load = fraction weight of Automobile on suspension system+ weight of person sitting on Automobile = 147+588 = 735 N.

Load on Rear Suspensions = 65 percentage of total load = 0.65*735 N = 441 N.
Approximating dynamic load as double of total static load W=2*441 N = 882 N.
Weight on Single Shock Absorber, W/2 = 441 N, Taking Factor of Safety = 1.2
So, design Load= 621.3N.
Magnetic flux Power per unit Area =2 N/mm^2 ; calculated based on Magnetic repulsion Force, as given in section 4.3. So, area of magnet required to bear suspension load of 621.3N.
2 N/mm^2 = 621.3/A A = 310.65 mm^2
Therefore, diameter of magnet required d_m = 48.1mm = 50mm.

3.3 Calculation of magnetic repulsion force.

Repulsive force between two magnets is found using Gilbert model of magnetic theory. Equation to find repulsive force F between two similar cylindrical magnets is given below.

$$F= [B_o^2 A^2 (L^2+r^2) / (\pi\mu_{0}/L^2)]*[1/d^2+ 1/(1+2L)^2-2/(d+L)^2]$$

B_O is flux density at each pole of magnet in tesla. A is area of surface of magnet poles in m^2, L is length of each magnet in m, r is radius of each magnet in m , μ_o is permeability of free space=$4\pi*10^{-7}$ T-M /A , d is the separation distance between two magnets in m and it is considerably more than radius r of magnets.

4. Results and discussion

In the present work, radius of magnet adapted is r = 25 mm, length of magnet is L= 12.5 mm. In normal riding conditions of automobile, magnets are apart inside the shock absorber at a distance of 82 mm and minimum distance, when they come close during shock is 30 mm. So repulsive magnetic flux force varies from maximum to minimum and it is determined using equation as stated in section 4.3. So, when x = 30 mm, repulsive force will be maximum F_{max} = 5512N, when x = 82 mm, repulsive force will be minimum F_{min} = 200N. Results show that, maximum repulsive force F_{max} = 5512 N is greater than load to be bear by suspensions = 441N. So, newly devised magnetic suspension system can sustain and bear operating load on it.

5. Conclusion

In the present work, magnetic suspension system has been designed and analyzed. Magnetic repulsive flux force generated by Neodymium magnet confirms to be a good damping media and absorb the shock efficiently. It facilitates lesser maintenance, absence of leakage of oil or air as in other sort of shock absorbers and makes the ride comfort.

References

[1] John C Dixon, "The shock absorber handbook", SAE International, SAE Order No 176, 1999.
[2] Suvriti Dhawan and Ravi Nandu, "Magnetic Suspension in Automobiles", *Journal of Aeronautical and Automotive Engineering*, Print ISSN: 2393-8579, Online ISSN: 2393-8587, Vol. 1, No.1, pp. 20-21, September 2014.
[3] S Gopinath , R J Golden Renjith, J Dineshkumar, "Design and fabrication of magnetic shock absorber", *International journal of Engineering and Technology*, Vol.3, No.2, pp.208-211, 2014.
[4] V V Borole, K K Chaudhari, "A Review on Magnetic Shock Absorber", *E-Issn*: 2348-0831, Vol.2, Issue 3, pp.104-109, Mar-Apr 2015.
[5] Shende Vignesh ,Nimbalkar Hrishikesh , Pawar Sanjay , Thorat Vijay , Raut P.S , "Magnetic Suspension System for Two Wheeler", *International Journal of Recent Research in Civil and Mechanical Engineering,* Vol. 2, Issue 2, pp. 141-146, October 2015 – March 2016

IOP Conf. Series: Materials Science and Engineering **594** (2019) 012038 doi:10.1088/1757-899X/594/1/012038

Simulation and Experimental Studies of MPC for Level Control of Modified Quadruple Tank System

Tarun Varshney[1,*], Shekhar Gehlaut[2] and Sanjay Kumar Bansal[3]

[1]Departmentof Electrical Engineering, PCE, Jaipur, India
[2]Departmentof Electrical Engineering, SGI, Sriganganagar, India
[3]Department of Electrical Engineering, GCET., Bikaner, India

Email-id: [*]t_varshney@yahoo.com

Abstract. Model Predictive Control (MPC) has been a powerful control strategy as it has already been proved with simulation and experimental studies. In this study, MPC has been explored for the level control of a nonlinear benchmark quadruple tank system. MPC has been applied as a real-time control strategy for set point control and disturbance rejection. Rigorous simulations and experiments have been done and fairy compared with those from the use of decentralized PI controller. To validate the results, real-time experimental studies have also been carried out for two possible configurations. The application offers a new look at the performance of the MPC for complex systems control.

1. Introduction

Mainly industrial real practices are multivariable in nature and have multi inputs and multi outputs. Such systems are likely to have interactions among process variables and control signals [1]. Because of these interactions among the variables, it is complex to design appropriate controller for multivariable systems. To overcome this issue a number of methods have been proposed and accepted in order to control the multivariable and MIMO systems, e.g. Relative Gain Array (RGA), decentralized techniques, decoupled control [1-5], mainly because these methods exhibit flexibility in operation, simplified design and tuning to improve the performance. The advantage of these methods lies in the fact that simpler linear SISO theory and methods can be applied on complex MIMO systems [6-7].

Decentralized control is simplest technique to deal MIMO systems in which a control for every input-output pair is designed as like SISO approach [7]. However, this mechanism is efficient only in presence of weak interactions between the loops [8]. If the system is highly interacted, a small change in the set point in one loop significantly affects the output in the other loop. Consequently, response may become unacceptable. Whereas in decoupling control strategy, the coupling between loops are considered and controller compensates the interactions which yields non-interacting SISO systems, where the set-point changes affect only the desired control variables [8]. However, applicability of these approximations depends on the neglected interaction dynamics, which can be viewed as modeling errors. The advantage of applying SISO methods on MIMO systems by using aforesaid methods, is that the analysis becomes easier, but one must compromise with the system dynamics which in many cases may change the system behavior. Moreover, the actual performance of the system may become significantly different from that of the expected one [9-10]. Despite the benefits of these methods, they cannot suppress the interactions among the variables in MIMO systems [10]. To overcome these limitations advanced controller structures are developed in which Model Predictive Controllers (MPC) are most popular because it is capable of handling multivariable systems online more effectively, while satisfying the constraints [11]. This has been the motivation for the author to use MPC. MPC is a collection of methods for optimal control, characterized by using a model to predict the behavior of the system being controlled and optimize control over a moving time horizon

[12-16]. The essence of MPC is to optimize forecasts of future behavior. The forecasting is accomplished with a system model and therefore the model is the essential element of the strategy.

In the present work, real-time experimental and simulation studies have been carried out on a benchmark control problem of Quadruple Tank System (QTS) and its modified version. The results are fairly compared with a decentralized Proportional-Integral (PI) controller. The rest of the paper is organized as follows, Section 2 illustrates the system dynamics, Section 3 describes the fundamentals and schematic of MPC, in section 4 results of experimental and simulation studies have been shown and fairly compared and finally conclusions are presented.

2. System description

The water level control of QTS is a challenging benchmark control problem owing to its multivariable nature, having well-built interaction among the variables, extremely nonlinear and non-minimum phase characteristics. QTS consists of four interconnected tanks and two pumps. The controlled inputs are voltages to the pumps and levels of lower two tanks are considered outputs. The outflows of every pump are separated into two tanks by using valves 1& 2. Pump-1 feeds Tank-1 & 4 whereas pump-2 feeds Tank-2&3. Tanks 3 & 4 discharge into Tanks-1 & 2 respectively while Tanks-1 & 2 discharges straight into the reservoir placed at the base, as given in Fig.1(a). The main objective is here to control the level of lower two tanks with a minimum mean squared error (MSE).

To test performance of MPC controller, author modified the structure of QTS by introducing an additional inter-connection valve between lower two tanks, as shown in Fig.1(b), which increases the complexity of system. This configuration is considered as Modified Quadruple Tank System (mQTS).

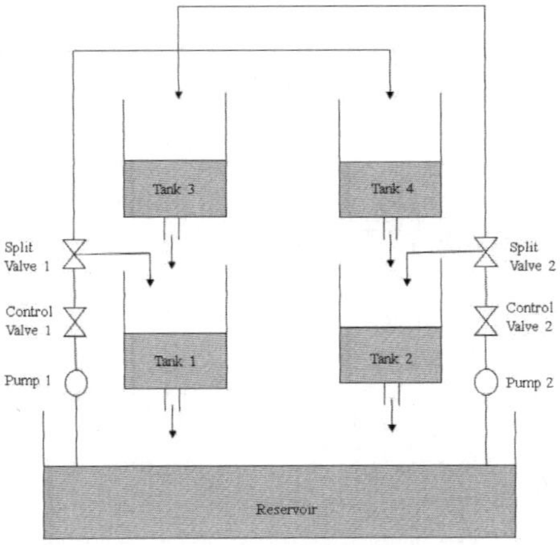

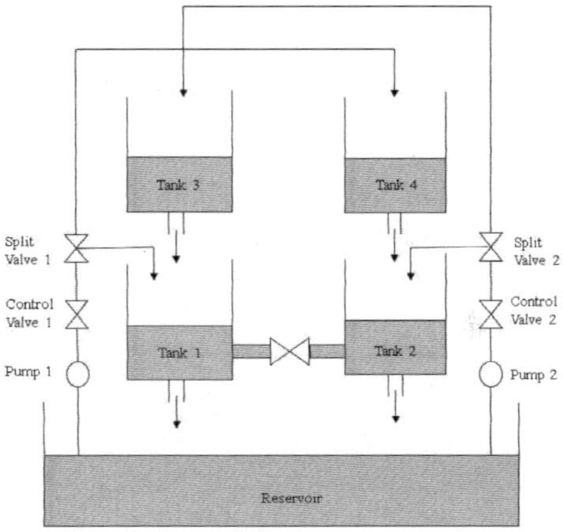

(a) QTS (b) mQTS

Fig. 1. The schematic of Quadruple Tank System

2.1. Mathematical Modeling for simulation study

2.1.1. Mathematical modeling of QTS. The dynamics of the system are described by the following equations [9].

$$\left.\begin{aligned}
\frac{dh_1}{dt} &= \frac{\gamma_1 k_1 u_1}{A_1} + \frac{a_3\sqrt{2gh_3}}{A_1} \quad \frac{a_1\sqrt{2gh_1}}{A_1} \\
\frac{dh_2}{dt} &= \frac{\gamma_2 k_2 u_2}{A_2} + \frac{a_4\sqrt{2gh_4}}{A_2} \quad \frac{a_4\sqrt{2gh_4}}{A_2} \\
\frac{dh_3}{dt} &= \frac{(1-\gamma_2)k_2 u_2}{A_3} - \frac{a_3\sqrt{2gh_3}}{A_3} \\
\frac{dh_4}{dt} &= \frac{(1-\gamma_1)k_1 u_1}{A_4} - \frac{a_1\sqrt{2gh_1}}{A_4}
\end{aligned}\right\} \quad (1)$$

System parameters are tabulated in Table 1.

Table 1. System parameters

Symbol	Parameters	Nominal values
Ai	Area of ith tank in cm^2 i=1,2,3,4	28,32,28,32
ai	Area of drain in cm^2 i=1,2,3,4	0.071,0.057,0.071,0.057
a12	Area of additional interconnection cm^2	0.064
hi	Maximum height if i^{th} tank in cm, i=1,2,3,4.	25
g	Gravitational constant	981
ki	Valve proportionality constant	3.14, 3.15
ui	Voltage ranges of ith control valve	[0-5 volts]

2.1.2. Mathematical modeling of mQTS. As there an interaction has been introduced in between Tank-1 and Tank-2 only, therefore the dynamics of Tank-3 and Tank-4 remains the same as of QTS and dynamics of Tank-1 and Tank-2 are modified as below:

$$\left.\begin{aligned}
\frac{dh_1}{dt} &= \frac{\gamma_1 k_1 u_1}{A_1} + \frac{a_3\sqrt{2gh_3}}{A_1} - \frac{a_{12}\sqrt{2g|h_1-h_2|}}{A_1} * sign(h_1-h_2) - \frac{a_1\sqrt{2gh_1}}{A_1} \\
\frac{dh_2}{dt} &= \frac{\gamma_2 k_2 u_2}{A_2} + \frac{a_4\sqrt{2gh_4}}{A_2} - \frac{a_{12}\sqrt{2g|h_1-h_2|}}{A_2} * sign(h_1-h_2) - \frac{a_2\sqrt{2gh_2}}{A_2} \\
\frac{dh_3}{dt} &= \frac{(1-\gamma_2)k_2 u_2}{A_3} - \frac{a_3\sqrt{2gh_3}}{A_3} \\
\frac{dh_4}{dt} &= \frac{(1-\gamma_1)k_1 u_1}{A_4} - \frac{a_1\sqrt{2gh_1}}{A_4}
\end{aligned}\right\} \quad (2)$$

*2.1.3. Linearization of QTS and mQTS.*If the nonlinear process is operated near a specified operating point then linearized model of the process may be sufficiently accurate. Using Taylor series and truncating after first order terms a linear approximation of the differential equations can be easily obtained. The linearization is performed around the normal steady-state points. Eq. (1) for QTS can be expressed in state-space model.

$$\begin{aligned} \dot{x} &= Ax + Bu \\ y &= Cx + Du \end{aligned} \quad (3)$$

where

$$A=\begin{bmatrix} -\frac{a_1}{A_1}\sqrt{\frac{g}{2h_{10}}} & 0 & \frac{a_3}{A_3}\sqrt{\frac{g}{2h_{30}}} & 0 \\ 0 & -\frac{a_2}{A_2}\sqrt{\frac{g}{2h_{20}}} & 0 & -\frac{a_4}{A_2}\sqrt{\frac{g}{2h_{40}}} \\ 0 & 0 & -\frac{a_3}{A_3}\sqrt{\frac{g}{2h_{30}}} & 0 \\ 0 & 0 & 0 & -\frac{a_4}{A_4}\sqrt{\frac{g}{2h_{40}}} \end{bmatrix} \quad (4)$$

$$B=\begin{bmatrix} \frac{\gamma_1 k_1}{A_1} & 0 \\ 0 & \frac{\gamma_1 k_1}{A_1} \\ 0 & \frac{(1-\gamma_2)k_2}{A_3} \\ \frac{(1-\gamma_1)k_1}{A_4} & 0 \end{bmatrix}, C=\begin{bmatrix} k_c & 0 & 0 & 0 \\ 0 & k_c & 0 & 0 \end{bmatrix}, \quad D=[0], \dot{x}=[\frac{d\Delta h_1}{dt}\ \frac{d\Delta h_2}{dt}\ \frac{d\Delta h_3}{dt}\ \frac{d\Delta h_4}{dt}]$$

$X=[\Delta h_1\ \Delta h_2\ \Delta h_3\ \Delta h_4]$; $U=[\Delta u_1\ \Delta u_2]$; $Y=[\Delta y_1\ \Delta y_2]$

Similarly, A, B, C, D matrices for mQTS can also be derived and it is observed that the dynamics of mQTS differs from QTS only in the composition of matrix A. A matrix for mQTS is shown below.

$$A=\begin{bmatrix} -\frac{a_1}{A_1}\sqrt{\frac{g}{2h_{10}}}-\frac{a_{12}}{A_1}\sqrt{\frac{g}{2h_{10}-h_{20}}}*sign(h_{10}-h_{20}) & -\frac{a_{12}}{A_1}\sqrt{\frac{g}{2h_{10}-h_{20}}}*sign(h_{10}-h_{20}) & \frac{a_3}{A_3}\sqrt{\frac{g}{2h_{30}}} & 0 \\ \frac{a_{12}}{A_1}\sqrt{\frac{g}{2h_{10}-h_{20}}}*sign(h_{10}-h_{20}) & \frac{a_2}{A_2}\sqrt{\frac{g}{2h_{20}}}-\frac{a_{12}}{A_1}\sqrt{\frac{g}{2h_{10}-h_{20}}}*sign(h_{10}-h_{20}) & 0 & \frac{a_4}{A_2}\sqrt{\frac{g}{2h_{40}}} \\ 0 & 0 & -\frac{a_3}{A_3}\sqrt{\frac{g}{2h_{30}}} & 0 \\ 0 & 0 & 0 & -\frac{a_4}{A_4}\sqrt{\frac{g}{2h_{40}}} \end{bmatrix} \quad (5)$$

3. Model Predictive Controller

Model predictive control represents a class of control algorithm that utilizes an explicit process model to forecast the expected response of a system. At each control horizon, algorithm optimizes future process behavior by calculating a sequence of expected manipulated variable manipulation. The first input in the optimal sequence is fed to the process, and the entire practice is repeated at succeeding control horizon. Ability to handle constraints, online process optimization, simplicity in design formulation are the major key factor that make it attractive tool to researchers and practitioners [17-20]. To design MPC it is mandate to define possible constraints, cost function, predictive model and actual process model. The control scheme is shown in Fig.2.

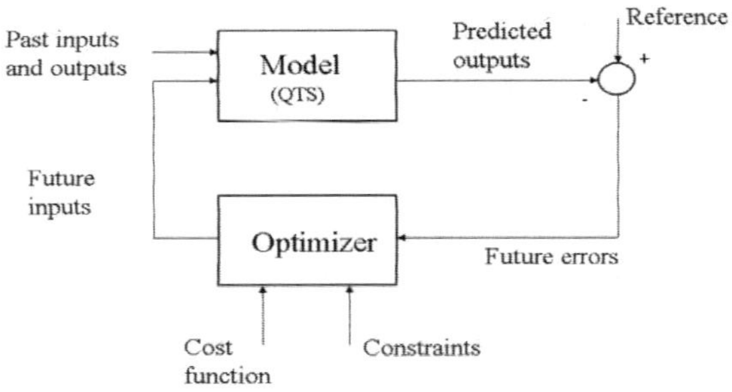

Fig. 2. Configuration of MPC

Let a state-space model of linear dynamic system

$$X_{k+1} = AX_k + Bu_k \tag{6}$$

$$Y_k = DX_k \tag{7}$$

for $k = k+1$, $Y_{k+1} = DX_{k+1}$

Substituting the value of X_{k+1} from Eq. (6), we get,

$$Y_{k+1} = D(AX_k + Bu_k) \tag{8}$$

$$Y_{k+1} = DAX_k + DBu_k) \tag{9}$$

for $k = k+2$, $Y_{k+2} = D(AX_{k+1} + Bu_{k+1})$

$$Y_{k+2} = DA^2X_k + DBu_k + DBu_{k+1} \tag{10}$$

for $k = k+3$, $Y_{k+3} = D(AX_{k+2} + Bu_{k+2})$

$$Y_{k+3} = DA^3X_k + DA^2Bu_k + DABu_{k+1} + DBu_{k+2} \tag{11}$$

$$\underbrace{\begin{bmatrix} Y_k \\ Y_{k+1} \\ Y_{k+2} \\ Y_{k+3} \end{bmatrix}}_{Y_{k|4}} = \underbrace{\begin{bmatrix} D \\ DA \\ DA^2 \\ DA^3 \end{bmatrix}}_{O_4} x_k + \underbrace{\begin{bmatrix} 0 & 0 & 0 \\ DB & 0 & 0 \\ DAB & DB & 0 \\ DA^2B & DAB & DB \end{bmatrix}}_{H_d^4} \underbrace{\begin{bmatrix} u_k \\ u_{k+1} \\ u_{k+2} \end{bmatrix}}_{u_{k|3}} \tag{12}$$

where,

$y_{k|4} =$ output variables matrix, $u_{k|3} =$ input variables matrix

O_4 = extended observability matrix for pair (*D*, *A*) ,H_4^d = lower block triangular Toeplitz matrix for *D*, *A*, *B* matrices. Subsequently Eq. (12) becomes

$$Y_{k|4} = O_4 x_k + H_4^d u_{k|3} \tag{13}$$

The above equation when the prediction horizon is equal to L becomes,

$$Y_{k|L} = O_L x_k + H_L^d u_{k+1|L-1} \tag{14}$$

Eq. (14) when $k = k+1$, becomes

$$Y_{k+1|L} = O_L x_{k+1} + H_L^d u_{k+1|L-1} \tag{15}$$

From Eq. (7), $X_{k+1} = AX_k + Bu_k$, substituting in the above equation,

$$Y_{k+1|L} = O_L Ax_k + O_L Bu_k + H_L^d u_{k+1|L-1} \tag{16}$$

Writing the Eq. (16) in matrix form gives,

$$Y_{k+1|L} = O_L Ax_k + (O_L B \;\; H_L^d) \begin{bmatrix} u_k \\ u_{k+1|L-1} \end{bmatrix} \tag{17}$$

where, $p_L = O_L Ax_k F_L$, $F_L = (O_L B \;\; H_L^d)$, $u_{k|L} = \begin{bmatrix} u_k \\ u_{k+1|L-1} \end{bmatrix}$, $Y_{k+1|L} = p_L + F_L u_{k|L}$

Eq. (17) is the prediction model and MPC calculates the predicted output by using this model. Substituting prediction model into the cost function,

$$J_k = (Y_{k+1|L} - r_{k+1|L})^T Q (Y_{k+1|L} - r_{k+1|L}) + u_{k|L}^T R u_{k|L} \quad (18)$$

$$J_k = (p_L + F_L u_{k|L} - r_{k+1|L})^T Q (p_L + F_L u_{k|L} - r_{k+1|L}) + u_{k|L}^T R u_{k|L} \quad (19)$$

$$J_k = u_{k|L}^T (F_L^T Q F_L + R) u_{k|L} + 2F_L^T Q (p_L - r_{k+1|L}) \; u_{k|L}^T R u_{k|L}$$

$$J_k = u_{k|L}^T H u_{k|L} + 2F_L^T Q (p_L - r_{k+1|L}) + (p_L - r_{k+1|L})^T Q (p_L - r_{k+1|L}) \quad (20)$$

where, $H = (F_L^T Q F_L + R) + 2f^T u_{k|L}$, $f^T = F_L^T Q (p_L - r_{k+1|L})$, $J_0 = (p_L - r_{k+1|L})^T Q (p_L - r_{k+1|L})$

Minimizing the cost function with respect to $u_{k|L}$,

$$\frac{\delta J_k}{\delta u_{k|L}} = u_{k|L}^T H u_{k|L} + 2f^T u_{k|L} + J_0 \quad (21)$$

$$u_{k|L}^* = -H^{-1} * f \quad (22)$$

The Eq. (22) becomes the future optimal control, where simply the first element of the vector is used for further process..

4. Experimental and simulation results

A simple negative feedback control topology has been implemented which minimizes the objective function as in Eq. (18) for both set point changes and disturbance rejection. To test the effectiveness of MPC, simulation studies have been carried out on MATLAB R2017a with Intel Pentium dual CPU E2220 @ 2.40 GHz.

4.1. Simulation results & discussion. Step responses of Tank-1 & 2 are shown in Fig. 3 & 4. The sampling period is 0.1 second and the length of the prediction horizon for MPC is chosen 15. Control efforts are also shown. The results are fairly compared with decentralized PI controllers whose parameters are chosen by trial and error method e.g. Kp1 =11, Ti1= 2.35 sec-1, Kp2 =11 and Ti2 = 2.5sec-1. From the simulation results it is clear that while satisfying input constraint (0 to 5 V), MPC is capable to track the set point changes efficiently and offers lesser steady state errors, peak overshoot and settling time. The Mean Squared Error (MSE) of 0.0057 and 0.0022 has been calculated for PI and MPC based scheme respectively which shows MPC has 61.4 % lesser MSE than that of PI controller.

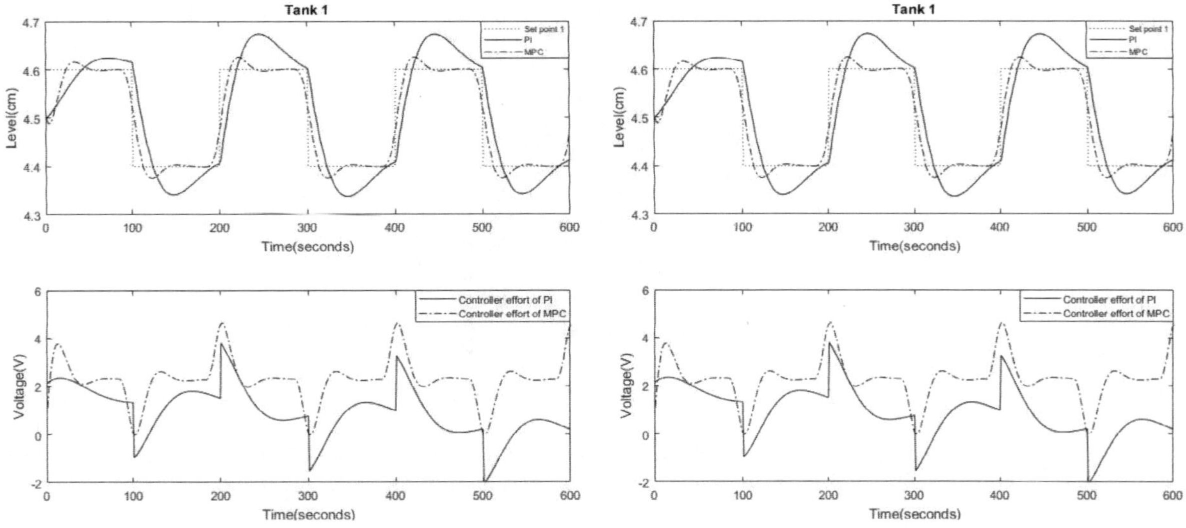

Fig. 3. Simulation results of QTS for Tank-1 **Fig. 4.** Simulation results of QTS for Tank-2

With the same objective and simulation parameters as for QTS, the whole scheme has been simulated for mQTS.

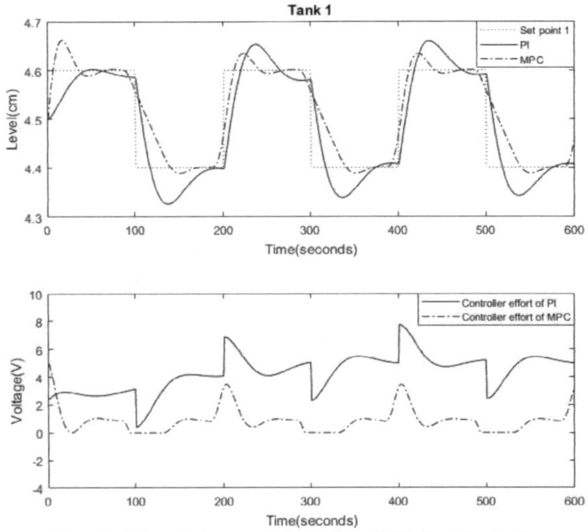

Fig. 5. Simulation results of mQTS for Tank-1

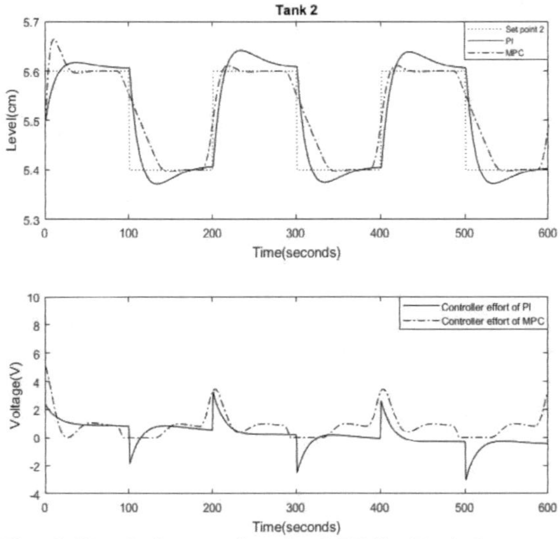

Fig. 6. Simulation results of mQTS for Tank-2

The step responses of Tank-1, Tank-2 and corresponding control efforts are shown in Fig. 5 & 6. From the simulation results it is clear that with the permissible inputs (0 to 5 V), the MPC is able to track the set point changes faster. MSE of 0.0056 and 0.0019 have been calculated for PI and MPC respectively which points towards the fact that MPC has 66.07% lesser MSE as compared to PI.

4.2. Experimental results & discussion. The experimental study has been carried out at Control laboratory of National Institute of Technology, Calicut, India. The real-time implementation of controller was carried out on LabVIEW platform which was interfaced by LTD_DAQ_013. The snap shot of QTS is depicted in Fig.7.

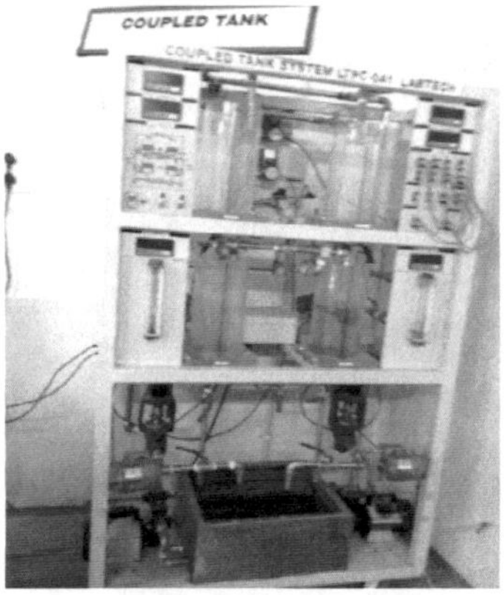

Fig. 7. Snapshot of QTS

The parameters of PI controller and MPC have been chosen as adopted in simulation studies. Experiments for both set-point changes and disturbance rejection have been carried out. The responses of PI controller and MPC are shown in Fig. 8 to 11. From the experimental results, it is noted that MPC is able to track set-point changes more smoothly. For the PI control scheme, the disturbances have been introduced by opening of outlet valves (more than 25 % of the previous opening) for Tank-1 at instant of 920 second for the duration of 40 seconds and for Tank- 2 at instant of 1290 second for the duration of 50 seconds. Whereas in MPC scheme disturbances have been introduced by closing the outlet valve for Tank-2 and by opening the valve for Tank-1, for the period of 30 secs starting at instant of 850 seconds and 750 seconds in Tank-1 & 2 respectively.

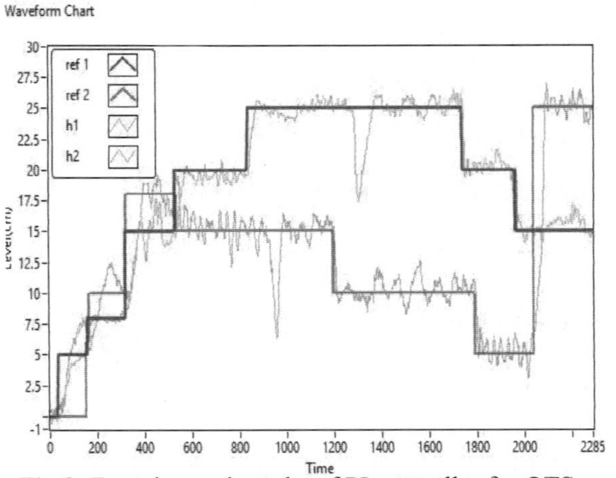

Fig 8. Experimental results of PI controller for QTS

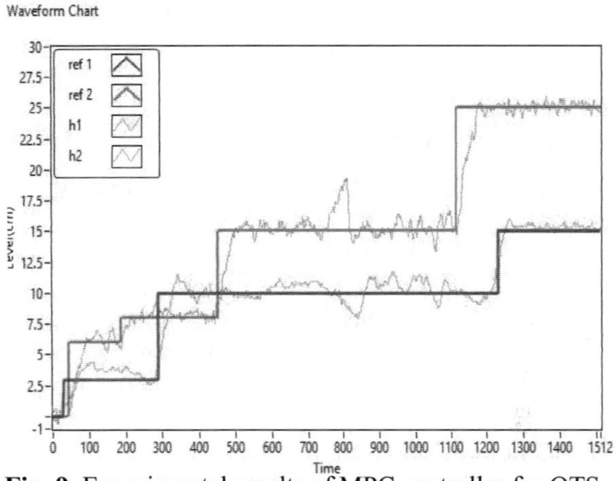

Fig. 9. Experimental results of MPC controller for QTS

It is important to note that MPC shows better results in trajectory tracking and disturbance rejection because the changes in level makes changes in dynamics of process, and a equivalent linearized model of the tank system can be utilized in real time by the MPC. However, the PI controller needs to have its parameters adjusted for optimal performance for every distinct case or when the dynamics of the process are changed by the changes in level

Furthermore, this study has been extended to mQTS. The experiment results of PI controllers and MPC are shown in Fig. 10 & 11. All results are anticipated as discussed for QTS. In spite of strong coupling between Tank -1 and Tank- 2, MPC is still capable of tracking the set point changes rapidly as compared to PI controller because PI controllers are not able to handle the strong coupling among the variables as shown in Fig. 10 at 425 seconds. In PI controller based scheme, disturbances are introduced at instant of 550 second for the duration of 30 seconds in Tank-1 and at the instant of 560 seconds for Tank-2 by opening the outlet valves of both tanks whereas the disturbances are introduced at instant of 620 second for the duration of 30 seconds in Tank-1 and at the instant of 1150 seconds for Tank-2 by opening the outlet valves of both tanks in MPC based scheme.

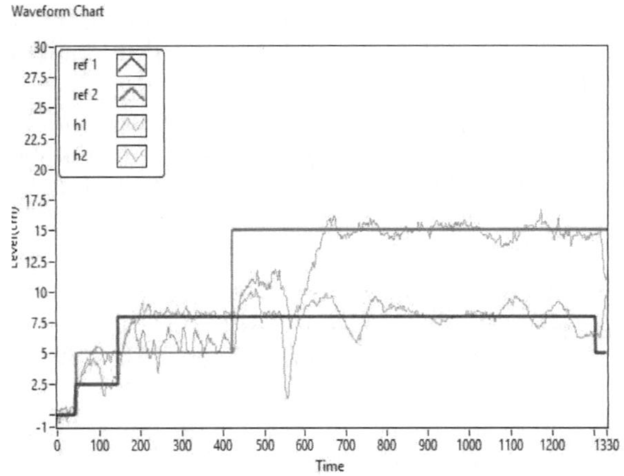

Fig. 10. Experimental results of PI controller for mQTS

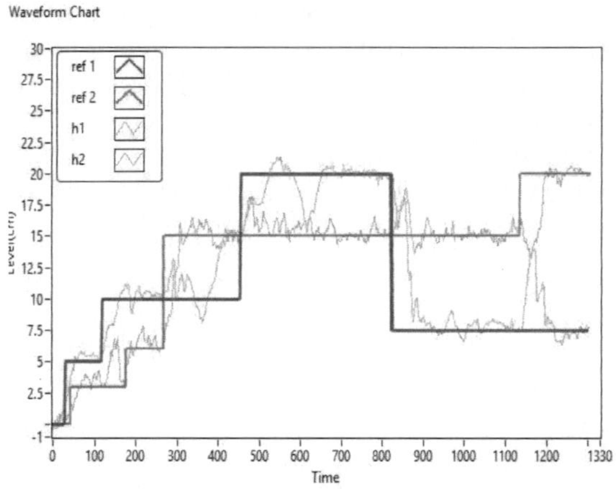

Fig. 11. Experimental results of MPC controller for mQTS

From the results, it is clearly observed that whenever a change in the set points of either of tanks occurs, the other tank gets significantly affected due to strong coupling among the variables. This effect is remarkable reduced in case of MPC.

5. Conclusions

In this work MPC based scheme has been formulated and implemented for nonlinear MIMO systems. Firstly, mathematical model of QTS has been derived and then it's linear state space model is developed. The illustrated scheme has been simulated for level control of benchmark problem of QTS. Thereafter, to test more effectiveness of the scheme, author has introduced complexity into the QTS by connecting a valve in between lower two tanks which leads to a strong interaction among the variables. The scheme has been simulated for this modified system referred as mQTS. The simulation results have been validated with real time experimental studies. The simulation results are compared within line with decentralized PI controller and it has been established that MPC is capable to track the set-points faster and also able to reject disturbances efficiently while satisfying the input constraints. MSE for MPC based scheme are found to be significantly lower than with PI controller. MPC offers an efficient technique because of it's ease use for wide application in process industry. To deal with the more complex situations, MPC strategy can be improved by using nonlinear process models and multiple objective functions.

References

[1] Seborg et al. (2011) Process dynamics and control. Wiley, India, pp 414-438

[2] P. Campo, M. Morari (1994) Achievable closed-loop properties of systems under decentralized control: conditions involving the steady-state gain. IEEE Transactions on Automatic Control vol. 39(5):392-394.

[3] K. Åström, K. Johansson, Q. Wang (2001) Design of decoupled PID controllers for MIMO systems. Proceeding of the American Control Conference, Arlington, 2015-2020.

[4] J. Freudenberg, R. Middleton (1996) Design rules for multivariable feedback systems. Proceedings of the Conference on Decision and Control, Kobe, Japan 1980-1985.

[5] S. Skogestad, I. Postlethwaite (1996) Multivariable Feedback Control: Analysis and Design. Wiley, New York.

[6] F. Garrelli, R.J. Mantz, H. De Battista (2006) Limiting interaction in decentralized control of MIMO systems. Journal of Process Control 16(5):437-483.

[7] E. Bristol (1996) On a new measure of interaction for multivariable process control. IEEE Transactions on Automatic Control 11:133– 134.

[8] W.J. Cai, W. Ni, M.J. He, C.Y. Ni (2008) Normalized Decoupling-A New Approach for MIMO Process

IOP Conf. Series: Materials Science and Engineering **594** (2019) 012038 doi:10.1088/1757-899X/594/1/012038

Control System Design. Ind. Eng. Chem. Res. 47 (19):7347-6356.
[9] K.H. Johansson (2000) The Quadruple Tank process: a multivariable Laboratory process with an adjustable zero. IEEE Transactions on Control system technology 8:456-465.
[10] W. Bequette (2003) Process Control: Modeling, Design and Simulation, Prentice Hall professional.
[11] García, C. E., D. M. Prett, M. Morari (1989) Model predictive control: theory and practice – a survey. Automatica 25:335.
[12] Eduardo F Camacho, Carlos Bordons (2003) Model predictive control, 3rd edn. Springer, London.
[13] David Di Ruscio (2013) Model Predictive Control with Integral Action: A simple MPC algorithm, Modeling, Identification and Control. 34:119–129.
[14] Liuping Wang (2009) Model Predictive Control system design and implementation using MATLAB. Springer, London.
[15] Lars Grune, Jurgen Pannek (2011) Nonlinear Model Predictive Control, Theory and Algorithms. Springer, London.
[16] Masood Askari, Mahmoud Moghavvemi, Hasider A.F. Almurib (2016) Multivariable Offset-free model Predictive controller for Quadruple tank system. IEEE transactions on industry applications. 52: 2.
[17] B. Vatankhah, M. Farrokhi (2014) Offset-free adaptive nonlinear model predictive control for pneumatic servo system. 2nd RSI/ISM Int.Conf. Robot. Mechatron. 896–901.
[18] K. Horváth, E. Galvis, M. G. Valentín, J. Rodellar (2015) New offset free method for model predictive control of open channels. Control Eng.Pract 41:13–25.
[19] V. Kirubakaran, T. K. Radhakrishnan, N. Sivakumaran (2014) Distributed multiparametric model predictive control design for a quadruple tank process Measurement. 47:841–854.
[20] A. Abdullah and M. Zribi (2014) Control schemes for a quadruple tank process. Int. J. Computer, Communication and Control 7:594–605.

IOP Conf. Series: Materials Science and Engineering **594** (2019) 012039 doi:10.1088/1757-899X/594/1/012039

Influence of antioxidant on performance of CI engine using waste fried oil methyl ester

Jagannath Balasaheb Hirkude*

Goa College of Engineering, Goa, India,
E-mail: *jbh@gec.ac.in

Abstract. The future of biodiesel is bright. However, biodiesel is vulnerable to oxidative degradation due to autoxidation. Adding antioxidant is a probable answer to achieve clean burning and improved combustion. This paper addresses experimental study the influence of antioxidant with WFO biodiesel on diesel engine performance. WFOME was produced through transesterification process. BHA and BHT have been used as antioxidants. Increased oxidation stability of B40 (60% diesel + 40% Biodiesel) was observed with accumulation of antioxidants. A 5HP engine was preferred for this study. The outcomes illustrate that antioxidant-treated B40 produced 2.15% to 4.78% higher brake thermal efficiency and BSFC by 2.08% to 4.74% compared to untreated B40. Reduction in nitrogen emissions in tune of 4 ppm to 8 ppm was observed during this research. Increased CO in tune of 8.54 to 17.08% has been pointed out. HC emissions were higher by 7.73 to 14.95%.

1. Introduction

So as encounter the forthcoming energy a crisis there is needed to discover the biode-gradable and environmental friendly alternate fuel. Several methodologies and regulations were followed to reduce NOx emission and soot particles because of use of biodiesel. Compression Ignition engines are a major source NOx and particulate matter. Many investigations have been carried out over past few years to achieve standard emission norms. To reduce the NOx emission effectively, Anti-Oxidants (AO) were preferred [1-2]. Additionally, a carbon dioxide emission from diesel combustion is responsible for damage of environment [3]. Antioxidants decelerate the biodiesel squalor process considerably.

Fattah et al. tested antioxidants and observed reduction of NOx emission accompanied by higher emission of CO and unburned HC [4]. Sajith et al. investigated with cerium oxide nanoparticle at 20, 40 and 60 ppm addition in jatropha biodiesel and observed significant reduction of NOx and HC [5-6]. Selvan et al. mixed cerium oxide nanoparticle of 25 ppm in diesel-biodiesel-ethanol blends and observed improved BSFC with reduction in HC, CO, NOx [7]. Hess et al. observed drop in NOx emissions because of substitution of antioxidants [8].

In the present study, the effect of antioxidant additives BHA and BHT on CI engine with B40 blends was analyzed.

2. Methodology

The biodiesel produced through transesterification process from WFO was blended with diesel. Oxidation stability was measured with help of Biodiesel Rancimat. To confirm uniform mixture 40% WFOME is mixed with 60 % diesel (B40) at the beginning of testing. The accumulation of antioxidant to B40 increased viscosity by 0.18% (Table 1).

Table 1 Fuel characterization.

Properties	B100	B40	B40+BHT	B40+BHA	Diesel	Method
Viscosity at 40°C (cSt)	6.8	5.56	5.57	5.57	4.320	ASTM D445
Specific Gravity	0.87	0.846	0.850	0.850	0.830	ASTM D941
Calorific Value (kJ/kg)	39000	41400	41250	41380	43000	ASTM D240
Oxidation stability(h)	7.4	15	24	25	59	ASTM D675
Flash Point (°C)	140	120	122	122	70	ASTM D93
Cetane Number	55	-	-	-	-	-

A typical 3.8 kW (5HP) diesel engine with a bore of 85 mm and 110 mm stroke length was used for test. Engine setup (figure 1) consists of Engine (1), Alternator (2), Electrical load bank (3), Fuel tank (4), Burette (5), Two way control valve (6), Air box (7), Orifice plate (8), U tube manometer (9), Exhaust gas analyser (10) and Exhaust gas thermocouple (11). The exhaust emissions were measured using AVL 444 Di-gas analyser. All the tests are carried out for three times under steady state condition. Table 2 presents error analysis of equipment.

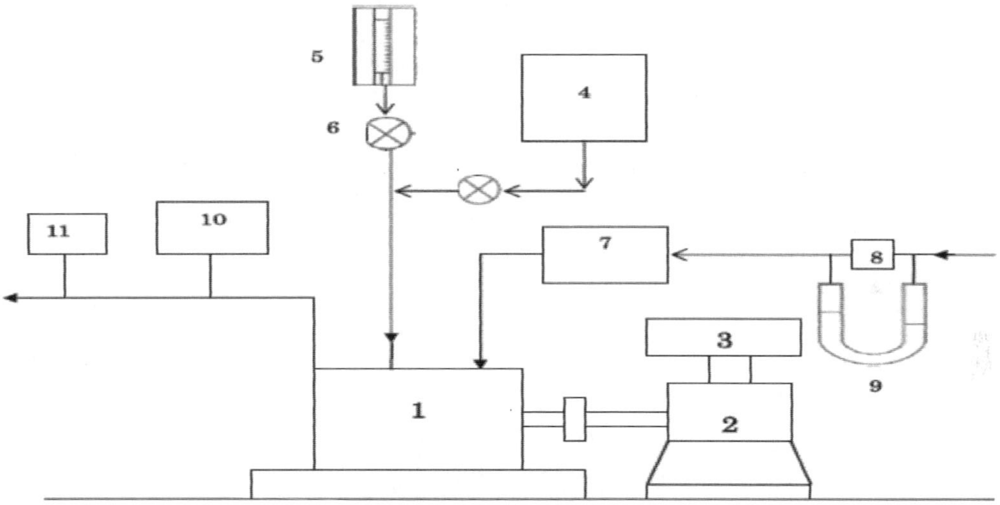

Fig. 1. Engine setup

Table 2 Error analysis

Instrument	Parameter	Accuracy	Parameter	Uncertainty
Dynamometer	Load	± 2.2 N	Torque	± 1.35 %
Digital counter	Speed	± 2 rpm	Power	± 1.35 %
Stop Watch	Time	± 0.5 %	BSEC	± 1.75 %
Temperature indicator	Temperature	± 1 °C	BTE	± 1.75 %
			NO_x	± 1 ppm
			HC	± 1 ppm
			CO Vol	±0.01 %

3. Results

3.1. Brake Thermal Efficiency (BTE)
The maximum values of BTE observed at 3.5 kW for B0, B40, B40 + BHA and B40 + BHT were 30.79%, 28.03%, 29.04% and 28.63% respectively. Antioxidants increased the mean BTE by 4.78% to 2.15% (Figure 2).

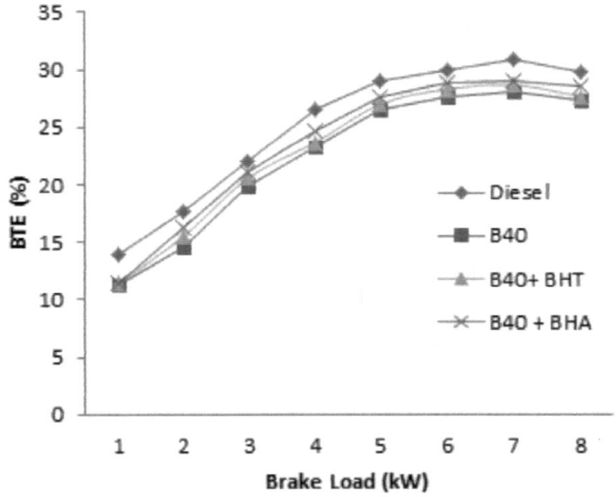

Fig. 2. BTE vs. Brake Load

3.2. Brake Specific Fuel Consumption (BSFC)
The maximum values of BSFC observed at 3.5 kW for B0, B40, B40 + BHA and B40 + BHT were 0.272, 0.310, 0.304 and 0.299 kg/kWh respectively. Antioxidants decreased the mean BSFC by 2.08 % to 4.74% (figure 3).

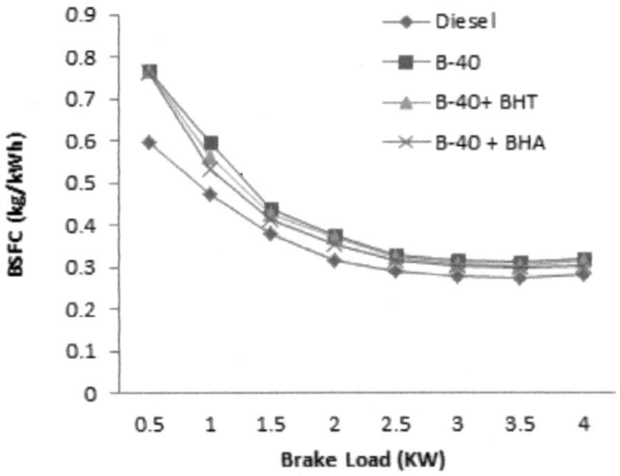

Fig. 3. BSFC vs Brake Load

3.3. Oxides of Nitrogen (NOx)
Accumulation of Antioxidant into the blends showed an encouraging impact on decreasing of emissions NOx [10]. BHA and BHT produced average decrease in NOx emission of 4 ppm to 8 ppm (figure 4). It could be due to the phenolic hydroxyl groups existent in these antioxidants interfere with the quick NOx mechanism [8].

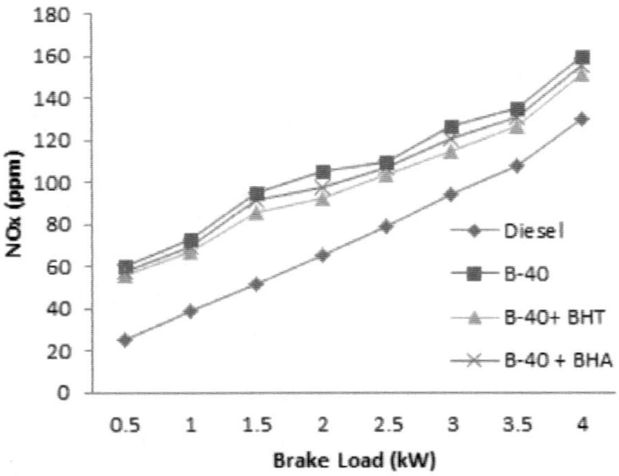

Fig. 4. NO_x Vs Brake Load

3.4. Carbon monoxide (CO)

The average CO emission reductions for were B40, B40 + BHA, B40 + BHT, 31.85%, 26.03% and 20.21%, respectively. Lesser CO emissions can be credited to higher cetane number [11]. Antioxidants increased CO emission by 8.54% and 17.08% (figure 5).

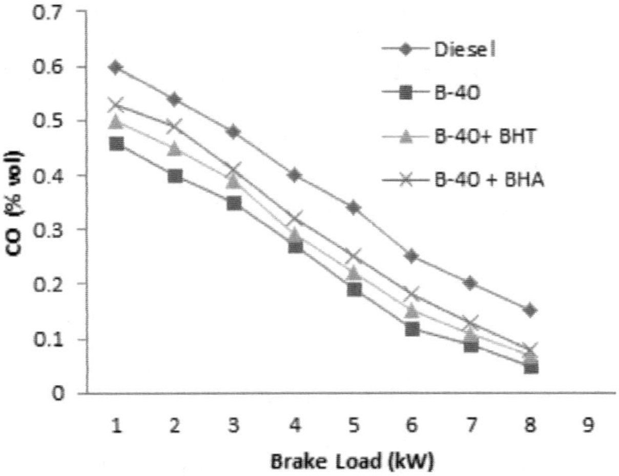

Fig. 5. CO Vs Brake Load

3.5. Hydrocarbon (HC)

Fuel spray characteristics and engine parameters setbacks HC emissions whereas improved oxygen content in the fuel can significantly reduce emissions of HC [12-13]. Decrease of oxidative free-essential materialization could be the cause of increase in HC in comparison with B40 with addition of AO [14]. B40 + BHA and B40 + BHT increased mean HC by 14.95% and 7.73%, respectively, compared to B40 (Figure 6).

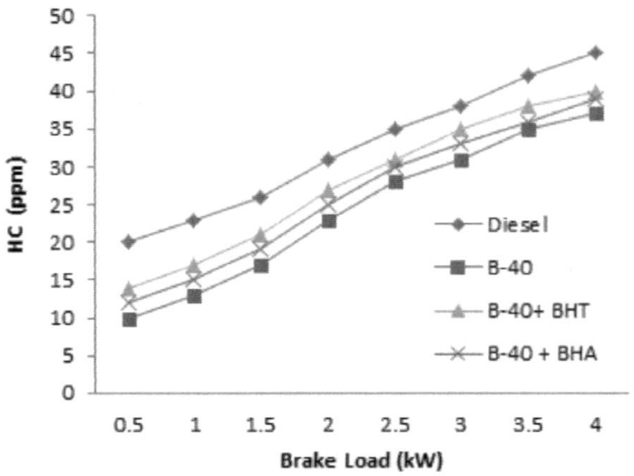

Fig. 6. HC Vs Brake Load

4. Conclusions

The following conclusions can be drawn based on the experimental investigations.

1. BHA produced better stabilization compared to BHT in B40.
2. The addition of antioxidants BHA and BHT to B40 increased the mean BTE by 4.78% and 2.15%, respectively compared to B40.
3. The accumulation of BHA and BHT to B40 decreased the mean BSFC by 2.08 % and 4.74%, respectively compared to B40.
4. Antioxidant-alleviated blends formed reductions in mean EGT of 3–8% com-pared to B40.
5. Accumulation of Antioxidant into the blends showed an encouraging impact on decreasing of emissions NOx. BHA and BHT producing a mean decrease in measured NOx emission of 4 ppm to 8 ppm in comparison to B40.
6. Mixing of BHA and BHT produced mean increases in CO emission of 8.54% and 17.08% compared to B40.
7. B40 + BHA and B40 + BHT produced mean HC reductions of 14.95% and 7.73%, respectively, compared to B40.

References

[1] Prabh A and Ramchandran B 2015 *Periodica Polytechnica Mechanical Engineering* **59**(2) pp. 88–93.

[2] Liaquat A M, Kalam M A, Masjuki H H and Jayed M H 2010 *Atmospheric Environment* **44** pp. 3869-3877.

[3] Pereira R G, Piamba Tulcan O E, Fellows C E, de Jesus Lameira V, Gonçalves Quelhas O L, Elias de. Aguiar M and Santo Filho D M d E 2012 *Journal of Cleaner Production* **29**(30) pp. 269-276.

[4] Fattah I M, Masjuki H H, Kalam M A, Mofijur M and Abedin M J 2014 *Energy Conversion and Management* **79** pp. 265–272.

[5] Sajith V, Sobhan C B and Peterson G P 2010 *Advances in Mechanical Engineering* **2** pp. 1-6.

[6] Arul Mozhi Selvan V, Anand R B and Udayakumar M 2009 *Journal of Engineering and Applied Sciences* **4**(7) pp. 1-6.

[7] Kalam M A and Masjuki H H 2008 *Biomass Bioenergy* **32** pp.1116-1122.

[8] Hess M A, Haas M J, Foglia T A and Marmer W N 2005 *Energy Fuels* **19** pp.1749-1454.

[9] Hirkude J B and Padalkar A S 2014 *Fuel* **119** pp. 266-273.

[10] Palash S M, Kalam M A, Masjuki H H, Masum B M, Fattah I M and Mofijur M *2013 Renew. Sustain. Energy Rev.* **23** pp.473-490.

[11] Hirkude J B and Padalkar A S 2012 *Applied Energy* **90** pp.68-72.

IOP Conf. Series: Materials Science and Engineering **594** (2019) 012039 doi:10.1088/1757-899X/594/1/012039

[12] Valente O S, Pasa V M D, Belchior C R P and Sodré J R 2012 *Science of the total environment* **431** pp. 57-61.
[13] Ozsezen A N, Canakci M, Turkcan A and Sayin C 2009 *Fuel* **88** pp. 629-636.
[14] Varatharajan K, Cheralathan M and Velraj R 2011 *Fuel* **99** pp. 2721-2725.

IOP Conf. Series: Materials Science and Engineering **594** (2019) 012040 doi:10.1088/1757-899X/594/1/012040

Mathematical model development of modified flow dispersion stress tensor in 2-D curvilinear flow domain

M.P. Akhtar[1], Nayan Sharma[2] and C. S. P. Ojha[3]
[1]Department of Civil Engineering, Manipal University Jaipur, India 203007.
[2]Department of Water Resources Development & Management, Indian Institute of Technology Roorkee, Roorkee-247667, India.
[3] Department of Civil Engineering, Indian Institute of Technology Roorkee, Roorkee-2467667, India
E-mail: parwez.akhtar@jaipur.manipal.edu

Abstract. For 2-D simulation of curvilinear flow field, use of momentum equations involves flow dispersion stress terms. Dispersion Stress terms take into account the effect of secondary flow variation arisen due to integration of the product of discrepancy between depth averaged velocity and the true velocity distributions. The objective of this paper is to present empirical mathematical functions to evaluate these terms. These terms can be incorporated in the 2D depth averaged flow equations as an additional source/sink term. In this work, the derivation is done to get revised set of empirical relations are later used in development of enhanced 2D numerical model. When compared with earlier investigations, the proposed formulations are simplified and numerically compatible. It is expected that modified formulation for flow dispersion stress tensor will lead to more realistic and improved simulation of flow field in curved flow domain.

1. Introduction

The empirical relations for dispersion stress terms in 2D curvilinear flow field have been given by numerous researchers. Towards this, Engelund and Skovaard [1] and Shimizu and Itakura [2] predicted the transverse velocity, however it was valid only near the bed. Lien et al. (1) used orthogonal curvilinear coordinate system and incorporated the dispersion terms derived from stream-wise and transverse velocity profile [4]. Duan [5] employed the Cartesian Co-ordinate to facilitate model application in meandering and non-meandering channels. Duan [5] found that in meandering channel, mass diffusion coefficient is much larger than turbulent diffusion coefficient. Mathematical expressions for components of dispersion coefficient tensor have been deduced by integrating the product of discrepancy between the depth averaged and actual velocity. Duan [5] deduced the dispersion terms with the assumption that the stream-wise velocity satisfies the logarithmic law. It seems that integration by Duan [5] ignored the role of boundary sub-layer formation at the bed. However, this assumption may not always hold well in many situations. For example, for very mild bed gradient with highly sub-critical flow zones in alluvial river flow case, the boundary sub-layer is rationally assumed to be intact to satisfy logarithmic law of velocity distribution. This paper attempts to deal with the derivation of flow dispersion tensor in general curved channels which are common features in braided and dynamic alluvial streams. The objective of this paper is to derive the appropriate set of mathematical expressions for dispersion stress terms for depth averaged *2D* model to be used for complex non-orthogonal curvilinear flow domain with mild bed slope.

2. Governing Equations

The governing equations for flow simulation are (Reynold's Averaged Navier Stokes) *RANS* equations with depth averaged approximation of continuity and momentum equation [Eqs.(1), (2) and (3)] in Cartesian coordinate system.

$$\frac{\partial \rho h}{\partial t}+\frac{\partial}{\partial x}(\rho h U_x)+\frac{\partial}{\partial y}(\rho h U_y)=0 \quad (1)$$

$$\frac{\partial}{\partial t}(\rho h U)+\frac{\partial}{\partial x}(\rho h U_x^2)+\frac{\partial D_{xx}}{\partial x}+\frac{\partial}{\partial y}(\rho h U_x U_y)+\frac{\partial D_{xy}}{\partial y} \quad (2)$$
$$=-\rho g h\frac{\partial H}{\partial x}-C_d \rho U_x\sqrt{U_x^2+U_y^2}+\rho h v_t\left(\frac{\partial^2 U_x}{\partial x^2}+\frac{\partial^2 U_x}{\partial y^2}\right)$$

$$\frac{\partial}{\partial t}(\rho h U_y)+\frac{\partial}{\partial x}(\rho h U_x U_y)+\frac{\partial D_{xy}}{\partial x}+\frac{\partial}{\partial y}(\rho h U_y^2)+\frac{\partial D_{yy}}{\partial y} \quad (3)$$
$$=-\rho g h\frac{\partial H}{\partial y}-C_d \rho U_y\sqrt{U_x^2+U_y^2}+h\rho v_t\left(\frac{\partial^2 U_y}{\partial x^2}+\frac{\partial^2 U_y}{\partial y^2}\right)$$

Where U_x and U_y= depth-averaged velocity components in x and y directions; t=time; ρ=density of water (kg/m^3); H=water surface elevation; h=depth of the flow; g=acceleration of gravity; Cd=frictional stress coefficient (for friction shear stress at the bottom in x and y directions); and equals $n^2\rho g / h^{\frac{1}{3}}$ with n=Manning's roughness coefficient; v_t=eddy viscosity;

2.1. Dispersion stress tensor

Components of dispersion stress terms in Cartesian Coordinate which can be included in momentum transport equations areD_{xx}, D_{xy} and D_{yy} . These terms can be expressed as follows, [5];

$$D_{xx} = \int_{z_0}^{h+z_0} \rho(u_x - U_x)^2 dz$$
$$D_{xy} = \int_{z_0}^{h+z_0} \rho(u_x - U_x)(u_y - U_y)dz \quad (4)$$
$$D_{yy} = \int_{z_0}^{h+z_0} \rho(u_y - U_y)^2 dz$$

Where z_0=zero velocity level.

For open channel free surface gravity flow, cohesive terms are non-significant and can be neglected. The depth averaged parabolic eddy viscosity model (Zero equation model) is adopted for the turbulence term. The depth averaged eddy viscosity is computed as given in Eq.(5) [6][11].

$$v_t = \frac{1}{6}\kappa U_* h \quad (5)$$

Where κ = *Von Karman' coefficient* and

U^*= *Shear velocity* = $\left[C_d\left(U_x^2+U_y^2\right)\right]^{1/2}$.

2.2. Transformed Governing Equations with dispersion stress tensor

The transformed depth averaged governing equations in generalized curvilinear coordinate system (ξ,η,τ) from continuity and momentum equation [Eqs.(1),(2) and (3)] are derived as follows.

$$\frac{\partial}{\partial \tau}(\rho h J)+\frac{\partial}{\partial \xi}(\rho h J \hat{u}_\xi)+\frac{\partial}{\partial \eta}(\rho h J \hat{u}_\eta)=0 \quad (6)$$

$$\frac{\partial}{\partial \tau}(\rho hJU_x)+\frac{\partial}{\partial \xi}\left[\rho hJ\hat{u}_\xi U_x\right]+\frac{\partial}{\partial \eta}\left[\rho hJ\hat{u}_\eta U_x\right]$$
$$-\rho hJv_t\left(\alpha_{11}\frac{\partial^2 U_x}{\partial \xi^2}+\alpha_{22}\frac{\partial^2 U_x}{\partial \eta^2}\right) \tag{7}$$
$$=-\rho ghJ\left(\xi_x\frac{\partial H}{\partial \xi}+\eta_x\frac{\partial H}{\partial \eta}\right)-C_d\rho J(U_x)\sqrt{(U_x)^2+(U_y)^2}$$
$$+\rho hJv_t\alpha_{12}\frac{\partial^2 U_x}{\partial \xi \partial \eta}-\left(\xi_x\frac{\partial D_{xx}}{\partial \xi}+\eta_x\frac{\partial D_{xx}}{\partial \eta}+\xi_y\frac{\partial D_{xy}}{\partial \xi}+\eta_y\frac{\partial D_{xy}}{\partial \eta}\right)$$

$$\frac{\partial}{\partial \tau}(\rho hJU_y)+\frac{\partial}{\partial \xi}\left[\rho hJ\hat{u}_\xi U_y\right]+\frac{\partial}{\partial \eta}\left[\rho hJ\hat{u}_\eta U_y\right] \tag{8}$$
$$-\rho hJv_t\left(\alpha_{11}\frac{\partial^2 U_y}{\partial \xi^2}+\alpha_{22}\frac{\partial^2 U_y}{\partial \eta^2}\right)$$
$$=-\rho ghJ\left(\xi_y\frac{\partial H}{\partial \xi}+\eta_y\frac{\partial H}{\partial \eta}\right)-C_d\rho J(U_y)\sqrt{(U_x)^2+(U_y)^2}$$
$$+\rho hJv_t\left(\alpha_{12}\frac{\partial^2 U_y}{\partial \xi \partial \eta}\right)-\left(\xi_x\frac{\partial D_{xy}}{\partial \xi}+\eta_x\frac{\partial D_{xy}}{\partial \eta}+\xi_y\frac{\partial D_{yy}}{\partial \xi}+\eta_y\frac{\partial D_{yy}}{\partial \eta}\right)$$

Where $\xi_x = {\partial \xi}/{\partial x}$, $\eta_x = {\partial \eta}/{\partial x}$, $\xi_y = {\partial \xi}/{\partial y}$, $\eta_y = {\partial \eta}/{\partial y}$, $\alpha_{11}=\xi_x^{\ 2}+\xi_y^2$, $\alpha_{22}=\eta_x^{\ 2}+\eta_y^2$, $\alpha_{12}=2(\xi_x\eta_y+\xi_y\eta_x)$, $J=x_\xi y_\eta - x_\eta y_\xi$,

In Eqs.(6b) to (8b) , $\widehat{u_m}$ (m=ξ, η) are the velocity components in the curvilinear coordinate (ξ,η,τ) which relate to U_x , U_y as

$$\begin{pmatrix}\hat{u}_\xi \\ \hat{u}_\eta\end{pmatrix}=\begin{pmatrix}\xi_x & \xi_y \\ \eta_x & \eta_y\end{pmatrix}\begin{pmatrix}U_x \\ U_y\end{pmatrix} \tag{9}$$

3. Derivation of dispersion stress terms in Momentum Equation

The dispersion terms resulting from the integration of the product of the discrepancy between the *mean* velocity and actual vertical velocity distribution were included in the momentum equations to take into account the effect of secondary current. Free surface flow in natural rivers is generally classified as turbulent-subcritical within the ranges of corresponding values of *Reynolds numbers* and *Froude numbers*. One of the important aspects of the free surface flow is shear velocity parameter which causes variation in velocity in different layers of fluid flow. So from the literature, one can readily assume that the stream wise velocity profile satisfies the logarithmic distribution law, *i. e.*

$$\frac{u_s}{U^*}=\frac{1}{\kappa}\ln\left(\frac{z}{z_o}\right) \tag{10}$$

Where z=vertical coordinate level (See Fig. 1), u_s=velocity in stream-wise direction and z_o is calculated according to flow Reynolds number as follows,

$$z_0 = 0.11\frac{\nu}{U^*}\text{.........,}\ \frac{U^*k_s}{\nu}\leq 5;$$
$$z_0 = 0.033k_s\text{..........,}\ \frac{U^*k_s}{\nu}\geq 70; \tag{11}$$
$$z_0 = 0.11\frac{\nu}{U^*}+0.033k_s\text{..,}5\geq\frac{U^*k_s}{\nu}\geq 70$$

Where in Eq.(11), ν = Kinematic Viscosity;

k_s = Roughness height(m); U^* = Shear Velocity

IOP Conf. Series: Materials Science and Engineering **594** (2019) 012040 doi:10.1088/1757-899X/594/1/012040

Ideally, at the bed boundary, *u* (stream-wise actual velocity) is zero; but for developing numerical scheme, value of base velocity should judiciously be taken non-zero value to ensure feasible solutions. Hence, it is well justified to exclude boundary sub-layer thickness (depth up to which boundary sub-layer is formed) and assign non-zero base velocity to achieve numerical solution close to experimental results. In other words, solid physical boundary is replaced with fluvial boundary and corresponding fluvial boundary condition has to be taken into consideration when analyzing the velocity profile vertically.

3.1. Modified approach with respect to Duan's [5] approach

Depth averaged stream wise velocity can be expressed as,

$$U_s = \frac{\int_{z_o}^{z_o+h} u_s dz}{\int_{z_o}^{z_o+h} dz} \tag{12}$$

Assuming velocity profile to be logarithmic. For the simplifying the derivation to obtain the analytical solution, one can assume

For zo<<h

$$z_o + h \cong h \tag{13}$$

In the first step, Duan(2004) apparently integrated the denominator without any approximation in Eq.(12) as follows.

$$U_s = \frac{\int_{z_o}^{z_o+h} u_s dz}{\left(z_o + h - z_o\right)} = \frac{1}{h}\int_{z_o}^{z_o+h} u_s dz \tag{14}$$

Then she approximated $\boldsymbol{z_0 + h \cong h}$ for simplifying the derivation for arriving the analytical solution as follows,

$$U_s = \frac{1}{h}\int_{z_o}^{h} u_s dz \tag{15}$$

In this way Duan [5] approximated to $\boldsymbol{z_0 + h \cong h}$ in the numerator only, however, in denominator she included zo in the upper bound of the integral in Eq.(12).

Here modification can be proposed through computing the depth averaged stream wise velocity through approximating upper bound of the integral both in the numerator and the denominator of Eq (12) through considering $\boldsymbol{z_0 + h \cong h}$ to obtain a consistent solution with mathematically symmetric approximation. Using this, one can simplify considerably in mathematical representation for dispersion stress tensor to be used in momentum transport equations in Eqs. (7) and (8). The derivation of modified approach and comparison with Duan's approach is discussed in detail in subsequent sections. Putting the value of the expression in Eq.(13) in Eq.(12), one can have,

$$U_s = \int_{z_0}^{h} u_s dz \Big/ \int_{z_0}^{h} dz \tag{16}$$

Integrating the logarithmic velocity profile along the depth, from Eq.(10), one can finally deduce an expression as follows.

$$U_s = \frac{\int_{z_o}^{h} u_s dz}{h - z_o} = \frac{1}{h}\frac{\int_{z_o}^{h} u_s dz}{\left(1 - \frac{z_o}{h}\right)} \tag{17}$$

IOP Conf. Series: Materials Science and Engineering **594** (2019) 012040 doi:10.1088/1757-899X/594/1/012040

Let $\eta_0 = {z_0}/{h}$, combining Eq.(10) and Eq.(17), one gets the following expression,

$$\frac{U_s}{U^*} = \frac{1}{\kappa h(1-\eta_0)} \int_{z_0}^{h} \ln\left(\frac{z}{z_0}\right) dz, \tag{18a}$$

or,

$$\frac{U_s}{U^*} = \frac{1}{\kappa} \frac{z_0}{h(1-\eta_0)} \left[\frac{z}{z_0} \ln\left(\frac{z}{z_0}\right) - \frac{z}{z_0} \right]_{z_0}^{h}, \tag{18b}$$

or,

$$\frac{U_s}{U^*} = \frac{1}{\kappa} \frac{z_0}{h(1-\eta_0)} \left[\frac{h}{z_0} \ln\left(\frac{h}{z_0}\right) - \frac{h}{z_0} - \frac{z_0}{z_0} \ln\left(\frac{z_0}{z_0}\right) + \frac{z_0}{z_0} \right], \tag{18c}$$

or,

$$\frac{U_s}{U^*} = \frac{1}{\kappa(1-\eta_0)} \left[\ln\left(\frac{h}{z_0}\right) - 1 + \frac{z_0}{h} \right] = \frac{1}{\kappa(1-\zeta_0)} \left[\frac{z_0}{h} - 1 + \ln\left(\frac{h}{z_0}\right) \right], \tag{18d}$$

or,

$$\frac{U_s}{U^*} = \frac{1}{\kappa(1-\eta_0)} \left[-\ln(\eta_0) - 1 + \zeta_0 \right] = \frac{1}{\kappa(1-\eta_0)} \left[\eta_0 - 1 - \ln(\eta_0) \right] \tag{18e}$$

Dividing Eq. (18e) with Eq. (10), one obtains,

$$\frac{u_s}{U_s} = (1-\eta_0) \frac{\ln\left(\frac{z}{z_0}\right)}{\eta_0 - 1 - \ln(\eta_0)} \tag{19}$$

Rearranging the above expression, one has

$$\frac{u_s}{U_s} - 1 = \frac{(1-\eta_0)\ln\left(\frac{z}{\eta_0 h}\right) - \eta_0 + 1 + \ln \eta_0}{\eta_0 - 1 - \ln(\eta_0)} \tag{20a}$$

or,

$$\frac{u_s - U_s}{U_s} = \frac{(1-\eta_0)\ln\left(\frac{z}{h}\right) - (1-\eta_0)\ln \eta_0 - \eta_0 + 1 + \ln \eta_0}{\eta_0 - 1 - \ln(\eta_0)} \tag{20b}$$

As $1 - \eta_0 \cong 1$, one assumes this term as 1 in the second term of the numerator for the ease and simplification, *i.e.*,

$$u_s - U_s = \frac{U}{\eta_0 - 1 - \ln(\eta_0)} \left((1-\eta_0)\ln\left(\frac{z}{h}\right) - \eta_0 + 1 \right) \tag{21a}$$

or,

$$u_s - U_s = \frac{U_s(1-\eta_0)}{\eta_0 - 1 - \ln(\eta_0)} \left(\ln\left(\frac{z}{h}\right) + 1 \right) \tag{21b}$$

The transverse velocity profile is assumed to be linear. As proposed by Odgaard [7][8], following relation is adopted for this model.

$$u_n = U_n + 2v_s\left(\frac{z}{h} - \frac{1}{2}\right) \tag{22}$$

Where u_n , U_n, and v_s are transverse velocity, depth averaged transverse velocity and the transverse velocity at the water surface. Engelund and Skovgaard [1] derived the deviation angle of the bottom shear as follows

$$\left(\frac{\tau_n}{\tau_s}\right)_b \approx \left(\frac{u_n}{u_s}\right)_b = 7.0\frac{h}{r}, \tag{23}$$

Where r = radius of channel curvature and the secondary flow at the surface and the bottom are equal. Therefore Eq. (20) is used to express transverse velocity at the surface. Thus,

$$(v_s)_b = 7.0(h/r)(u_s)_b \tag{24}$$

or,

Substituting Eq.(21) in Eq.(19), one obtains (same as Duan's approach),

$$u_n = U_n + 7.0\frac{h}{r}U_s\left(\frac{z}{h}-\frac{1}{2}\right) \tag{25}$$

Let us define

$$\frac{(1-\eta_0)}{\eta_0 - 1 - \ln(\eta_0)} \equiv \gamma \tag{26}$$

Substituting Eq.(26), Eq. (21b) and Eq. (25) convert to ;

$$u_s - U_s = \gamma U_s\left(\ln\left(\frac{z}{h}\right)+1\right) \tag{27a}$$

$$u_n - U_n = 7.0\frac{h}{r}U_s\left(\frac{z}{h}-\frac{1}{2}\right) \tag{27b}$$

3.2. Expressions for Dispersion stress Tensors

The dispersion stress terms at the stream-wise and transverse directions can be expressed as.

$$D^c{}_{xx} = \int_{z_0}^{h+z_0} \rho(u_s - U_s)^2 dz \tag{28a}$$

$$D^c{}_{xy} = \int_{z_0}^{h+z_0} \rho(u_s - U_s)(u_n - U_n)dz \tag{28b}$$

$$D^c{}_{yy} = \int_{z_0}^{h+z_0} \rho(u_n - U_n)^2 dz \tag{28c}$$

In Eqs. (28a), (28b) and (28c) D^c_{xx} , D^c_{xy} and D^c_{yy} are dispersion stress terms in curvilinear coordinate system. Substituting Eqs.(27a, b) in Eqs.(28a,b and c), one can deduce dispersion stress tensor presented in following steps.

3.2.1. The First Dispersion Term D_{xx}^c. Substituting Eq.(24a) in Eq.(25a), one can get,

$$D^c_{xx} = \gamma^2 U_s U_s \int_{z_0}^{h}\left(\ln\left(\frac{z}{h}\right)+1\right)^2 dz \tag{29}$$

Now consider $z/h=m$, then, $dz=h.dm$, $m_1=z_0/h=\eta_0$ lower integral bound, $m_2=h/h=1$ upper integral bound. With these substitution Eq.(29) becomes

$$D^c_{xx} = \gamma^2 U_s U_s h\left[\int_{\eta_0}^{1}(\ln(m))^2 dm + 2m\int_{\eta_0}^{1}\ln(m)dz + (1-\eta_0)\right] \tag{30a}$$

or,

$$D^c_{xx} = \gamma^2 U_s U_s h\left[\begin{array}{l}\left(m(\ln m)^2 - 2m\ln m + 2m\right)_{\eta_0}^{1} \\ + 2(m\ln m - m)_{\eta_0}^{1} + (1-\eta_0)\end{array}\right] \tag{30b}$$

or,

$$D^c_{xx} = \gamma^2 U_s U_s h\left[\begin{array}{l}2-\eta_0(\ln\eta_0)^2 + 2\eta_0\ln\eta_0 - 2\eta_0 \\ +2(-1-\eta_0\ln\eta_0+\eta_0)+1-\eta_0\end{array}\right] \tag{30c}$$

One gets the final expression as,

$$D^c_{xx} = \rho\gamma^2 U_s U_s h\left[-\eta_0(\ln\eta_0)^2 - \eta_0 + 1\right] \tag{30d}$$

3.2.2. The Second Dispersion Term D_{xy}^c. Similar to the above, one can get expression for the Second Dispersion Stress Term

$$D^c_{xy} = \int_{z_0}^{h}\gamma U_s\left(\ln\left(\frac{z}{h}\right)+1\right)\times 7.0\frac{h}{r}U_s\left(\frac{z}{h}-\frac{1}{2}\right)dz \tag{31a}$$

or,

$$D_{xy}^{c} = -7.0\gamma\frac{h}{r}U_sU_s\int_{z_0}^{h}\left(\ln\left(\frac{z}{h}\right)+1\right)\left(\frac{1}{2}-\frac{z}{h}\right)dz \quad (31b)$$

or,

$$D_{xy}^{c} = -7.0\gamma\frac{h}{r}U_sU_s\left(\frac{1}{2}\int_{z_0}^{h}\ln\frac{z}{h}dz-\int_{z_0}^{h}\frac{z}{h}\ln\frac{z}{h}dz+\frac{1}{2}\int_{z_0}^{h}dz-\int_{z_0}^{h}\frac{z}{h}dz\right) \quad (31c)$$

Again taking $m=z/h$ and integrating and transforming the upper and lower bound as done earlier, and taking h common, one has ,

$$D_{xy}^{c} = -7.0\gamma\frac{h^2}{r}U_sU_s\begin{pmatrix}\frac{1}{2}[m\ln m-m]_{\eta_0}^{1}-\frac{1}{4}[2m^2\ln m-m^2]_{\eta_0}^{1}\\+\frac{1}{2}(1-\eta_0)-\frac{1}{2}[m^2]_{\eta_0}^{1}\end{pmatrix} \quad (32a)$$

$$D_{xy}^{c} = -7.0\gamma\frac{h^2}{r}U_sU_s\frac{1}{4}\begin{pmatrix}2[-1-\eta_0\ln\eta_0+\eta_0]\\-[-1-2\eta_0^2\ln\eta_0+\eta_0^2]\\+2(1-\eta_0)-2[1-\eta_0^2]\end{pmatrix} \quad (32b)$$

or,

$$D_{xy}^{c} = -1.75\gamma\frac{h^2}{r}U_sU_s\left(\begin{bmatrix}-2-2\eta_0\ln\eta_0+2\eta_0\\+1+2\eta_0^2\ln\eta_0-\eta_0^2\\+2-2\eta_0-2+2\eta_0^2\end{bmatrix}\right) \quad (32c)$$

or,

$$D_{xy}^{c} = -1.75\gamma\frac{h^2}{r}U_sU_s\left([-2\eta_0\ln\eta_0+2\eta_0^2\ln\eta_0+\eta_0^2-1]\right) \quad (32d)$$

or,

or,

$$D_{xy}^{c} = -1.75\gamma\frac{h^2}{r}U_sU_s\left(\begin{bmatrix}-2\eta_0\ln\eta_0(1-\eta_0)\\-(1+\eta_0)(1-\eta_0)\end{bmatrix}\right) \quad (32e)$$

or,

$$D_{xy}^{c} = 1.75\rho\gamma\frac{h^2}{r}U_sU_s(1-\eta_0)(2\eta_0\ln\eta_0+\eta_0+1) \quad (32f)$$

3.2.3. The Third Dispersion Term D_{yy}^{c}. Using the similar procedure as above, one can obtain the expression for D_{yy}^{c}

$$D_{yy}^{c} = (7)^2\frac{h^2}{r^2}U_sU_s\int_{z_0}^{h}\left(\frac{z}{h}-\frac{1}{2}\right)^2 dz \quad (33a)$$

or,

$$D_{yy}^{c} = 49.0\frac{h^2}{r^2}U_sU_s\left(\int_{z_0}^{h}\left(\frac{z}{h}\right)^2 dz-\int_{z_0}^{h}\left(\frac{z}{h}\right)dz+\frac{1}{4}\int_{z_0}^{h}dz\right) \quad (33b)$$

Again taking $m=z/h$ and integrating and transforming the upper and lower bound as done earlier, one can obtain,

$$D_{yy}^{c} = 49.0\frac{h^3}{r^2}U_sU_s\left(\left[\frac{m^3}{3}\right]_{\eta_0}^{1}-\left[\frac{m^2}{2}\right]_{\eta_0}^{1}+\frac{1}{4}(1-\eta_0)\right) \quad (34a)$$

or,

$$D_{yy}^{c} = 49.0\frac{h^3}{r^2}U_sU_s\left(\frac{1}{3}-\frac{\eta_0^3}{3}-\left(\frac{1}{2}-\frac{\eta_0^2}{2}\right)+\frac{1}{4}-\frac{\eta_0}{4}\right) \quad (34b)$$

or,

$$D_{yy}^{c} = 49.0\frac{h^3}{r^2}U_sU_s\left(\frac{1}{3}-\frac{\eta_0^3}{3}-\frac{1}{2}+\frac{\eta_0^2}{2}+\frac{1}{4}-\frac{\eta_0}{4}\right) \tag{34c}$$

or, finally one may obtain,

$$D_{yy}^{c} = 49.0\rho\frac{h^3}{r^2}U_sU_s\left(-\frac{\eta_0^3}{3}+\frac{\eta_0^2}{2}-\frac{\eta_0}{4}+\frac{1}{12}\right) \tag{34d}$$

The relation between depth averaged velocities in curvilinear coordinates and Cartesian coordinate can be given as [5]

$$U_x = U_s\cos\theta_s + U_n\cos\theta_n \tag{35a}$$

$$U_y = U_s\sin\theta_s + U_n\sin\theta_n \tag{35b}$$

Where θ_s and θ_n are angles between stream-wise, transverse directions pointing outward and positive x-axis respectively. Similarly the dispersion terms in Cartesian coordinates can be related to that in curvilinear coordinates as follows[5].

$$D_{xx} = D_{xx}^{c}\cos^2\theta_s + 2D_{xy}^{c}\cos\theta_s\cos\theta_n + D_{yy}^{c}\cos^2\theta_n \tag{36a}$$

$$D_{yy} = D_{xx}^{c}\sin^2\theta_s + 2D_{xy}^{c}\sin\theta_s\sin\theta_n + D_{yy}^{c}\sin^2\theta_n \tag{36b}$$

$$D_{xy} = D_{xx}^{c}\cos\theta_s\sin\theta_s + 2D_{xy}^{c}(\cos\theta_n\sin\theta_s + \sin\theta_n\cos\theta_s + D_{yy}^{c}\sin\theta_n\cos\theta_n \tag{36c}$$

The dispersion stress terms finally obtained in Eqs.(30d, 32f and 34d) can be transformed by Eqs.(32) and Eqs.(33) to get modified dispersion stress tensor in Cartesian coordinate system.

4. Results and Discussion

The correlations by Duan [5] for dispersion stress tensor in curvilinear coordinate system are as follows.

For the statistical comparison, theoretical data for a wide rectangular channel is analysed. A qualitative comparison of variations of dispersion stresses for varying sinuosity with the modified formulation (Eqs.30d, 32f and 34d) and Duan's formulations (Eqs. 37 a, 37b and 37c) have been compared. Four configurations (*Curvature* 0.34, 0.72, 1.00 and 1.05) were chosen to cover low, moderate, high and very high sinuosity curved channels[9]. The width ratio ($\beta=B/h$, where h=water depth, B=channel width) were chosen as 10, 15, and 20. Longitudinal slope is kept as 0.001 and 0.025 for creating sub-critical and supercritical condition, respectively. Average velocities are estimated using Manning's equation (Manning'n is kept 0.025). z_o is taken as D_{50}/30 (Roughness height (k_s) is kept equal to D_{50} =0.44 mm).

It can be seen that expressions of dispersion stress terms as obtained in the present work are not in complete agreement of Duan's formulations as given in Eq. (37a). From Eq.(37b), it is apparent that any comparison between the two approaches, the value of '*C*' should have been available. However, in their paper [10] there is no *C* in Eq. (37b) .Thus, there is lack of enough insight into the adoption of any appropriate value of *C*. However, in view of Duan and Julien [10], the value of '*C*' is taken unity for comparative purposes only.

Assuming *C* as one, an attempt is made to relate Eq.(37b) with Eq. (32f), developed in the present work. Following empirical relation is obtained.

$D_{xy}^{c}=A_0 + A_1\times D_{xy}^{c}$(Duan) (R-Square=0.9854; Adjusted R-Square=0.9847)

Where A_0=-3.147 and A_1=-103.676

To appreciate the difference between the two expressions for two approaches of dispersion terms (as given in Eqs.30d, 32f, 34d and 37), certain computations are done for a variety of conditions (Table 1).

The differences do appear in the formulations of the first term (D_{xx}^{c}) and second terms (D_{xy}^{c}). The formulation for D_{xx}^{c} are different for the present approach and Duan's approach yet computed values of this term is similar and close valued, as shown in Table 1. However, the modified model (Eq. 30d) has much simpler mathematical representation than Duan's model (Eq. 37a).

Table 1 Computations of dispersion stress tensor by modified and Duan's (2004) expressions

S No.	Curve Type	Curvature	Width Ratio (β)	Longitudinal Slope	Velocity (m/s)	Froude Number	Modified Terms			Duan(2004)-Terms		
							D^c_{xx}	D^c_{xy}	D^c_{yy}	D^c_{xx}	D^c_{xy}	D^c_{yy}
1	Mild	0.34	10	0.001	0.57	0.3304	1.204	1.90	4.00	1.204	-0.0182	4.00
2		0.34	15	0.001	0.43	0.2522	0.512	0.52	0.69	0.512	-0.0068	0.69
3		0.34	20	0.001	0.36	0.2082	0.280	0.20	0.20	0.280	-0.0034	0.20
4		0.34	10	0.025	2.83	1.6521	30.098	47.56	99.88	30.101	-0.4540	99.88
5		0.34	15	0.025	2.16	1.2608	12.800	12.89	17.23	12.802	-0.1697	17.23
6		0.34	20	0.025	1.79	1.0408	6.994	5.11	4.95	6.996	-0.0844	4.95
7	Medium	0.72	10	0.001	0.57	0.3304	1.204	4.06	18.24	1.204	-0.0388	18.24
8		0.72	15	0.001	0.43	0.2522	0.512	1.10	3.15	0.512	-0.0145	3.15
9		0.72	20	0.001	0.36	0.2082	0.280	0.44	0.90	0.280	-0.0072	0.90
10		0.72	10	0.025	2.83	1.6521	30.098	101.61	455.88	30.101	-0.9699	455.88
11		0.72	15	0.025	2.16	1.2608	12.800	27.54	78.66	12.802	-0.3626	78.66
12		0.72	20	0.025	1.79	1.0408	6.994	10.92	22.61	6.996	-0.1802	22.61
13	High	1.00	10	0.001	0.57	0.3304	1.204	5.69	35.70	1.204	-0.0543	35.70
14		1.00	15	0.001	0.43	0.2522	0.512	1.54	6.16	0.512	-0.0203	6.16
15		1.00	20	0.001	0.36	0.2082	0.280	0.61	1.77	0.280	-0.0101	1.77
16		1.00	10	0.025	2.83	1.6521	30.098	142.16	892.42	30.101	-1.3570	892.42
17		1.00	15	0.025	2.16	1.2608	12.800	38.53	153.98	12.802	-0.5073	153.98
18		1.00	20	0.025	1.79	1.0408	6.994	15.28	44.26	6.996	-0.2522	44.26
19	very high	1.05	10	0.001	0.57	0.3304	1.204	5.95	39.09	1.204	-0.0568	39.09
20		1.05	15	0.001	0.43	0.2522	0.512	1.61	6.74	0.512	-0.0212	6.74
21		1.05	20	0.001	0.36	0.2082	0.280	0.64	1.94	0.280	-0.0106	1.94
22		1.05	10	0.025	2.83	1.6521	30.098	148.76	977.22	30.101	-1.4200	977.22
23		1.05	15	0.025	2.16	1.2608	12.800	40.32	168.62	12.802	-0.5309	168.62
24		1.05	20	0.025	1.79	1.0408	6.994	15.99	48.47	6.996	-0.2639	48.47

The variability of D^c_{xy} for both approaches are shown in Fig. 1. The trend of the variation is closely related (R-square=0.985), but values of D^c_{xy} differs considerably. Plot of D^c_{xy} against width ratio (β) for both approaches for variety of conditions are shown in Fig. 1.

The third term D^c_{yy} is identical in both cases as in present formulation and Duan's work. The trend of second terms is statistically similar with very low difference in mean and standard deviation (With using statistically determined '*C*' value). R-square (0.99 for D^c_{xx} and D^c_{yy}, 0.98 for D^c_{xy}) and Standard Error suggests high degree of goodness of fit for both models (For Eqs. 32f, and 37b). The inconsistency in the values of dispersion stress terms from Duan's model for different hydraulic conditions is evident in the Table 1. For example, in modified model's D^c_{xx}, D^c_{xy}, and D^c_{yy} is are varying consistently for different width ratio (β). Whereas, same terms show inconsistent variations with different β for Duan's Model. These models are developed for sub- critical flow condition; however, their trends are also analyzed for super critical flow condition. Plots of variation of different terms with width ratio for sub critical and super critical conditions are shown in Fig. s 2 and 3.

Fig. 3 shows that D^c_{xx} remains nearly constant with varying sinuosity and β. D^c_{yy} increases sharply with increasing sinuosity and reducing width ratio. For wider channels, variations are low. But for narrow bends transverse deviations in velocity is quite high. The trend remains same in case of supercritical flow condition (Fig. 3) except higher magnitude of three components of flow dispersion tensor. This is caused due to enhanced transverse mixing of the flow at high turbulence.

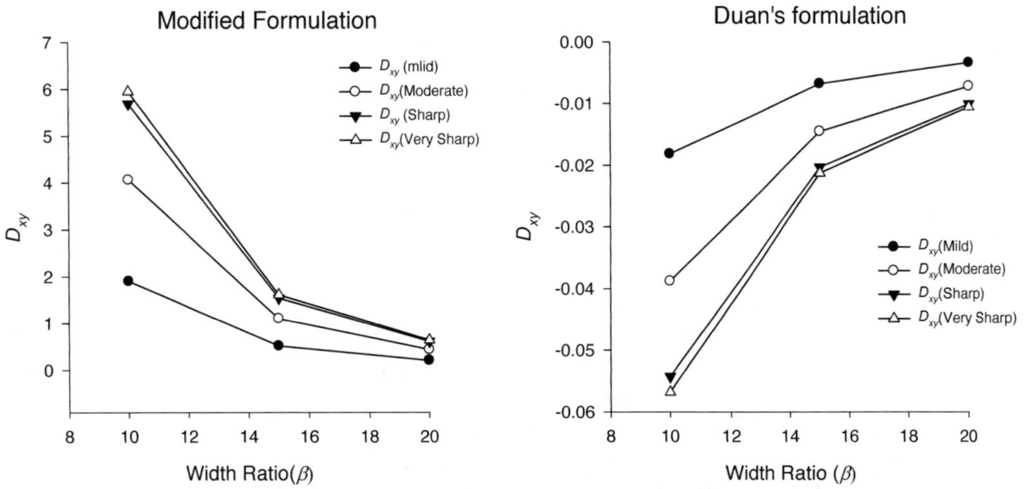

Fig. 1. Variation of D_{xy} with width ratio (β) for two approaches.

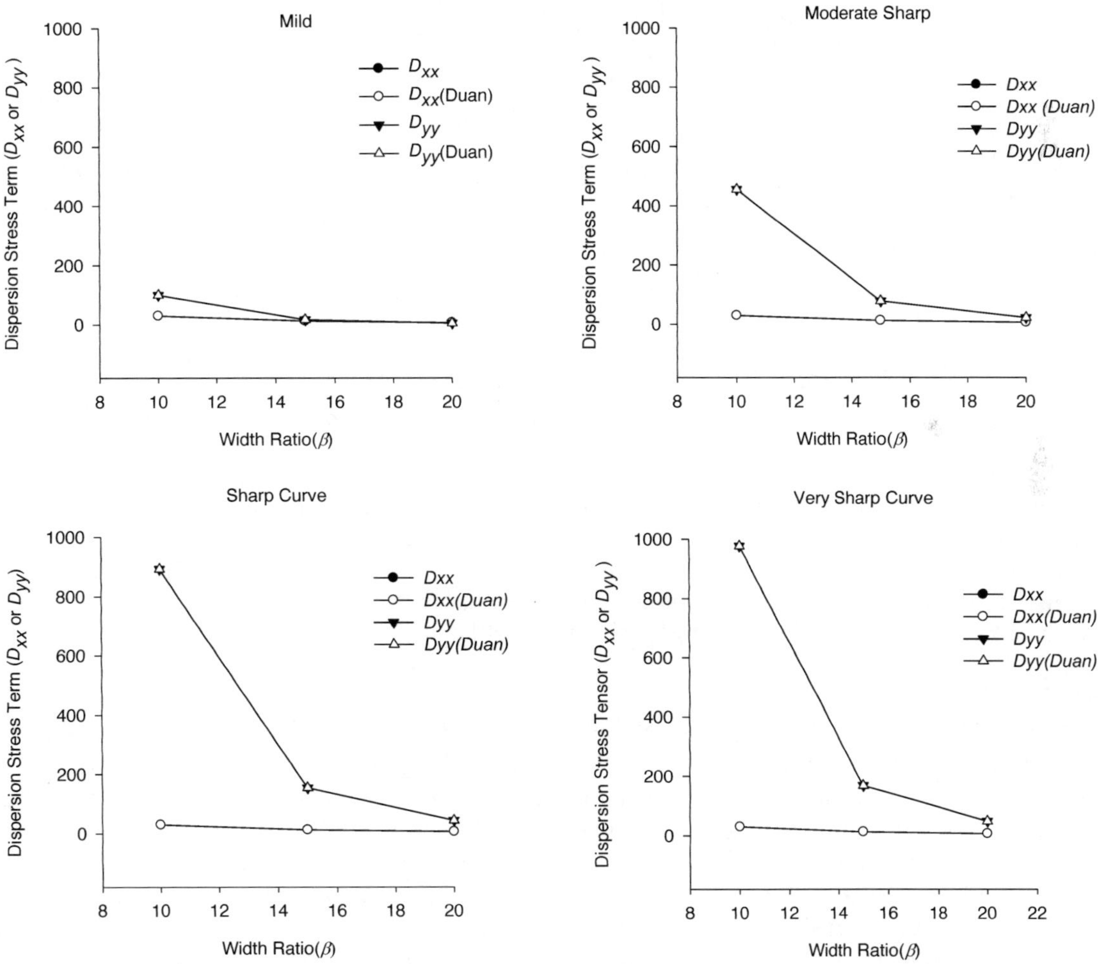

Fig. 2 Variation of D_{xx}, and D_{yy} with width ratio (β) for *sub- critical flow condition*

IOP Conf. Series: Materials Science and Engineering **594** (2019) 012040 doi:10.1088/1757-899X/594/1/012040

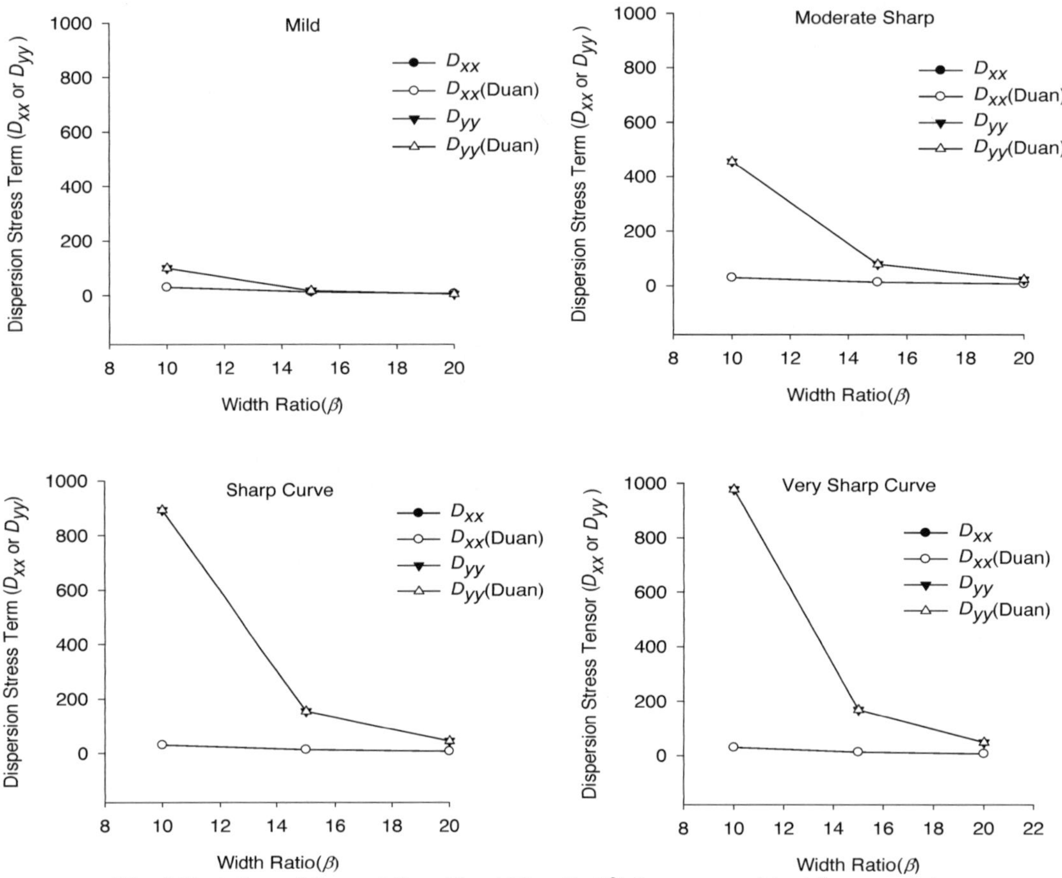

Fig. 3 Variation of D^c_{xx}, and D^c_{yy} with width ratio (β) for *super-critical flow condition*

5. Conclusions

New expressions for dispersion stress tensor are proposed. A comparison between these terms and one given by Duan, indicates the conditions in which there is a good agreement between the two. An insight is provided to estimate one of the unknown parameters in the Duan's dispersion stress tensor. Compared to Duan's model, two of the three components of dispersion stress tensor namely D^c_{xx}, D^c_{xy} are considerably simplified in the mathematical representation.

Notation

β = width ratio; C_d =frictional stress coefficient; D^c_{xx} , D^c_{xy} and D^c_{yy}=components of dispersion coefficient tensor in curvilinear coordinates;D_{xx}, D_{xy} and D_{yy} =components of dispersion coefficient tensor in Cartesian coordinates; g =acceleration of gravity; H =water surface elevation; h =flow depth; κ_s =roughness height; n =Manning's roughness coefficient; η_0=dimensionless zero bed elevation; ξ=stream-wise direction; η=transverse direction; θ_s ,θ_n=angles between the stream-wise and transverse directions towards outer bank and the positive x axis; $\varkappa$ = *Von Karman's* constant; ν=kinetic viscosity; ν_t =eddy viscosity; ϱ=density of flow; r=radius of curvature; t=time; $\widehat{u_m}$ =depth averaged velocity in curvilinear coordinate system; U =magnitude of depth averaged velocity; U^*= shear velocity; U_x, U_y =depth averaged velocities in Cartesian coordinates; u_n, U_n=transverse actual and depth averaged velocities; u_s, U_s= Stream-wise actual and depth averaged velocities; x, y = horizontal Cartesian coordinate; z=vertical coordinate; z_0=zero velocity level.

IOP Conf. Series: Materials Science and Engineering **594** (2019) 012040 doi:10.1088/1757-899X/594/1/012040

References

[1] Engelund F and Skovgaard O 1973 On the origin of meandering and braiding in alluvial streams J. Fluid Mech **57** 289–302

[2] Shimizu Y and Itakura T 1989 Calculation of bed variation in alluvial channels J. Hydraul. Eng. **115**(3) 367–384

[3] Lien H C Hsieh T Y Yang J C and Yeh K C 1999 Bend flow simulation using 2D depth-averaged model J. Hydraul. Eng. **125**(10) 1097–1108

[4] de Vriend H J 1977 A mathematical model of steady flow in curved shallow channel *J. Hydraul. Res.* **15**(1) 37–54

[5] Duan J G 2004 Simulation of flow and mass dispersion in meandering channels *J Hydraul. Eng* **130**(10) 964-976

[6] Kalkwijk J P T and De Vriend H J 1980 Computation of the flow in Shallow River bends J. Hydraul. Res. **18**(4) 327–342.

[7] Odgaard A 1989a River meander model I: Development. J. Hydraul. Eng. **115**(11) 1433–1450

[8] Odgaard A 1989b River meander model II: Application J. Hydraul. Eng. **115**(11) 1451–1464

[9] Abad J D and Garcia M H 2005 Hydrodynamics in kinoshita-generated meandering bends: importance for river-planform evolution *In Proceedings of the 4th IAHR Symposium on River, Coastal and Estuarine Morphodynamics* RCEM. 4–7 October 2005. Parker G Garcia MH (eds) Taylor and Francis Balkema Urbana, Il

[10] Duan J G and Julien Y P 2005 Numerical simulation of the inception of channel Meandering *Earth Surface Process Landforms* **30** 1093–1110

[11] Zhou J G 1995 Velocity-depth coupling in shallow-water flows *J. Hydraul. Eng.* **121**(10) 717–724

IOP Conf. Series: Materials Science and Engineering **594** (2019) 012041 doi:10.1088/1757-899X/594/1/012041

Design and Implementation of LiFi Communication system

D Ramananda[1], Amar Michael Sequeira[2], Sagar R Raikar[3] and Chandan Kumar Shanbhag[2]

[1]Department of Physics, Bhandarkars' Arts & Science College, Kundapura, Karnataka, India
[2]Department of Computer Science, Bhandarkars' Arts & Science College, Kundapura, Karnataka, India
[3]Department of Electronics and Communication Engineering, Indian Institute of Information Technology, Dharwad, Karnataka, India
E-mail: *msamar25@gmail.com

Abstract: Visible Light Communication technology is considered complimentary to Radio Frequency (RF) technology. More specifically, Light Fidelity (LiFi) based Visible Light Communication system appears to have enhanced wireless capability to realize IoT, 5G Cellular System conforming to requirements such as high data rate, low latency, connectivity etc. Recent developments in solid-state lighting technology have significantly influenced LiFi system Architecture. In view of these developments, we propose to carry out a study on feasibility of deploying LiFi technology along with the existing RF infrastructure. We emphasize to look for designing full-duplex LiFi Indoor communication module, utilizing multiple White Light Emitting Diode (WLED) based lighting equipment and integrate the module with network edge devices such as PC, smart phones for establishing efficient communication systems.

1. Introduction

We live in a modern society where several electrical modes of communication are in use, which includes among others landline telephone, television set, fax machine, mobile phone, computer with internet, personal digital assistant (PDA), smart phones and voice-assisted devices etc.

Due to miniaturization, most of these communication aids have become essential smart gadgets in the hands of current generation. It may be opt to call the significant progress in the communication area as Communication Revolution. Today everything in the communication world is connected wirelessly. It appears to be one of the fastest growing segments in modern world. Moreover, recent studies have shown that microwave radio links do not support high-speed data transmission due to complexity in establishment of backbone network and elevated cost factors.

In view of these developments, there arises a huge demand for large spectral bandwidth and an efficient alternative wireless communication technology, which adds extra capacity to existing RF infrastructure. Therefore, an efficient wireless technology combined with more spectral bandwidth as found in visible region of the electromagnetic spectrum is taken up for implementation. Among the upcoming communication technologies, the optical wireless communication (OWC) (1) is considered to be promising

IOP Conf. Series: Materials Science and Engineering **594** (2019) 012041 doi:10.1088/1757-899X/594/1/012041

one which can avoid the congested RF spectral bandwidth. Primarily, Communication system will have five blocks, which include information source, transmitter, channel, receiver, and destination blocks.

Specifically, VLC technology has been gaining more attention in both academia & industry. Visible Light Communication Consortium (VLCC) and IEEE have published two standards in 2003 and 2011 respectively (2) for VLC. In 2011, Haas et al (3) successfully demonstrated Visible Light Communication using LEDs utilized in normal lighting systems and coined the new technology as Light fidelity (LiFi). The theme of the combined effort is to design LiFi USB Dongle Integrated with smart devices. A company 'pureLiFi' has released a customizable LiFi dongle as a proof of concept for upcoming LiFi technology with capacity of supporting data rates up to 11.5 M bits/sec. In the near future, LiFi based ICs will be integrated with mobile and LED lighting – to access high-speed internet. These devices with specialized LiFi chipsets will allow end user to access the data over internet through smart LED lights in workplaces. Existing LED light bulb technology requires only the addition of a tiny microchip to become a LiFi transmitter. Similar project works are underway in different parts of the world testing the technology. In addition, LiFi can be made to use existing infrastructure for wireless backhaul. For a community progressing towards adopting eco-friendly technology, LiFi is a promising alternative. It has a wide range of applications in the field of Health-care, Security, Underwater Communication, Augmented Reality, Intelligent Transport Systems etc. Due to several interesting properties such as large bandwidth and immunity against radio interference, VLC has attracted the attention of research community. Sometimes VLC is often referred to as green communication technology.

There are of course drawbacks with the proposed LiFi technology that it is proved robust in Line-Of-Sight environment. Under bright daylight conditions, the receivers would not be able to distinguish the signal from ambient light, and unlike Wi-Fi, LiFi signal cannot pass through wall. Of course, these limitations could be overcome with smart architecture technology where the light follows the user around the space (5). Here, Algorithms will determine lighting conditions and adapt the systems to obtain better efficiency.

2. LiFi in India

According to telecom regulatory TRAI, the average data usage per user appears to have increased 23 times 45GB/month in current year. Despite the raising data usage, mobile broadband penetration (which is over 95% of total broadband connection) remains low at 30% of the total mobile connection within these base 344.57 million broadband users, nearly 67%, access data through low speed 2G/3G networks. The LiFi technology is currently in its nascent stage across the world. There are few R&D organizations based in Europe, US and India working on LiFi technology standardization and deployment.

3. LiFi

Light Fidelity is a prime application of Visible Light Communication, a wireless communication technique that carries information by modulating light signals around 400-700 nanometers. The key advantage of LiFi technology is that it can reuse the existing lighting infrastructure for providing indoor wireless data transmission along with illumination. For data transfer, Li-Fi relies on the modulation of light at high frequencies imperceptible to the human eyes.

IOP Conf. Series: Materials Science and Engineering **594** (2019) 012041 doi:10.1088/1757-899X/594/1/012041

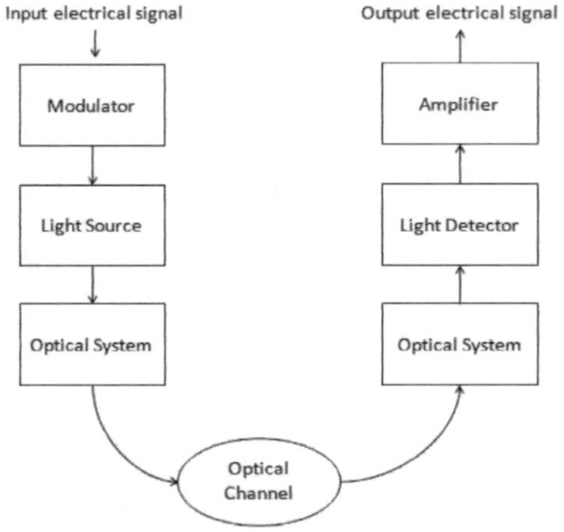

Figure 1: Block Diagram of VLC system

In Li-Fi communication system, a Digital Signal Processor integrated with an LED driver modulates digital data from a network/server for transmission through the optical medium. The LED driver in the fixture transduces digital data bit-stream into optical signal, which is transmitted at a very high frequency. These signal frequencies lie in the visible light spectra and it contributes to avoid congestion in existing RF spectrum. The decoder module in receiving terminal — say, a computer, smart phone, or any network edge device, translates the modulated signals into understandable data for end-user. The proposed LiFi module has to receive and decode data transmitted over downlink, modulate and upload signal over uplink via-LiFi.

VLC is a one-way, point-to-point optical communication technology while Li-Fi enables two-way communication over multiple frequencies for mobile and networked connections. Though VLC and Li-Fi are both "light communication technologies", the latter is a hybrid of VLC and IR communication. "The downlink—LED to photo detector—is VLC, but the uplink is infrared". In brief, LiFi module can broadly be classified into sections namely Transmitter, Optical Channel and Receiver. Modulator, LED driver and LEDs are the major components in Transmitter section. Similarly, the receiver section is composed of demodulator unit and Photodiode.

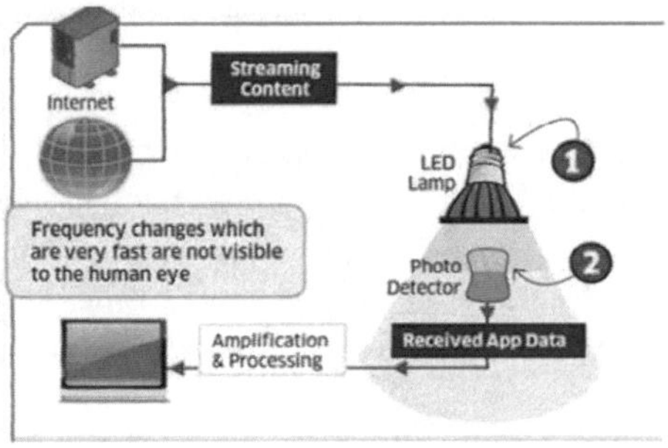

Figure 2: Typical LiFi System (6)

4. LiFi Architecture

IOP Conf. Series: Materials Science and Engineering **594** (2019) 012041 doi:10.1088/1757-899X/594/1/012041

LiFi communication system involves the full duplex communication mode. The downlink in LiFi technology uses the visible light spectrum for transmission of by intensity modulation of LEDs in solid-state lighting systems. However, uplink connection in LiFi system is designed using IR communication since uplink in Visible Light spectrum could cause interference with the downlink DataStream. In the current work, Downlink connection from Access point to Edge device of LiFi system is established.

Modulation: In VLC, the data to be transmitted is modulated in optical medium. VLC can be realized as an Intensity Modulation/Direct Detection (IM/DD) system with real valued - unipolar modulation signal. This limits the application of the modulation schemes to the techniques like On-Off Keying (OOK), Pulse Position modulation (PPM), Pulse Width Modulation (PWM) and Unipolar Pulse Amplitude Modulation (PAM). Since these modulation schemes introduce Inter Symbol Interference (ISI) in communication channel, Orthogonal Frequency Division Multiplexing (OFDM) (7) technique is considered in LiFi system. However, we have used OOK modulation technique for data modulation at the transmitter.

In this paper, a prototype design for short-range optical communication system with white LED source is proposed and demonstrated. Off-the-shelf array of white LEDs and single photodiode (PD) are utilized for analog circuitry design. Investigations into the influence of Signal-to-Noise Ratio (SNR) of communication signal and channel coding on the Bit-Error Rate (BER) performance is performed using the prototype. Further, the effects of incident LED beam angle, relative position of transmitter with respect to receiver on the horizontal coverage will be taken up for performance analysis.

4.1 Hardware/Software Setup:

In LiFi system the transmitter and receiver module constitute the main part of the system. A system controller is required for data pre-processing and modulation at the transmitter and for demodulation at the receiver end.

4.2. Transmitter Block:

In transmitter, Atmega328P microcontroller is used as system processor for pre-processing and modulation of data to be transmitted over downlink. The data to be transmitted may contain Image, Text, sound etc. Further conversion to the digital domain and data modulation/encoding is performed in controller. LED driver circuit is designed to convert the modulated data into optical signal to transmit through the optical channel.

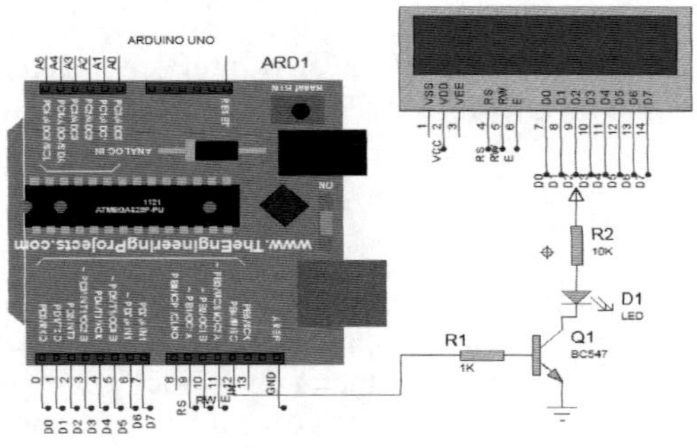

LiFi Transmitter Setup

Figure 3: LiFi Transmitter block

4.2.1 Optical Channel: In the present work we have chosen a theoretical model for optical channel given by Barry for analytical studies (8).

4.2.2 Receiver Block: At receiver end, Photodiode is used as transducer, which converts the light modulations into electrical impulses, which can be processed in system processor. Transistor based amplifier circuitry is designed to enhance the signal strength. Atmega328P microcontroller is used in the work as system processor for demodulation and processing of encoded data bits.

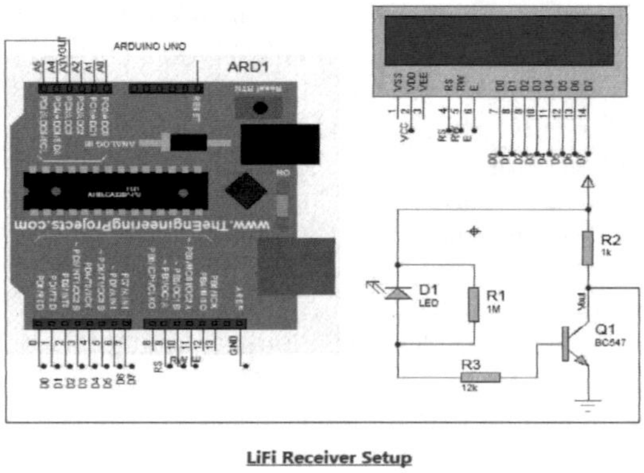

Figure 4: LiFi Receiver Block

4.2.3 Software block: In software block, the modulation and demodulation of data bits are performed. In this work, we have implemented Manchester coding (9) at the transmitter for encoding data bits and OOK modulation technique by continuous switching of LEDs. The encoding algorithm is implemented using C++ on Arduino IDE.

5. Experimental Results

We successfully designed a full duplex LiFi system where in user can request for data from server which in turn triggers a response from transmitter side. The transmitter broadcasts the data to the end user across the optical channel by modulating LEDs. We were able to achieve downlink data rates up to 10 Kbps under ambient lighting conditions. In addition, we are working to design an integrated circuitry for establishing uplink channel comprising IR LED, as it is robust towards interference caused by downlink channel. The LiFi system designed herewith could be integrated with existing Ethernet Backbone network making hybrid Wi-Fi - LiFi system.

6. Scope of Future Work

The proposed system design is very promising in terms of reliability and throughput for future high-speed wireless networks. Visible Light Communication appears to be a significant choice for a wireless indoor data transmission system. Earlier studies have shown that there does not exist a comprehensive comparison of RF and VLC communication channels. A qualitative analysis of transmission schemes with channel characteristics would help to assess the potential of LiFi technology, which can be a complement and/or alternative to RF communication systems. An effective numerical analysis and computer simulation study can be performed to validate the proposed system architecture.

7. References

IOP Conf. Series: Materials Science and Engineering **594** (2019) 012041 doi:10.1088/1757-899X/594/1/012041

[1] Wireless infrared communications. Kahn, J.R. Barry J.M. S.l.: Proceedings of the IEEE, Feb 1997, Vol. 85(2).

[2] 802.15.7, IEEE Std. IEEE standard for local and metropolitan area networks–part 15.7: Short-range wireless optical communication using visible light. 2011.

[3] What is LiFi? Haas, H. MARCH 15, 2016, JOURNAL OF LIGHTWAVE TECHNOLOGY, VOL.34

[4] Indoor visible light data transmission system utilizing white led lights. al., Y. Tanaka et. s.l.: IEICE Trans. Communication, 2003.

[5] Marr, Bernard. Will LiFi Take Big Data And The Internet Of Things To A New Level? www.forbes.com. [Online] Jan 12, 2016.

[6] Projects, Homemade Circuit. LiFi block diagram.

[7] Optical OFDM for visible light communications. Qi Wang Zhaocheng Wang, Tianqi Mao. s.l.: 13th International Wireless Communications and Mobile Computing Conference, 2017.

[8] Simulation of multipath impulse response for indoor wireless optical channels. Al, J.R.Barry et. s.l.: IEEE Journal on Selected Areas in Communications, 1993, Vol. 11(3).

[9] Maxim Integrated. Manchester Data Encoding for Radio Communications. www.maximintegrated.com. [Online] maxim integrated.

[10] Haas, Mohamed Sufyan Islim and Harald. LiFi Research and Development Centre, Institute for Digital Communications, University of Edinburgh, Edinburgh EH9: s.n., April 2016, Vol. Vol.14 No.2.

IOP Conf. Series: Materials Science and Engineering **594** (2019) 012042 doi:10.1088/1757-899X/594/1/012042

An inventory model with variable demand for non-instantaneous deteriorating products under the permissible delay in payments

Priya Mathur[1], A K Malik[2*] and Satish Kumar[3]

[1]Department of Mathematics, Poornima Institute of Engineering and Technology, Jaipur, Rajasthan, India
[2]Department of Mathematics, B. K. Birla Institute of Engineering and Technology, Pilani, Rajasthan, India
[3]Department of Mathematics, D.N. (PG), College Meerut, Uttar Pradesh, India
*Email: ajendermalik@gmail.com

Abstract. In today's competitive business deal the demand for any product is positively affected by its displayed stock level due to which the suppliers normally gave permission to his/her retailers for the delay in payment. We developed an inventory model to stem the auspicious conditions under which permissible delay in payment and displayed stock levels are beneficial for the manufacturing industry and business markets. In this decision model, for the permissible delay period; the retailer is permitted to postpone his/her payment for the products, which was bought without paying any interest. In particular, we analyze in this study the optimal renewal policy for non-instantaneous decaying products through variable demand. Furthermore, the results reveal that the optimum cost and time using the assistance of a numerical example which demonstrates the model with sensitivity analysis is carried out.

1. Introduction

In the current age the exposition of products/items in bulky quantities in the supermarkets appeals progressively consumers and produces the advanced demand. Thus the significance of inflation and multivariate demand cannot be overlooked for getting the optimal inventory policy. It has been studied that maximum number of the researchers researching on the inventory models do not assumed the case of trade credit and the inflation simultaneously. Trade credit and inflation play an important role in the optimal order policy inventory model and influences the demand of certain products/items. An EOQ model with inflation condition established first by [1]. During long ago lot of researchers discussed inventory models with the permissible delay in payment condition. First an EOQ constant demand inventory model through the environment of permissible delay in payments [2]. Expanded the inventory model [2], assuming an inventory model using fix deterioration rate [4]. First investigated the mathematical model system which optimized the total cost, cogitate the stock-dependent demand for utilization environment [3]. In the last two decades, most of the researchers have published numerous research articles with inventory-dependent demand, integrating some other key factors—such as decay, demand, permissible delay in payment, shortage, inflation; some of such articles are by [5-8]. In production and inventory management literature, stock-dependent level is an appropriate methodology to deal customers to increase the interest level to purchase more items and inventory models with demand and deteriorate rate were discussed [9-19]. Some studies were also conducted with two storage capacity with inventory model and deteriorate rate, see, for example [20-23].

2. Assumptions and Notations

IOP Conf. Series: Materials Science and Engineering **594** (2019) 012042 doi:10.1088/1757-899X/594/1/012042

For the developed inventory model, we use the subsequent notations and assumptions:

- Here the demand rate is considered as time and stock-dependent i.e., $D(t) = a + bt + cI(t)$, where $(a,b,c) \geq 0$ are positive constants and I(t) is the inventory level at time t.
- α : deterioration rate
- C_o : ordering cost per order
- C_h : inventory holding cost per unit time
- C_d : deteriorating cost per unit
- C_p : purchasing cost per unit
- C_s : sales revenue cost per unit
- R : net discount rate of inflation; $R = r - i$, where r is the discount rate which indicating the
- time value of money and i is the inflation rate
- t_m : permissible delay in payment offered by supplier in months
- I_p : interest charges per month
- I_e : interest earned per in stocks per month
- TVC_1 : total optimum inventory cost per unit time of the developed system ($t_1 \leq t_m \leq T$).
- TVC_2 : total optimum inventory cost per unit time of the developed system ($t_m \geq T$).
- $T = t_1 + t_2$: total time horizon for the developed inventory system.

3. Mathematical Model

The initial idea was to develop the model, the retailer collect the income and earn interest after consumers pay for the expanse of purchasing cost from the retailer in expectation of the conclusion of the allowable postponement in payment period accessible by the supplier. During the time interval $[0, t_1]$, the inventory level (I_1) decreases due to multivariable demand. The inventory level beads to 0 because of demand and the decay in the items during the interval is $[t_1, T]$; in time ($0 \leq t \leq t_1$) items has no deterioration and in time ($t_1 \leq t \leq T$) the items has deterioration. Thus, the proposed inventory level at some time t can be showed by the following differential equations:

$$\frac{dI(t)}{dt} = \begin{cases} -D(t), & 0 \leq t \leq t_1 \\ -D(t) - \alpha I(t), & t_1 \leq t \leq T \end{cases} \quad \text{.... (1)}$$

with the boundary conditions $I_1(0) = W$, $I_2(T) = 0$ respectively. After solving the system of equations (1), we have

$$I(t) = \begin{cases} (W + \gamma_1)e^{-ct} - \gamma_1 - \frac{b}{c}t, & 0 \leq t \leq t_1 \\ (\gamma_2 - t\gamma_3) + e^{(c+\alpha)(T-t)}(T\gamma_3 - \gamma_2), & t_1 \leq t \leq T \end{cases}, \text{ where } \gamma_1 = \left(\frac{ac-b}{c^2}\right), \gamma_2 = \left(\frac{b - a(c+\alpha)}{(c+\alpha)^2}\right), \gamma_3 = \left(\frac{b}{c+\alpha}\right) \quad \text{.... (2)}$$

Studying the continuity of inventory levels at $t = t_1$, from the system of equations (2) we have

$$W = \left(\frac{b}{c}t_1 + \gamma_1 + \gamma_2 - t_1\gamma_3\right)e^{ct_1} - \gamma_1 - e^{(c+\alpha)T}(\gamma_2 - T\gamma_3) \quad \text{...(3)}$$

For the total optimum inventory cost per cycle contains the subsequent components:

Inventory ordering cost (ICO) is CO= C_o. ...(4)

Inventory holding cost (ICH) is $ICH = C_h\left(\int_0^{t_1} e^{-Rt}\, I(t)dt + \int_{t_1}^{T} e^{-Rt}\, I(t)dt\right)$

$$= C_h\left[\begin{array}{l} \left(\frac{W+\gamma_1}{R+c}\right)\left(1 - e^{-(R+c)t_1}\right) + \left(e^{-Rt_1} - 1\right)\left(\frac{cR\gamma_1 + b}{cR^2}\right) + \frac{b}{cR}t_1 e^{-Rt_1} \\ + \frac{(\gamma_3 - \gamma_2 R)}{R^2}\left(e^{-RT} - e^{-Rt_1}\right) + \frac{\gamma_3}{R}\left(Te^{-RT} - t_1 e^{-Rt_1}\right) + \left(e^{(c+\alpha)t_2 - Rt_1} - e^{-RT}\right)\left(\frac{T\gamma_3 - \gamma_2}{R+c+\alpha}\right) \end{array}\right] \quad \text{...(5)}$$

Inventory deterioration cost (ICD) is

$$ICD = C_d \int_{t_1}^{T} \alpha\, e^{-Rt}\, I(t)dt = C_d\alpha\left[\frac{(\gamma_3 - \gamma_2 R)}{R^2}\left(e^{-RT} - e^{-Rt_1}\right) + \frac{\gamma_3}{R}\left(Te^{-RT} - t_1 e^{-Rt_1}\right) + \left(e^{(c+\alpha)t_2 - Rt_1} - e^{-RT}\right)\left(\frac{T\gamma_3 - \gamma_2}{R + c + \alpha}\right)\right] \quad \ldots(6)$$

Inventory purchasing cost (ICP) is

$$ICP = C_p \times W = C_p\left[\left(\frac{b}{c}t_1 + \gamma_1 + \gamma_2 - t_1\gamma_3\right)e^{ct_1} - \gamma_1 - e^{(c+\alpha)T}(\gamma_2 - T\gamma_3)\right] \quad \ldots(7)$$

In this developed model we have deliberate two cases with permissible delay in payment conditions:

Case-1: $t_1 \le t_m \le T$, in this instance, the interest payable is

$$IP_1 = C_p I_p \int_{t_m}^{T} I(t)\, dt = C_p I_p\left[\gamma_2(T - t_m) - \frac{\gamma_3}{2}\left(T^2 - t_m{}^2\right) - \left(\frac{T\gamma_3 - \gamma_2}{c + \alpha}\right)\left(1 - e^{(c+\alpha)(T - t_m)}\right)\right] \quad \ldots(8)$$

Interest earned is

$$IE_1 = C_s I_e\left[\int_0^{t_1} t.\left(a + bt + cI(t)\right)dt + \int_{t_1}^{t_m} t.\left(a + bt + cI(t)\right)dt\right]$$

$$= C_s I_e\left[\begin{array}{l}\frac{at_m{}^2}{2} + \frac{bt_m{}^3}{3} + \left(\frac{W + \gamma_1}{c}\right)\left(1 - e^{-ct_1}(ct_1 + 1)\right) + c\left(\frac{\gamma_2 t_m{}^2}{2} - \frac{\gamma_3 t_m{}^3}{3} - \frac{\gamma_2 t_1^2}{2} + \frac{\gamma_3 t_1^3}{3}\right)\\ -\frac{\gamma_1 c t_1^2}{2} - \frac{bt_1^3}{3} + c(T\gamma_3 - \gamma_2) \times \left\{\frac{e^{(c+\alpha)t_2}}{c + \alpha}\left(t_1 + \frac{1}{c + \alpha}\right) - \frac{e^{(c+\alpha)(T - t_m)}}{c + \alpha}\left(t_m + \frac{1}{c + \alpha}\right)\right\}\end{array}\right] \quad \ldots.(9)$$

The total optimum inventory cost (TVC₁ for $t_1 \le t_m \le$T) per unit time is

$$TVC_1 = \frac{1}{T}\left[ICO + ICH + ICD + ICP + IP_1 - IE_1\right] \quad \ldots(10)$$

For TVC_1 is minimum, the necessary and sufficient conditions are $\frac{dTVC_1}{dt_2} = 0$ and $\frac{d^2TVC_1}{dt_2^2} > 0$

Case-I1: $t_m \ge T$, in this instance, there is no interest charges are paid for the products, *i.e.*, $IP_2 = 0$ …. (11)

Interest earned is

$$IE_2 = C_s I_e\left[\int_0^{t_1} t.\left(a + bt + cI_1(t)\right)dt + \int_{t_1}^{t_m} t.\left(a + bt + cI_2(t)\right)dt + D(t)T(t_m - T)\right]$$

$$= C_s I_e\left[\begin{array}{l}\frac{at_m{}^2}{2} + \frac{bt_m{}^3}{3} + \left(\frac{W + \gamma_1}{c}\right)\left(1 - e^{-ct_1}(ct_1 + 1)\right) + c\left(\frac{\gamma_2 t_m{}^2}{2} - \frac{\gamma_3 t_m{}^3}{3} - \frac{\gamma_2 t_1^2}{2} + \frac{\gamma_3 t_1^3}{3}\right) - \frac{\gamma_1 c t_1^2}{2}\\ -\frac{bt_1^3}{3} + c(T\gamma_3 - \gamma_2) \times \left\{\frac{e^{(c+\alpha)t_2}}{c + \alpha}\left(t_1 + \frac{1}{c + \alpha}\right) - \frac{e^{(c+\alpha)(T - t_m)}}{c + \alpha}\left(t_m + \frac{1}{c + \alpha}\right)\right\} + aT(t_m - T)\end{array}\right] \quad \ldots.(12)$$

The total optimum inventory cost (TVC_2 for $t_m \ge$T) per unit time is

$$TVC_2 = \frac{1}{T}\left[ICO + ICH + ICD + ICP + IP_2 - IE_2\right] \quad \ldots(13)$$

For TVC_2 is minimum, the necessary and sufficient conditions are $\frac{dTVC_2}{dt_2} = 0$ and $\frac{d^2TVC_2}{dt_2^2} > 0$

4. Numerical Example and Sensitivity Analysis

To demonstrate the obtained solution procedure is optimal, discussed the following examples:

Ex.1. Let C_o=1000, R=0.020, C_s=200, C_p=100, t_m=0.50 month, I_p= 0.12, I_e=0.090, C_h= 0.40, C_d= 0.060, α=0.20, *a*=100, *b*=0.40 and *c*=0.20. With the help of Table 1, we note that the inventory cost (TVC_1=15298.2941) is minimum when t_1=1/2 and t_2=0.5552 month.

Table 1 Sensitivity analysis of various parameters with TVC_1

Change in		t_2	W	TVC₁	d^2TVC_1 / dt_2^2
C_o	1100	0.5691	139.2509	15402.2293	7705.7073
	1000	0.5552	137.1921	15298.2941	7774.2886
	900	0.5404	135.1027	15192.8129	7848.6454
R	0.022	0.5552	137.1925	15298.2826	7774.2000
	0.020	0.5552	137.1921	15298.2941	7774.2886
	0.018	0.5552	137.1917	15298.3054	7774.3772
C_s	220	0.5522	136.7620	15270.5827	7786.4059
	200	0.5552	137.1921	15298.2941	7774.2886
	180	0.5581	137.6206	15325.9392	7762.3329
C_p	110	0.5451	135.7286	16748.0929	8603.3540
	100	0.5552	137.1921	15298.2941	7774.2886
	90	0.5671	138.9587	13487.5674	6947.4834
t_m	0.55	0.5550	137.1722	15244.4920	7739.9378
	0.50	0.5552	137.1921	15298.2941	7774.2886
	0.45	0.5551	137.1834	15381.2366	7809.8509
I_p	0.132	0.5490	136.2906	15312.0302	7957.3070
	0.120	0.5552	137.1921	15298.2941	7774.2886
	0.108	0.5615	137.1234	15284.2563	7592.0012
I_e	0.099	0.5522	136.7620	15270.5827	7786.4060
	0.090	0.5552	137.1921	15298.2941	7774.2886
	0.081	0.5581	137.6206	15325.9392	7762.3329
C_h	0.44	0.5549	137.1494	15300.8626	7781.8833
	0.40	0.5552	137.1921	15298.2941	7774.2886
	0.36	0.5554	137.2349	15295.7248	7766.6955
C_d	0.066	0.5551	137.1911	15298.3148	7774.4802
	0.06	0.5552	137.1921	15298.2941	7774.2886
	0.054	0.5552	137.1932	15298.2733	7774.0970
α	0.22	0.5478	138.2805	15527.0320	8259.6184
	0.20	0.5552	137.1921	15298.2941	7774.2886
	0.18	0.5612	135.7944	15060.0169	7311.6776
a	110	0.5429	148.9246	16722.3020	8610.8538
	100	0.5552	137.1921	15298.2941	7774.2886
	90	0.5698	125.4287	13872.8845	6941.2343
b	0.44	0.5544	137.6409	15298.7756	7787.1988
	0.40	0.5552	137.1921	15298.2941	7774.2886
	0.36	0.5559	137.2934	15297.8085	7761.3831
c	0.22	0.5222	132.5867	15231.6321	8394.3737
	0.20	0.5552	137.1921	15298.2941	7774.2886
	0.18	0.5911	142.2234	15277.1672	7175.6456

The following graph (Fig. 1) shows that the relation among optimal inventory cost (TVC_1) and time period t_1 and t_2.

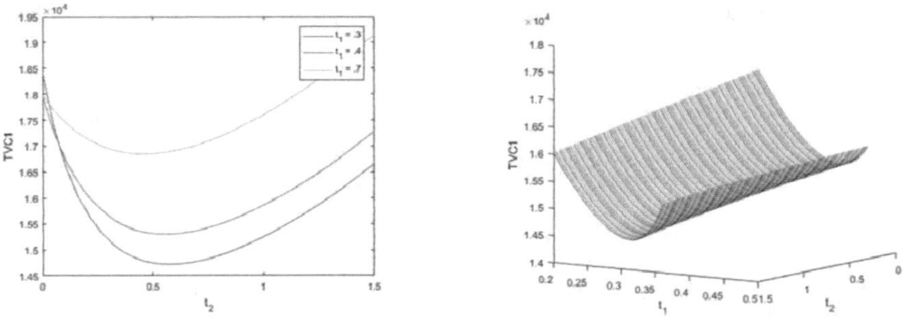

Fig.1 Graphical representation and 3-D view of optimal inventory cost TVC_1 and t_2, TVC_1 v/s t_2 and t_1

Ex.2. Let C_o=1000, R=0.020, C_s=200, C_p=100, t_m=1.20 month, I_p=0.12, I_e=0.090, C_h=0.40, C_d=0.060, α=0.20, a=100, b=0.40 and c=0.20. With the help of Table 2, we note that the inventory cost (TVC_2=13073.0029) is minimum when t_1=1/2 and t_2=0.2782 month.

Table 2 Sensitivity analysis of various parameters with TVC_2

Change in		t_2	W	TVC_2	d^2TVC_2/dt_2^2
C_o	1100	0.2945	100.8115	13360.1536	12814.6914
	1000	0.2782	98.6474	13214.4607	13073.0029
	900	0.2613	96.4355	13065.1483	13352.5294
R	0.022	0.2782	98.6475	13214.4541	13072.9157
	0.020	0.2782	96.6474	13214.4607	13073.0029
	0.018	0.2782	96.6473	13214.4673	13073.0902
C_s	220	0.2463	94.4840	12924.9591	14154.8114
	200	0.2782	98.6474	13214.4607	13073.0029
	180	0.3108	102.9664	13490.3799	12079.2127
C_p	110	0.2928	100.5801	14667.3831	13613.0830
	100	0.2782	98.6474	13214.4607	13073.0029
	90	0.2612	96.4226	11708.0814	12556.1212
t_m	1.32	0.2410	93.7991	12628.0473	13681.3124
	1.20	0.2782	98.6474	13214.4607	13073.0029
	1.18	0.3103	102.8951	13757.6501	12062.6865
I_p	0.132	0.2782	98.6474	13214.4607	13073.0029
	0.12	0.2782	98.6474	13214.4607	13073.0029
	0.108	0.2782	98.6474	13214.4607	13073.0029
I_e	0.099	0.2463	94.4820	12924.9591	14154.8114
	0.090	0.2782	96.6474	13214.4607	13073.0029
	0.081	0.3108	102.9664	13490.3799	12079.2127
C_h	0.44	0.2780	98.6299	13216.4684	13082.9031
	0.40	0.2782	98.6474	13214.4607	13073.0029
	0.36	0.2783	98.6649	13212.4527	13063.1037
C_d	0.066	0.2782	98.6470	13214.7678	13073.2505
	0.060	0.2782	98.6474	13214.4607	13073.0029
	0.054	0.2782	98.6471	13212.4636	13073.7553
	0.22	0.2833	00.8840	13442.1032	13482.8034

α	0.20	0.2782	98.6474	13214.4607	13073.0029
	0.18	0.2706	96.0150	12791.7989	12712.1172
a	110	0.2634	106.3896	14388.6666	14638.8403
	100	0.2782	98.6474	13214.4607	13073.0029
	90	0.2957	90.8586	12036.8735	11527.0669
b	0.44	0.2776	98.5609	13212.6777	13042.3132
	0.40	0.2782	98.6474	13214.4607	13073.0029
	0.36	0.2788	98.7338	13216.2310	13053.7142
c	0.22	0.2556	93.3361	13164.2535	13934.0374
	0.20	0.2782	98.6474	13214.4607	13073.0029
	0.18	0.3027	102.2805	13272.0046	12231.7768

The following graph (Fig. 2) shows the relation between optimal inventory cost (TVC_2) and time period t_1 and t_2.

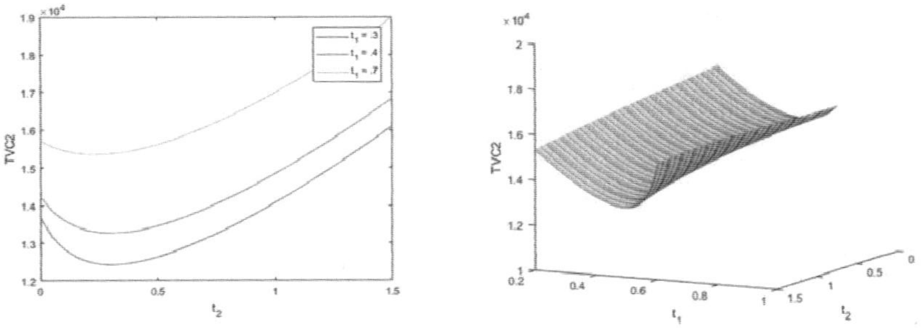

Fig.2 Graphical representation and 3-D view of optimal inventory cost TVC_2 and t_2, TVC_2 v/s t_2 and t_1

5. Conclusion

This paper deals with the idea of an inventory system beside the non-instantaneous decaying products under the inflation and trade credit conditions. For real conditions, maximum products degrade due to expiration of their maximum life time. In this proposed inventory model two examples have been taken to illuminate the significant features of the results and also sensitivity analysis of various parameters with the optimal solution are discussed. Further the promising future investigation for the study of an inventory model can be developed with production dependent model, partial backlogging, two warehouses etc.

References

1. Buzacott J A 1975 *Operation Research* **26**, 553–558.
2. Goyal S K 1985 *J. Oper. Res. Soc.* **36**, 35–38.
3. Gupta R, Vrat P 1986 *Operations Research* **24**, 41–42.
4. Aggarwal S P, Jaggi C K 1995 *J. Operat. Res. Soc.* **46**, 658- 662.
5. Liao H C, Tsai C H, Su C T 2001 *International Journal of Production Economics* **63,** 207–214.
6. Jaggi C K, Aggarwal K K, Goyal S K 2006 *International Journal of Production Economics* **103**, 707– 714.
7. Pal M, Ghosh S K 2006 *Aust. Soc. Operat. Res. Bull.* **25(3).**
8. Soni H, Shah NH 2008 *European Journal of Operational Research* **184(9)**, 1–10.
9. Chang C T, Teng J T, Goyal S K 2010 *International Journal of Production Economics* **123**, 62–68.
10. Khanra S, Ghosh S K, Chaudhuri K S 2011 *Appl. Math. Comput.* **218**, 1–9.
11. Malik A K, Singh Y 2011 *International Journal of Soft Computing and Engineering* **1(5)**, 317-321.

IOP Conf. Series: Materials Science and Engineering **594** (2019) 012042 doi:10.1088/1757-899X/594/1/012042

12. Gupta K K, Sharma A, Singh P R, Malik A K 2013 *International Journal of Soft Computing and Engineering* **3**, 279-281.
13. Singh Y, Malik A K, Kumar S 2014 *International Journal of Computer Applications* **96(25)**, 14-18.
14. Sarkar B, Sarkar S 2013 *Economic Modelling* **30**, 924–932.
15. Sarkar M, Sarkar B 2013 *Economic Modelling* **31**, 245-252.
16. Ghoreishi M, Mirzazadeh A, Weber G W 2014 *Optimization* **63**, 1785-1804.
17. Sarkar B, Mandal B, Sarkar S 2015 *Journal of Manufacturing Systems* **35**, 26-36.
18. Sharma A, Gupta K K, Malik A K 2013 *International Journal of Computer Applications* **67(25)**, 6-9.
19. Vashistha V, Tomar A, Shekhar C, Malik A K 2016 *Indian Journal of Science and Technology* **9(15)**, 1-6.
20. Singh S R, Malik A K 2008 *International Journal of Mathematical and Applications* **2(1-2)**, 141-149.
21. Singh S R, Malik A K 2009 *International Journal of Applied Mathematical Analysis and Applications* **4(1),** 59-70.
22. Sett B K, Sarkar B, Goswami A, *Scientia Iranica* **E 19**, 1969–1977.
23. Singh S R, Malik A K 2010 *International Journal of Mathematical Sciences* **9**, 513–528.

IOP Conf. Series: Materials Science and Engineering **594** (2019) 012043 doi:10.1088/1757-899X/594/1/012043

Analysis of an inventory model with both the time dependent holding and sales revenue cost

A. K. Malik[1*], Priya Mathur[2] and Satish Kumar[3]

[1]Department of Mathematics, B. K. Birla Institute of Engineering & Technology, Pilani, Rajasthan, India
[2]Department of Mathematics, Poornima Institute of Engineering and Technology, Jaipur, Rajasthan, India
[3]Department of Mathematics, D.N. (PG), College Meerut, Uttar Pradesh, India

E-mail: *ajendermalik@gmail.com

Abstract. The research paper explores the possibility of long-life products on inventory model with variable demand rate. We established a new scheme for maximum life time product; their deterioration rate is considered as a function of maximum life time. The main conception behind the planned model is: time dependent both holding and sales revenue cost for finding the optimal total profit with the optimal cycle. Finally, numerical example and sensitivity analysis demonstrate that the proposed mathematical model result and realize the effect of variation in optimal profit function with respect to several parameters used.

1. Introduction

The value of any commercial operation is contingent on deeply on the combining of demand and supply for sharing the messages form supplier to customers. Several articles appeared in dealing inventory models with constant deterioration rate but in real situations, the most of the items decaying due to finish according to their life time. Examined an inventory model with two types of decline rate: one is constant and other in variable form [1]. In early phases explored the inventory model with decaying items [5]. The works on inventory models with diminishing items at a constant decline rate were discussed by the eminent researchers [2-4], [11-14]. It is important for the inventory items, to maintain their freshness or original quality [6] provided a concept known as "non-instantaneous deterioration". A number of researchers have worked in the area of inventory control and planning in relationships of supply and demand and take care of customer satisfaction. In those scenarios; including linear deteriorating rate and time dependent function; [7-10]. Analysis an inventory model with quality improvement and controllable lead time under the backorder price discount policy [15].

2. Assumptions and Notations

For the developed inventory model the following assumptions are used:

- The demand rate, sales revenue cost and holding cost all are linear increasing function of time.
- The deterioration function $\alpha(t) = 1/(1+m-t)$ depends upon time, where m is the maximum life time of the product.
- Shortages are not allowed and lead time is considered as zero.

For the developed inventory model the following notations are used:

$a+bt$	:	the demand rate,
C_o	:	the ordering cost per order,
C_1	:	the purchasing cost per unit item,

$C_2(t)$	:	the sales revenue cost per unit item,
$C_h(t)$	:	the inventory holding cost per unit time per unit item,
$\alpha(t)$	:	the time varying deterioration rate at time t, $\alpha(t)=1/(1+m-t)$, where $0\leq\alpha(t)\leq 1$,
S	:	maximum inventory level
m	:	maximum life time of the deteriorating item,
TVP	:	the total profit per unit time of the inventory system.

3. Mathematical Model

In this section; we formulate the mathematical model for the proposed inventory system under consideration in the demand rate and sales revenue cost and holding cost all are linear. The inventory level $Q_1(t)$ represents the level in the time $[0, t_1]$ and no degradation in the product while the inventory level $Q_2(t)$ represents the level in the time $[t_1, T]$ with the degradation in the product. Thus the inventory levels in the time interval [0, T] can be symbolized by the following differential equations:

$$\frac{dQ_1(t)}{dt} = -(a+bt), \qquad 0 \leq t \leq t_1 \tag{1}$$

$$\frac{dQ_2(t)}{dt} + \alpha(t)Q_2(t) = -(a+bt), \quad t_1 \leq t \leq T \tag{2}$$

With the boundary conditions $Q_1(0) = S, Q_2(T) = 0$, respectively. Solving the above system of differential equations and using the boundary conditions, we get the stated result.

$$Q_1(t) = S - at - \frac{b}{2}t^2 \tag{3}$$

$$Q_2(t) = (1+m-t)\left[a_1 \log\left(\frac{1+m-t}{1+m-t_2}\right) - b(T-t)\right], \text{ where } a_1 = a + b(1+m) \tag{4}$$

For continuity of inventory levels at $t=t_1$, using equations (3) and (4), we have

$$S = at_1 + \frac{b}{2}t_1^2 + (1+m-t_1)\left[a_1 \log\left(\frac{1+m-t_1}{1+m-T}\right) - b(T-t_1)\right] \tag{5}$$

The optimum profit per cycle contains the following values:
The ordering cost per cycle is

$$\text{OC}= C_0 \tag{6}$$

The purchasing cost per cycle is given by

$$\text{PC}=C_1 \times S = C_1\left\{at_1 + \frac{b}{2}t_1^2 + (1+m-t_1)\left[a_1 \log\left(\frac{1+m-t_1}{1+m-T}\right) - b(T-t_1)\right]\right\} \tag{7}$$

The sales revenue cost per cycle is given by

$$\text{SRC} = \int_0^T C_2(t)(a+bt)dt = \left(C_{21}aT + \frac{T^2}{2}(C_{21}b + C_{22}a) + C_{22}b\frac{T^3}{3}\right) \tag{8}$$

The deterioration cost per cycle is cycle is given by

$$\text{DC}=C_3\int_{t_1}^{T}\alpha(t)Q_2(t)dt=C_3\left(a_1\left[-(T-t_1)(1+m-t_1)\log\left(\frac{1+m-t_1}{1+m-T}\right)-\frac{b}{2}(T-t_1)^2\right]\right) \tag{9}$$

The holding cost per cycle is given by

$$\text{HC}=\left(\int_0^{t_1}C_h(t)Q_1(t)dt+\int_{t_1}^{T}C_h(t)Q_2(t)dt\right)$$

$$=h_1\left\{\begin{array}{l}St_1-\frac{a}{2}t_1^2-\frac{b}{6}t_1^3+a_1\left(\frac{(T^2-t_1^2)}{4}+\frac{1}{2}(1+m-t_1)^2\log\left(\frac{1+m-t_1}{1+m-T}\right)-\frac{(1+m)}{2}(T-t_1)\right)\\-\frac{b}{2}(1+m)(T-t_1)^2+\frac{b}{2}T\left(T^2-t_1^2\right)-\frac{b(T^3-t_1^3)}{3}\end{array}\right\}$$

$$+h_2\left\{S\frac{t_1^2}{2}-\frac{a}{3}t_1^3-\frac{b}{8}t_1^4+a_1\left[\begin{array}{l}\frac{(T^3-t_1^3)}{9}+\frac{(1+m)}{12}\left(T^2-t_1^2\right)-\frac{(1+m)^2}{6}(T-t_1)\\+\frac{1}{6}\left((1+m)^3-3(1+m)t_1^2+2t_1^3\right)\log\left(\frac{1+m-t_1}{1+m-T}\right)\\-\frac{b}{2}(1+m)T\left(T^2-t_1^2\right)+\frac{b}{3}(1+m+T)\left(T^3-t_1^3\right)-\frac{b(T^4-t_1^4)}{4}\end{array}\right]\right\} \tag{10}$$

Thus the optimum profit (TVP) per cycle per unit time is given by

$$TVP=\frac{1}{T}[SRC-OC-PC-DC-HC] \tag{11}$$

To obtain the optimal cycle time T* we have to maximize the total present value of profit. To maximize the TVP, we have to differentiate with respect to T and equate it to zero $\frac{dTVP}{dT}=0$ and

$$\frac{d^2TVP}{dT^2}<0 \tag{12}$$

The profit function is highly non-linear, therefore; it cannot inventory system is maximized by a developed solution procedure. Also a numerical example, graphical design with the sensitivity analysis is utilized to illustrate this developed inventory model.

4. Numerical Example and Sensitivity Analysis

To demonstrate the above solution procedure, we take the following example: C_o=1000, C_{21}=200, C_{22}=0.05, C_1=140, h_1=0.40, h_2=0.040, C_3=0.06, m=25, t_1=0.5, a=600 and b=80. Using optimization techniques we obtain the optimal ordering cycle T*=5.1005 and corresponding TP*=36141.9801 and L*=4521.5008 are evaluated and tabulated in Table 1 for varying model's parameters:

Table 1 Variation of model's parameters

Change in		t_2	L	TP	d^2TP/dT^2
a	660	4.4720	4689.7099	38790.7931	-1496.0787
	600	4.6005	4521.5008	36141.9801	-1409.8421
	540	4.7401	4350.0192	33388.4626	-1328.2882
b	88	4.7244	4799.0941	37020.5683	-1468.5760
	80	4.6005	4521.5008	36141.9801	-1409.8421

	72	4.4614	4240.8804	35287.7381	-1355.7550
m	27.5	5.1669	5186.2209	37389.4365	-1258.8338
	25	4.6005	4521.5008	36141.9801	-1409.8421
	22.5	4.0249	3876.7480	35012.9895	-1610.5600
C_{21}	220	5.2106	5301.5199	52460.8665	-1230.3321
	200	4.6005	4521.5008	36141.9801	-1409.8421
	180	4.0778	3899.6065	20276.5951	-1672.8774
C_{22}	0.055	4.6026	4523.9975	36153.1026	-1408.7882
	0.05	4.6005	4521.5008	36141.9801	-1409.8421
	0.045	4.5985	4519.0071	36130.8635	-1410.8968
C_1	154	4.0137	3826.1399	24026.6064	-1771.7820
	140	4.6005	4521.5008	36141.9801	-1409.8421
	126	5.3674	5511.8895	48903.6149	-1134.3382
C_o	1100	4.6033	4524.8524	36122.3796	-1410.2905
	1000	4.6005	4521.5008	36141.9801	-1409.8421
	900	4.5978	4518.1457	36161.5911	-1409.3930
C_3	0.066	4.6004	4521.3577	36141.4854	-1409.9129
	0.06	4.6005	4521.5008	36141.9801	-1409.8421
	0.054	4.6007	4521.6439	36142.4749	-1409.7713
h_1	0.44	4.7313	4683.6746	36359.0730	-1434.3244
	0.40	4.6005	4521.5008	36141.9801	-1409.8421
	0.36	4.4599	4350.0099	35950.8155	-1383.9323
h_2	0.044	4.5620	4474.1780	35891.7277	-1427.2673
	0.04	4.6005	4521.5008	36141.9801	-1409.8421
	0.036	4.6397	4569.7190	36394.3591	-1427.2673

The sensitivity analysis of the proposed inventory model with cost parameters have been discussed in above table 1. The following features are observed, on the basis of the sensitivity analysis of the various parameters:

1. The higher value of a implies more and more demand i.e., which improve the total profit.
2. The increasing value of R implies more life time of the product which implies more profit of the system.
3. The increasing value of Sales revenue cost S_l implies more profit of the system.
4. The increasing values of purchasing cost C_1, holding cost $C_h(t)$, ordering cost C_o and deteriorating cost C_d implies higher values of the total cost which reduce the total profit.

The following graphs (Fig.1 & Fig.2) show the relation between the total profit TP with time duration t_1 and T.

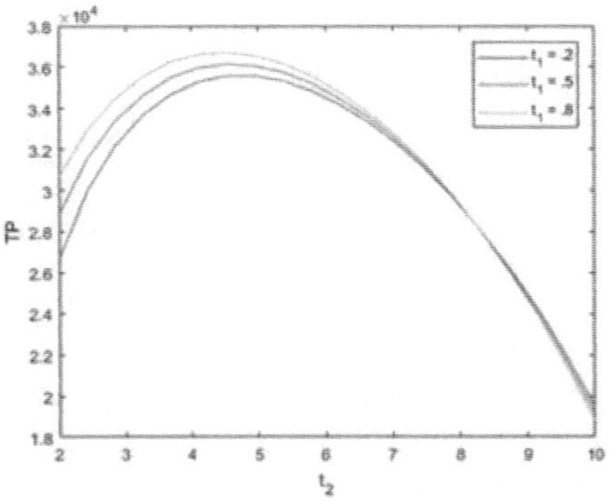

Fig. 1: Total Profit w.r.t. t_2

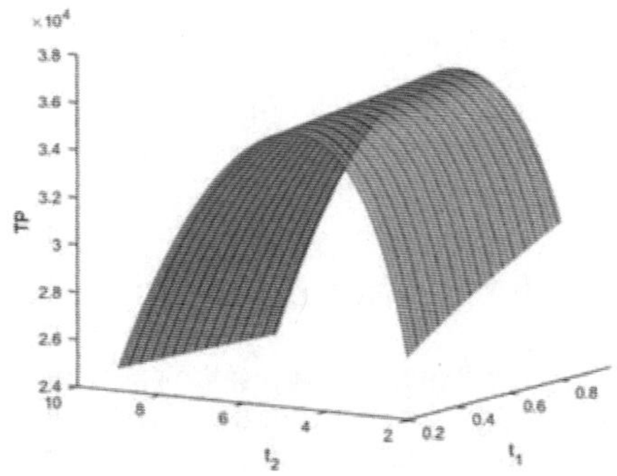

Fig. 2: Graph representation of Total Profit w.r.t. t_1 and t_2

5. Conclusion

The projected inventory model delivers the optimal solutions vigorously over analysis of the limitations of constraints involved which is time dependent and direct connected with the present market scenario. The proposed inventory system having some certain limitations such as: maximum life time of the product, variable demand rate and time dependent sales revenue cost. These important limitations could be addressed in further studies on inventory management system. Finally, we can also investigate the improved inventory model using stochastic holding cost, stock and price dependent demand, permissible delay in payment, inflation, probabilistic deterioration, reliability, two storages etc., and also soft computing techniques methods such as fuzzy logic, particle swarm optimization, etc.

References

[1]. Mishra R B 1975 *International Journal of Production Research Society* **13**, 495-505.
[2]. Shah Y K, Jaiswal M C 1977 *Opsearch* **14**, 174-184.
[3]. Aggarwal S P 1978 *Opsearch* 15, 184–187.
[4]. Chowdhury R, Chaudhuri K S 1983 *Opsearch* **20**, 99-106.
[5]. Dave U 1986 *Opsearch* **23**, 244-249.
[6]. Wu K S, Ouyang L Y, Yang CY 2006 *International Journal of Production Engineering* **101**, 369-384.
[7]. Singh S R, Malik A K 2008 *International Journal of Mathematical and Applications* **2(1-2)**, 141-149.
[8]. Singh S R, Malik A K 2010 *International Journal of Mathematical Sciences* **9**, 513–528.
[9]. Sett B K, Sarkar B, Goswami A, *Scientia Iranica* **E 19**, 1969–1977.
[10]. Sarkar B 2012 Mathematical and Computer Modelling **55**, 367–377.
[11]. Malik A K, Singh Y 2011 *International Journal of Soft Computing and Engineering* **1(5)**, 317-321.
[12]. Gupta K K, Sharma A, Singh P R, Malik A K 2013 *International Journal of Soft Computing and Engineering* **3**, 279-281.
[13]. Singh Y, Malik A K, Kumar S 2014 *International Journal of Computer Applications* **96(25)**, 14-18.
[14]. Vashistha V, Tomar A, Shekhar C, Malik A K 2016 *Indian Journal of Science and Technology* **9(15)**, 1-6.
[15]. Sarkar B, Mandal B, Sarkar S 2015 *Journal of Manufacturing Systems* **35**, 26-36.

IOP Conf. Series: Materials Science and Engineering **594** (2019) 012044 doi:10.1088/1757-899X/594/1/012044

Gain Characteristics of InGaAsN Quantum Well Heterostructures with GaAs and InP Substrates

M. I. Khan[1], Sandhya K.[2], A. M. Khan[3] and P. A. Alvi[1*]

[1]Department of Physics, Banasthali Vidyapith, Banasthali-304022, Rajasthan, India
[2]Higher Colleges of Technology, Abu Dhabi, UAE
[3]Department of Electronics & Communication Engineering, Integral University, Lucknow-226026, Uttar Pradesh, India
E-mail:*drpaalvi@gmail.com

Abstract. The present paper reports the effect of substrate on the optical gain characteristics of InGaAsN material based quantum well (QW) heterostructure. To analyze the substrate effect on the optical gain, two substrate, namely, GaAs and InP have been chosen. Taking in to account both of the substrates, the quantum mechanical approach (k.p method) has been adopted to calculate the confinement of the carriers related with quantum well and the optical gain. In addition, for InGaAsN/InP heterostructure, the cladding effect has also been observed. The simulated results exhibit that the substrate has played a very important role in modifying the optical gain characteristics due to the presence of strain which comes into play due to the lattice discrepancy.

Keywords: optical gain, strain, lattice mismatch, InP, GaAs, InGaAsN

1. Introduction

III-Nitride semiconducting heterostructures have paying attention researchers belonging to electro-optic society due to their probable utilizations in FET (field effect transistors), LEDs (light emitting diodes) [1-4], solar cells, detectors and laser diodes [5, 6]. The reason behind their vast applications is that these semiconductors based heterostructures have been found to contribute exceptional features and exclusive electronic, electrical, mechanical and optical properties which can be exploited for designing of future optoelectronic and photonic devices [7-9]. So far, III-nitrides founded photodetectors running in ultraviolet regime have exhibited promising output. Additionally, III-nitrides established LEDs had been commercialized because of emission of tremendous vibrant blue and inexperienced gentle. Additionally, these substances have some fine bodily residences which make the contraptions ready for applications past the info storage and imaging [10]. These materials are additionally valuable in high temperature electronics and area functions on the grounds that these are bodily and chemically powerful. This feature makes them ultimate for operation in crucial environment.

Keeping in views the interesting properties of III-N QW heterostructures, a thorough study has been carried out on InGaAsN QW heterostructure, in the following sections, by optimizing its valence band mechanism and optical gain characteristics taking into account two different substrates (i.e. GaAs, and InP). In addition, a slight attention has also been paid towards the role of claddings on the gain characteristics of InGaAsN/InP heterostructure.

2. Theory

In any quantum well devices which can absorbor produce photonic energies, the confinement of the carriers and radiations is a big issue. The confinement of the carriers and radiations should be larger as much as possible to show a good output. Hence, to estimate the confinement of the carriers allied with

IOP Conf. Series: Materials Science and Engineering **594** (2019) 012044 doi:10.1088/1757-899X/594/1/012044

the quantum well, 6 × 6 Luttinger-Kohn Hamiltonian has been solved [11-13]. By deliberating the carrier's confinements and their localities, the optical transitions elements have been calculated followed by gain computations.

Since the proposed heterostructures are symmetrical (i.e. heterostructures containing the QW sandwiched by barriers followed by claddings on both the sides) and, consequently, for symmetrical structure, the optical gain can be expressed as [12];

$$G(\hbar\omega) = \frac{2.\pi e^2}{nc\epsilon\omega Lm^2} \sum_{\sigma=U,L} \sum_{n,m} \int \left|\left(\hat{e}.M_{nm}^{\eta\sigma}(k_t)\right)\right|^2 \times \frac{\left(f_n^c(k_t) - f_{\sigma m}^v(k_t)\right)\left(\frac{\gamma}{\pi}\right)}{\left(E_{\eta,\sigma nm}^{c,v}(k_t) - \omega\hbar\right)^2 + \gamma^2} \frac{k_t dk_t}{2\pi}$$

In the above expression, the detail of all the terms can be seen in ref. [12].

3. Results and Discussion

Generally, any opto-electronic device's output performance is calculated in terms of optical inverse loss or optical features since it is most significant and primary condition for the understanding the mechanism of a semiconducting materials-based laser. It is due to the stimulated emission that take place by electron and hole pairs (EHPs) recombination. For InGaAsN/GaAs (GaAs is substrate) heterostructure, the simulated optical gain and modal gain is shown in figure 1. In figure 1, it can be seen that the both the gain i.e. peak optical gain and the peak modal gain (which is determined by multiplication of optical gain with confinement factor) is achieved at a single wavelength ~ 1.30 μm. Moreover, in figure 1, two peaks can be observed. Appearance of these two peaks are due to two transitions which occurs between first electronic state (lying in conduction band) and heavy and light hole states (lying within the valence band).

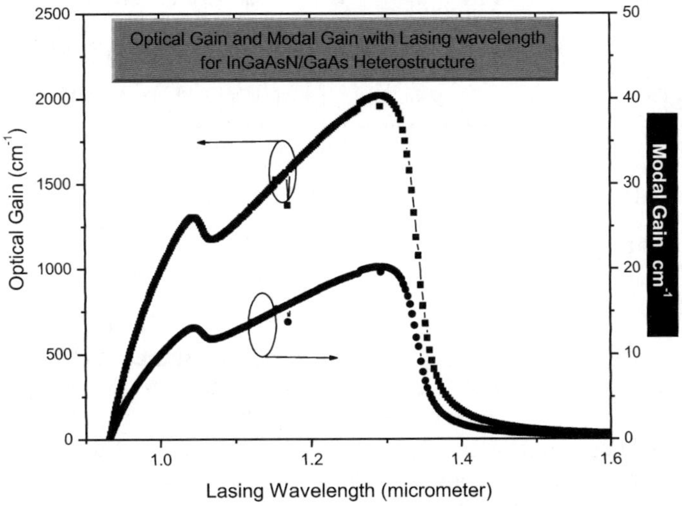

Fig.1. Gain characteristics (on right scale) and modal gain (on left scale) as a function of wavelength for InGaAsN/GaAs QW heterostructure

Refer to figure 2, the optical gain for InGaAsN/InP (InP is substrate) heterostructure taking into account claddings is simulated and shown. In figure 2, it can be observed that the peak optical inverse loss with claddings is achieved at energy 1.11 eV (at wavelength ~1.11 μm); while the optical gain without claddings are achieved at wavelength ~1.14μm (see inset figure). Further, form figure 2, it can be seen that the presence of claddings not only affects the magnitude of optical gain, it also affects the wavelength and photonic energy. Positively, the inclusion of claddings in the heterostructure enhances the strength of the optical inverse loss by confining the much light wave in the quantum well. The numerically outcome data of the simulation of InGaAsN QW heterostructure with substrates GaAs and InP is shown and compared in table1.

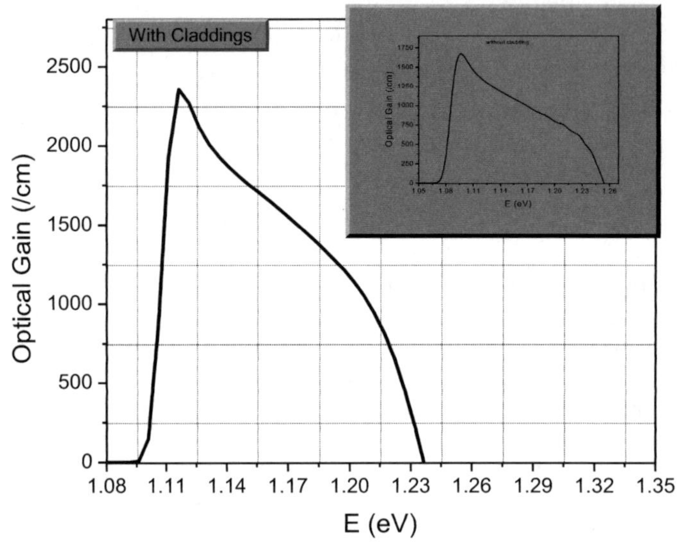

Fig.2. Optical gain for InGaAsN/InPStraddled Heterostructures (with claddings).

Table 1: Comparative Study of InGaAsN QW heterostructure with GaAs and InP substrate

Quantum well Heterostructure Detail	Substrate	Optical Gain (/cm)	Wavelength (μm)
InGaAsN (with claddings)	GaAs	2100	1.30
InGaAsN (without claddings)	InP	1675	1.14
InGaAsN (with claddings)	InP	2353	1.11

Conclusion

For InGaAsN quantum well heterostructure with the substrates GaAs and InP,The quantum mechanical approach (k.p method) has been adopted to calculate the confinement of the carriers associated with quantum well and the optical gain. In addition, for InGaAsN/InP heterostructure, the cladding effect has also been observed. The simulated results exhibit that the substrate has played a very significant role in modifying the optical gain features due to the presence of strain which comes into play due to the lattice difference.

References

[1]. A. Kuramata, K. Domen, R. Soejima, K. Horino, S. Kubota and T. Tanahashi, “InGaN Laser Diode Grown on 6H–SiC Substrate Using Low-Pressure Metal Organic Vapor Phase Epitaxy”, *Japan. J. Appl. Phys.* **36** L1130 (1997).

[2]. S. Nakamura, S. Pearton and G. Fasol, *The Blue Laser Diode* (Berlin: Springer) 2000.

[3]. J. K. Sheu, F. W. Huang, C. H. Lee, M. L. Lee, Y. H. Yeh, P. C. Chen and W. C. Lai, “Improved conversion efficiency of GaN-based solar cells with Mn-doped absorption layer”, *Appl. Phys. Lett.* **103** 063906 (2013).

IOP Conf. Series: Materials Science and Engineering **594** (2019) 012044 doi:10.1088/1757-899X/594/1/012044

[4]. F. Y. Wu, P. B. Keller, D. Kapolnek, P. S. Denbaars and U. K. Mishra, "Measured microwave power performance of AlGaN/GaN MODFET", *IEEE Electron. Dev. Lett.* **17** 455–7 (1995).
[5]. F. Y. Wu, P. B. Keller, P. Parikh, D. Kapolnek, S. P. Denbaars, and U. K. Mishra, "Bias dependent microwave performance of AlGaN/GaN MODFET's up to 100 V", *IEEE Electron Dev. Lett.* **18** 290–2 (1997).
[6]. Z. I. Alferov, "The history and future of semiconductor heterostructures", *Semiconductors* 32, 1 (1998).
[7]. P. A. Alvi, Sapna Gupta, Meha Sharma, Swati Jha, F. Rahman, "Computational modeling of novel InN/$Al_{0.30}In_{0.70}N$ multilayer nano-heterostructures", *Physica E: Low-Dimensional Syst. Nanostruct.*, 44, pp. 49-55 (2011).
[8]. P. A. Alvi, "Strain Compensated InN based Superlattice", *Adv. Sci. Lett., Vol. 5, No.17,* pp. 101-107(2012).
[9]. Sapna Gupta, F. Rahman, M. J. Siddiqui, P. A. Alvi, "Strain Profile in III-Nitride based Multilayer Nano-heterostructures", *Physica B: Condensed Matter*, Vol. 411, pp. 40-47 (2013).
[10]. Y. S. Park, "Wide bandgap III-nitride semiconductors: opportunities for future optoelectronics", *Optoelectronics Review* 9(2), 117-124 (2001).
[11]. S. L. Chuang, Physics of optoelectronic devices (Wiley, New York, 1995).
[12]. H. K. Nirmal, Nisha Yadav, F. Rahman, P. A. Alvi, "Optimization of High Optical Gain in Type-II$In_{0.70}Ga_{0.30}As/GaAs_{0.40}Sb_{0.60}$Lasing Nano-heterostructure for SWIR Applications" *Superlattices Microstruct.*, 88, 154-160 (2015).
[13]. P. A. Alvi, "Transformation of type-II InAs/AlSb nanoscale heterostructure into type-I structure and improving interband optical gain", *Phys. Status Solidi B,* **254**, No. 5, 1600572 (2017).

IOP Conf. Series: Materials Science and Engineering **594** (2019) 012045 doi:10.1088/1757-899X/594/1/012045

Finite Element Analysis of Circular Silicon Diaphragm

Samridhi[1], Manish Kumar[2], Kulwant Singh[3], P. A. Alvi[1*]

[1]Department of Physics, Banasthali Vidyapith, Banasthali-304022, Rajasthan (INDIA)
[2]School of Engineering, Indian Institute of Technology Mandi- 175005, Himachal Pradesh (INDIA)
[3]Department of Electronics and Communication Engineering, Manipal University Jaipur, Jaipur 303007, Rajasthan (INDIA)

*E-mail corresponding author: drpaalvi@gmail.com

Abstract. Miniaturization of sensing devices not only reduces the size, but also enhances the performance and reliability in sector of automation, space, weather monitoring and forecasting. It has become important for scientist and researchers working in the field of micro-electro-mechanical system area to simulate the structure before actual fabrication. This paper reports about the performance of proper meshing required to lead the exact results of simulated circular silicon diaphragm with the help of finite element method. In this paper, we have shown that the proper meshing of the circular silicon diaphragm has shifted the vibrating frequency of the diaphragm under different pressures applied.

Keywords: Sensors, Meshing, Frequency, Displacement

1. Introduction

Circular plates are employed in many applications like biological analysis, energy harvesting and silicon sensing device [1]. Therefore, silicon is still proving itself to be a good quality material for the sensing device [2]. MEMS sensing devices based on this material were fabricated, firstly in form of transducer to generate a signal as a function of pressure. These sensing devices have undergone a significant growth in the area of industries as well as in the commercial sector [3, 4]. They are cheap due to batch fabrication and are in small size, consume low power and gives enhanced performance. The era of MEMS fabrication allows miniaturization of complicated systems in a way of controlling, actuating and has integrated sensing capabilities on a single chip. These sensors have changed the mindset of designers and engineers the way they measure the sensitivity of the pressure sensor. Therefore, it is necessary to understand the complex structure of the sensor on a single microchip before its actual fabrication. Finite Element Analysis is a valuable tool for the designing purpose of a sensor since it gives more accurate and precise results up to high level of accuracy and it tells the physical behaviour like the structural mechanics of a sensor [5].In this article, the simulated results show how frequency response of diaphragm is changing with meshing size. The proper meshing is carried out by finite element analysis which enables us to obtain the displacement of the diaphragm under a different range of applied pressure. Finer the meshing, the solution converges near to the real behaviour of a diaphragm with a very high level of accuracy.

IOP Conf. Series: Materials Science and Engineering **594** (2019) 012045 doi:10.1088/1757-899X/594/1/012045

2. Modelling and Theory

According to the literature review, it has been found that the maximum displacement of circular diaphragm occurs at the center, while stress is minimal at the edges of circular diaphragm. The maximum stress of the diaphragm is given by the formula [6].

$$\sigma_{max} = 1.25Pa^2/t^2 \quad (1)$$

Also, the maximum deflection at the center of circular diaphragm can be given as:

$$\boldsymbol{w_{max} = (1 - v^2)Pa^4/4.13Et^3} \quad (2)$$

Where t is thickness, ν represents Poisson's ratio, and P is pressure, a refers the dimension, and E is young modulus. In the present paper, we have performed displacement and stress analysis of circular diaphragm geometry with Silicon as a substrate material using COMSOL package.

3. RESULT AND DISCUSSION

The finite element analysis has been performed for the silicon circular diaphragm and the simulated results have been shown in figure 1. Fig.1 (a) and (b) shows the geometry of a circular diaphragm with radius of 1mm and diaphragm thickness 50 μm, which is supported by a fixed region having radius of 2mm. A fine and normal meshing is carried out by FEM (finite element method) on the same substrate with same dimension and geometry.

Modes [mn]	**Fine Meshing**	**Normal Meshing**
	Displacement of Diaphragm at 275KHz frequency	**Displacement of Diaphragm at 350 KHz frequency**
	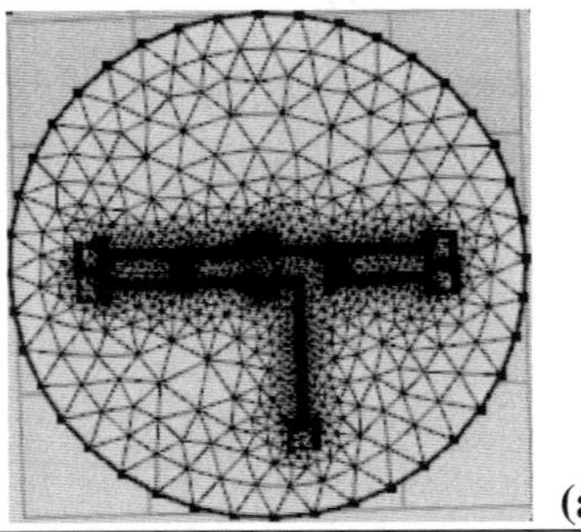 **(a)**	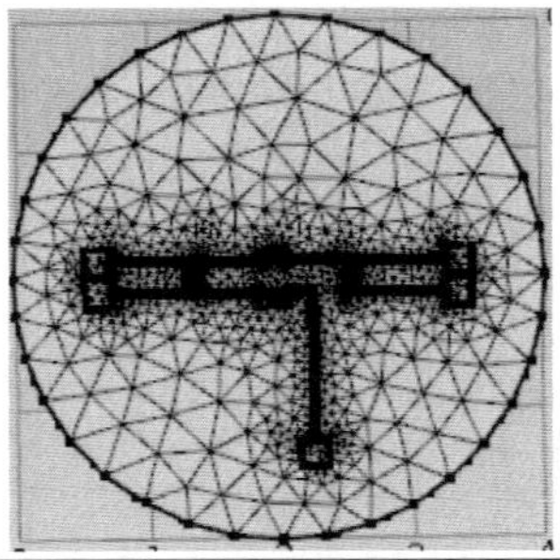 **(b)**
[11]	 **(c)**	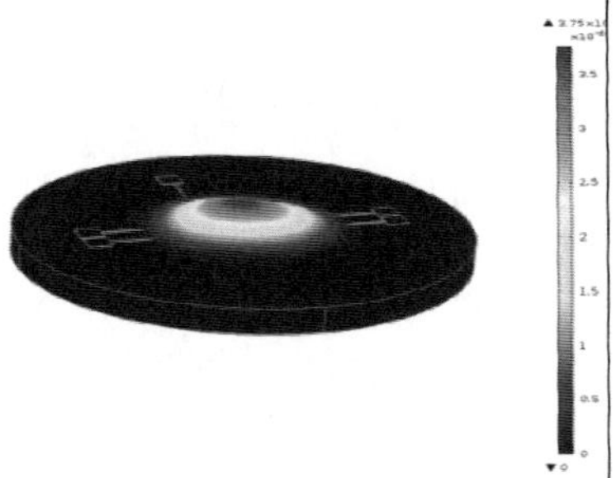 **(d)**

Fig.1 Displacement Profile under Fine meshing and Normal Meshing

For fine meshing, the vibrating frequency of circular silicon diaphragm is noted ~ 275KHz, as shown in Fig.2 (a) at pressure about 60 KPa with step size of 10 KPa. The maximum displacement of the silicon diaphragm is about ~ 16μm under this pressure. The deflection so induced by this pressure is shown in Fig.1(c) and it shows that the displacement is maximum at the centre of circular diaphragm. Similarly, for normal meshing, the vibrating frequency of circular silicon diaphragm is noted as ~350KHz, as shown in Fig.2 (b) at pressure about 26 KPa with step size of 5KPa and. Here, the maximum displacement of the silicon diaphragm is about ~ 6.5 μm under this pressure. The deflection so induced by this pressure is shown in Fig.1(d).

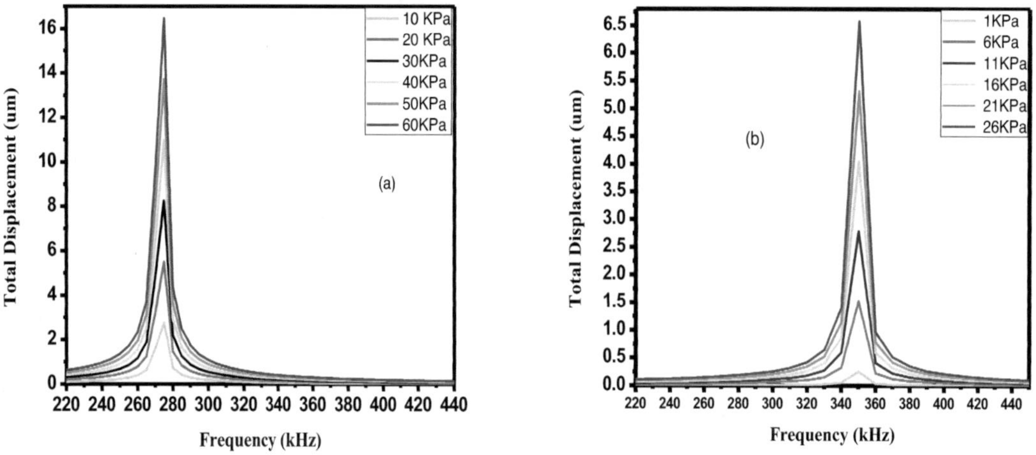

Fig. 2 Frequency Vs Displacement of Si diaphragm at (a) Fine Meshing having ~275KHz frequency and (b) Normal Meshing having ~350KHz frequency

Conclusion

A proper meshing has been carried out by finite element analysis within the framework of the COMSOL package to find the real behaviour of the vibrating diaphragm which can be applicable for the various sensing devices and micro-structures. Finer the meshing, the solution converges near to the real behaviour of a diaphragm with a very high level of accuracy.

Acknowledgement

Samridhi and P. A. Alvi are grateful to DST, Government of India, New Delhi, for providing the facilities thorough "Banasthali Center for Research and Education in Basic Sciences" under CURIE programme.

References

[1]. W.H. Duan, C.M. Wang, C.Y. Wang" Modification of fundamental vibration modes of circular plates with free edges" *Journal of Sound and Vibration*, *317*(3-5), 709-715.
[2]. Nisanth.A, Suja.K.J, Rama Komaragiri "Performance Analysis of a Silicon Piezoresistive Pressure Sensor Based on Diaphragm Geometry and Piezoresistor Dimensions", International Conference on Circuit, Power and Computing Technologies, pp. 1273-1278. IEEE, 2014.

IOP Conf. Series: Materials Science and Engineering **594** (2019) 012045 doi:10.1088/1757-899X/594/1/012045

[3]. P. A. Alvi, B. D. Lourembam, V. P. Deshwal, B. C. Joshi and J. akhtar, A process to fabricate Si_3N_4 and SiO_2 micro-membrane using front side lateral etching technology, *Sensor Review*, Vol. 26, No. 3, pp. 179-185, 2006.
[4]. P. A. Alvi, P. A. Alvi, J. Akhtar, K. M. Lal, S. A. H. Naqvi, and A. Azam, "Design and Fabrication of Micromachined Absolute Micro Pressure Sensor", Sensors & Transducers Journals, 96, No. 5, 47-55, 2008.
[5]. U. Sampath Kumar, N. Jagadesh Babu "Design and Simulation of MEMS Piezoresistive Pressure Sensor to Improve the Sensitivity" International journal of innovative research in electrical, electronics, instrumentation and control engineering 3, no. 3, 2015.
[6]. K. Y. Madhavi, K. A. Sumithradevi, M. Krishna, & M. N. Vijayalakshmi, Analysis of Square and Circular diaphragms for a MEMS pressure sensor using a Data Mining tool, *Communication Systems and Network Technologies (CSNT),* pp 258-261, 2011.

IOP Conf. Series: Materials Science and Engineering **594** (2019) 012046 doi:10.1088/1757-899X/594/1/012046

Generic IoT Platform for Analytics

Pradeep B[1]. and Balasubramani R[2,*]
[1]Accenture Solutions Private Limited, Bangalore
[2]Department of Information Science and Engineering, NMAMIT, Nitte
E-mail: *balasubramani.r@nitte.edu.in

Abstract: Internet, since its root has reclassified the manners by which the world collaborates today, countless devices speak with each other which make our life simple and thus improve this world a place to dwell. The Internet of Things (IoT) is the system of physical gadgets, vehicles, home machines and different things implanted with hardware, programming, sensors and network which empowers these objects to interface and trade information. Today the world has swung to a place where each and everything should be associated. This paper discusses, how a generic IoT platform has been developed to get the data from various sensors from agricultural lands and conversely control the devices from distant places using Internet. For e.g., a smart plug can help you save and monitor your energy consumption by showing energy statistics. Sensors fitted in agricultural lands has been chosen to test the quality of the IoT platform.

Index-Terms: Node Red, Raspberry Pi, DHT11, Node JS.

1. Introduction

The Internet of things (IoT) is a processing idea that illustrates the possibility of ordinary physical items being associated with the web and having the capacity to recognize themselves to different gadgets. The term is firmly related to RFID as the strategy for correspondence, despite the fact that it likewise may incorporate other sensor innovations, remote advancements or QR codes.

The IoT is huge on the grounds that a protest that can speak to itself carefully moves toward becoming an option that is more prominent than the question independent from anyone else. Never again does the question relate just to its client, however it is currently associated with encompassing articles and database information. At the point when numerous articles demonstration as one, they are known as having "surrounding knowledge". Internet of Things (IoT) is the setup of all the physical gadgets that we can envision in our everyday life for gathering and transmission of information through different sensors. This makes life simple with negligible human interruption.

Creation of a Generic platform for data transmission and reception from various sensors and plotting the graphs on a dashboard was developed using JavaScript. It also includes the process to setup a Smart plug in which a device can be plugged in and controlled using a dashboard application after authentication. The data transmission is secured so that no external user can see the information that is being exchanged between an authenticated client and server.

The data exchange is done using MQTT protocol and ESP8266. ESP8266 is the WiFi module which includes a microcontroller unit that can be programmed using Arduino IDE. ESP8266 connects to the active WiFi connection and supports the data transmission using MQTT. Various libraries have been included in the project and corresponding links have been provided in this documentation.

The project is divided into two sections

- **Hardware section:**

Hardware section involves arranging the setup and making the circuits required for the project. It also includes setting up the ESP8266 module and programming it is using Arduino IDE.

- **Software section:**

The Software section includes creating a Web Application for the user using NodeJS and WebViewer in Android Studio. It provides many functionalities to the user like adding a device, controlling it etc.

IOP Conf. Series: Materials Science and Engineering **594** (2019) 012046 doi:10.1088/1757-899X/594/1/012046

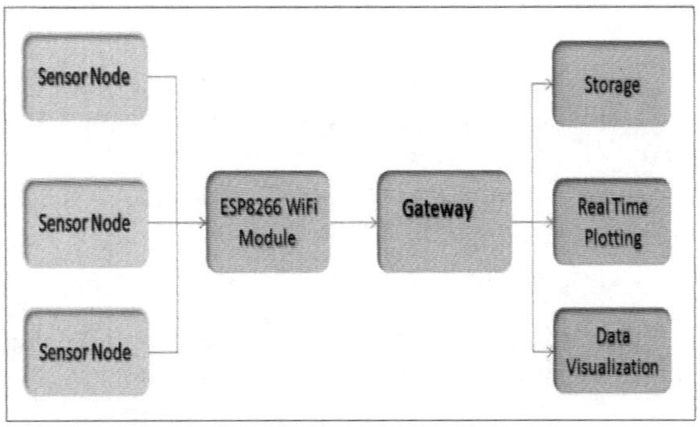

Fig. 1 Generic IoT Platform for Sensors flow diagram

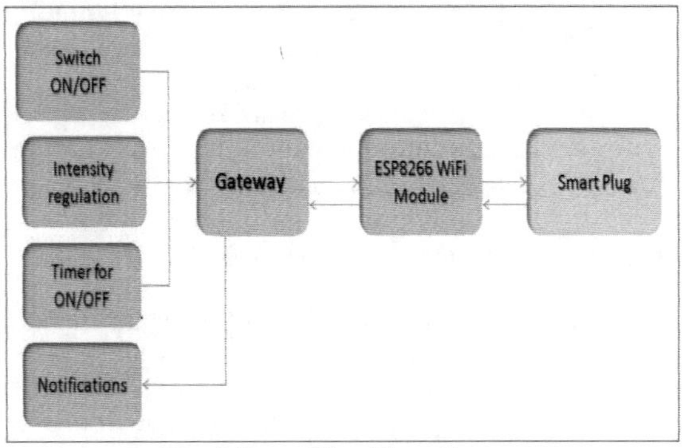

Fig. 2 Smart Plug Flow diagram

2. Literature Review

Hamdan Hejaziin [1] explained an overview on IoT stages, talking about their models and basics of IoT building components and the corresponding conventions between them. This helps the per user to pick a reasonable and sufficient IOT platform for own demands. M. Nelwin.Rajkumarin [2] outlines about IOT technologies and applications interconnected to agriculture with contrast to pumping motor on and off. Also live streaming of farm on android mobiles by using WiFi can be watched. Priyanka Padalalu in [3] explained the idea which helps the farmer in irrigating the farms efficiently using automated irrigation system which is based on values from soil, moisture and pH sensors. The data collected is recorded from time to time to the server and can be acquired via android app. Rajalaxmi.P in [4] explained that in order to automate the irrigation systems, sensors are installed in the field for supervision. Wireless transmission is used to conceal data in database of server. If the moisture and temperature of the field falls below the threshold, the irrigation is automated. The farmers mobile are sent with notification regularly. Farmers being situated in any part of the world can observe the status of his field.

In areas where water is inadequate this system is useful. Compared to traditional approach this system is 92% more effective. The system is cost effective. Also Power consumption is benefit able. The system can be used in Green houses to control the light intensity. There is increase in Crop Production and wastage of crops is less compared to traditional farming. Gives feasible results. Guruprasadh JP in [5] explained information about soil moisture and Nutrient deficiency obtained from Intelligent sensors. The information obtained can be utilized to automate. Modern electronic technology with automatized features which makes use of microcontroller can detect the dampness content of the earth and turns the pumping motor on and off.

IOP Conf. Series: Materials Science and Engineering **594** (2019) 012046 doi:10.1088/1757-899X/594/1/012046

3. Objectives:

A generic IoT platform helps us to get the data from various setups and conversely control the devices from distant places using internet. The readings of temperature and humidity sensor can be used to control an air conditioner.

A smart plug can help you save and monitor your energy consumption by showing you energy statistics. Moreover, it will help you regulate the intensity at which your device operates which will indeed help you save power.

This project is beneficial for budding IoT enthusiasts as it would help them to gain an in-depth understanding of IoT and make a project which they can implement in their daily lives.

4. Experimental Setup

A. *Hardware components*
1.DHT11: Output will be temperature in degree Celsius and relative humidity in percentage. Data pin connected to digital pin gives digital output. Capacitive Humidity sensor to measure the relative humidity and thermistor to measure temperature of surrounding.
2.ADXL335-Accelerometer: Consists of three axes for sensing
(x, y, z). Output will be signal conditioned voltage output for each axis proportional to acceleration.
3.Light Intensity Sensor: Voltage Proportional to light Intensity in the output, use analog Read to read it.
As light intensity increases, resistance of LDR decreases therefore voltage on Analog pin also decrease as light intensity increases.
4.RFID: Radio-frequency identification (*RFID*) uses electromagnetic fields to automatically identify and track tags attached to objects.
The tags contain electronically stored information. The tags are further classified into two types.

a. Passive Tags: Operate from short distances as they collect energy from a nearby RFID reader interrogating radio waves.
b. Active Tags: They have a local power source such as a battery and may operate at hundreds of meters from the RFID reader.RFID can be used for security purposes.

5.Node MCU: NodeMCU development board is Open Source IoT platform. NodeMCU is a small board, based on ESP-12E WiFi Module from AI-Thinker, itself containing a single-chip ESP8266 WiFi-SoC.
6.Relay
7.Motor
8.Tungusten Bulb
9.MQTT protocol: MQTT (Message Queue Telemetry Transport) is a lightweight publish subscribe protocol for IoT based application. MQTT is preferred over HTTP protocol because HTTP protocol requires handshaking for every request whereas, on other hand, MQTT is a simple protocol where data gets published and subscriber receives it by subscribing to that particular topic.
10.Raspberry Pi: The Raspberry Pi is a low cost computer and can function as a proper monitor. Raspberry Pi can be used to build smart devices where it acts as a Gateway. Raspberry Pi is usually slow than our normal laptop, but is still a Linux computer. Though Raspberry is slow it is preferred over computers as its power consumption is quite less.

B. Software Requirements

1. Adding NodeMCU/ESP8266 board to Arduino IDE
2.Installing Mosquitto Broker for MQTT in Ubuntu
3.Installing node.js for Node-Red
4.Installing Node-Red
5.Installing UI Dash Board on Node Red

C. Technology used

1. Canvas Gauges: are open source minimalist HTML5-based components for web applications.

IOP Conf. Series: Materials Science and Engineering **594** (2019) 012046 doi:10.1088/1757-899X/594/1/012046

2. JQuery: 3.1.1: JQuery is one of the most popular JavaScript library used to make the client-side scripting of HTML easier.
3. Node JS:Node.js is a platform built on Chrome's JavaScript runtime for easily building fast and scalable network applications. It is an open source server framework that allows us to use JavaScript on the server.
4. Node Red: Node-Red is a software tool built on Node.js developed by IBM which makes it easier to wire together hardware devices. It provides simple browser based editing facility which makes it easier to create flows just by a simple drag of the mouse. The flows can then be deployed at runtime.

5. Generic IoT Platform for Sensors

- *Schematic for Sensor Setup:*

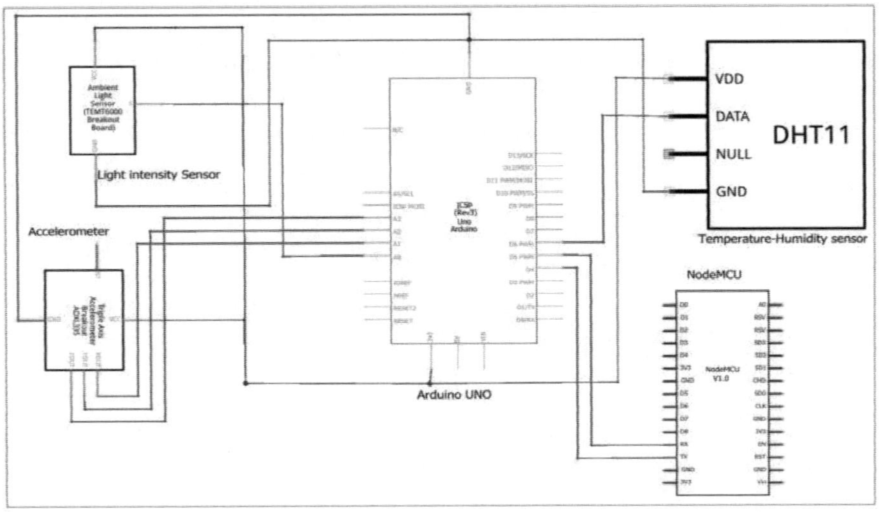

Fig. 3 Schematic for sensors setup

Above Figure shows the Hardware setup for Node MCU and sensors.Sensors like DHT11, Light Intensity sensor,Accelerometer are connected via Aurdino, Node MCU.

B. *Algorithm for Generic IOT Platform*

Above flowchart shows the step by step process of how the data is collected from sensors and visualization is done on node red dash board. First step code is to upload the code to aurdino-UNO and Node MCU.As soon as the data is received by the aurdino, serial communication between the aurdino and Node MCU starts. Node MCU is connected to the wifi and Mosquitto broker. The Encrypted data is published by the client to the broker and other client subscribes the data from the broker. Inturn, the visualization of data is done on the Node Red Dash Board.

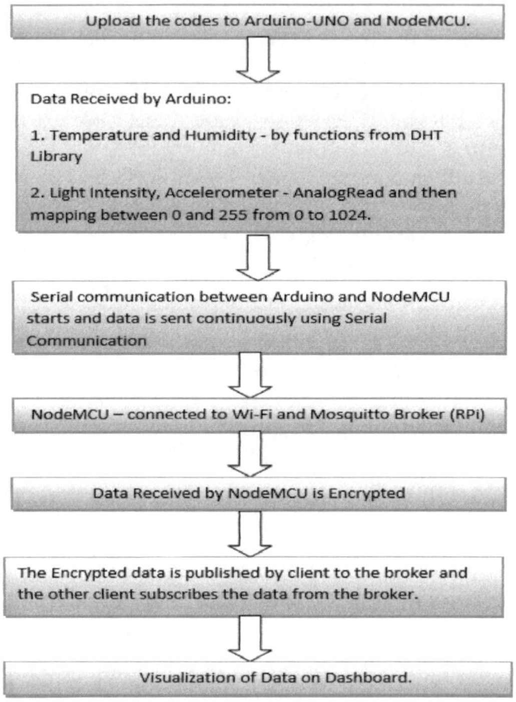

Fig.4 Flow for sensors

C.Flow in Node Red

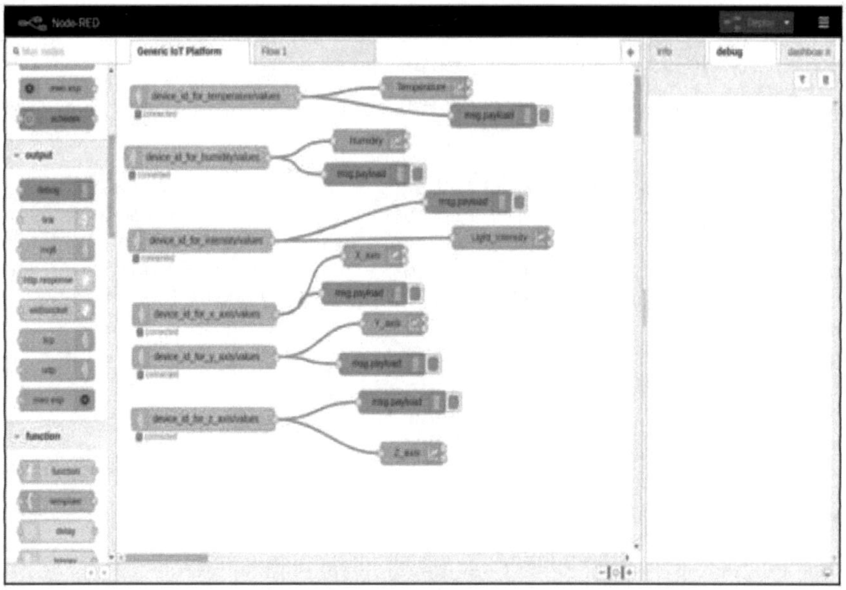

Fig.5 Flow in node red for changing sensor values

IOP Conf. Series: Materials Science and Engineering **594** (2019) 012046 doi:10.1088/1757-899X/594/1/012046

Figure above shows the flow of data among the sensors in node red where visualization entire setup is showed.

D.Results obtained on Node Red

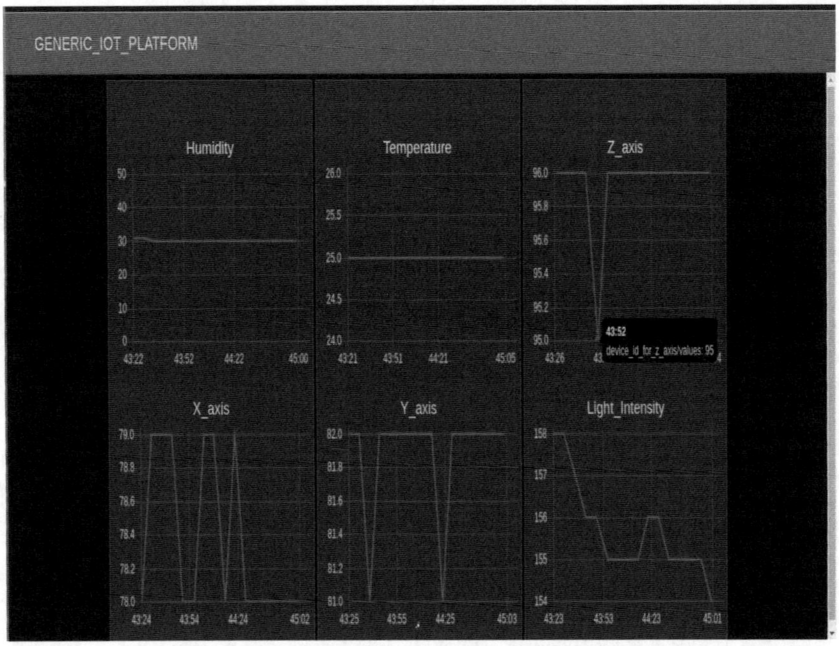

Figure.6 Results obtained on Node Red Dashboard

Figure above shows the graph where time versus value is plotted for values of temperature and humidity during different time intervals.

- The user can view the current reading of the sensors through the gauge chart.
- Time vs Value data of the device is plotted through line chart in real time.
- The user can see day-to-day analysis of the device readings through a line chart.

6. Conclusion

A platform for data transmission and reception from various sensors are created (Node MCU and Temperature sensor) and the graphs are plotted on dashboard developed, the gauge chart provides current readings of the sensor which can be viewed by the user. Every day analysis of device readings through line chart can be seen by the user through line chart.

References

[1] Hamdan Hejazi, Husam Rajab, Tibor Cinkler, László Lengyel, "Survey of Platforms for Massive IoT".

[2] Dr. M Nelwin Rajkumar, S.Abhinaya, Dr.V Venkatesh Kumar,"Intelligent Irriagation system-IOT based approach".

[3] Priyanka Padalalu, Sonal Mahajan, Kartikee Dabir, Sushmita Mitkar & Deepali Javale Department of computer Engineering," Smart Water Dripping System for Agriculture".

[4] Rajalakshmi. P, Mrs.S.Devi Mahalakshmi,"IOT based crop field Monitoring and Irrigation Automation".

[5] Guruprasadh JP, Harshananda A, Keerthana IK, Rachana,Yadu Krishnan K,Murali Rangarajan,"IntelligentSoil Quality Monitoring system for Irrigation Automation".

[6] N. Alexandratos and J. Bruinsma, "World agriculture towards2030/2050: The 2012 revision", Global Perspective StudieTeam,FAO Agricultural Development Economics Division; Accessed May20, 2017.

[7] D. Piementel, B. Berger, D. Filiberto, M. Newton, B. Wolfe, E.Karabinakis, et al, “Water resources: Agricultural and Environmental issues”, BioScience, vol. 54, pp. 909-918, October 2004.
[8] G.V. Satyanarayanasnd S. D. Mazaruddin, “Wireless sensorBased remote monitoring system for agriculture using ZigBee and GPS”,Conf. Adv. Commun. Contr. Syst, pp. 110-114, 2013.
[9] H.C. Punjabi,S. Agarwal, V. Knithani, V. Muddaliar, And M.Vasmatkar, “Smart farming using IoT”, Int. J.Electr.Commun.Engg.Tech, vol. 8, pp. 58 – 66, February 2017.
[10] V.I. Adamchuk, J.W. Hummel, M.T. Morgan, and S.K.Upadhayaya,“On-the-go soil sensors for precision agriculture”,Comput. Electron. Agr, vol. 44, pp. 71-91, March 2004.
[11] J.V. Stafford, “Implementing precision agriculture in the 21stcentury”,J. Agr. Eng. Res, vol. 76, pp. 267-275, July 2000.
[12] D.L. Corwin and S.M. Lesh, “Apparent of soil electricconductivity in precision agriculture”,Comput. Electron.Agr, vol.46,pp. 1-10, March 2005.
[13] P. Divvya, S. Sonkiya, P. Das, V.V. Manjusha, and M.V. Ramesh, “CAWIS: Context Aware Wireless Irrigation System”,Int. Conf. Comput. Commun. Contr. Tech., pp. 310 – 315, 2014.
[14] S. Harishankar, R..S. Kumar, K.P.Sudharsan, U. Vignesh, and T. Viveknath, “Solar powered smart irrigation system", Adv. Electron. Electr.Engg, pp.341 -346, 2014.
[15] M.R.M .Kassim, I. Mat, and A.N. Harun, “Wireless sensor network in precision agriculture application”IEEEInt.Conf.Comput.Information Technol., August 2014.
[16] R.F.M. Huerta, A.D.J.O. Melendez, I.R.G.G. Gonzalez, I.I.T Pacheco, I.G.H.Ruiz, I.L.M.C. Medina,et al, “An analysis of electrical impedance measurements applied for plant N status estimation in lettuce (Lactuca sativa)”, Sensor, vol. 14, pp. 11492 11503, July 2014.
[17] Z. Chik, O.F. Murad, and Rahmad, “Dependency of dry density ofsoil on water content in the measurement of electrical resistivity ofsoil”, J. Eng. Res. Tech., vol. 2, pp. 141-145, June 2015.
[18] L. Meiqing, L. Jinyang, M. Hanping, and W. Yanyou, “Diagnosis anddetection of phosphorus nutrition level for solanumlycopersicumbased on electrical impedance spectroscopy”, Biosys. Eng, vol. 143,pp. 108-118, January 2016.
[19] A. Lasia, “Electrochemical impedance spectroscopy and itsapplications”, Modern Aspects of Electrochemistry”, vol. 32, B.E.Conway, J. Bockris, and R.E. White, Eds.New York: KluwerAcademic/Plenum Publishers, 1999, pp. 143-248.
[20] T. Islam, Z. Chik, M.M. Mustafa, and H. Sanusi, “Modeling ofelectrical resistivity and maximum dry density in soil compactionmeasurement”, Environ. Earth Sci., vol. 67, pp. 1299 – 1305,November 2012.
[21] Adding NodeMCU board to Arduino IDE:https://github.com/esp8266/Arduino
[22] Installing and Testing Mosquitto: http://wingsquare.com/blog/setting-up-mqtt-mosquitto-broker-in-ubuntu- linux/
[23] Installing Node-Red: http://nodered.org/docs/getting-started/installation.html [24]RFID: https://en.wikipedia.org/wiki/Radio-frequency_identification’
[25] MQTT: Mqtt.org
[26] Node JS: https://www.tutorialspoint.com/nodejs/nodejs_introduction.htm
[27]. Navarro-Hellín, H., Martínez-del-Rincon, J.,Domingo-Miguel, R., Soto-Valles, F. and Torres-Sánchez, R., 2016. A decision support system formanaging irrigation in agriculture.Computers andElectronics in Agriculture, 124, pp.121-131.
[28]. Roussel, N., F. Frappart, G. Ramillien, J.Darrozes, F. Baup, and C. Ha. "Detection of soilmoisture content changes by using a single geodeticantenna: The case of an agricultural plot." In 2015IEEE, International Geoscience and Remote SensingSymposium (IGARSS), pp. 2008-2011. IEEE, 2015.
[29]. Bing, Fu. "Research on the agriculture intelligentsystem based on IOT." In 2012 InternationalConference on Image Analysis and SignalProcessing, pp. 1-4. IEEE, 2012.
[30]. Alnaimi, Firas B. Ismail, Yee Chaw Chu, andKhairulSalleh Mohamed Sahari. "Hybrid renewablepower system for agriculture irrigation system."In System Integration (SII), 2014 IEEE/SICE International Symposium on, pp. 736-742. IEEE,2014.
[31]. Kavianand, G., V. M. Nivas, R. Kiruthika, and S.Lalitha. "Smart drip irrigation system for sustainableagriculture."In Technological Innovations in ICT forAgriculture and Rural Development (TIAR), 2016IEEE, pp. 19-22.IEEE, 2016.
[32]. Wu, Hao, Fangpeng Chen, Hanfeng Hu, Qi Liu,and Sai Ji. "A Secure System Framework for anAgricultural IoT Application."In InternationalConference on Computer Science and its Applications, pp. 332-341.Springer Singapore, 2016.
[33]. Abrisqueta, Isabel, WenceslaoConejeroMercedes Valdés-Vela, Juan Vera, Ma FernandaOrtuño, and Ma Carmen Ruiz-Sánchez. "Stem waterpotential estimation of drip-irrigated early-maturingpeach trees under Mediterranean conditions." Computers and Electronics inAgriculture 114 (2015): 7-13.
[34]. Campos, Isidro, Claudio Balbontín, JoseGonzález-Piqueras, Maria P. González-Dugo,Christopher MU Neale, and Alfonso Calera."Combining a water balance model with evapotranspiration measurements to estimate totalavailable soil water in irrigated and rainfed vineyards." Agricultural Water Management 165 (2016): 141-152.

[35]. Giusti, Elisabetta, and Stefano Marsili-Libelli. "A Fuzzy Decision Support System for irrigation and water conservation in agriculture." Environmental Modelling & Software 63 (2015): 73-86.
[36]. Zwart, Sander J., and Wim GM Bastiaanssen. "Review of measured crop water productivity values".

Institute of Physics
Dirac House, Temple Back
Bristol BS1 6BE UK

ISSN: 1757-8981
ISBN 978-1-5108-9815-8

CURRAN ASSOCIATES INC.
proceedings
.com